RFID Security

Techniques, Protocols and System-on-Chip Design

Paris Kitsos • Yan Zhang
Editors

RFID Security

Techniques, Protocols
and System-on-Chip Design

 Springer

Editors

Paris Kitsos
Hellenic Open University
School of Science & Technology
Computer Science
Tsamadou 13-15
262 22 Patras
Greece

Yan Zhang
Simula Research Laboratory
Martin Linges v 17, Fornebu
P.O. Box 134
1325 Lysaker
Norway

ISBN: 978-0-387-76480-1   e-ISBN: 978-0-387-76481-8
DOI: 10.1007/978-0-387-76481-8

Library of Congress Control Number: 2008929331

Preface

Radio Frequency Identification (RFID) technology is gaining a recent explosion of development in both industry and academia. RFID is believed to be an indispensable foundation to realize ubiquitous computing paradigm. In RFID systems, RFID tag consists of three primary components: RFID transponder, RFID transceiver, and an application system. The RFID transponder is composed of a small microchip with data storage, limited logical functionality, and an antenna. The RFID transceiver can be distinguished based on the operation frequency (HF or UHF) or on the powering techniques (active, passive, or semipassive). Communications between a transceiver and a transponder involve interrogating the transceiver to obtain data, writing data to the transponder, or delivering commands to the transponder. The application system is used to collect data through the transceiver and the database utilizes the data for a variety of purposes. RFID tags have been deployed for several years. Though tagging shipping containers is the largest business space for RFID, there are a number of emerging applicable scenarios. For instance, electronic payment, RFID passports, office folders, microsensors, intelligent labels, port management, food production control, animal identification, and so on. It is strongly believed that many more scenarios will be identified when RFID principle is thoroughly understood, cheap components are available, and RFID security is guaranteed.

RFID security is a prerequisite to enable wide applications of RFID systems. R&D efforts are in progress to propose secure system architecture, secure protocols, and secure self-configuration. Due to unique characteristics of RFID systems, all components shall be sufficiently secured, including RFID tags, personal data, transaction, middleware, back-end system, and RFID readers. There are a number of challenges in designing efficient security schemes in RFID systems. First, the air interface of RFID systems poses an inherent problem as regards data confidentiality. Any data transferred over the air could be easily subject to eavesdropping if the transaction is unencrypted. Second, it is often desired to use RFID tags as cryptographic tokens, e.g., in a challenge–response protocol. In this case, the tag must be able to execute a secure cryptographic primitive. However, the low-cost demand for RFID tags (0.05–0.1$) forces the lack of resources to perform true cryptographic operations. Typically, these systems can only store hundreds of bits and have 5K to 10K logic gates, but only 250 to 3K can be devoted to security tasks. Finally, for RFID users, the foremost protection involves confidentiality, privacy, and non-reputability. The process of unique identification involves collecting large amounts of personal data, which needs to be highly confidential. Special care must be also taken to ensure that the tag–reader communications is adequately encapsulated and shielded. To address these issues, the security subjects are being explored in various scenarios and emerging standards. The topics include authentication, access control and authorization, attacks, privacy and trust, encryption, dynamic privacy protection, and hardware implementation of algorithms, case studies, and applications.

RFID security. Techniques, Protocols and System-on-Chip Design focuses on the security issues in RFID systems, recent advances in RFID security, attacks and solutions, security techniques, and practical hardware implementation of cryptography algorithms. This book consists of 16 chapters, which are organized into three parts as follows:

- Fundamentals
- Security protocols and techniques
- Encryption and hardware implementations

Part I presents RFID fundamentals, system architectures, and application. In addition, a general discussion on RFID security is introduced. Part II explores the latest security protocols and techniques. This part includes a comprehensive collection of the recent state-of-art protocols and techniques to secure RFID and avoid all potential security forces and cracks. Part III deals with the hardware implementations of cryptography algorithms and protocols dedicated to RFID platforms and chips. This part is very useful for developers to develop practical systems.

This book has the following salient features:

- Identifies the basic concepts, key technologies, and cutting-edge research outcomes of RFID security
- Provides comprehensive references on state-of-the-art technologies for RFID security
- Contains a sufficient number of illustrative figures for easy reading and understanding of the materials
- Details the hardware implementation for the algorithms in RFID security
- Exploits the selected techniques for enhancing RFID security and performance

This book represents a useful and comprehensive reference for RFID basics and RFID security. The book is written for people interested in wireless networks and mobile communications at all levels, especially the researchers and engineers worrying about the security issues in wireless communications. The principal audiences include students, educators, engineers, VLSI developers, scientists, researchers, and research strategists. It can also be used as textbook for an advanced selected topic course on RFID security for graduate students.

This book could not be possible without the great efforts and time invested by all the contributors. They were extremely professional and cooperative, and did a great job in the production of this book. Our reviewers provided valuable comments/feedback, which, we believe, greatly helped to improve the quality of this book. Special thanks go to Ward, Jason, and Caitlin L. Womersley of Springer for their continued support, patience, and professionalism from the beginning to the final stage. Last but not least, we thank our families and friends for their constant encouragement, patience, and understanding throughput this project, which was a pleasant and rewarding experience for us.

Patras, Greece *Paris Kitsos*
Oslo, Norway *Yan Zhang*

Editors' Biographies

Paris Kitsos received the BS degree in Physics in 1999 and a PhD in 2004 from the Department of Electrical and Computer Engineering, both at the University of Patras. From June 2005, he is a research fellow with the Digital Systems & Media Computing Laboratory, School of Science & Technology, Hellenic Open University (HOU), Greece (http://dsmc.eap.gr/en/main.php). He is an associate editor of Computer and Electrical Engineering, an International Journal (Elsevier) and member on the editorial board of International Journal of Reconfigurable Computing (Hindawi). Except this book he is serving as co-editor for one more book: Security in RFID and Sensor Networks to be published by Auerbach Publications, Taylor & Francis Group. He has participated as program and technical committee member in more than 40 conferences and workshops in the area of his research. Also he has participated as guest co-editor in the following special issues: Computer and Electrical Engineering, An International Journal (Elsevier Ltd) with subject Security of Computers & Networks; Wireless Personal Communications, An International Journal (Springer) with subject Information Security and Data Protection in Future Generation Communication and Networking; and Security and Communication Network (SCN), Wiley Journal with subject Secure Multimedia Communication. His research interests include VLSI design, hardware implementations of cryptographic algorithms and security protocols for wireless communication systems, and hardware implementations of RFID cryptography algorithms. Dr. Kitsos is adjunct assistant professor in the Department of Computer Engineering & Informatics, University of Patras and adjunct lecturer in the Department of Computer Science and Technology, University of Peloponnese. He has published more than 60 publications in international journals, books, and technical reports, as well as is reviewing manuscripts for books, international journals, and conferences/workshops in the areas of his research. Also, he is a member of the Institute of Electrical and Electronics Engineers (IEEE) and Institution of Electrical Engineers (IEE).
Email: pkitsos@eap.gr
http://dsmc.eap.gr/en/members/pkitsos/

Yan Zhang received the PhD degree in School of Electrical and Electronics Engineering, Nanyang Technological University, Singapore. From August 2006, he works with Simula Research Laboratory, Norway (http://www.simula.no/). He is associate editor of Security and Communication Networks (Wiley); on the editorial board of International Journal of Network Security; International Journal of Ubiquitous Computing, Transactions on Internet and Information Systems (TIIS); International Journal of Autonomous and Adaptive Communications Systems (IJAACS) and International Journal of Smart Home (IJSH). He is currently serving the Book Series Editor for the book series on "Wireless Networks and Mobile Communications" (Auerbach Publications, CRC Press, Taylor and Francis Group). He serves as guest co-editor for Wiley Security and Communication Networks special issue on "Secure Multimedia Communication"; guest co-editor for Springer Wireless Personal Communications special issue on selected papers from ISWCS 2007; guest co-editor for Elsevier Computer Communications special issue on "Adaptive Multicarrier Communications and Networks"; guest co-editor for Inderscience International Journal of Autonomous and Adaptive Communications Systems (IJAACS) special issue on "Cognitive Radio Systems"; guest co-editor for The Journal of Universal Computer Science (JUCS) special issue on "Multimedia Security in Communication"; guest co-editor for Springer Journal of Cluster Computing special Issue on "Algorithm and Distributed Computing in Wireless Sensor Networks"; guest co-editor for EURASIP Journal on Wireless Communications and Networking (JWCN) special Issue on "OFDMA Architectures, Protocols, and Applications"; guest co-editor for Springer Journal of Wireless Personal Communications special Issue on "Security and Multimodality in Pervasive Environments".

He is serving as co-editor for several books: Resource, Mobility and Security Management in Wireless Networks and Mobile Communications; Wireless Mesh Networking: Architectures, Protocols and Standards; Millimeter-Wave Technology in Wireless PAN, LAN and MAN; Distributed Antenna Systems: Open Architecture for Future Wireless Communications; Security in Wireless Mesh Networks; Mobile WiMAX: Toward Broadband Wireless Metropolitan Area Networks; Wireless Quality-of-Service: Techniques, Standards and Applications; Broadband Mobile Multimedia: Techniques and Applications; Internet of Things: From RFID to the Next-Generation Pervasive Networked Systems; Unlicensed Mobile Access Technology: Protocols, Architectures, Security, Standards and Applications; Cooperative Wireless Communications; WiMAX Network Planning and Optimization; RFID Security: Techniques, Protocols and System-On-Chip Design; Autonomic Computing and Networking; Security in RFID and Sensor Networks; Handbook of Research on Wireless Security; Handbook of Research on Secure Multimedia Distribution; RFID and Sensor Networks; Cognitive Radio Networks; Wireless Technologies for Intelligent Transportation Systems; Vehicular Networks: Techniques, Standards and Applications; and Orthogonal Frequency Division Multiple Access (OFDMA).

He serves Track Co-Chair for ITNG 2009, Publicity Co-Chair for SMPE 2009, Publicity Co-Chair for COMSWARE 2009, Publicity Co-Chair for ISA 2009, General Co-Chair for WAMSNet 2008, Publicity Co-Chair for TrustCom 2008, Workshop General Co-Chair for COGCOM 2008, Workshop Co-Chair for IEEE APSCC 2008, Workshop General Co-Chair for WITS-08, Program Co-Chair for PCAC 2008, Workshop General Co-Chair for CONET 2008, Workshop Chair for SecTech 2008, Workshop Chair for SEA 2008, Workshop Co-Organizer for MUSIC'08, Workshop Co-Organizer for 4G-WiMAX 2008, Publicity Co-Chair for SMPE-08, International Journals Coordinating Co-Chair for FGCN-08, Publicity Co-Chair for ICCCAS 2008, Workshop Chair for ISA 2008, Symposium Co-Chair for ChinaCom 2008, Industrial Co-Chair for MobiHoc 2008, Program Co-Chair for UIC-08, General Co-Chair for CoNET 2007, General Co-Chair for WAMSNet 2007, Workshop Co-Chair FGCN 2007, Program Vice Co-Chair for IEEE ISM 2007, Publicity Co-Chair for UIC-07, Publication Chair for IEEE ISWCS 2007, Program Co-Chair for IEEE PCAC'07, Special Track Co-Chair for "Mobility and Resource Management in Wireless/Mobile Networks" in ITNG 2007, Special Session Co-organizer for "Wireless Mesh Networks" in PDCS 2006, a member of Technical

Program Committee for numerous international conference, including ICC, PIMRC, CCNC, AINA, GLOBECOM, ISWCS etc. He received the Best Paper Award and Outstanding Service Award in the IEEE 21st International Conference on Advanced Information Networking and Applications (AINA-07).

His research interests include resource, mobility, spectrum, energy and security management in wireless networks and mobile computing. He is a member of IEEE and IEEE ComSoc.

Email: yanzhang@ieee.org

http://home.simula.no/~yanzhang/

Contents

Part I
Fundamentals

RFID: Fundamentals and Applications

Andreas Hagl* **and Konstantin Aslanidis**

Abstract This section of the book will give you an overview of the basic RFID functionality as well as current and future applications, without going into deep technical details. The information may also help developers from other technology areas, to understand the basics of RFID and to develop efficient concepts and techniques suitable for RFID systems.

1 Introduction

Radio Frequency Identification (RFID) is the most reliable way to electronically identify, data capture, control, track, and inventory items using RF communication. Today RFID is ubiquitous having a very broad use but most of the time such systems are invisible or are not recognized by the users.

The basic RFID system consists of a *Reader* and a *Transponder*.

The *Reader* or *Transceiver* is the unit acting as the master and supplies the RFID transponder with energy and triggers the communication signals to force the transponder to execute the requested action. The reader control can be either via a computer terminal or the automatic execution of program scripts. In stationary installations, fixed readers are connected to power and communication lines, whereas in mobile applications, hand held readers (not connected to main power or communication lines) are used. For further data exchange, the reader may be connected to a host computer or database as shown in Fig. 1.

The *Transponder* or *Tag* is the identification device which is located on the item to be identified. Most RFID transponders are without an internal power source (battery) and are called *passive* transponders. The power supply of a tag is the RF field generated by the reader. The tag generates its own supply voltage by rectifying the

A. Hagl

Texas Instruments Deutschland GmbH, Haggertystrasse 1, 85356 Freising, Germany

e-mail: a-hagl@ti.com

P. Kitsos, Y. Zhang (eds.), *RFID Security: Techniques, Protocols and System-on-Chip Design,* © Springer Science+Business Media, LLC 2008

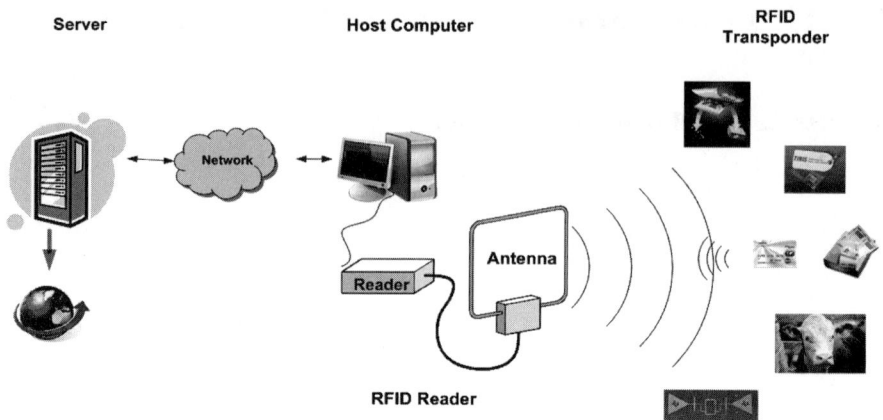

Fig. 1 RFID system

induced voltage from the Reader's RF signal. *Active* transponders have an integrated power source (internal battery) and behave the same way as passive devices but with increased performance. These tags are using the battery to supply the circuitry and to generate the response data. Their activation is mostly triggered by the reader signal.

2 History

RFID technology developments started back in the 1950s where these systems were developed specifically for Governmental and Military use in USA and USSR. Semiconductor technologies at that time were in their infancy and devices were large in size with high-current consumption and expensive, which did not recommend their use for passive RFID systems. The real explosion of passive RFID technology was at the end of the 1980s and was made possible by the improved size, current consumption of the circuitries, and price of semiconductor technologies. This enabled an acceptable RFID performance (communication distance) for passive systems under acceptable investment.

The first generations of RFID tags were only used as identification devices, having only a fixed identification code stored into the tag's memory. There was mainly a one way communication with the tag communicating back its memory content when triggered by reader activation. An example of an early RFID system patent from Mario Cardullo [2] is shown in Fig. 2.

Now RFID systems are widely used in applications with the primary task to identify items, but there are also new applications where higher security and computation as well as integrated sensors and actors are required. Due to the current cost structure of RFID systems, new application fields can be justified based on Return of Investment (ROI).

United States Patent [19]

Cardullo et al.

[11] **3,713,148**

[45] **Jan. 23, 1973**

[54] **TRANSPONDER APPARATUS AND SYSTEM**

[75] Inventors: **Mario W. Cardullo**, Rockville; **William L. Parks, III**, Bethesda, both of Md.

[73] Assignee: **Communications Services Corporation, Inc.**, Rockville, Md.

[22] Filed: **May 21, 1970**

[21] Appl. No.: **39,309**

[52] U.S. Cl. ....343/6.5 R, 343/6.8 R
[51] Int. Cl..G01s 9/56
[58] Field of Search ....343/6.5 R, 6.5 LC, 6.5 SS, 343/6.8 R, 6.8 LC

[56]   **References Cited**

UNITED STATES PATENTS

3,541,257   11/1970   McCormick et al.........343/6.5 LC X
3,144,645   8/1964   McIver et al...................343/6.5 R X

Primary Examiner—T. H. Tubbesing
Attorney—Jacobi, Lilling & Siegel

[57]   **ABSTRACT**

A novel transponder apparatus and system is disclosed, the system being of the general type wherein a base station transmits an "interrogation" signal to a remote transponder, the transponder responding with an "answerback" transmission. The transponder includes a changeable or writable memory, and means responsive to the transmitted interrogation signal for processing the signal and for selectively writing data into or reading data out from the memory. The transponder then transmits an answerback signal from the data read-out from its internal memory, which signal may be interpreted at the base station. In the preferred inventive embodiment, the transponder generates its own operating power from the transmitted interrogation signal, such that the transponder apparatus is self-contained.

7 Claims, 3 Drawing Figures

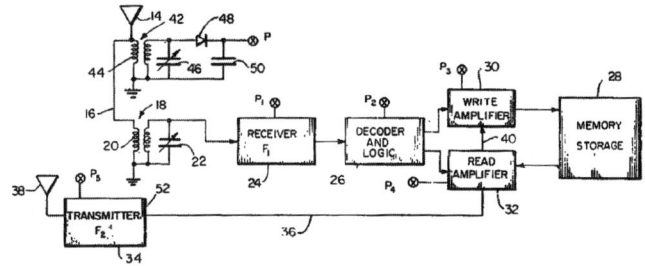

Fig. 2 RFID patent

3 RFID System Basics

Transponders basically operate as active or passive devices. The functionality of both types is similar; the main difference is the increased performance in view of communication distance and computation capabilities of the active vs. the lower cost of the passive transponders. The integrated battery increases the cost of the transponder, limits the tag's life time, causes environmental issues over disposal, and limits the form factor and thickness of the tag. These disadvantages of the active transponders limit the applications where these tags can be used. Due to the very high market share of the passive technology, only this technology will be presented in the following sections.

The Tag mostly acts as a slave and relies on the reader to activate it using the "Reader Talks First" (RTF) concept. The reader supplies energy via the RF field and transmit requests/commands to instruct the tag about the action to be executed. The

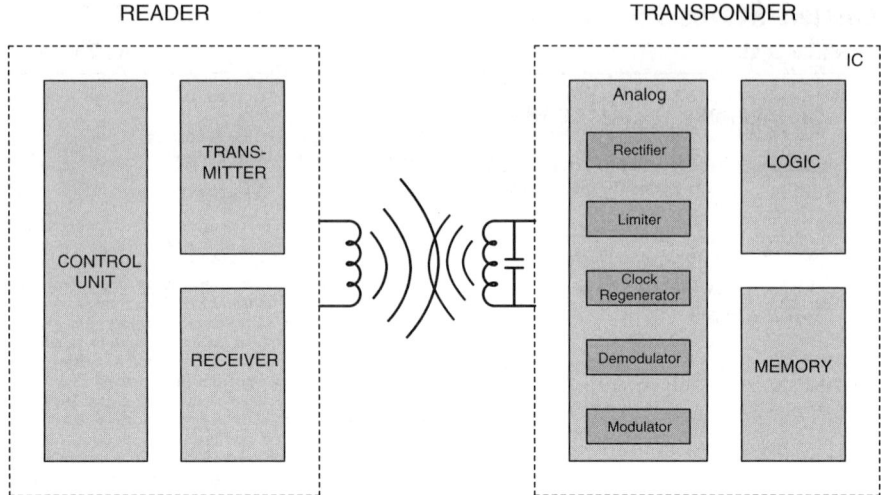

Fig. 3 RFID functional blocks

tag receives and decodes RF signals coming from the reader, executes the instructed action, and may respond with data or status information. The cost structure of the tag can be roughly split in costs for IC, antenna, assembly, and test. The electronics part (IC, Integrated Circuit) of the tag consists of some basic functional modules which are used to enable certain functionality as shown in Fig. 3:

- The induced voltage is rectified by the *Rectifier* to supply the IC with energy.
- The *Limiter* limits the RF voltage at the inputs pins to avoid over voltage which would destroy the circuitry.
- The *Clock Regenerator* extracts the frequency signal from the RF signal which is used as an internal clock.
- The *Demodulator* decodes the incoming data signal and generates a binary bit stream representing the command and data to be executed. These data are used by the IC to execute the requested activities.
- The *Modulator* modulates the decoded response data.
- The *Logic* part represents the microcontroller or digital circuitry of the tag.
- The *Memory* unit (mostly EEPROM) contains the tag specific data as well as additional memory where application specific data can be programmed.

The Reader consists of a control unit and the radio frequency (RF) unit containing the transmitter and the receiver modules. In the control unit, the firmware and the hardware is implemented to control the reader activities such as communications with a host computer and the tag, as well as data processing. The transmitter generates the RF signal (frequency and power level) which is connected to the antenna resonance circuit. The receiver part receives the RF signal generated by the tag, demodulates and decodes the data, and sends the binary data to the control unit for further processing.

3.1 Why RFID

In the past, the most used identification system has been the barcode. The main reason for the wide usage of this system is the low cost of a barcode by simply printing it on the items and the improved performance (detection rate, and reliability) of the new generation of scanners. There are still some disadvantages of this technology though:

- Data cannot be modified or added
- Requires line of sight for operation (label must be seen by the reader)
- High maintenance effort for the complex scanner optics

Modern application processes like item tracking, require extended capabilities of the ID system which cannot be achieved by the barcodes. In these applications, RFID systems can add value through extended functionality. This should not be misunderstood to imply the complete replacement of barcodes by RFID. RFID is an alternative to barcodes which will lead to a coexistence of both technologies based on the performance and capability requirements and the specific investment to use the RFID technology for these applications. Most applications will require the use of both, barcode and RFID in parallel.

Summarizing the advantages of the RFID systems in relation to other identification systems currently in use and especially barcode:

- Battery-less. Supply voltage derived from the RF field
- No line-of-sight required for the communication
- Large operating and communication range
- Read and Write capability of the transponder memory
- High communication speed
- High data capacity (user memory)
- High data security
- Data encryption/authentication capability
- Multiple tag read capability with anticollision (50–100 tags)
- Durability and reliability
- Resistant to environmental influence
- Reusability of the transponder
- Hands free operation
- Miniaturized (IC size $<1\,\mathrm{mm}^2$)
- Very low power

3.2 RFID System Selection

"The Universal" RFID system which covers all application requirements does not exist. The selection of the right RFID system for a specific application is difficult and depends on a number of factors to be considered. The final solution may be

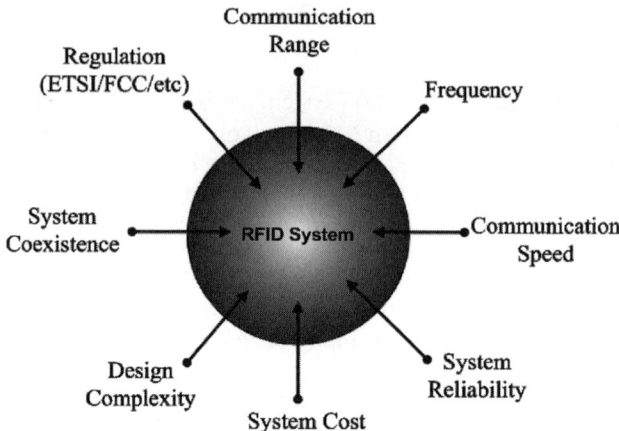

Fig. 4 RFID technology selection parameters

something of a compromise between the various parameters and will represent the "best fit." The most important system parameters are frequency, communication range, and cost.

The choice of the right frequency mainly dictates the communication range and data rate based on the existing regional or worldwide regulations. The right frequency choice in combination with the required features (e.g., memory size, security, etc.) will dictate the design complexity. This complexity will be translated in chip size on the tag side, and associated cost. Due to the high ratio between readers and transponder used in the application, low transponder costs are essential.

For the implementation of RFID technology, additional parameters like standards and system coexistence must be considered (Fig. 4). The use of application and technology standards enables, a multiple sourcing of the system components and the use of lower cost "of the shelf" products. In cases where an installed base already exists, the coexistence (operation in parallel without interfering each other) of the systems must be considered.

3.3 RFID ISO Standards

From the beginning of RFID and until a few years ago, most of the companies involved developed their own systems, trying to improve the technology performance to cover specific customer/application needs. All these solutions were based on similar concepts, but smaller differences especially in the frequency and data coding lead to interoperability issues of these technologies in the field and a "monopoly" for the component sourcing. Users were bound to a single supplier with all the disadvantages related to a single source.

Users and RFID manufactures recognized that this situation could not help to enter new high volume applications where competition, multiple source, and interoperability are the main requirements. The best way out of this situation was to define International Organization for Standardization (ISO) standards [11]. Standards may not provide the optimized cost and performance and mostly represent a compromise over all the available systems, but can guarantee the minimum functionality and interoperability at least for the majority of the applications.

The most important RFID Standards defining communications cover the following applications:

- Animal identification
- Item management
- Logistics
- Access control
- Smart cards
- Payment
- e-Passports
- Waste management

Low Frequency Standards 134.2/125 kHz:

- ISO11784 – Animal ID code structure
- ISO11785 – Animal ID technical concept
- ISO14223 – Animal; Advanced transponders
- ISO18000-2 – Item management; Air interface and protocol structure for LF

High Frequency Standards 13.56 MHz:

ISO/IEC15693 – Vicinity cards;
ISO/IEC14443 – Proximity cards;
ISOIEC18000-3 – Item management; Air interface and protocol structure for HF

Ultra High Frequency Standards 868/915 MHz:

ISO/IEC18000-6 – Air interface and protocol structure
EPC Gen 2 – Air interface and protocol structure (ISO18000-6C) [5]

There are also some additional standards available defining the Air interface communication at other ISM frequency bands which mainly cover some niche applications.

3.4 Frequency Regulation

To allow the use of radio frequency communication without interference from other services, the frequency spectrum has been split in spectrum bands and assigned to specific services which are allowed to operate within that spectrum. These are

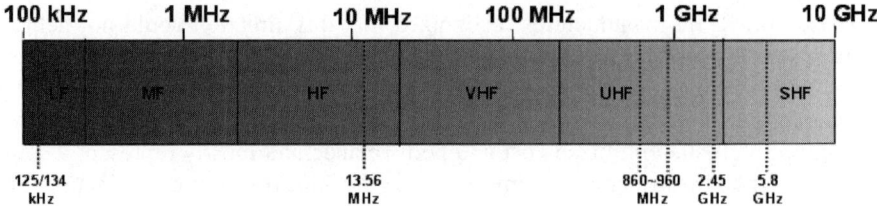

Fig. 5 Frequency spectrum

called primary services, like cell phone, astronomy, broadcast, etc. The frequency spectrum is managed by the different national and international organizations like ITU [12], FCC [8], ETSI [7], CEPT [3], ERO [6] etc. In addition to these primary user bands there are some frequency bands defined which can be used by different services. These bands are called industrial, scientific and medical (ISM). RFID is not assigned as a primary user of a particular frequency band, and is only allowed to operate within these general use ISM bands. The most important ISM bands for RFID systems are following frequency bands:

125/134 kHz, 13.56 MHz, 868/915 MHz, 2.45 GHz, and 5.8 GHz (Fig. 5).

There is not an ideal operating frequency, which can cover all application requirements. Every frequency has its pros and cons, which are given by the properties of the frequency itself but also given from the regulation restrictions (level and bandwidth).

4 RFID Systems and Applications

RFID systems [10,15] operate mainly in three frequency bands which are worldwide available.

- LF – Low frequency
- HF – High frequency
- UHF – Ultra high frequency

4.1 Low Frequency (LF)

A large variety of applications are using LF frequency for the RF communication (Fig. 6). Two basic systems dominate this band, FDX-B (Full Duplex) and HDX (Half Duplex). The communication concept of these systems is different, but due to their properties, both systems can coexist without interfering with each other. LF systems use the magnetic component of the electromagnetic field.

At low frequency, the antenna volume and the form factor mean that tag volumes may be higher than for tags in higher frequencies but this allows for higher flexibility

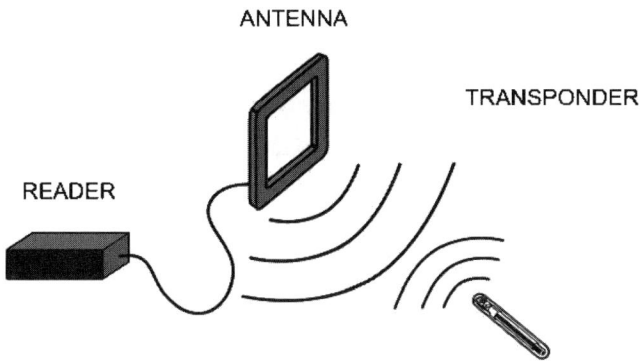

Fig. 6 LF communication

and variations in the antenna design. The typical operating distance under normal operation is around 1 m.

The applications with the highest volumes are animal identification, vehicle immobilizers, access control, and leisure applications. The main tag form factors are cards for access control, discs for industrial process control, and ear tags and glass encapsulated transponders for animal identification. Glass transponders are used for injection into pets and fish, as well inside boluses for ruminant animals.

4.1.1 FDX Technology

FDX systems use a communication concept, where, at the same time as the tag is supplied with energy by the readers' RF field, the RF field is modulated by the tags modulator circuit. The field modulation is in accordance with the data to be transmitted to the reader. Typically the FDX systems use amplitude shift keying (ASK) as their modulation concept. The communication concept is shown in Fig. 7.

The characteristics of the FDX technology are:

- RF frequency is used as system clock. No internal oscillator is needed
- Simple, low cost implementation
- Optimization of "Reader to Tag" and "Tag to Reader" communications cannot be done. The final adjustment of the system is a compromise between the two communication directions
- The system may be further compromised by using lower quality factors for the resonance frequency circuits
- High signal to noise (S/N) ratio of the ASK modulation can result in reduced performance and communication reliability
- The simultaneous presence of the strong reader RF signal and the weak tag response signal (about $-60\,\text{dB}$) means that readers have to use sophisticated methods to extract the response signal from the field. The bit error rate of this communication concept and data modulation is quite high
- Minimum number of components for the transponder (antenna and IC)

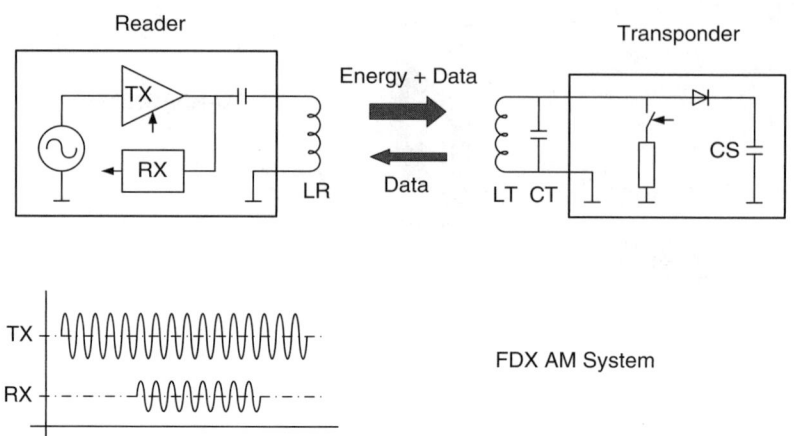

Fig. 7 FDX system operation

Due to the continuous RF signal during communication, the system does not need an additional external storage capacitor (CS) for operation, but some of the systems use an external resonance capacitor (CT) parallel to the antenna (LT). Due to the low capacitance of the resonance capacitor, this capacitor can be integrated in the IC.

4.1.2 HDX Technology

HDX systems [13] use a communication concept, where the tag energy supply and the response data communication take place sequentially. HDX systems use frequency shift keying (FSK) modulation for the data communication from the tag to the reader. The main difference between FDX and HDX systems is that the HDX system needs a storage capacitor (CS) to store its energy for operation, before the response of the data. During the so called "Charge Phase," the readers' RF field induces energy in the tag which is stored in the charge capacitor. After the charge phase, the reader switches off the RF field, and the transponder sends the response using the charge capacitor as an "internal" power source. The communication concept is shown in Fig. 8.

The characteristics of the HDX technology are:

- RF frequency cannot be used as system clock. An internal oscillator or clock generation is needed.
- Sophisticated implementation concepts are needed.
- Optimized performance for the Reader to Tag communication, which means High Quality factor resonance circuits can be used on the reader and tag side, allowing higher induced voltages on the tag side.

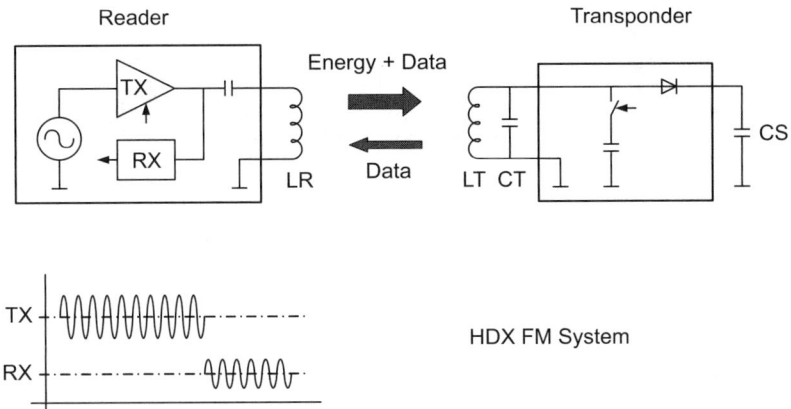

Fig. 8 HDX system operation

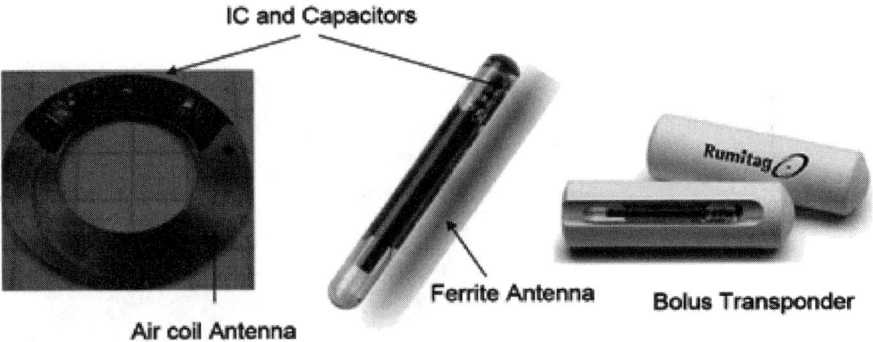

Fig. 9 HDX transponders [16, 17]

- During Tag to Reader communication (RX) the quality factor of the resonance circuit of the reader is switched to a lower value, optimizing the bandwidth for higher data rate communication. The adjustment of the system can be optimized for both communication directions.
- The system performance can be improved by use of higher quality factors for the resonance frequency circuits.
- The S/N ratio of the FSK modulation is lower and therefore the reliability of the reception of the return signal is higher.
- The absence of the strong reader RF signal and the associated noise during the response means a simple reader can be used and the bit error rate is quite low.
- It needs external storage capacitor, which due to the required capacity of 100–300 nF cannot be integrated (Fig. 9).

Examples of LF transponders with different form factors are shown in Fig. 10 and Fig. 11.

LF readers Fig. 12 are available as mobile readers with internal antenna and stationary readers with external antennae.

Fig. 10 LF transponder form factor overview [17]

Fig. 11 Eartag transponders [1]

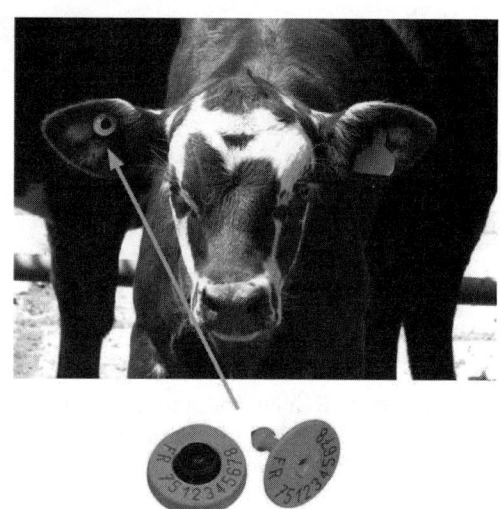

4.2 High Frequency (HF)

HF systems operate at 13.56 MHz, using mainly the FDX communication concept as used for LF systems. HF systems also use the magnetic component of the electromagnetic field to transfer energy (Fig. 13).

Due to the higher frequency, there are some advantages compared with LF, such as the ability to have higher data rates. Due to higher resonance frequency of the antenna circuit, the inductance and the capacitance values used in the resonance circuit can be reduced. The lower antenna inductance of the transponder means a fewer number of turns (5–10) (Fig. 15) (compared to 200–300 turns for the LF

Mobile readers

Low power reader module

High power RF module

LF antennae

Fig. 12 LF readers and antennae [16, 17]

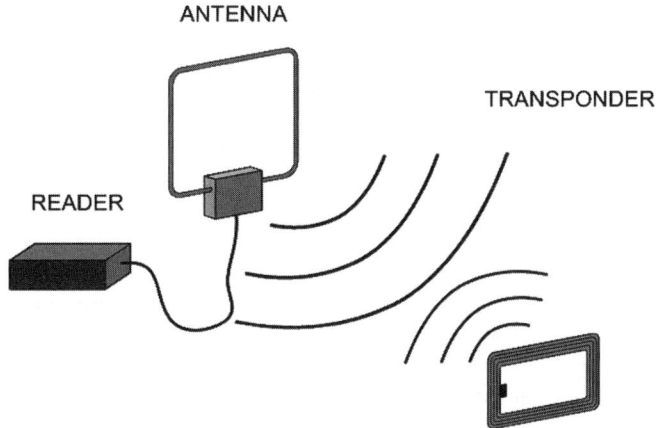

ANTENNA

TRANSPONDER

READER

Fig. 13 HF communication

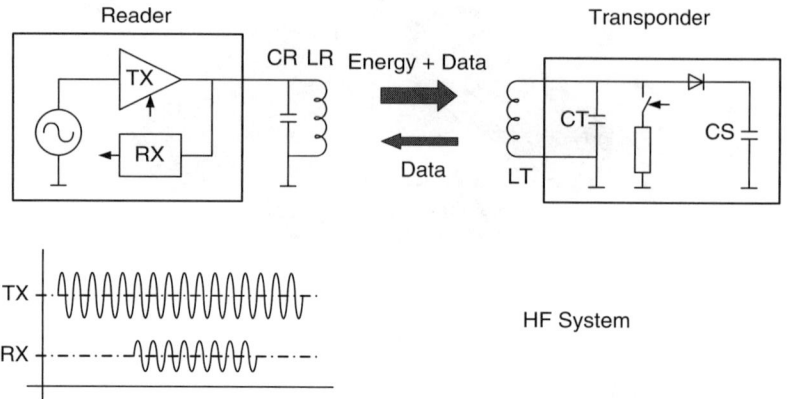

Fig. 14 HF system operation

systems), gives the possibility of being produced not only from copper wire but also printed or etched on foil. These antennas are flexible and can be easily laminated in a credit card or label form factor. These basic advantages lead to a lowered cost for the system.

There is a variety of HF application. The higher volumes applications are banking/payment, [4] credit cards, access control, ticketing, item tracking, library, laundry, drug pedigree (Fig. 16) and authentication, and supply chain management. The typical operating distance under normal operation is about 0.7 m.

The main tag form factors are PVC cards, paper labels, and encapsulated tags. The credit card sized tags (Fig. 15) are used for payment and payment applications as well as access control and ticketing. Tags laminated in paper are mostly used for item tracking and ticketing. Encapsulated tags (Fig. 17) are used for applications like Laundries, where these tags have to withstand in the rough environmental conditions such as humidity, temperature, and chemicals. The communication concept of HF systems is shown in Fig. 14.

The characteristics of the HF technology are:

- The carrier frequency is used as system clock. No internal oscillator needed.
- Simple, low cost transponder.
- Higher data rate capability.
- Support of simultaneous identification of transponders (see Fig. 16) based on collision avoidance protocols.
- The simultaneous presence of the strong reader RF signal and the weak tag response signal (about $-60\,dB$) means that reader has to use sophisticated methods to extract the response signal from the field. The bit error rate of this communication concept and data modulation is quite high.
- Minimum number of external components (only antenna and IC, Fig. 15).

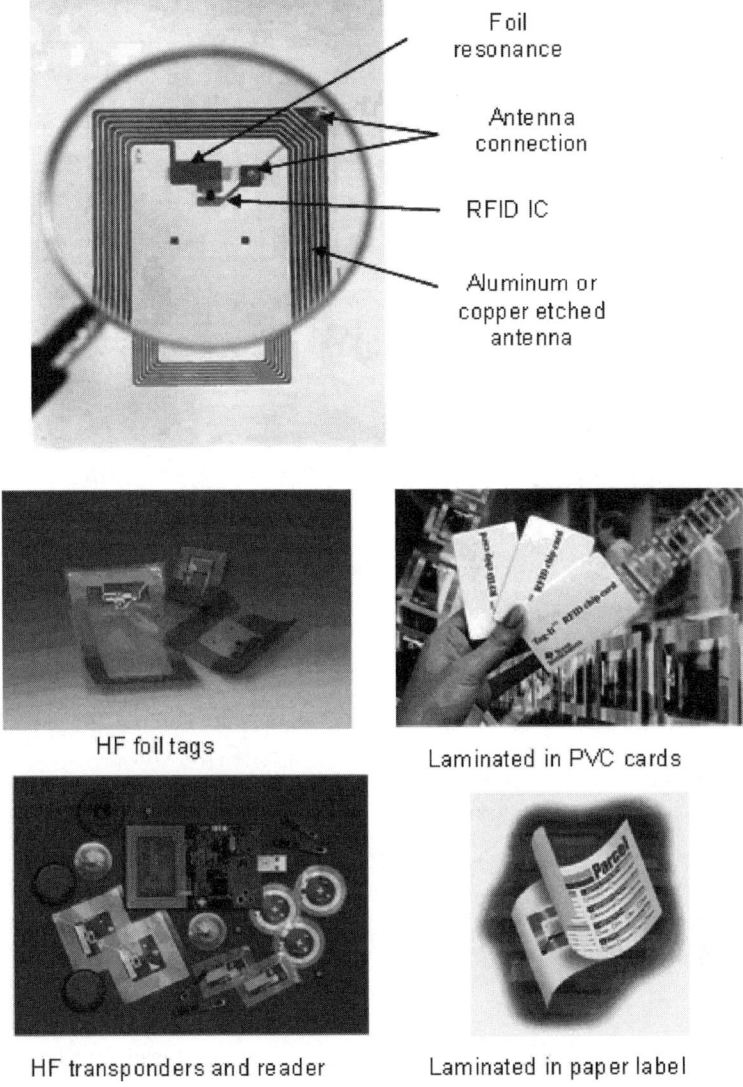

Foil
resonance

Antenna
connection

RFID IC

Aluminum or
copper etched
antenna

HF foil tags

Laminated in PVC cards

HF transponders and reader

Laminated in paper label

Fig. 15 HF transponders [17]

- Flexible tags on foil, which can be laminated (PVC or paper, Fig. 15).
- Worldwide regulation of frequency and power levels.

The tag is supplied with energy by the continuous presence of the RF signal during communication; therefore the system does not need an additional external capacitor to store energy for operation (Fig. 14). Due to the low capacitance of the transponder resonance capacitor (CT), this can be integrated in the IC. In that case, the components needed are an IC and an antenna (Fig. 14).

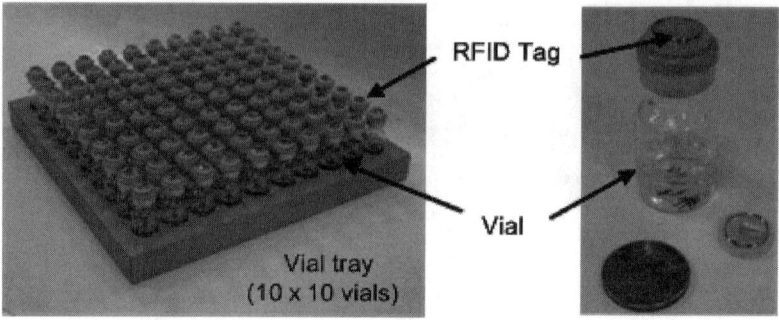

Fig. 16 Pharmaceutical vial identification samples [17]

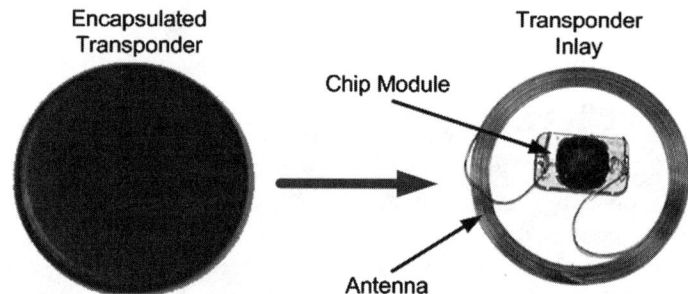

Fig. 17 Encapsulated transponder [17]

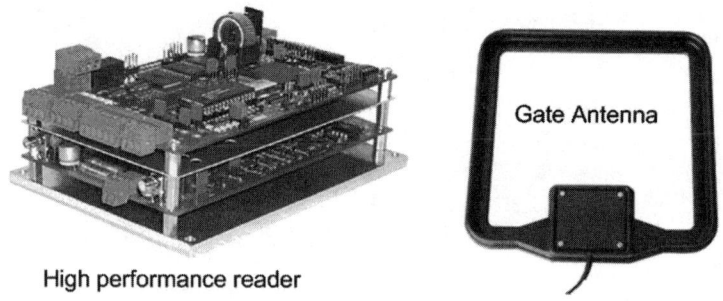

Fig. 18 HF reader and antenna [9, 17]

4.3 Ultra High Frequency (UHF)

UHF systems operate in the frequency bands of 866–868 MHz in Europe, 902–928 MHz in USA, and 952–954 MHz in Japan. The electric component of the electromagnetic field is used for energy propagation with backscatter modulation used for the tag's response (Fig. 19).

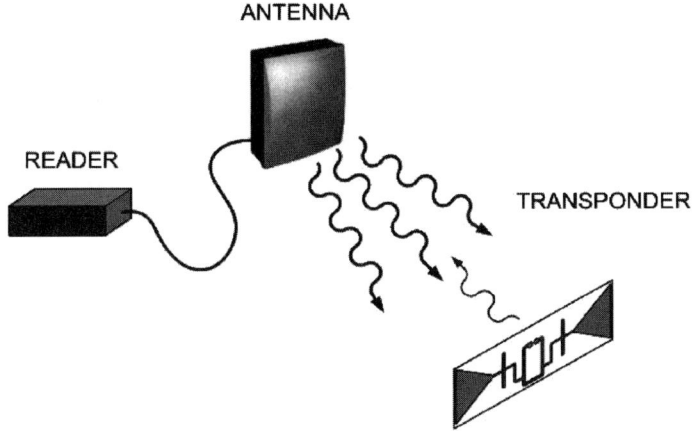

Fig. 19 UHF communication

The regulation of the UHF frequency band and power levels is different depending on the region of operation. This complicates the design of the systems which must have the ability to operate in different frequencies.

The antennas used in UHF systems are typically dipole antennas. These antenna can be linear or circular polarised depending on the application. Dipole antennas are directional and do not radiate uniformly in all directions. These antennas are normally etched on foil or printed and can be made very small, at a lowest cost.

One of the characteristics of UHF systems is the so-called "Multipath signal transmission": The signal radiated by a reader may arrive at the tag by different routes. Depending on the object materials (floors, roof, walls, etc) in the operating range, the signal may be reflected, and may arrive at the tag at a slightly different time to the direct signal. This effect may cause field nulls (signal disappears) and field peaks (signal amplified) which influences the system operation. In addition, the signal may be absorbed (signal disappears) by the environment.

At the UHF frequency, tags are also sensitive to materials on which they are attached. This influence can affect the system operating performance, and can be mostly overcome by tuning/selecting the tags for the particular material.

The operating range of the UHF systems depends mainly on the regional regulation (frequency, bandwidth, channels, and power level). The typical operating distance under normal operation is in the range of 1–6 m. The operating concept is shown in Fig. 20.

The UHF system became very popular after the implementation of the "Gen2" system protocol defined by the Electronic Product Code (EPC) organization [5]. These activities are supported by large worldwide operating retailers like Wal-Mart, Metro, Tesco, Marks & Spencer, etc. The main applications targeted by the Gen2 system are the Supply Chain applications, Item management, and Logistics. Currently the growth in Gen2 protocol implementations is only modest partly because the cost expectation <\$0.05 for a complete tag cannot be achieved within today (Figs. 21).

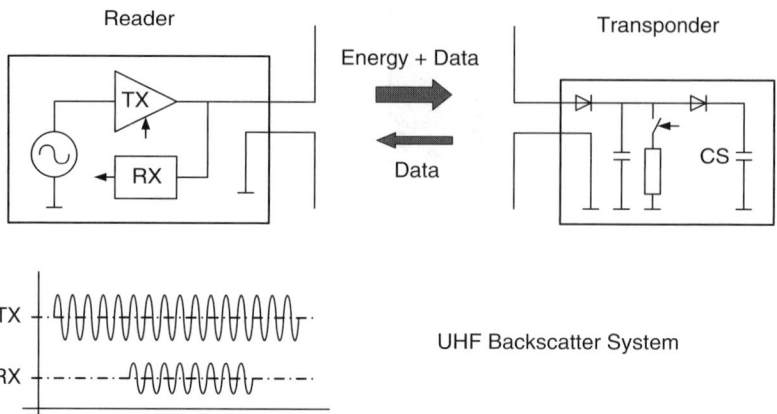

Fig. 20 UHF system operation

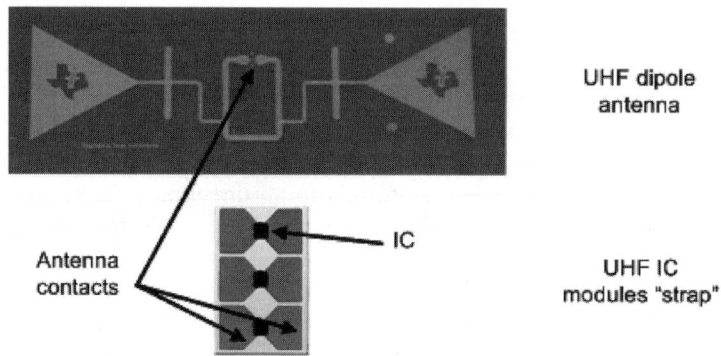

Fig. 21 UHF tags [17]

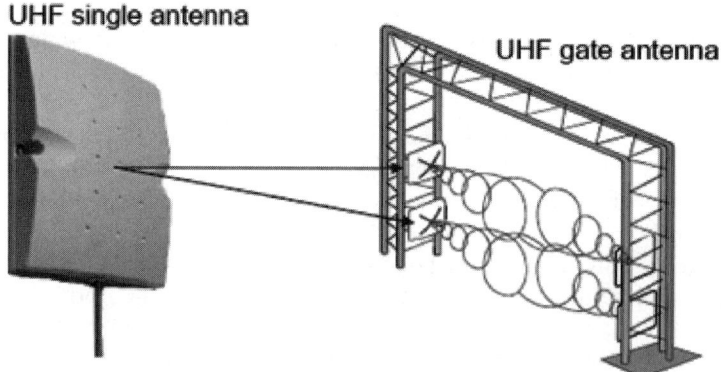

Fig. 22 UHF single antenna and gate antenna [9]

5 RFID in Automotive

Automobile theft has increased year by year creating demands for extra vehicle security systems beyond the mechanical key. Especially after boarders with Eastern Europe fell in the 1990s, automobile theft has increased dramatically. European insurances companies required immobilizer systems to decrease theft rates.

Different systems appeared on the market such as locks with pin codes and electronic contact sticks, besides the mechanical key. Systems with RFID transponders appeared to be most convenient for the user because they did not require an additional process to start the engine, they were transparent to the user, and the RFID transponder operated battery-less. Since the adoption of the immobilizer system with RFID transponders the automobile theft rate decreased continuously in Europe. Immobilizer systems are now used worldwide and the majority of new cars are equipped with immobilizer systems based on RFID technology. Immobilizer systems were the first high volume application for RFID.

5.1 Immobilizer System

An immobilizer system (Fig. 23) consists of an RFID transponder (Fig. 25) and an RFID transceiver to communicate with the transponder. The transponder is encapsulated in the mechanical key and the transceiver is located typically close to the mechanical key lock at the steering column. The communication between transponder and transceiver is typically controlled by the dash board control unit or engine control unit.

Immobilizer systems operate at low frequency of 120–140 kHz. The transceiver module is mounted at the ignition lock cylinder and the air coil antenna is mounted around the ignition lock cylinder (Fig. 24). The transceiver module contains mostly just an RFID reader ASIC and a few external components. Because of the short communication range requirements of 1–5 cm, both systems FDX and HDX are commonly used. The communication time is unrecognized by the driver because it is well below 100 ms.

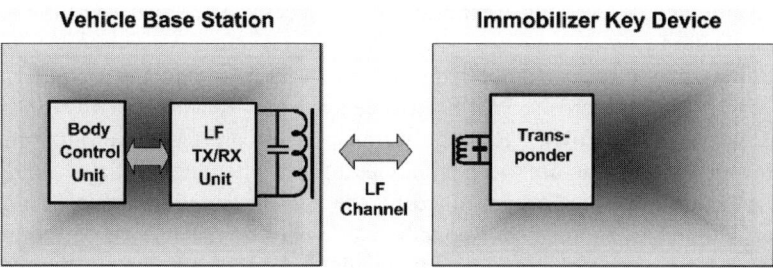

Fig. 23 Immobilizer system

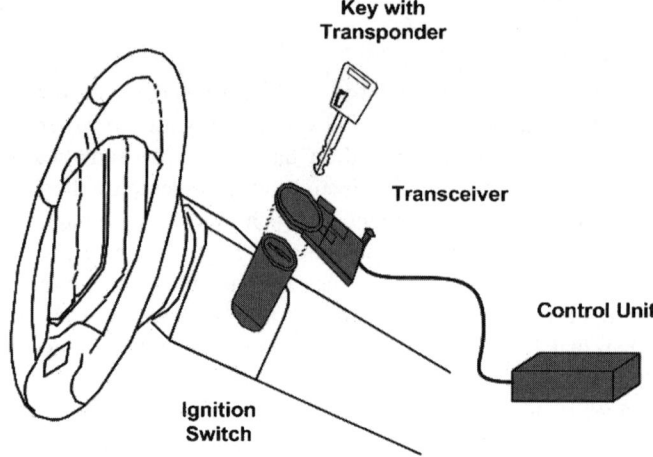

Fig. 24 Immobilizer system with key lock

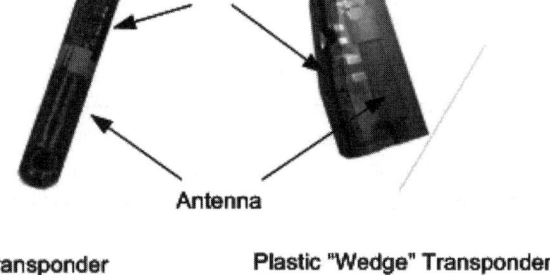

Fig. 25 Transponders used for immobilizer [17]

5.1.1 System Operation

The driver places the ignition key into the ignition lock cylinder and turns the key. Not recognized by the driver the communication between transponder and transceiver is initiated by the control module. After successful verification of the key the engine is started.

The first generation of immobilizer systems used fixed code transponders. Two types of transponder were used for fixed code systems – Read Only (R/O) and Write Once Read Many (WORM). Read-only transponders are factory programmed by the transponder manufacturer with a unique number where no duplicate is allowed. WORM transponders are programmed by the system supplier with a non-unique number. Duplicates of such transponders can easily be made by programming of "blank" transponders with commercially available RFID readers.

Some immobilizer applications used rolling code technology. A read–write transponder with EEPROM is reprogrammed after each successful immobilization with a new number. At the next read cycle the new number is checked for a valid immobilization. Advantage of this approach is that a transponder number is only valid for one immobilization.

A different approach to increase the security level of the immobilizer system is to use a password-protected transponder. The transponder responds only after the reception of valid password. Typical passwords used have a length of 8–32 bits. Advantage of this approach is that a transponder is secured against activation away from the vehicle. An attacker may attempt to guess the password until the transponder responds. This is quite easy if the password is just 8 bits but gets time consuming if a password is 32 bits when each attempt has a communication time of 100 ms.

Today's immobilizer systems use challenge–response authentication technology with encryption key length of 40–96 bits. A random challenge number is sent to the transponder. After calculation with an encryption module the transponder replies with the encrypted response. The control module verifies the correctness of the response signal before starting the engine.

In future, immobilizer systems require longer encryption keys. Across all car manufacture it seems that the Advanced Encryption Standard (AES) with 128-bit key length is the preferred encryption algorithm. Furthermore AES will also be used for authentication of modules inside the vehicle.

5.2 Remote Keyless Entry System

First generation Remote Keyless Entry (RKE) systems (Fig. 26) were introduced before immobilizers were used. The RKE function was implemented as convenience feature unlocking and locking doors within 10 m around the vehicle. They were battery operated and in case of an empty battery the door could be opened with the mechanical key. Frequencies used in the UHF band are 315 MHz in the USA and 433 MHz in Europe. The first generation of RKE systems used fixed code transmission with lowest security. They were replaced by rolling code systems with higher security level.

With implementation of the first immobilizers into vehicles the RKE system and Immobilizer system were totally separated. The transponder was used for immobilization and the RKE key was used for unlocking and locking of the car. Because of cost reasons, both systems are today combined in a single integrated circuit [14]. A low frequency ferrite coil antenna is connected to the IC for the transponder function and an UHF transmitter is utilized for RKE function. In case of an empty battery, RKE does not work but the transponder is still functional to guarantee the immobilizer function. Access to the vehicle is guaranteed by at least one mechanical lock at the driver door.

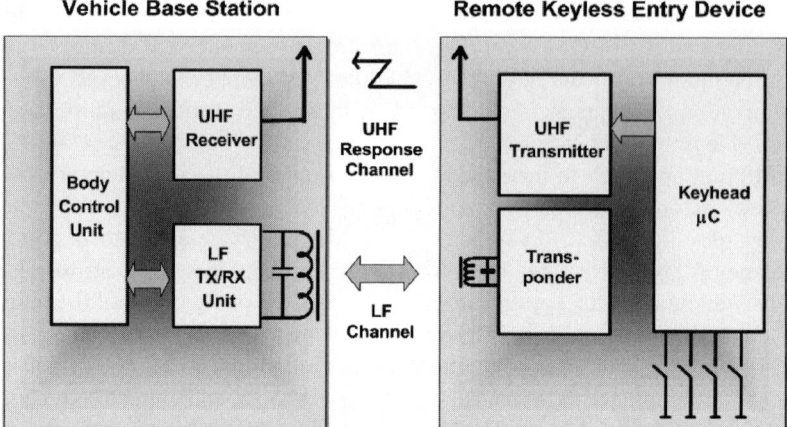

Fig. 26 Remote keyless entry system

5.3 Passive Entry System

Passive entry systems (Fig. 27) were introduced as convenience feature for the driver as it is not necessary to press any button as with RKE devices. The driver only has to go with the passive entry device (electronic key) to the vehicle and pull the door handle. Once inside, the driver has just to push a starter button for the engine to operate. The passive entry device needs to be inside the vehicle for this operation. There are basically two systems on the market: triggered and polling. Triggered systems have detection switches at the door handles to initiate activation/readings, whereas polling systems perform repeated reads with a repetition rate of about 500 ms.

Systems such as these require a precise detection of the key location – it is essential to determine if a key is inside or outside the vehicle. Starting the engine should only be possible if the passive entry device is inside the vehicle and locking of the vehicle should only be possible if the key is outside the vehicle.

The characteristics of the magnetic field are utilized to determine the passive entry key location. In the near field, the magnetic field strength declines with 60 dB per decade, while in the far field it declines with 20 dB per decade (Fig. 28). Therefore, low frequency is used for communication from vehicle to passive entry key. Several low-frequency transmit antennas are located at the vehicle for proper system operation. The key itself consists of three channels to be independent of orientation. If one channel receives a sufficiently strong low-frequency signal from the transmitter it will wake the controller for data processing. The response signal from the passive entry device to the vehicle is sent via UHF similar to RKE systems.

Immobilizer function is also guaranteed in case of an empty battery. The driver has to either put the key in a slot and push the starter button or push the starter button while holding the passive entry key a few centimeters from the starter button. Access to the vehicle is guaranteed by one mechanical key lock at the driver door.

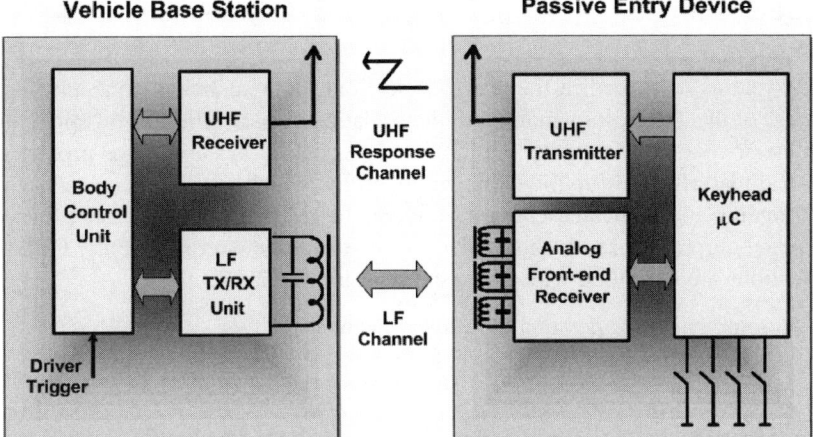

Fig. 27 Passive entry system

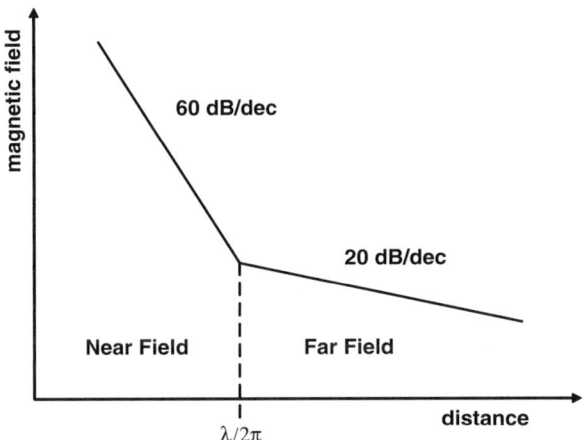

Fig. 28 Characteristics of magnetic field

Both Immobilizer and Passive entry systems typically use challenge–response authentication because of the availability of two-way communication. Remote keyless entry typically uses rolling code encryption because of one-way communication. Today there are many different encryption methods used. Most of them are proprietary to a car or semiconductor manufacturer. Next generation systems will use standardized encryption for immobilizer, passive entry, and remote keyless entry function.

6 Conclusion

RFID systems described in the previous sections show technical concepts and the wide use of this technology mainly in identification. Most of the current applications have no or moderate security implemented. As RFID systems have proven their technical maturity, new applications are identified which require increased security requirements and dedicated encryption technologies. RFID will be used to extend or replace currently used technologies to satisfy the new requirements. Following applications are identified with increased security requirements:

- Electronic passport, personal ID, divers license
- Payment, credit cards
- Drug pedigree
- Automotive

However, technical maturity and technology improvements should not divert the attention from privacy issues. Privacy will be a big challenge for all parties participating in the development of the new applications. Solutions to solve the privacy issue will enable the success of the RFID systems in the new high security applications.

References

1. Allflex. http://www.allflexusa.com
2. Cardullo et al., United States Patent 3,713,148
3. CEPT. http://www.cept.org
4. C. Cook, K. Aslanidis, H. Vollbrecht, Texas Instruments, Inc., Improving Consumers' Contactless Payment Experience White Paper, 2007.
5. Electronic Product Code (EPC), http://www.epcglobalinc.org
6. ERO. http://www.ero.dk
7. ETSI, http://www.etsi.org
8. FCC. http://www.fcc.gov
9. Feig Electronic. http://www.feig.de
10. K. Finkenzeller, RFID Handbook: Radio-Frequency Identification Fundamentals and Applications, Wiley, New York, 1999
11. ISO, http://www.iso.org
12. ITU, http://www.itu.int
13. U. Kaiser and W. Steinhagen, A low-power transponder IC for high-performance identification systems. IEEE Journal of Solid-State Circuits, 30: 306–310, 1995
14. H. Meier, P. Mittertrainer, U. Kaiser, and M. Knebelkamp, A High Security Transponder with Integrated Remote Keyless Entry Function, Proceedings of ISATA, Duesseldorf, June 1998
15. D. Paret, RFID Radiofrequency Identification and Smart Card Applications, Wiley, New York, 2005
16. Rumitag. http://www.rumitag.com
17. Texas Instruments. http://www.ti.com/rfid

RFID: An Anticounterfeiting Tool

Brian King* and Xiaolan Zhang

Abstract In this paper, we describe several applications concerning how an RFID system can be used as an anticounterfeiting tool. We survey the security services that are necessary for such RFID applications. We introduce potential threats to such an RFID system and the necessary security services. We discuss security models for the security services in an anticounterfeiting RFID system and construct a generic protocol that can be used (i.e., slightly modified to fit the necessary application).

1 Introduction

RFID computing system is a technology whose use has become more and more diverse and has become pervasively applied in many environments. Initially the use of RFID systems was limited to low-impact applications like inventory and antitheft in the retail area [12, 27, 31]. However, today RFID computing systems have been applied in sophisticated areas, such as in IDs and passports [28] and pharmaceutical pedigrees [32], where their impact is critical to the success. There are multiple proposals of the use of RFID in future complex systems that will greatly impact the efficiency of these systems.

In many cases, RFID has been implemented in a "behind-the-scene" scenario such as in a distribution center. In some other cases, RFID has been used in a merchant "antitheft" application while these tags are used in a manner that is directly

B. King
Department of Electrical and Computer Engineering, Indiana University
Purdue University Indianapolis, Indianapolis, IN 46202, USA
e-mail: briking@iupui.edu

X. Zhang
Department of Electrical and Computer Engineering, University of Illinois,
Urbana, IL 61801 USA
e-mail: xyzhang29@crhc.uiuc.edu

P. Kitsos, Y. Zhang (eds.), *RFID Security: Techniques, Protocols and System-on-Chip Design,* © Springer Science+Business Media, LLC 2008

interacting with consumers, these tags are removed once a purchase is completed. Such applications of RFID do not impact privacy. However, as RFID is applied in more sensitive applications, the RFID computing system will become more pervasive, in terms of consumer effect, and so privacy and privacy-related issues will become more and more important. In order to successfully apply RFID to sensitive applications within the consumer mainstream, privacy preserving protocols need to be developed. The RFID system should reveal information only to those authorized and unauthorized parties should not be able to extract side information from the RF communications.

Along with the identification function of RFID tags, anticounterfeiting is one of the functions that tags will be able to provide [26]. Examples of its use in an anticounterfeiting system include embedding RFID tags in Euro banknotes [35], pharmaceutical products [13], or passports [19]. Anticounterfeiting is usually implemented by authenticating an item remotely or semiremotely. The party who authenticates the item is able to obtain its tag information as well. However, many RFID systems will operate in complex environments, so adversaries may tamper the tag information in many ways to fool the legitimate users of the wrong item. Preserving integrity of tag data is key to trusting an RFID authentication system. Thus any protocol should consider integrity. Much of the past work on integrity mainly focused on the trust of operations. Due to the complexity of transactions and the number of custodians of a tag during its lifetime, an important focus should be on how to evaluate the trust of tag information that has been obtained and modified by many parties. For example in [37], we constructed a model, which we call *perfect integrity of tag information*. This model can be used to evaluate protocols that provide integrity of information and/or anticounterfeiting protocols. In an integrity (anticounterfeiting) application, passive RFID tags could possibly be embedded into product to reduce the cost. The chip may only be capable of preliminary computing like exclusive-or and does not have a lot of memory space, perhaps about 1 Kbits. Although the tag may have memory access control through a key, it should not be relied on, since the key is usually transmitted clear in air, which will be stolen easily. So a more complicated scheme is needed to provide a secure tracing against eavesdropping and spoofing. In Sect. 3, we discuss more integrity problems and applications for which, an anticounterfeiting RFID protocol can be applied.

The focus of this work concerns the use of RFID as an integrity tool, in particular its use in an anticounterfeiting system. In addition, we assume that the applications will require consumer/bearer privacy. That is any successful anticounterfeiting/integrity RFID protocol must preserve tag owner/bearer (those parties that "possess" the tagged product) privacy. An outline of this chapter is as follows: First, we provide an introduction. We then provide a brief discussion RFID technology. Next we discuss the potential of using RFID as an anticounterfeiting tool and applications. We then discuss security design criteria. After that we discuss re-encryption, an anonymity cryptographic tool that can be used for privacy. We then construct a generic integrity (anticounterfeiting) protocol, which satisfies the integrity model while preserving bearer privacy. We then conclude.

2 RFID Technology

RFID stands for Radio Frequency IDentification. RFID tags are small integrated circuits connected to an antenna, which can respond to an interrogating RF signal with simple identifying information, or with more complex signals depending on the size of the IC. They usually have very little memory (around several Kbits), some of which are keyed read or write enabled such as Atmel e5561 [1]. One classification is by source of power. Passive tags derive all their transmission and computation power from the RF signal. It is inexpensive and less powerful. Active tags have batteries and more complex. They are able to communicate over a longer distance (over 10 ft.) than passive ones (just over a foot).

RFID tag is a tiny chip that can be stimulated by radio frequency queries which cause the tag to respond with the required data [34]. It has a pair of antennas and very limited computational power and memory. A tag reader or RFID transceiver wireless stimulates a tag and communicates with it. The data stored in the memory of a tag can be accessed by read or write. RFID tags have been used for many years to track and trace remote objects. Low-frequency tags are commonly used for animal identification and automobile key-and-lock, antitheft systems. High-frequency tags are used in library book or bookstore tracking, pallet tracking, building access control, airline baggage tracking, and apparel item tracking. Ultra-high-frequency RFID tags are commercially used for pallet and container tracking, or for truck and trailer tracking in shipping yards. Microwave tags are used in long-range access control for vehicles.

A typical RFID system consists of a tag (transponder), a reader (transceiver), and some means to process information, such as a computer. The reader queries the tag for some information. The tag then responds with the corresponding information. The reader then forwards such information to the data processing device via reader's network. The reader may be a handset device or a computer, which is capable of complicate computation, such as public key algorithm. An RFID system usually operates on 868–956 MHz or 13.56 MHz frequency band. The higher frequency ones have higher transmission range and smaller size. But they are easily blocked by the presence of liquid intensive mass, even human beings. An RFID tag can respond to multiple readers and a reader can talk to thousands of tags. Their communication, in some applications, should be authenticated and confidential.

One typical use of an RFID system today is the Electronic Produce Code (EPC) system [25]. A unique code is assigned to each object by means of a tag so that it can be identified and traced remotely. The EPC system classifies tags [4] into six classes

Class 0 which is read only.

Class 1 write once and read many.

Class 2 read/write.

Class 3 read/write with on-board sensors

Class 4 read/write with integrated transmitters can communicate independent of readers

Class 5 read/write with integrated transmitters can communicate with passive devices

Clearly as we move toward higher class of tags, not only do the tags possess greater communication capabilities, but also the cost of the tags increase. Cost could hinder the ability to pervasively employ tags. Today most tags used in practice are Class 0 or Class 1 tags. As we move toward more complex applications requiring greater functionality by the tag, we will see the use of higher class of tags. The resulting increase in communication capabilities implies a greater risk due to some security vulnerability.

3 RFID Anticounterfeiting Applications and Potential Threats

Traditional RFID systems were developed as a replacement for bar-code, deployed and managed by one or few trusted entities so little or no security was required. Today's RFID applications have become more diverse and open. Anticounterfeiting has become a fast growing field for RFID applications. For example, people have begun to tag valuable items, such as banknotes, passports, or drugs, for the purpose of additional authentication and trust. Anticounterfeiting is also a demanding security feature for new RFID applications, such as access control key cards, credit cards, tollway payment, to prevent fraud and identity theft. The widespread use of RFID inflames an already heated debate concerning privacy vs. RFID identification technology. Privacy advocates may view the wireless tracing of personal items as a violation to privacy. We could be unwarily tracked by other people for personal information, like the amount of money we carry, who we are, what we buy, where we are, etc. Directly or indirectly, the potential use of RFID tag to spy on us is very real.

For RFID anticounterfeiting applications, preserving integrity as well as privacy becomes a serious topic to study. The problem becomes difficult when future RFID systems may be operated in complex environment that multiple parties of a variety of trustfulness jointly manage the same system. In this section, we present some typical applications, potential threats, and the challenges.

3.1 Motivation for RFID Anticounterfeiting Applications

In general, for an RFID anticounterfeiting system, the reader interrogates a tag, not only for identification but also for verifying the identity. In some cases, the reader may already know the tag identity prior to RF contact. RFID tag is used as a means for additional authentication. The identity to be protected could be any information associating with the property of the tag bearer. RF information provides a digital fingerprint for these credentials, complicating the forgery process. In the cases when a reader contacts a tag for both identification and authentication, tag integrity consists of an important part for trust in a system.

3.1.1 Financial Credentials

Euro banknotes, which are issued by European Central Bank (ECB), have been circulated by the European Union for the last 10 years. The unification of the currency system brings to Europe many commercial convenience and economic benefits. However, the widespread use of Euros has increasingly made money management more difficult. Counterfeiting banknotes is one of the most serious crimes. According to the biannual report of ECB [9], a total of 311,925 counterfeit banknotes were identified and removed from circulation in the second half of 2003. Although many antiforgery technologies have been adopted for the Euro, the Euro has become a criminal favorite for money laundering. First, it is a valuable currency with a value similar to the US dollar. Second, it has high denomination banknotes like the €200 and €500 (€ stands for Euros). Third, many countries accept Euros, so they facilitate the circulation of counterfeit banknotes through various channels. The circulation of Euros between the many different countries of the European Union makes it hard for the law enforcement agencies to trace the counterfeit money. These crimes, related to the use of the Euros, have become so challenging that new technology has been demanded to fight against them. The European Central Bank plans to put RFID tags into Euro banknotes to defense against a variety of monetary crimes, such as forgery, money smuggling, money laundering, etc. These tagged banknotes will be harder to forge and easier to trace.

Visa, MasterCard, and American Express started experimenting contactless credit card system empowered by RFID technology [33]. They expect that the new system would speed up the checkout process compared with traditional magnetic strip swapping card. Signature authorization may be waived for transactions below $20. Researchers [14] found that RFID credit cards are vulnerable to many integrity attacks. They are able to skim a credit card by an unauthorized reader, eavesdrop authorized RFID sessions and clone credit cards. Using a tampered reader and a credit card emulator, a "relay" attack could be performed to place a transaction on a nearby innocent credit card. The credit card emulator first initiates a transaction with a real reader. The credit card emulator relays any communication received from the real reader to the tampered reader so it could use them to initiate a fraudulent session with a nearby innocent credit card. In a similar manner, the reader relays information received from the innocent credit card to the credit card emulator that responds to the real reader. Effectively, a transaction is placed on the innocent card by the real reader via communication redirection. To prevent these attacks, a card should be able to recognize authorized readers and a reader should be able to identify real cards.

3.1.2 Merchandise Chains

Gillette[1] began to use RFID tags to label commodities and ship to Wal-Mart[2] warehouse [3, 27, 31]. Logistically the product may transition through many entities that

---

[1] Global Gillette is a business unit of Procter & Gamble that mainly manufactures safety razors.

[2] Wal-Mart is currently the world's largest retailer.

are individually operated, from Gillette manufacture, third-party warehouse, transportation services, and Wal-Mart. Certainly Wal-Mart will be concerned that the RFID tag information is altered by an unauthorized party after they have shipped out from Gillette. In future, Gillette may extend tag life to collect user feedback. Integrity of tagged product should be maintained to ensure that the data obtained by Gillette from tags are authentic.

The United States Food and Drug Administration (FDA) has been considering to use RFID tags to prevent counterfeit and adulterated pharmaceutical products [8]. Mostly due to its high development and manufacturing cost, pharmaceutical products are amongst the most expensive retail merchandise. The price of pharmaceutical products varies significantly due to variance of standards and regulations applied on drugs in different countries. Some pharmaceutical products made in Canada, are known to be much cheaper than their counterpart in USA. However, the United States is reluctant to allow importing drugs due to lacking of method for integrity check. On the other hand, high profit behind a drug price gap constantly fuels the incentive for international drug trafficking and forgeries [20]. Drug counterfeiting is increasingly threatening public health and consumer rights. It becomes an urgent demand to authenticate pharmaceutical products in an effective way. RFID technology can automate the identification of pharmaceutical products in a shorter time and higher level of accuracy so applying RFID in inventory and prescription is expected in the near future. It would be attractive when drug validation, anticounterfeiting, prescription, and consumer inventory are solved by an all-in-one RFID tag throughout manufacture, distribution, and postretail stages [15].

3.1.3 Personal Identification and Access Control

Many companies and hotels have used RFID contactless keycard for building access. Compared with traditional keys, digital key cards have many security advantages, that they are more expensive to copy and easy to disable after lost. Contactless keycard provides fast access and is mostly referred to as a "wave." However, security is still a concern when RFID key cards are to be used for accessing very valuable assets. People may still try to counterfeit RFID cards even at a higher cost, driven by incentives to steal high profit from controlled assets. Enhancing integrity for RFID key cards would be very desired.

As early as 20 years ago, US Federal officials had estimated that at least 30,000–60,000 passports are fraudulent among the then 3 million US passport applications received everyday [10]. Only 1,000 fraudulent passports (typically very obvious ones) have been detected. Record shows that 80% illegal drug dealers and about 300,000 fugitives and terrorists are aided with bogus passports and visas, travel freely over the world. To improve the integrity of passport, the US government has been issuing passports including a 64-kilobyte RFID chip that contains the passport holder's personal information [30]. The contents on chip will include the name, nationality, gender, date of birth, place of birth, and a digital photo of the passport holder. This data will match the data printed on the paper of the passport. RFID technology is in use to improve the security of passports, making them difficult to forge by criminals. However, wireless passports make people more concerned about their identity safety. Especially since people carry passports where they are traveling to foreign countries, into crowded international airports and sites within public areas, which are usually unfamiliar places so people have little or no control over the environment. If the passport is ready to broadcast chip contents to any receiver, it would be a real danger to traveler's privacy when some people try to gather passport data for unauthorized background checking, identity theft, illegal tracking, or investigation. A Dutch company has already successfully cracked a weakly encrypted Dutch-prototype RFID passport within 2 hours [5, 36], obtained all plaintext information perfectly for a clone passport. So it would be equally important to study how RFID provide integrity as well as keeping personal information safe.

3.1.4 Sensoring Network

Telepathx[3] plans to integrate RFID tags, crash sensors, and wireless mesh networks to detect and report highway crashes [24]. They install RFID tags on guardrails, utility pools, and other roadside structures. With a sensor inside, these RFID tags are able to detect an impact and send it to the nearby remote transmission unit (RTU), which is a part of a wide area wireless mesh network that has been already implanted throughout Australia. The density of RTUs are about one every 250 m. Each RFID tags can communicate to RTUs within range of at least 125–250 m. The severity of an impact is accessed by the number of RTUs that are informed. A small collision may only trigger a few RFID sensors while a large one may trigger most of them along the guardrails. Currently the company is trying to improve the sensitivity of the system to reduce false alarms. Besides infrastructure limits, other sources of a false alarm may result from intentional tag tampering or counterfeiting. The next step is to determine how much trust one can put on each RFID tag. Anticounterfeiting tags would improve system trust and receive wide use.

3.2 Potential Threats and Challenges

As we see the utilization of RFID for anticounterfeiting becomes a demanding feature for many applications, the security of RFID system becomes a vital factor to provide certain level of protection and trust. Moreover, these applications require extension of tag life and active range that a tag may be accessed by many parties at different time and location, with various levels of trust. For example, an RFID enabled Euro bill may be accessed by your local bank, supermarket and even your neighbor with a reader. In an open access RFID system, tags may operate in unknown or untrusted environments, exposing to attacks from various purposes. Malicious readers

---

[3] An Australian firm operates remote monitoring systems.

may intercept an authentic RF communications between a tag and a reader, attempt to contact an authentic tag, or even tamper tag data. Tampered tags may spoof an authentic reader with falsified tag information. Later, we will discuss the threats and challenges to design an RFID anticounterfeiting system. Many of them are common to most RFID systems but these threats become more exacerbated for systems specialized for anticounterfeiting.

3.2.1 Integrity

Integrity is a foremost concern, especially for anticounterfeiting systems. Leaving "dirty" tags undetected aids the smuggling of "dirty" tag bearers. Even in regular RFID systems, serious integrity attacks could be everywhere. Example 1 and 2 shows some integrity attacks and their severe consequence.

Example 1 (Tag Integrity) *Flex tag provides a user read and write memory space on tag. For privacy protection, the tag is able to write many times. Tag-Pharmacy programs the serial number of each product into attached tag memory space. A Flex tag reader is installed at each checkout door to provide scanning-free check out service. The cashier uses RF reader to read the serial number of each tagged item in a cart without taking out individual items and sums the total payment. Alice hides a Flex tag reader in her backpack when she was in store. She replaced the on-tag serial number of the LuxLife hotpot that is worth $50 with the serial number of the SimpLife hotpot that is worth $10. Then she placed LuxLife hotpot in an obscured place in her cart during checkout. The cashier was not able to see the hotpot and checked it out with the price of the SimpLife one.*

Example 2 (Deny of Service) *Tag-Mart becomes popular in the neighborhood after they adopted RFID technology. Bob was fired by his boss and holds a grudge. He decided to do some destruction to Tag-Mart business. The next day after he was fired, Bob brought a reader in his bag and reprogrammed the serial numbers of all tags into random junk while he pretended to walking around the store. Soon, all scanning-free checkout points were overcrowded with impatient customers who were unable to checkout. Eventually Tag-mart had to close the store for a few days to correct the tags' serial numbers.*

Hardware integrity attacks may involve tag cloning, memory tampering, reproducing tagged items, and physical tag damage. These attacks are very threatening but also expensive. Specialized hardware equipment has to be made to perform these attacks. It is difficult to falsify a tag than readers mostly because tags are usually embedded and encapsulated into tag bearers, such as in banknote papers. The consistency of the physical appearance of tagged items should always be a big concern, due to the potential of tampering resulting in forgery. One could conceive that software attacks on a tag by a subverted reader would be easier and more prevalent since readers are large in size and easy to modify both in software and hardware. *For these reasons, we focus mainly on software tag integrity throughout this chapter*. Most tags currently in use are read-only or write-once, such as EPC Class I and II tags. These tags are subject only to hardware integrity attacks. However, tags without or with limited functionality as well as tags that are reused by different entities will be at a disadvantage concerning privacy protection. Therefore, software integrity is a concern for tags with general modification capability.

3.2.2 Privacy

Privacy becomes a real concern [3] when tags are massively implemented in the mainstream, for example on banknotes, passports, credit cards, medicines, and clothing. Further, in most anticounterfeiting applications, tags may carry and broadcast information that are very sensitive to individuals. People are worried when the contents of an RFID tag are disclosed to unauthorized third parties, or even the existence of tag provides some traces of personal information. In addition to personal privacy, some companies may worry about industrial espionage via unauthorized interrogating of their RFID tagged packages or cargo. Adversaries may attack on-tag data confidentiality or illegally track individuals by RF fingerprint.

Confidentiality was not a concern in traditional RFID systems, such as Auto-ID that has been used in manufacturing, transportation, and inventory. The functioning of tags are localized. Each company may have their own RFID identification system and the system/data specification is not shared among companies. However, their internal RFID system is unsecured. Each RFID tag broadcasts identification information, usually a data stream of unique sequence number, in plaintext to the air, a reading device supporting such tag type is able to receive the identification within the range that the signal strength of such tag would be able to reach.[4] In lack of authentication and communication secrecy, any reader is able to identify the tag and further the item that this tag represents for. An unsecured system simplifies tag design and lowers the cost to manufacture. However, potential privacy problems rise, as illustrated in Example 3, when an insecure system is directly applied within an open environment, a person's privacy is violated via unsecured tagging.

Example 3 (Confidentiality) *Remote identification may infringe the legal privacy of a person [21]. Tag-Library uses EPC Class I to tag library books. The tag memory stores the ISBN code of a book. Tags are not killed after checked out because the library plans to reuse the tag after Alice returns the books. Alice loaned some English classic novels from Tag-Library and kept them in a locked wood cabinet in her office. Alice's boss Bob is interested in the reading preference of his employees so he used an EPC Class I reader to remotely query Alice's cabinet that contains her books. Bob successfully retrieved all the ISBN codes and found the book titles via Google search. Tag-Library unintentionally discloses to Bob the reading interests of Alice that is legally protected personal privacy.*

---

[4] The reader's range varies depends on the type of tags, from a few meters to hundreds of meters. Both tags and readers will affect the reader's ranges. Limiting reader's range is one way for privacy protection. However, unstandardized readers may boost the signal power of some tags and reach them farther away. Privacy may still be a concern for some applications that requires long range RFID tag.

In general, the nature of remote access of RF tags makes privacy protection more difficult. Even if the information that a tag contains is not sensitive or private, privacy violation may still occur. For example, when a person is identified and tracked remotely via detecting RF signals from tags that are embedded in personal items. Remote traceability does not produce a particular privacy problem for RFID systems. That is, other mobile devices, such as cell phones or WiFi computers, have similar security concerns. However, unlike other devices, bearing active RFID tag may be mandatory if tags are embedded into necessities and tag activities, like power on or off, are out-of-control of the bearer. Sometimes people may not even be fully aware of the existence of tags. Providing security against illegal remote tracking in this sense is very important for RFID enabled merchandises gain general acceptance. Example 4 shows that tracking the location of a person without permission becomes easier when RFID tags are widely used on the vehicles. The accuracy of location tracking could be guaranteed by the limitation of RF signal broadcasting range.

Example 4 (Tracking) *An RFID toll card could be used to disclose some information about the location of an individual's car at a particular time. In a divorce case [3], the wife claimed that her husband had an affair with the record from his toll card as one of the evidence. The toll card was charged by a reader at a certain location at a certain time where her husband never confessed being present. Unless there was another person driving his car at that time, he had to come up with an explanation for this errant.*

3.2.3 Challenges

Modern cryptography provides many solutions to confidentiality and authentication. However, hardware resources is the main limiting factor to secure an RFID system by applying PC's security tools. For communication between readers and the rest of the network, we could consider using some algorithms that are currently used on hand-held devices. The real challenge is to secure tags and their RF channels. A typical RFID tag may only have a few thousand gates, less than one kilobyte of memory, and limited and irregular power supply [16]. Since an RFID system is an asymmetric system the proper computational burden on tag and reader does not need to be equal. The proper computation load assignment between tags and readers to perform expensive cryptographic algorithms, and a key distribution system may vary depending on the application. In an open environment where multiple parties may access the same tag at different times, the problem of how to distribute keys between readers that are loosely connected can be a hard problem. Further, we should prevent illegal tracing, which can be done by changing the RF appearance of a tag. Clearly the more security and privacy we have on a tag, the more complex of a protocol between tags and readers will be. If a protocol takes too much time, it may be infeasible to implement due to the number of tags a reader needs to communicate in a limited period of time.

We should be aware that solutions for integrity and privacy may not be compatible with each other in many cases. To prevent illegal tracing of tag data, there are several existing solutions. One is to change the RF data (RF identifiers) frequently so that a malicious party cannot identify a tag by recording the tag information in advance. As shown in Examples 1 and 2, tags with modification capability are prone to integrity attacks if data is written into tags in a wrong way. An immediate solution for integrity is to use a password for write access. Password protected write has been already implemented in EPC Class I version 2, and EPC Class II tags. Class I version 2 tags also generate a one-time pad to obscure authentication process. However, the adversary may still be able to acquire the password if s/he eavesdrops the whole session. As for a stronger solution, one could use public key encryption authentication protocol that has been commonly used in a computer system, but this may require too much computational resources for a resource constrained RFID tag to handle.

Here we summarize some design challenges to secure an RFID system. Due to diversity and complexity of real world applications, it is impossible to find a panacea for all problems. However, finding some generic solutions for a set of problems that receives general interests is feasible:

1. Level of security required in some applications needs to be strong.
2. Due to limitation in physical size and cost, an RFID tag may not be able to perform expensive computation, such as public key encryption schemes used on personal computers.
3. Tags may be subject to intensive physical attacks, such as memory tampering, brute force password attacking or cloning.
4. Current communication channels between tags and readers are open and insecure. Many common computer network attacks, such as eavesdropping, impersonating or DOS, could be translated to RFID systems.
5. Because of the pervasive and invisible nature of tags, privacy protection becomes an important issue. The consequences of a violation of privacy could be serious.
6. Multiparty, multilevel trust access may coexist in one system.
7. The solutions for security, privacy, and performance may not be compatible.
8. Specific requirement for each application varies in a way that it is hard to find a one-fits-all solution for all problems.

4 Designing a Secured System

We have shown that an RFID anticounterfeiting system is prone to many attacks. In this section, we will discuss concrete design factors and steps toward securing an RFID system. First of all, we need to understand the functionality and levels of protection that the proposed system is to provide, and the cost to provide them. It requires a study of cost, software/hardware limitations, trustiness of access parties, and potential vulnerabilities for the application. Then, we build a security model based on our requirements to describe our goal in a precise and accountable fashion. The security model will be used to evaluate the competence of each implementation. The model should include security requirements and cost for each functionality to be implemented. The security model should be general enough to be implementation independent and provide enough flexibility for various engineering decisions. An iterative process of reviewing requirements and model refinements may be needed for a complex system. To preserve integrity, completeness, and neutrality of the model, it is recommended to have a stable model before actual implementation and any modification of model should require a review of system requirements. The next step is to actually design the system, specifically a suite of secured RFID communication protocol(s) to implement desired functions. Customizing existing cryptographic tools, RFID system architecture and technology or inventing new ones may be needed for certain functionality. In the end, the system should be evaluated against the security model iteratively to verify if it provides the necessary RFID services with the necessary level of security.

4.1 Understand Requirement and Cost

For each targeted application, we should understand specific requirements of functionality, performance, security, and privacy in detail to find the best possible solutions within resource limitations.

The foremost questions are what data should each tag possess, who owns them and the what is the level of sensitivity. We should distinguish the concept between tag data and tag memory contents. Note the tag data is application oriented. In most current applications, it is a bare serial number owned by the manufacturer. However, the actual tag physical memory may contain other data besides an identification number, such as the access password in an EPC Class I Version 2 tag. Such data is not used for the purpose of the application but is used for auxiliary security functions so we do not need to consider them at the first step. The data owners could be the creator, modifier, or maintainer. Manufacturer is usually the owner of tag identifier. Other tag data, such as logistic information, could be owned by distributors or other parties. Data sensitivity describes the severity of damages if the data is disclosed or tampered. Determining ownership and sensitivity is to understand security assets to protect and is the first step toward a secured design.

Secondary, we determine the access privilege of each party that may contact the tag. In a pharmaceutical chain, the manufacturer, distributors, retailers, and even customers may access the same tag at a different time. It is very important to determine who should access which portion of tag data. Access functionalities can include read, write, and append. One may limit the maximum number of accesses that a party is allowed for a particular memory asset over a given time period. If necessary, we need to build an access model to classify trust levels and elaborate access modes.

Then the next step is to understand system limitations that are determined by either technology or budget. From a security perspective, some constraints may enhance security but others may cause a security vulnerability. Each component in the system has its own limitations. Here we focus on tag resource, tag access, and reader's range. Tag resource limits include *physical removeability*, the number of gates (computational power and memory size), and power source. Physical removeability, an important integrity property, describes how easily a tag could be removed or replaced from the bearer without damaging the bearer. Current VLSI process technology determines the maximal number of gates that could be crammed into a certain area. The more gates available, the more computational unit and memory space will be available on tag. However, more working components requires more power supply and advanced chip design, which increases the cost of the tag. Since tags are to be attached into products in a high volume, people try to reduce per tag cost while providing necessary functions. Tag access is limited by reader's range and memory access type. For different type of tags, reader's range can vary. A short reader's range uses less power and provides more privacy protection. But a long-range reader requires less readers and reduces infrastructure investments on readers and reader's network. Determining tag memory access type is another important aspect for system functionality and security. *Readability* and *writability* are two access types. Readability tells the portion of memory contents that are able to be accessed by RF read. Writability is the part of memory that is allowed to modify or append via RF contact. For most tags, reader's range varies from a few meters to hundreds of meters. Physically tags are unremoveable in most applications, like tagged banknotes. The number of gates are about 400–4,000 whose computational power hardly can support symmetric key encryption [23]. Tags that cost a few dollars could have memory around 1 Kbits with read and write access.

System reliability is another concern. Encryption errors, communication errors, and hardware errors affect the quality of services, including security services. One important part of a requirement study is the need to understand the probability of system errors, faults, and failures. And we should learn the cost to improve reliability in different ways and the tolerance thresholds for our particular application.

Lastly, we should build an adversary model for possible attacks. Adversaries are considered as parties (readers) performing operations that they are not authorized for. Such operations can be interleaving RF communication, querying a tag (try to act as an authorized reader), spoofing a reader (try to act as a valid tag), tampering tag physically, or performing DOS attacks by any means. An adversary may attempt to eavesdrop RF signals, initiate or intercept a session. The most important part in our adversary model is to decide the amount of resource, money, and time, available for adversary to conduct each attack and the cost associated with them. For example, we need to answer the following questions in order to analyze how an unauthorized reader attempts to identify a tag:

1. How easy will it be for an adversary to acquire a compatible reader for this particular type of tag?
2. How often will a reader be able to enter within the communication range of the tag?

3. How many successful RF communication sessions does the reader need to make to identify the tag?
4. How much computing is required for the adversary to process data? Computational resource can be an issue when one wants to hack an encrypted data.

Note that always assuming an adversary has unlimited resources is unrealistic and it even prevents us to understand the practical security needs for a system. Due to tag limitations, it is impossible to design a perfect secured protocol but it is highly likely to have a secure one with reasonable adversary assumptions if we understand the actual protection requirements and the implementation cost well enough.

An RFID system may need security services for confidentiality, integrity, and availability to prevent various attacks. In some systems one must make sure the communication between tags and readers are confidential and authenticated, in other systems the information provided by the tags needs to be authenticated. Sometimes the access (read or write) to the RFID systems, including tags, readers, and other related equipment, should be classified for access parties. One of the design decisions is to determine the necessary security to be included and the level of protection provided.

4.2 Proposed Security Models

Security models provide a formal definition of security features that we wish to accomplish. They must be precise, accountable, and requirement oriented. They provide a tool for us to evaluate the security of real implementations so they must also be stable, general, and neutral. In some past constructions, security models were developed from the idea and parameters of a particular protocol. And then the model was applied to the protocol to assess the quality of security services implemented. Juels' model [17] and Ohkubo's model [22] were developed in this way, and are among the first security models proposed for RFID systems. Their models, together with the protocols, provide a security solution as well as a proof for security. However, confining the scope of a model to one particular implementation intrinsically limits the ability of the model to review a protocol as an outsider or verify alternative implementations. Some security vulnerabilities within a protocol may not be able to be found by verifying itself with models that are built from it. Avoine [2] proposed a protocol-independent adversary behavior model that could be applied on any existing protocols. By using his neutral model, Avoine successfully demonstrated the vulnerability of many protocols. Disconnection between models and protocols brings the model a neutral and critical texture. On the other hand, a security model should be specific to one security requirement. A close connection to security requirements improves the focus and precision of a model that could be used to proof or disproof the security features that it was built for. Several security models may be required for applications that requires several security services. The models they choose should address security vulnerabilities that are specific to the application. Security models should also be general enough for different applications to adjust their desired security levels. As we have shown in previous sections, RFID applications vary in security requirements decided by a particular balance of demand and budget. The model should allow users to adjust security parameters to fit their practical needs. In an accountable and rigid fashion, the model provides reasons for us to trust a protocol to a certain extent. A good security model should answer the following questions about a security protocol:

1. What kind of attacks can the protocol resist?
2. How well does the protocol provides protection against these attacks?
3. In what situations does the security protocol fail?
4. How much does it cost an adversary, to break the protocol in various situations?
5. How does one specify the level of security, in terms of adversary advantage, failure tolerance, or both?
6. Does the protocol allow one to adjust to different levels of security, if one wants to upgrade or downgrade the level of security?

An anticounterfeiting RFID application may require one or more security services. They are privacy of tag data, privacy of bearers, integrity of tag data, and availability of tag identity. Security models are to be developed to address these requirements. Below is a list of concepts that a model may define for:

1. Availability of tag identities
2. Intrinsic integrity of tag data
3. Observable integrity of tag data
4. Confidentiality of tag identities
5. Confidentiality of tag owners
6. Indistinguishability of tag identities
7. Indistinguishability of tag owners
8. Forward and backward security

Remote identification is one of the most important functionalities of an RFID system. RF readers identify the identities of a tagged item by making RF interrogations to that tag. Availability of identities means an authorized reader of item should always be able to obtain the identity from RF contact. They should be able to verify and distinguish the identity of this item. Any approach to block or spoof the RF communication between a tagged item and an authorized reader is an attack on the availability of the RFID system.

Intrinsic integrity is maintained on a tag if tag data can only be modified by authorized reader in an authentic way. Many physical attacks, such as tag cloning and memory tampering, break tag intrinsic integrity. Tampering tag memory space via RF channel is a common violation of intrinsic integrity on unprotected writable tags. Usually, it is expensive to maintain intrinsic integrity.

Observable integrity is preserved if there is no way that an unauthorized party can alter tag data that goes unnoticed by parties that are authorized for that data. Observable integrity is weaker but more practical for RFID integrity. Usually it is easier to implement observable integrity and it is also the minimum integrity that an integrity-sensitive RFID application should provide.

Confidentiality of identities is to protect the identity of tagged item from revealing the identity to an unauthorized reader via RF access. A tagged item is identifiable only to a reader authorized for it. If a tag contains multiple fields of data that are owned by different authorized users, each field must be separately available to the parties authorized for that field. For RFID enabled personal items that the owner always bears, such as eye glasses, the identity of owners should also be protected. Confidentiality of tag owners (bearers) is to protect the identity of the person who carries an RF tagged item. Confidentiality of tag identity and owners are essential to privacy protection.

Indistinguishability of tag identities and owners prevents a tag or a bearer being tracked remotely. More strict than confidentiality, tag identity should not only being unidentifiable to unauthorized readers, but also the RF signature of that tag are indistinguishable to those from other tags. Otherwise, an unauthorized reader will be able to trace and track a tag even without knowing its identifier. Tag owners should also be protected from being traced illegally by RFID tags they bear.

Forward security [22] requires that past history of a tag is untraceable and invisible to an adversary even if it acquires the current cleartext data on tag. Past history could be RF read and write transactions made by other parties. A more general version of forward security is that the past k's transaction, where k is a finite number, on a tag is untraceable and invisible to an adversary.

Backward security requires that past access history of a tag does not help an unauthorized reader identify an tag better than guessing. The "history" is a finite collection of pairs of information and results obtained from prior remote accesses.[5] A reader's membership could change with respect to time (i.e., within one's history).

Any canonical language can be chosen to formalize a security model. For example, Zhang and King's model [37] described security using a probabilistic mathematical definition. Using formal language for definitions provides provability of security when the model is to apply on protocols.

4.3 Designing Security Protocols and Review

During the design process, we should often ask the following questions to improve our protocol iteratively:

1. What function do we need to implement?
2. Does this solution solve our problem? How well does it solve the problem?
3. Do we actually need this function? How can we implement it in a effective way?

Selecting suitable cryptographic tools for our protocol is a core design decision. According to our security models, our aim is to choose the right security mechanism that provides adequate protection at a cost within our expectation. Although

---

[5] A access record may be acquired by eavesdropping other parties active session or this party's own query.

cryptography provides many tools for confidentiality, authentication, and integrity of different security levels, most of them are very computational resource intensive. Tag's physical limitation is the biggest challenge. When passive RFID tags are used, the chip may only be capable of preliminary computing like exclusive-or and does not have much memory space. Following are a list of concerns that need to be considered in the design an RFID cryptosystem:

1. How much computing and memory resources can be utilized for encryption and decryption, respectively? Where will encryption and decryption be performed, on the tag or the reader or both? Readers have more resources compared to tags. So it is recommended to place computationally intensive tasks on readers as much as possible. This is especially the case when public key encryption is employed. In cases where tags may not have the ability to perform public key cryptographic computations, a suitable symmetric-key cryptosystem should be selected.
2. How are plaintext and ciphertext transmitted and stored? One should avoid transmitting cleartext via RF. Even ciphertext should be cautiously transmitted in the air because the eavesdrop of a static ciphertext may provide RF signature to trace a tag or tag owner.
3. How is the key distributed? If key distribution is a problem, then we could use a public key cryptosystem. But public-key cryptography is computationally resource intensive. Thus typically a symmetric key cryptosystem is chosen. The security of symmetric key encryption relies on the security of both encryption and decryption keys. Tags can share a key with reader when the membership of readers is obvious. But it becomes difficult when the system scales.

In a large system, several protocols may be needed for transactions for many purposes. For example, in a secured merchandise chain network, we need a protocol for identifying tags, one for modifying tags, and one for verifying tags. Each may be individually designed and used for an access party.

After we have a protocol, we scrutinize the security performance of our protocol against our model. We should either be able to prove a security function or find a security vulnerability. We then improve the protocol by fixing security holes. After every fix, we need to check all security functions again to make sure new modification does solve the problem without introducing another problem.

Example 5 illustrates an m-history backward secure RF protocol for indistinguishability of tag identities with some security assumptions, and also discusses the security briefly for different security models.

Example 5 *Suppose a cryptosystem is secure against adaptive chosen ciphertext attack (CCA2) if it is impossible to recover the plaintext from the corresponding ciphtertext without the secret key even if at most m ciphertext-plaintext pairs are obtained. However, if more than m pairs are obtained by the adversary, the cryptosystem can be broken with some probability p. Here we have a security protocol for a simple RFID system. Tag identification data is stored in tag via ciphertext encrypted by the cryptosystem. The decryption key is assumed to be delivered to authorized readers safely. When a reader interrogates a tag, a tag simply responds* *with its ciphertext. After each access, the ciphertext is re-encrypted. Assume that re-encryption process is secure and ciphertext never collide with previously used ones. We could show that the protocol satisfies m-history backward security for identification indistinguishability. With any access history of entries less or equal to m that contains pairs of ciphertext and corresponding plaintext, an unauthorized readers cannot decrypt the current ciphertext to identify the tag because of the CCA2 cryptosystem. This protocol is weakened if the adversary acquires a history of length more than m. She could hack the ciphertext with a success rate p. For a security model that tolerates failure rate q and adversary advantage r, such that $q + r < p$, the protocol does not satisfy the requirement. For the case $q + r \geq p$, the protocol is secure enough.*

5 Re-encryption: An Anonymity Tool

Re-encryption is an important tool to ensure anonymity, and is used in many applications for which anonymity is a necessary security service, for example e-voting and mix networks.

In general, a ciphertext C can provide static information. For example, suppose that banknotes were RFID tagged, information concerning the note's serial number and denomination were encrypted with some bank-regulatory public-key and the resulting ciphertext information was available via public query. Even though unauthorized parties cannot decrypt, this static ciphertext provides a "tracing signal" to the unauthorized parties. That is, if Alice has RFID enabled-banknote which will transmit ciphertext C, and Bob, who possesses a reader, comes in contact with Alice then he will be aware of the ciphertext C. Later at a different location, utilizing his reader if he observes this same ciphertext then he knows that Alice is located nearby.

By re-encrypting the ciphertext, the message that was encrypted does not change, but the ciphertext does. Thus preserving anonymity.

Re-encryption is performed in various ways, for example if one uses the discrete-log based cryptosystem El-Gamal [11] (or the elliptic curve variant) then any user with knowledge of the public-key can re-encrypt an El-Gamal ciphertext without knowing the original message.

Obviously any party who knows the cryptosystem, the message, and the public-key can encrypt. Further, by using a random value can make the appearance of a ciphertext look random and nonstatic. For example if we encrypt message m with public-key PK we have $C = \text{ENC}(\text{PK}; m)$. This forms a static ciphertext of the message m. If anyone encrypts m, they would have the same ciphertext. For example in a voting application, if m represents a Clinton vote then all Clinton votes would have the same ciphertext. One can make the encryption nonstatic by selecting a random value r and concatenating it to m and encrypting it,

$$C = \text{ENC}(\text{PK}; m||r). \tag{1}$$

The value r is called the re-encryption factor. In the context of this work, we assume that the encryption of content is padded by random value as illustrated in (1), here $||$ represents a concatenation. A re-encryption is generated by a party with access to m by having them select a new random value r' and computing $C' = \text{ENC}(\text{PK}; m||r')$, thus giving the ciphertext of message m a new appearance.

6 A Generic Protocol to Demonstrate Integrity of RFID Information

In [38, 39] we introduced an anticounterfeiting protocol that was first introduced to protect the integrity of currency. This protocol was developed by enhancing and improving the integrity features of the protocol described in [18].

As discussed earlier one can see that software and human attacks are attacks that can easily be distributed over the network and replicated by others. These attacks are a major hindrance toward using RFID as an anticounterfeiting/integrity tool. For this reason, we focus mainly on software tag integrity throughout this chapter and our generic integrity protocol is constructed to withstand software attacks.

We assume that the protocol is to be applied to demonstrate the validity of product $\mathcal{P}$. Further each product $\mathcal{P}$ will have a unique serial number S associated with it. We denote that information that is written to the tag by the manufacturer by I_0. Of course there will be other information, denoted by I, associated with each product. Initially I will be set equal to $S||I_0$. As the product moves through the *supply chain*, modification may occur, when modification occurs, in order to provide integrity of information I, one would need to make changes to the information I. This modification to I would be limited to append only, any modification would need to be signed, and the signing party would need to be identified. Thus at different times the information I available on the tag is such that $I_0 \subset I$. Serial number S and information I_0 will be signed by the manufacturer $\Sigma_{\text{MA}} = \text{Sig}(\text{SK}_{\text{MA}}; S||I_0)$ where SK_{MA} is the signing key of the manufacturer.

We assume that an authorized party needs to monitor the existence of product $\mathcal{P}$ and its information from remote sources. The concerns of this party revolves around the "authenticity of the product," is it valid or counterfeit. Such an authorized party could be a customs/border agent or law enforcement agent or some type of regulatory official. We denote the authorized party by $\mathcal{L}$. The intent is to provide $\mathcal{L}$ with serial number S and relevant information which is denoted by I^* where I^* is a subset of the information I satisfying $I_0 \subseteq I^* \subseteq I$. Authorized party $\mathcal{L}$ may wish to access S and I^*, further in order to determine that this information is valid, $\mathcal{L}$ needs Σ_{MA}. This will be transmitted over the RF channel, but if transmitted as cleartext this would violate privacy. Then it must be encrypted with the public-key of law enforcement. However if we encrypt $S, I^*, \Sigma_{\text{MA}}$, then this would form a static ciphertext and could violate bearer's (consumer) privacy. Thus, as recommended in [18], we use a random factor r and encrypt S, I^*, and r, this is denoted by $C = \text{Enc}(\text{PK}_{\mathcal{L}}; S||I^*||\Sigma_{\text{MA}}||r)$, where $\text{PK}_{\mathcal{L}}$ is the public-key of authorized party $\mathcal{L}$. As the product $\mathcal{P}$ travels though the *supply chain* and *consumer environment* the ciphertext can be refreshed by selecting a new r and re-encrypting it.

Our protocol will need to provide a cryptographic link between the serial number printed on the product and its RF ciphertext. We assume that the adversary is not be able to remove/replace, physically clone, tamper or block a tag. Recall our focus concerns software attacks and not on physical attacks. The tool we developed creates a cryptographic binding between the RF signal and the Serial Number (optical key). This provides a way for the authorized party $\mathcal{L}$ to verify the serial number remotely. In order to protect privacy we use re-encryption where the static signature that is available to $\mathcal{L}$ (encrypted with $\mathcal{L}$'s public key) is re-encrypted by parties who encounter the product via the *supply chain* or by consumers who wish to protect privacy.

On the product label we place the serial number S, the manufacturer information I_0, and the manufacturer's signature Σ_{MA} of S and I_0. Thus this information is available optically. The RF access key is used for access control of cell functions that are password protected, this key is denoted by d. Here d is computed using the optical information Σ_{MA}, so that $d = h(\Sigma_{\text{MA}})$ where h is a cryptographic secure publicly known hash function. This mechanism of constructing the RF access key from optical access was first utilized in [18]. Note that not all cell functions are password protected. Thus some RF cell functions are public and other RF cell functions are "keyed."

The organization of the RF cells is to support several security services. One security service we need to provide is the integrity of the product, that is provide some type of pedigree of the product, where has it been, who manufactured it, is the product valid or is it counterfeit. A second security service is to ensure remote RF access of the product information to authorized parties and to deliver this information in a manner that allows the authorized party to ascertain the authenticity of this information. In the first security service it may be assumed that the party checking on the integrity has physical control so that the RF access key is readily available. To achieve the second security service capability, RF communication is necessary, but the RF communication cannot compromise privacy. A primary concern of our work is to ensure owner/bearer privacy while providing the necessary security services. As noted earlier, encrypting RF communication will not protect the tag bearer's privacy. In order to successfully allow authorized parties to access information remotely there must exist a publicly available RF channel. But the information available must be private, hence encrypted. By using public-key cryptography authorized parties possessing the secret key can receive this information, however, to ensure bearer privacy a static ciphertext cannot be used, so re-encryption will be necessary. At the same time, "nonstatic ciphertext" is not sufficient to provide authorized parties the necessary information, in that counterfeit information could be placed in the ciphertext. What is needed is mechanisms that will allow the authorized party $\mathcal{L}$ the means to determine the authenticity of the information.

Some aspects of the RF cell organization was inspired by Juels and Pappu [18]. Some of the improvements were first discussed in [38] and [39].

As the authorized party $\mathcal{L}$ need to be provided the necessary product information remotely, there must exist a RF cell for which this information is available publicly. As discussed earlier, this need to be encrypted, thus there exists some RF cell, cell γ, for which C is available publicly. Here $C = \text{ENC}(\text{PK}_{\mathcal{L}}; S||I^*||\Sigma_{\text{MA}}||r)$ where $\text{PK}_{\mathcal{L}}$ is the public-key of $\mathcal{L}$ and r is a random value which is periodically changed (re-encrypted) to refresh the ciphertext. Those parties interacting with the product will need to follow Algorithm 2 to refresh C by selecting a new value r. A weakness exists that some malicious party could modify the product information so that information revealed by the remote access to $\mathcal{L}$ is incorrect. For example suppose product $\mathcal{P}$ has serial number S and manufacturer information I_0. A malicious party may wish to dupe $\mathcal{L}$, and attempt to pass $\mathcal{P}$ as product $\mathcal{P}'$ by placing an encryption of S', $I^{*'}$, and Σ'_{MA} the respective serial number and information of $\mathcal{P}'$. Without optical access to the product $\mathcal{P}$, $\mathcal{L}$ would not be able to determine the authenticity of the serial number and information that they decrypted from cell γ. Thus other information must be available to $\mathcal{L}$, to determine the authenticity of the decrypted information. It is a necessity that there exists some mechanism that provides $\mathcal{L}$ the capability of verifying the authenticity of the decrypted ciphertext. This is achieved by the following.

We require that the random value r, used as the re-encryption factor, is available via keyed read (see cell δ). But in order for the authorized party $\mathcal{L}$ to retrieve it they would have to compute the access password $d = h(\Sigma_{\text{MA}})$ and transmit it in the clear. Rather than doing this, since this could violate bearer's privacy, we require the following. The hash of the serial number S is stored internally. Thus $h(S)$ is stored internally where we assume it cannot be (physically) altered. We denote this place by cell ω. During the process of re-encryption the value $W = h(r)$ is computed and transmitted to the tag and written to the tag via a keyed write into cell ϕ. We then allow a public function called "compare" to the RFID tag cell ε. A reader may transit a binary string b to the tag, the tag will XOR the values in cell ϕ and cell ω to compute the value V. If b equals the XOR value V, then the tag sends a "1" otherwise the tag sends a "0." Observe that the value in cell ω MUST BE $h(S)$, and that the value stored in cell ϕ SHOULD BE $h(r)$ where r is the re-encryption factor.[6] Thus the XOR value should be $V = h(r) \oplus h(S)$. After decrypting the ciphertext, $\mathcal{L}$ will have S, I^*, Σ_{MA}, and r. They compute b to be $h(S) \oplus h(r)$. If the tag responds with a "1" then they know with a very high degree of probability that the information revealed to $\mathcal{L}$ is valid. Note that though $h(S)$ is static, the value r is random, thus $h(r)$ is pseudorandom, hence $h(S) \oplus h(r)$ is pseudorandom. Therefore from a privacy point of view, little side information is revealed during the compare request, which is a public RF function.

---

[6] It is possible that a malicious party may have placed an incorrect value in cell ω in an attempt to fool $\mathcal{L}$. However by using a cryptographic hash function, it is infeasible that a malicious party can find $h(r')$ and S', such that $h(r') \oplus h(S') = h(r) \oplus h(S)$.

The integrity of the product $\mathcal{P}$ is enhanced by providing those in contact with the product additional information. As discussed earlier the serial number S, initial information I_0 and manufacturer signature $\Sigma_{MA} = \text{Sig}(SK_{MA}; S||I_0)$ is placed optically on the product. The information I is placed in cell ρ. Information is placed in an append manner. Any appending of information will be signed by the party making the appending. The details of this is achieved is discussed in Sect. 6.4. This information is available via RF cell ρ, in order to ensure privacy of tag owner (bearer), this is keyed read and keyed append. Recall that the RF access key d is computed as $d = h(\Sigma_{MA})$. The party which has physical access to the product thus can read and append to cell ρ. Another aspect of checking the integrity of the information is to check that the ciphertext in cell γ is correct. The party with physical access cannot decrypt the ciphertext since they do not possess the authorized party's secret key $SK_{\mathcal{L}}$, but both the serial number S and I are available with physical access (so they can compute I^*) and that the random value r is available in cell δ. Consequently they can compute the ciphertext C. Cell δ is both keyed read and keyed write. The algorithm which describes the modification of cell δ is given in Algorithm 2.

The organization of the cells for the RFID tag is provided in Table 1.

Once a product completes the manufacturing process the following algorithm Algorithm 1 is enacted to set up the optical and RF cell information of the tagged product.

Table 1 RF cell organization

	Internal	
Hash of serial number	$h(S)$	
	Optical	
Serial number	S	
Signature	$\Sigma_{MA} = \text{Sig}(SK_{MA}; S||I_0)$	
Manufacturer information	I_0	
	RFID tag	Mem.[a]
Ciphertext	$C = \text{Enc}(PK_{\mathcal{L}}; S||I^*||\Sigma_{MA}||r)$	cell γ: r$\bar{\text{w}}$
Encryption factor	r	cell δ: $\bar{\text{r}}\bar{\text{w}}$
Hash of encryption factor	W	cell ϕ: $\bar{\text{w}}$
Exclusive-or	$h(S) \oplus W$	cell ε: c[b,c]
Information	I	cell ρ: $\bar{\text{r}}\bar{\text{a}}$
Hash of serial number	$h(S)$	cell ω no access

Note that the RF access key is static, but the chance of tracking tags by a static key is far less than tracking by static tag responses.

[a] There are six kinds of access control for each memory cell: Normal read r, keyed read $\bar{\text{r}}$, normal write w, keyed write $\bar{\text{w}}$, keyed append $\bar{\text{a}}$, compare c. Key is $h(\Sigma_{MA})$

[b] We denote the value in this cell by V

[c] This value is not "stored in memory," merely computed from cells ϕ and ω

Algorithm 1 Initializing the tag

1: **for all** product $\mathcal{P}$ **do**
2:      Choose and print a unique $S_{\mathcal{P}}$
3:      Collect manufacturer information $I_{0,\mathcal{P}}$
4:      Print $\Sigma_{MA} \leftarrow \text{Sig}(SK_{MA}; S_{\mathcal{P}} || I_{0,\mathcal{P}})$
5:      Compute RFID key $d_{\mathcal{P}} \leftarrow h(\Sigma_{MA})$
6:      Compute $h(S_{\mathcal{P}})$ and burn into ω
7:      Compute $I_{\mathcal{P}}^* = I_{0,\mathcal{P}}$
8:      Set $I_{\mathcal{P}} = S_{\mathcal{P}} || I_{0,\mathcal{P}}$
9:      Keyed write $I_{\mathcal{P}}$ to cell $\rho_{\mathcal{P}}$
10:     Randomly select $r_{\mathcal{P}}$ and key write $r_{\mathcal{P}}$ to cell $\delta_{\mathcal{P}}$
11:     Compute $h(r_{\mathcal{P}})$ and key write it to cell $\phi_{\mathcal{P}}$
12:     Compute and write $C_{\mathcal{P}} \leftarrow \text{Enc}(PK_{\mathcal{L}}; S_{\mathcal{P}} || I_{\mathcal{P}}^* || \Sigma_{MA} || r_{\mathcal{P}})$ into $\gamma_{\mathcal{P}}$

6.1 How Re-encryption is Applied

As noted earlier, static ciphertext can be used to trace and so it can violate bearer's privacy. One solution is to periodically change/refresh the ciphertext. Observe that any party with physical access to the product possesses the secret key d (since this is obtained by the optical information). Thus any party with physical access can update the ciphertext, so third parties with physical access apply Algorithm 2 to re-encrypt the information.

Algorithm 2 Re-encryption algorithm

1: Randomly select $r'_{\mathcal{P}}$ and RF keyed write $r'_{\mathcal{P}}$ into cell $\delta_{\mathcal{P}}$
2: Compute $h(r'_{\mathcal{P}})$ and RF keyed write $h(r'_{\mathcal{P}})$ into cell $\phi_{\mathcal{P}}$
3: Compute and RF keyed write $C'_{\mathcal{P}} \leftarrow \text{Enc}(PK_{\mathcal{L}}; S_{\mathcal{P}} || I_{\mathcal{P}}^* || \Sigma_{MA} || r'_{\mathcal{P}})$ into cell $\gamma_{\mathcal{P}}$

6.2 User Verification of Product $\mathcal{P}$

When a party $\mathcal{UV}$ comes in contact with product $\mathcal{P}$, they should use the information available to verify the authenticity of the product. Since this party has physical access to the product they possess the optical information so that they can compute the secret key d. Once they have the secret key they can obtain all information from all of the RF cells. This party $\mathcal{UV}$ should verify that all information on the tag is valid, even those cells that are intended for other purposes. In particular, $\mathcal{UV}$ must verify the manufacturer's signature Σ_{MA}. Further $\mathcal{UV}$ can obtain information I which contains I_0, any additional information that has been written to I was appended and this information must have been signed by the third-party modifier. The party $\mathcal{UV}$ should verify all of the signatures that have been placed on I. Algorithm 3 illustrates the complete verification process, if at any time the algorithm is aborted then the product's authenticity is invalid.

Algorithm 3 User verification algorithm

1: **for all** product $\mathcal{P}$ to be verified **do**
2:     Optically read $\Sigma_{MA}, S_{\mathcal{P}}, I_{0,\mathcal{P}}$
3:     **if** $\mathrm{Ver}(\mathrm{PK}_{MA}; \Sigma_{MA}; S_{\mathcal{P}} || I_{0,\mathcal{P}})$ is false **then**
4:         abort.
5:     **if** RF read $C_{\mathcal{P}}$ or RF keyed read $r_{\mathcal{P}}$ fails **then**
6:         abort.
7:     **if** $V_{\mathcal{P}} \neq h(S_{\mathcal{P}}) \oplus h(r_{\mathcal{P}})$ **then**
8:         abort.
9:     **if** any of the signatures contained in information I does not verify **then**
10:        abort
11:     Compute $I_{\mathcal{P}}^{*}$ from $I_{\mathcal{P}}$
12:     **if** $C_{\mathcal{P}} \neq \mathrm{Enc}(\mathrm{PK}_{\mathcal{L}}; S_{\mathcal{P}} || I_{\mathcal{P}}^{*} || \Sigma_{MA} || r_{\mathcal{P}})$ **then**
13:        abort.
14:     Randomly select $r_{\mathcal{P}}'$ and RF keyed write $r_{\mathcal{P}}'$ into cell $\delta_{\mathcal{P}}$
15:     Compute $h(r_{\mathcal{P}}')$ and RF keyed write $h(r_{\mathcal{P}}')$ into cell $\phi_{\mathcal{P}}$
16:     Compute and RF keyed write $C_{\mathcal{P}}' \leftarrow \mathrm{Enc}\,(\mathrm{PK}_{\mathcal{L}}; S_{\mathcal{P}} || I_{\mathcal{P}}^{*} || \Sigma_{MA} || r_{\mathcal{P}}')$ into cell $\gamma_{\mathcal{P}}$

6.3 Authorized Tracing of Products

Authorized party $\mathcal{L}$ can track product $\mathcal{P}$, provided they have access to the secret key $\mathrm{SK}_{\mathcal{L}}$. To be considered authorized, it is assumed that they possess or have the means to possess $\mathrm{SK}_{\mathcal{L}}$. The manner in which the secret key $\mathrm{SK}_{\mathcal{L}}$. is distributed may vary depending on the application. Further, to gain access of the key some type of legal procedure may have to be conducted, the exact procedure is outside the scope of our work. The algorithm that describes how authorized party $\mathcal{L}$ verifies the authenticity of product $\mathcal{P}$ is described in Algorithm 4. Again, if at any time the algorithm is aborted, then $\mathcal{L}$ has detected that $\mathcal{P}$ is invalid.

Algorithm 4 Authorized tracing algorithm

1: **for all** product $\mathcal{P}$ to be traced **do**
2:     **if** RF read γ (value will be $C_{\mathcal{P}}$) fails **then**
3:        abort.
4:     $S_{\mathcal{P}} || I_{\mathcal{P}}^{*} || \Sigma_{MA} || r_{\mathcal{P}} \leftarrow \mathrm{Dec}(\mathrm{SK}_{\mathcal{L}}, C_{\mathcal{P}})$
5:     Compute $I_{0,\mathcal{P}}$ from $I_{\mathcal{P}}^{*}$
6:     **if** $\mathrm{Ver}(\mathrm{PK}_{MA}; \Sigma_{MA}; S_{\mathcal{P}} || I_{0,\mathcal{P}})$ is false **then**
7:        abort.
8:     Compute $b = h(S_{\mathcal{P}}) \oplus h(r_{\mathcal{P}})$
9:     Compare b to RF cell ε (value is denoted by $V_{\mathcal{P}}$)
10:     **if** $V_{\mathcal{P}} \neq b$ **then**
11:        abort.

6.4 Modification of Information I

Depending on the product $\mathcal{P}$, additional information (beyond the manufacturer information) may need to be collected. If additional information is collected, it should be appended to the manufacturer information I_0 to form I. Every time I is modified, the modifying party appends it to the previous information and then signs the current information I. The signature of the modification would also need to be appended to I. In addition, the modifying party needs to identify themselves. So

$$I \leftarrow I||\text{data}||\text{Sig}(\text{SK}_{\text{MP}};I||\text{data})||\text{ID}_{\text{MP}}$$

here SK_{MP} and ID_{MP} represents the signing key and identification of the modifying party and $||$ represents concatenation. In order to conserve memory, ID_{MP} will represent some type of efficient encoding of the identification of the modifying party.

Counterfeit products have proven to be a great problem, causing great financial woes as well as health hazards. In several situations, counterfeit products have emerged from import countries into the consumer mainstream. These violations of the marketplace form a real danger to consumers. For example, Proctor and Gamble has found numerous instances of counterfeit toothpaste [7]. In addition to the financial impact of this counterfeit toothpaste, it has become a serious health risk causing several deaths. This problem can be relieved by applying an integrity protocol as illustrated in the following example.

Example 6 *The BriteWash toothpaste company has decided to tag their toothpaste products. In order to demonstrate authenticity of the product they will apply the RFID integrity protocol. Initially when manufactured they set the manufacturer information I_0 to include the company's name (CN), day, time, place of manufacturing and product name (PN). Also included in I_0 is the country of destination for which the product is to be sold (denoted by DEST).*

$$I_0 \leftarrow CN||day/time||PLACE||PN||DEST.$$

Then I is computed as

$$I \leftarrow S||I_0.$$

Later when the product is imported to the country of destination, the importer will note the receivership of the product by signing it and update information I to include this signature.

$$I \leftarrow I||IMPORTER||day/time||Sig(SK_{IMP};I_0||IMPORTER||day/time)$$

The information I^*, which is provided to $\mathcal{L}$, would be I.

Another example of a potential application of using RFID to determine authenticity is in the area of pharmaceutical drugs. Numerous documented instances of counterfeit drugs have been observed. For example in [29] counterfeit versions of the products Viagra and Cialis were imported into the market. The results of counterfeiting can be quite lucrative. To demonstrate how to reduce the amount of counterfeit pharmaceutical products we provide the following example.

Example 7 *The RexWay pharmaceutical company has decided to tag pharmaceutical products to assure its authenticity. The manufacturer will initialize I_0 to include the company's name (CN), day, time, place of manufacturing and drug name (DN).*

$$I_0 \leftarrow CN||day/time||PLACE||DN. \tag{2}$$

Then I is computed as $I \leftarrow S||I_0$. Afterwards since pharmaceuticals represent high-cost and valuable products, each party of the supply chain needs to denote who they are and sign when they are delivered the drug (become custodian, labeled CTD). Also to ensure that the supply chain is clear, each party who possesses the drug will include in its information the party to which they will be delivering the product to (PTDT, party to deliver to). So information I will have the following form

$$I \leftarrow I||CTD||day/time||PTDT||Sig(SK_{CTD};I||CTD||day/time||PTDT). \tag{3}$$

The potential members of the supply chain include (and their role):

1. Law enforcement/Regulatory agency.
2. Manufacturer. An entity within the manufacturing system, they produce, label, and package drugs. They often determine the supply chain (routing) and verify that each drug is delivered to the right pharmacy.
3. Internal distribution channel. An entity within the manufacturing system, they store, and sell drugs. May often be the manufacturer.
4. External distributor. Registered merchants or firms (wholesaler, warehouse, etc.) that deliver products from manufacture system to pharmacies. In some situations manufacturer directly deal with pharmacies.
5. Pharmacy. Registered retailers or dispensers that sell drugs to end consumers. A medical clinic is also a form of pharmacy in this sense.
6. Consumer.
7. Recycling center.

A memory estimate for the RF cell organization in Example 7 is as follows. Although we indicated that there could be seven members of the supply chain, we will assume that there is no requirement for consumers to designate ownership, otherwise we would require a huge PKI. Further we will assume that the recycling center does not need to designate itself as a custodian. Lastly, observe that manufacturer could represent more than one member of the supply chain. So we really require memory for at most four members of the supply chain (one of which is the manufacturer). We assume that a short signature is used to sign, such as the scheme described by Boneh in [6]. Thus we estimate the size of a signature to be 160 bits. Assume S requires 64 bits. To estimate the size of I_0, see (2), we assume an efficient encoding of company name (CN) and drug name (DN). We assume CN to be 20 bits, which allows for 2^{20} different members. We assume day/time to be 6 bits and

PLACE to be 12 bits. Lastly assume DN to be 20 bits. Thus I_0 is $2*20+6+12 = 58$ bits. Initially I is $64+58 = 122$ bits. Assuming only four custodians of the drug, then there are at most threes appendings, as illustrated by (3). An estimate for the size of the final I is 122 bits $+ 3*(20+6+20+160)$ bits which is 740 bits. Let I^*, the information that is provided to the authorized party, equal I_0. We assume that the re-encryption factor is on the order of 160 bits, thus the ciphertext C stored in cell γ is approximately $64+122+160+160$ which is 506 bits. The re-encryption factor r is stored in cell δ is 160 bits. The value $W = h(r)$ is stored in cell ϕ, so 160 bits are needed, As estimated above, we require 740 bits to store I in cell ρ. The value $h(S)$ is stored internally. Thus as a very rough estimate, we estimate 1,566 bits would be needed. Current RFID technology does not support this amount of RF memory, but as technology develops the cost of constructing such tag would decrease.

6.5 Security of the Protocol

As mentioned earlier, the goal of the protocol is to defeat a software and/or human attack and not physical attacks on the tag. Physical attacks can always be created given one has the expertise and resources. However, software attacks represent attacks that can be distributed over the internet. Assessing the security of this protocol from software-based attacks we assume the following. First we use a cryptographic digital signature scheme, thus it is infeasible to create forgeries. We also assume that each party protects the secret key and has constructed suitably strong keys. We utilize a cryptographic hash function, which is preimage and collision resistant. Further, we utilize the hash function to generate serial numbers, so they are unique. Thus forgeries can be attempted but it is infeasible to generate a forgery without the secret key. Further collisions of values (like serial numbers) can be attempted but they too are infeasible. Now a malicious party could clone information from product $\mathcal{P}'$ and place in onto product $\mathcal{P}$. However, such attempts will be discovered. In [39] we established that when applying an integrity protocol as a banknote integrity tool, $\mathcal{L}$ would be able to detect fraudulent behavior. More formally, in [37], we constructed a model for integrity of RFID tag information, described as *perfect integrity of tag information*. In [39], we established that a banknotes application of the integrity protocol for authorized tracing satisfied perfect integrity of tag information for $\mathcal{L}$.

7 Conclusion

We have discussed several of the security challenges concerning the use of RFID as an anticounterfeiting tool. Further, we have discussed several of the security threats and vulnerabilities. Moreover, we have discussed the security design criteria, as well as several security models. Lastly, we have provided a generic integrity protocol, and illustrated the model with two examples that utilize the integrity protocol.

References

1. Atmel Corporation: Atmel e5561 data sheet (2003)
2. Avoine, G., "Adversarial model for radio frequency identification". *Cryptology ePrint Archive, Report 2005/098*, http://eprint.iacr.org/ (2005)
3. Baard, M., "Watchdogs push for RFID laws". *Wired News* (2004)
4. Baudin, M, Rao, A., "RFID applications in manufacturing" http://www.mmt-inst.com/RFID%20applications%20in%20manufacturing%20_Draft%207_.pdf
5. Blass, E., "Dutch rfid e-passport cracked – us next?" *Engadget* (2005)
6. Boneh, D., Lynn, B., Shacham, H., "Short signatures from the Weil pairing." *ASIACRYPT '01* **2139** (2001) 514–532
7. "Contaminated Counterfeit Toothpaste Now Found in 6 States, Canada" http://www.foxnews.com/story/0,2933,287544,00.html
8. Davey, M., "Illinois to help residents buy drugs from Canada, and Afar." *The New York Times* (2004)
9. ECB *Biannual Information on the Counterfeiting of the Euro*. ECB Press, Germany (2004)
10. "Fake passports". Time (1981) http://www.time.com/time/magazine/article/0,9171,925044,00.html
11. Gamal, T.E., "A public key cryptosystem and a signature scheme based on discrete logarithms", *IEEE Transactions on Information Theory*, **31** (1985) 469–472
12. "Gillette confirms RFID purchase". *RFID Journal* (2003)
13. Harris, G., "Tiny antennas to keep tabs on U.S. drugs". *New York Times* (2004)
14. Heydt-Benjamin, T.S., Bailey, D.V., Fu, K., Juels, A., O'Hare, T., "Vulnerabilities in first-generation rfid-enabled credit cards". In: *Eleventh International Conference on Financial Cryptography and Data Security, Lowlands*, Scarborough, Trinidad/Tobago (2007)
15. James, J.S., "FDA, companies test rfid tracking to prevent drug counterfeiting". *The Body* (2006)
16. Juels, A., "Privacy and authentication in low-cost RFID tags" http://www.rsasecurity.com/rsalabs/staff/bios/ajuels/ (2003)
17. Juels, A., "Minimalist cryptography for low-cost RFID tags". In: *The Fourth International Conference on Security in Communication Networks – SCN 2004*. Lecture Notes in Computer Science, Amalfi, Italia, Springer, Berlin (2004), pp. 149–164
18. Juels, A., Pappu, R., "Squealing euros: Privacy-protection in RFID-enabled banknotes". In: *Financial Cryptography*, Springer, Berlin (2003) pp. 103–121
19. Kanellos, M., "E-passports to put new face on old documents". CNET News.com (2004)
20. Koh, R., Schuster, E.W., Chackrabarti, I., Bellman, A., "Securing the pharmaceutical supply chain". Technical Report MIT-AUTOID-WH-021, AUTO-ID Center (2003)
21. Molnar, D., Wagner, D, "Privacy and security in library RFID: Issues, practices, and architectures." In Pfitzmann, B., Liu, P., eds., *Conference on Computer and Communications Security – ACM CCS, Washington, DC, USA*, ACM Press, New York, NY (2004) pp. 210–219
22. Ohkubo, M., Suzuki, K., Kinoshita, S., Cryptographic approach to "privacy-friendly" tags. In: RFID Privacy Workshop, MIT, Cambridge, MA, USA (2003)
23. Ranasinghe, D., Engels, D., Cole, P., "Low-cost RFID systems: Confronting security and privacy," in *Auto-ID Labs Research Workshop*, Zurich, Switzerland (2004)
24. "Rfid sensor system promoted for highway safety". *RFID Update* (2007)
25. Sarma, S.E., Weis, S.A., Engels, D.W., RFID systems and security and privacy implications. In: *Workshop on Cryptographic Hardware and Embedded Systems, Lecture Notes in Computer Science* (2002) pp. 454–470
26. Staake, T., Thiesse, F., Fleisch, E., Extending the EPC network – the potential of RFID in anti-counterfeiting. In: *Auto-ID Labs Research Workshop*, Zurich, Switzerland (2004)
27. Symbol Technologies: RFID Technology and EPC in Retail. (2004)
28. United States Department of Homeland Security "United States Visitor and Immigrant Status Indicator Technology Program (US-VISIT)" http://www.dhs.gov/xlibrary/assets/privacy/privacy_pia_usvisit_adis_i94.pdf

29. United States Department of Justice, "Pharmacist Sentenced to Prison for Ordering and Receiving Counterfeit Pharmaceutical Drugs" `http://www.usdoj.gov/criminal/cybercrime/georgeSent.htm`

30. United States Department of State, "Department of State Begins Issuing Electronic Passports to the Public" `http://www.state.gov/r/pa/prs/ps/2006/70433.htm`

31. "Wal-Mart details RFID requirement". *RFID Journal* (2003)

32. Wasserman, E., "A Prescription for Pharmaceuticals", *RFID Journal*, `http://www.rfidjournal.com/magazine/article/1739`

33. "Wave the card for instant credit". *Wired News* (2003)

34. Wikipedia contributors, "RFID," In: *Wikipedia, The Free Encyclopedia*, Dec. 14, 2004, 18:42 UTC.

35. Yoshida, J., "Euro bank notes to embed RFID chips by 2005". *EE Times* (2001)

36. Zappone, C., "e-passports: Ready or not here they come". *CNNMoney* (2006)

37. Zhang, X, King, B., "Modeling RFID Security", *CISC 2005. Information Security and Cryptology, First SKLOIS Conference, CISC 2005*, Beijing, China, December 15–17, 2005. Lecture Notes in Computer Science 3822 (Springer 2005) pp. 75–90.

38. Zhang, X., King, B., "Integrity improvements to an RFID privacy protection protocol for anticounterfeiting", *ISC 2005. Information Security, Eighth International Conference, ISC 2005*, Singapore, September 20–23, 2005, Proceedings. Lecture Notes in Computer Science 3650 (Springer 2005) pp. 474–481.

39. Zhang, X., King, B., "Applying Integrity to an Anticounterfeiting RFID Privacy Protection Protocol", *Journal of Computer Science and Technology*, 22(3) (2007) 438–448.

RFID Security and Privacy

Tassos Dimitriou

Abstract Radio Frequency IDentification (RFID) is a method of remotely storing and retrieving data using small and inexpensive devices called RFID tags. Products labeled with such tags can be scanned efficiently using readers that do not require line-of-sight. This form of identification, often seen as a replacement of barcode technology, can lead to improved logistics, efficient inventory management, and ultimately better customer service.

However, the widespread use of radio frequency identification also introduces serious security and privacy risks since information stored in tags can easily be retrieved by hidden readers, eventually leading to violation of user privacy and tracking of individuals by the tags they carry.

In this chapter, we will start by building some background on the types, characteristics, and applications of RFID systems. Then we will describe some of the potential uses and abuses of this technology, discuss in more detail the attacks that can be applied to RFID systems and, finally, review some of the countermeasures that have been proposed to date.

1 Introduction

Radio Frequency Identification (RFID) is a new technology for automated object identification. An RFID tag is an electronic device that consists of an antenna and an inexpensive chip, often smaller than a grain of rice, that can be read from distance by a nearby reader. This device is typically attached to an object and upon request it can return information related to the tagged item, such as product characteristics, date of manufacture, date of purchase, and so on.

T. Dimitriou
Athens Information Technology, 19.5km Markopoulo Ave., 19002 Athens, Greece
e-mail: tdim@ait.edu.gr

P. Kitsos, Y. Zhang (eds.), *RFID Security: Techniques, Protocols*
and System-on-Chip Design, © Springer Science+Business Media, LLC 2008

RFID tags can be *passive* or *active*. Passive or semiactive tags get their power directly from the signal broadcasted by a reader. This ability to draw power from a nearby reader is what makes passive tags attractive; they do not need batteries, so they can be smaller and cheaper opening a new way of interesting applications. Active tags, on the other hand, have their own power source but are typically more expensive and are used only in specialized applications.

Tags, of the "passive" variety, are often envisioned as a next-generation bar-code technology, automating inventory procedures, thus cutting costs for manufacturers and retailers. Their two most important characteristics are small size, which allows them to be implanted within objects, and their ability to be read inside boxes, pallets, etc. which does not require line-of-sight. Passive tags are consequently less expensive and offer an unlimited operational lifetime. The tradeoff is that they have shorter read ranges and memory capable of holding a very small amount of information.

Despite these limitations, however, RFID tagged items can have remarkable applications. One can imagine a future where passive RFID tags are in every human-made object and even in some natural ones (such are animals or even people). This would allow better tracking of items in complex automated chains and revolutionize distribution networks, thus permitting goods to be traced from manufacturers to retail stores. This may help companies combat theft or improve management of stock and inventories in shops or warehouses (many industries and government agencies in the US, including the Department of Defense, already mandate the use of RFID tags by all of their suppliers [1, 2]).

However, the introduction of RFID tags in all objects could also directly benefit the consumer: One could imagine refrigerators issuing warnings about expired food or about remaining bottles of milk. Laundry machines could select washing cycles based on color and sensitivity of clothes. Waiting times at checkout lines may be drastically reduced since RFID readers can scan tags at rates of hundreds per second. Pharmaceutical products may be checked for being counterfeit or expired and animals (or more controversially children) could be retrieved in case they are lost.

Despite this increased productivity and convenience, one must wonder about the social consequences of a world full of tagged items. Will this pervasive use of RFID tags open up the possibility for violating user privacy? Consider for example the communication between tag and reader. The mere fact that this communication is wireless and does not require physical contact opens up the possibility for abuse. Currently, RFID tags respond to any reader request within range. Consequently, a person carrying a tagged item effectively broadcasts a fixed identifier to nearby readers. Thus anyone with a reader can read the information in the tag, potentially violating the owner's privacy.

To see how this new technology can be (ab)used, under today's bar-code technology, an ABC wrist watch sold in Athens has the same bar-code as a watch sold in Paris. With RFID, however, each watch carries a *unique* identification number which could be tied to a particular person, the buyer of that watch. The person could then be tracked if he/she ever entered the same store, or if they entered *any* other store with an RFID reader at the premises. Eventually, the reader may be able to identify that wrist watch, the time and date it was bought, where it was bought, and how frequently one visits a particular store.

In general, violation of privacy can have two forms: information leakage and location tracking. The first form deals with direct information obtained from a tag that may help in identifying the owner's preferences and physical condition. For example, information about medication may point to a particular disease, clothing information may reveal a particular life style, and so on. One of the major worries of privacy advocates is that purchased tagged items would link buyers to these specific items in central databases. Marketers could then use this information to build personal profiles and target individuals with specialized sale offers. However, even if tag responses are not tied to a particular product, static data can help in tracking the whereabouts of a person. This second form of privacy violation can be achieved by correlating tag "sightings" from multiple readers at fixed locations. Thus people can be tracked by the tags they carry!

The use of RFID tags in products and everyday items offers many benefits for both industry and consumers alike. The concern, however, is that this technology can potentially be abused in numerous ways. Unless changes are incorporated at a time when this technology is still developing, we may suffer the consequences later on. In this chapter, we are going to describe some of the potential uses and abuses of this technology and review in more detail some of the countermeasures that have been proposed to date.

The remainder of this chapter is organized as follows: Section 2 aims to build some background on RFID technology by giving more details about the types, operational characteristics and applications of RFID systems. Section 3 is the heart of the chapter; it focuses on the security and privacy concerns introduced by the use of this technology and presents possible solutions and countermeasures. It starts by listing a set of requirements that should be true of any secure solution and then presents in more detail the various approaches that have been proposed to date. While this list by no means is complete, it serves to highlight the issues that need to be taken into account when providing for secure and anonymous RFID transactions.

2 RFID Primer

At the highest level, RFID tags can be characterized as either *active* or *passive*. Active tags require a power source and they may have limited lifetime if they are powered by integrated batteries. Passive tags are of more interest to retailers for various reasons: they exhibit indefinite operational lifetime, they require no battery, and they can be made small enough to fit almost everywhere. They consist of an antenna, that is used to receive power from a reader and send back information to it, and a chip, whose simplest operation can be that of retrieving a unique ID that identifies the tag.

The simplest approach for implementing a passive tag is the use of *near-field coupling*. In near-field coupling, a reader first creates a magnetic field in its locality. If the tag is placed in such a field, an alternative voltage will be generated in the antenna's coil, which can be used to power the tag's chip. The data sent back to the reader typically uses load modulation, a technique in which a varying load is applied in the tag's antenna that eventually can be detected by the reader as a small change in the current flowing through the reader's coil. Different modulation techniques can be used depending on the number of bits required in the tag ID as well as the rate of data transfer. In near-field coupling, however, the range of communication between tag and reader is proportional to $c/2\pi f$, where c is the speed of light and f is the operational frequency. As f must be increased to accommodate for larger IDs and higher data rates, this technique poses certain limitations and new tags have been developed that are based on the concept of *far-field coupling*.

In far-field communications, tags are equipped with a dipole antenna which allows the tags to be outside the reader's near field, thus allowing for longer communication ranges. The technique used to send data back to the reader is called *back scattering* in which some of the reader's incoming signal is reflected back, thus allowing the tag to communicate its ID. Far-field tags, by means of a larger antenna, can achieve better data rates and longer read distances.

Passive tags can operate using a number of frequencies. Low-Frequency (LF) tags operate in the 125–135 kHz range; they have a typical read range of less than half a meter and a data rate of a few kbps. High-Frequency (HF) tags operate at 13.56 MHz, may have ranges up to a couple of meters and typical data rates of tens of kbps. Ultra-High-Frequency (UHF) tags operate in the 860–960 MHz range (also in 2.45 GHz), they can communicate in distances of up to tens of meters and achieve rates up to a few hundred kbps. Typically, tags operating in UHF use far-field communications and back scattering while LF and HF tags use load modulation for tag-to-reader communications.

2.1 Standards

There have been many standards in the world of RFID, however, two important families include those developed by EPCGlobal and those developed by the International Standards Organization (ISO).

EPCGlobal is a consortium of several companies and universities that created the standards for Electronic Product Code or EPC. The most important standard coming out from EPCGlobal is the Generation-2 standard that was created to mitigate many of the issues that limited the success of its Generation-1 standard (mainly interoperability issues for tags of the Gen-1 variety). Four classes of tags are distinguished within Generation-2 that progressively build upon the properties of lower classes. Class 1 refers to write-once read-many passive tags that carry unique ID, password-based access control, and a kill switch that can be used in deactivating the tag at a point-of-sale. Class 2 extends Class 1 passive tags mainly by allowing rewritable memory and authenticated access control. Class 3 refers to semipassive tags that carry an integral power source to supplement captured energy. Finally, Class 4 refers to active tags that enable tag-to-tag communication, more complex protocols, and ad hoc networking.

ISO 18000 is a multistandard that specifies protocols for a number of frequencies, including LF, HF, and UHF. One meeting point between the two families is the incorporation of EPC Gen-2 Class 1 standard in the new ISO 18000-6 standard. In addition to that, two other RFID standards are ISO 14443 and ISO 15693. ISO 14443 was created for proximity cards and RFID tags that operate at short distances, typically in the order of several centimeters. ISO 15693, is a more recent standard for both vicinity cards and RFID tags that typically operate at distances of about 1 m. Both standards operate in the HF band, while UHF is covered by EPC Gen-2 Class 1.

Another important development in the world of RFID standards is the establishment of the Near-Field Communications (NFC) forum that aims in integrating mobile phones with existing passive RFID products based on near-field coupling. The NFC standard is compatible with both ISO 14443 and ISO 15693 and allows devices to operate as either a reader or a tag, thus allowing both transmission and reception of data. One application of this technology is secure device pairing since, for example, a mobile phone can communicate securely with others in the vicinity by exchanging keys without the fear of eavesdropping or person-in-the-middle attacks (physical proximity enhances security since an attacker that is present runs the danger of being discovered). A complication of the wide-scale adoption of the NFC standard is its incompatibility with the EPCGlobal standards which are based on far-field communications.

2.2 System Architecture

An example of system architecture is shown in Fig. 1 consisting of a tag, a reader, and a back-end database. RFID tags are relatively cheap nowadays but in order to achieve greater penetration their cost must drop to a few cents per tag [3]. Upon a

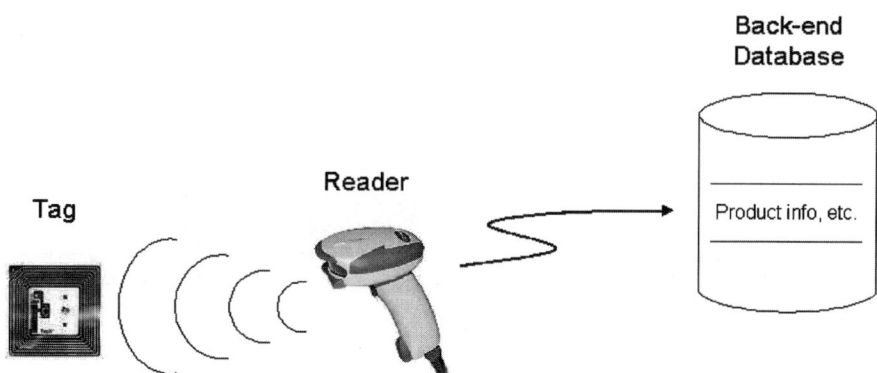

Fig. 1 System architecture

scan request by a reader, the tag responds with a unique ID that is transferred to a back-end infrastructure for further processing. The tag is identified and particular info about the tagged product can be retrieved by means of accessing a system database containing all possible tag identifiers.

Communication between tags and readers is wireless and is therefore subject to eavesdropping. Thus most protocols aim to secure this part either by securing tag-to-reader communications or by making tag responses indistinguishable from random data (we will have to say more about this in Sect. 3). On the other hand reader-to-database communications can be assumed to be secure since both reader and back-end systems are more powerful devices and can handle the overhead introduced by encryption. One issue, however, that needs to be remembered is that the use of encryption should not affect the end-system's performance. If the back-end database does not have an efficient way to disambiguate the tag among (perhaps) millions of concurrent encrypted tag responses, the database will become the bottleneck and performance will be lost. Thus protocols must not only be secure but be efficient as well. *Scalability* is therefore an important issue that affects protocol design and choice of cryptographic tools.

Each tag carries a unique number called Electronic Product Code (EPC) as shown in Fig. 2. The *Header* field allows for an expansion of the tag format to 64-, 96-, and 256-bit versions. For the 96-bit format shown in Fig. 2, the *EPC Manager* field defines the domain manager for the remaining fields. *Object Class* describes the generic type of tagged product and *Serial Number* denotes an individual item number of 36-bits length.

This EPC serves as an identifier for the physical object carrying the tag, which can now be recognized, identified, and tracked by the underlying IT-infrastructure. Since information about objects should not in general be stored on the tag itself, EPCGlobal has developed a lookup system by which such information can be supplied by distributed servers on the Internet [4]. This system is called the Object Name Service (ONS) and is similar in spirit to the Domain Name System for resolving Internet names. By using this system it will be possible to let parties dynamically register any kind of EPC Information Service for the objects, e.g., the main manufacturer, suppliers, shops, or after-sale service providers, thereby opening the way for new business ideas (however, for some security considerations regarding the ONS, see [5]).

Fig. 2 Example of an 96-bit EPCGlobal tag

2.3 Example Use

In this section, we review some of the basic applications of RFID technology along with some privacy concerns regarding its use. Although this list is by no means complete, it serves as indicator of how the technology can eventually affect our every day life activities.

2.3.1 Supply Chain Management and Inventory Control

Perhaps the driving force behind the widespread adoption of RFID technology is supply chain management and manufacturing logistics. RFID technology presents a major improvement over bar-code technology in a number of ways. In addition to their small size which allows them to implant within objects, RFID tags can be scanned in large numbers, without the need for line-of-sight as in optical readers. RFID technology can improve product "visibility" and help combat theft at various stages in the supply chain: Readers can gather information about the location of goods as they travel from manufacturers, to warehouses or distribution centers, and eventually to stores. Access to more accurate information about the location of products in the various stages of the distribution chain allows retailers to keep what they need in stock, thus avoiding costly delays in delivery/manufacture of items, as well as helping increase sales by ensuring that shelves are always stocked.

This potential for lowering the operational cost in the supply chain is what motivates major retail stores such as Wall-Mart [1] or government agencies such as the US Dept. of Defense [2] to adopt UHF RFID tags and require their suppliers to put electronic tags in the pallets and packing cases they deliver to them. In summary, there is considerable momentum behind RFID adoption by major retailers which will also have an impact on smaller companies and customers as well.

2.3.2 Access Control

Another important area of application of the RFID technology is authentication and access control. Replacing keys with electronic cards or budges has a number of advantages. The primary one is that cards are more difficult to forge and can be revoked more easily when compromised or lost than having to change the lock as is the case for mechanical keys. Additionally, this method provides for better control and knowledge of the people present in a particular area, which may be helpful in emergency situations.

RFID-based keys are also used in automobiles where the goal is to make theft harder. A car equipped with a reader will only start if both the mechanical key and the unique RFID tag are present. The lack of the correct electronic tag will prevent the vehicle from starting up, thus serving as an immobilizer. The car can start only when the tag correctly replies to a challenge sent by a reader. Thus the requirement that both a physical key and an RFID interaction are present is what makes theft a harder task.

2.3.3 Automatic Payments

There are many instances where RFID tags can be used for making electronic payments. Perhaps the simplest one is their use in automated toll collection. A small plaque placed in the windshield of a car is interacting with a reader at the tollbooth to automatically collect fares and debit a prepaid account every time the tag ID is detected. The system is designed so that it can operate at relatively high speeds, thus helping reduce traffic jams, especially at peak commuting hours.

Electronic tickets is another application of RFID technology. An RFID-based ticket is a prepaid token that provides access to some facility or a resource. The ticket can be validated as the user passes through a reader and even be renewed electronically, thus allowing for greater convenience and eliminating the need for printed tickets.

2.3.4 People and Pets

Human implantation of RFID tags dates back to at least 1998, when Kevin Warwick, professor of Cybernetics at the University of Reading, implanted an RFID tag above his left elbow, which he used to control doors, computers around his office. In 2004, VeriChip had a tag approved for implantation in people. The idea is to offer rapid, secure patient identification, helping at-risk patients get the right treatment when needed most. By scanning the tag, doctors can identify a patient and access personal medical information even if a person is admitted to a hospital unconscious. Other uses include controlled access to restricted areas. Such access had already found application in certain clubs that use them to let their members bill drinks directly to their accounts and gain access to VIP areas.

While human implantation has some profound ethical and privacy implications (for example, Scott Silverman, CEO of VeriChip, proposed in an interview implanting chips in immigrants to assist the US government in later identifying them [6]), animal implantation is already a reality. Dairy farms use RFID tags to monitor the status of each animal (amount of food, medication, identity) as well as the ownership and medical history to protect against infectious diseases. Owners of pets tag their animals so that when lost, authorities can scan the tag and identify the owner's name and address. While stray animals often lose their collars, this seems to be unlikely with subdermal tags (but do you see how this can lead to possible violation of the owner's privacy?).

2.3.5 Authenticity of Money and Drugs

Adding RFID to banknotes is seen as an automatic way of validating their authenticity and helping reduce counterfeiting (the European Central Bank allegedly had such plans for high denomination Euro banknotes [7]). However, adding a unique ID that is readable everywhere increases the chances of tracking and violation of privacy.

This is because tagged banknotes can be easily tracked between transactions, thus providing information about where people spend their money.

Another area of using RFID tags to enhance authenticity is in pharmaceutical products. But in today's era where anybody can purchase medication over the Internet, can we be sure that the products we buy are authentic and safe for use? Antitamper proof packaging together with hard-to-forge RFID tags may be a solution to the problem. However, again many privacy issues arise since a person can be scanned for the medicine he/she carries thus revealing possible medical conditions and health problems. This information then can be of value to insurance companies and employers.

2.3.6 Passports

The International Civil Aviation Organization (ICAO), a body run by the United Nations recently set a mandate for incorporating RFID tags into passports [8]. These electronic passports have already been deployed in many European countries, USA, Japan, and several others. The goal of course is to provide strong authentication through documents that strongly identify their bearers and cannot be forged.

Unfortunately, as demonstrated in [9] and throughout Europe [10], e-passports fall short of their goals as they can be used for clandestine scanning and tracking. Even if a mechanism called Basic Access Control is used to encrypt tag-to-reader communications, the cryptography is very weak since the secret key can be brute-forced very easily (the secret key really provides about 50-bits of security).

3 Security and Privacy

As we have already highlighted, RFID technology poses unique risks to personal privacy and security. RFID tags come in a wide variety of shapes and sizes. While some tags are easy to spot, others can be very small (the μ-chip by Hitachi [12] measures 0.4 mm including the antenna) or they can be embedded inside boxes, clothes, etc. and thus be invisible to the human eye. Since tags can be interrogated by hidden readers, the data transmitted by the tag may provide identification and/or location information as well as specifics about the tagged product, such as price, color, date of purchase, etc.

This threat to privacy is magnified further by the fact that readers can scan multiple tags at the same time. This means that if a person enters an area carrying several tagged items, it takes just one RFID reader to collect the information emitted by all the tags! This information can then be used in numerous ways. For one, retailers can use RFID readers to build complete profiles and target specific individuals. For another, even if tag responses do not make sense because they are encrypted, a collection of *fixed* identifiers may still help identify the whereabouts of a person by matching these identifiers either through some form of payment that reveals the person's identity (e.g. credit card) or through a snapshot taken by a hidden camera.

Although our focus here is personal privacy, RFID tags can also pose a threat to corporate security because many different entities can read the tags [13]. For example, competitors can remotely gather confidential supply chain data or more frighteningly deny this information to the company by performing radio frequency jamming attacks. For a more complete list of threats, the authors in [14] offer a taxonomy of attacks across the different stages of a typical industrial supply chain. Finally, several interesting books (other than the current one) have already been published: [15], which is dedicated to RFID security and privacy and [16] that focuses on consumer privacy violations.

Thus incorporating security to RFID technology is not only in the interest of individuals but companies as well. In the remainder of this section, we will list in more details the various threat and security requirements we expect from secure RFID protocols and then present an overview of existing privacy enhancing techniques.

3.1 Security Requirements

Most threats to personal privacy arise from the fact that RFID tags come equipped with unique identifiers that can be read by anyone with a reader. In a typical system (Fig. 3), a reader makes a scan request and the tag responds with an identifier that helps identify the object carrying the tag. We list a set of general security goals that should be true for any RFID protocol as follows:

- *Privacy enforcement.* No secret information should leak from the tag that can help in identifying tag contents or the bearer of the tag. Current RFID tags aim to be cheap [17], so they emit constant identifiers that may reveal personal and sensitive information. Examples include money, medicine (which may link to a particular disease), books (which may indicate a particular political preference), and so on.
- *Protection against tracking.* Since tags can be read from inside wallets, backpacks, suitcases, etc. and readers can be everywhere (hidden in walls, entrances, . . .) current RFID deployments can be used to track people by the tags

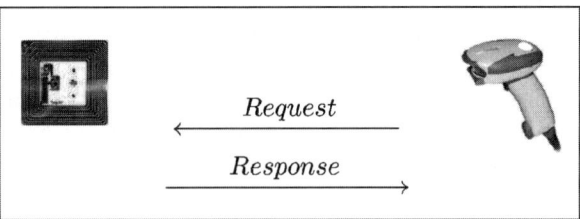

Request

Response

Fig. 3 A reader querying a tag

they carry. Additionally, a collection of tagged objects may pinpoint to a particular person, even if a personal identity is not associated with the tags. To solve this problem no *fixed* identifiers should be emitted by the tags.

In addition to the above basic requirements, a list of derived requirements follows:

- *Efficiency of identification.* Once a tag response is received by a reader, it is forwarded to a back-end database that helps identify the tag from a list of valid tag identifiers. Since this database may consist of millions even billions of items and readers can be used to read multiple tags at the same time, *scalability* must be a necessary requirement of any well-designed protocol. We feel that security solutions that are not coupled with efficient identification eventually will be dropped out.
- *Protection against tag spoofing or cloning.* A tag cloning attack allows an adversary to either install a replacement tag or simply query the tag and forward its response to a nearby reader. This would fool the system into believing the product is still on the shelf, or alternatively, an expensive item could be purchased for the price of a cheap one. These *person-in-the-middle* type of attacks can have serious consequences especially when RFID tags are used for access control. Examples of such attacks have been demonstrated in [18], where serious security weaknesses in SpeedPass and many automobile immobilizer systems have been identified, in [19], where a device was built that could clone RFID-based entry systems made by several vendors, and in [20], where an attacker can trick a reader into communicating with a victim smartcard that is very far away despite the fact that contactless smartcards have an alleged range of just a few centimeters. Unfortunately, all protocols that have the form of Fig. 3 fall prey to these relay attacks since tag responses cannot be authenticated without further involvement of the reader. One possible defense is to consider *distance-bounding* techniques that are designed so that the round-trip delay time of a challenge–response exchange is measured accurately and is protected against manipulation [21]. However, the method of [21] does not quite fit in the RFID paradigm since many messages have to be exchanged to reduce the probability of error.
- *Forward privacy.* RFID tags are inexpensive devices that offer no tamper resistance, hence they suffer from physical attacks that attempt to expose their memory contents, in particular secret values (keys) stored in the tag. An attacker upon compromising a tag may be able to link this tag with past actions performed on the tag. Forward privacy ensures that messages transmitted today will still be secure in the future, even after compromising the tag.

3.2 Policy and Other Nonprotocol Approaches

Some simple solutions to mitigate the privacy problems introduced by the use of RFID tags include "killing" the tags or alternatively putting them to "sleep."

Killing the tags can be enforced by a password protected kill operation. A reader at a point of sale may instruct the tag to kill itself by sending a small tag-specific pin, thus rendering the tag inoperable. While this measure certainly enforces user privacy, a killed tag may lose all its postsale benefits for the consumer. For example, in a home equipped with this ubiquitous technology, refrigerators, ovens, washing machines, and other devices could not longer interact with tagged objects and act on behalf of the consumer based on the exchanged information. People with Alzheimers disease could not maintain their independence by getting assistance and guidance in everyday activities [22]. For these reasons and others that require tags to remain operable through their lifetime (libraries, rental shops, etc.), one should look for less "devastating" approaches to security.

One such alternative would be to put the tags to "sleep." A user could disable the tag at the point of sale and then wake it up again at the convenience of her home. However, this approach poses certain management problems; if users do not want anybody with a reader to reactivate their tags, these "wake-up" commands must be protected by a password known only to the user. But then consumers would lose all the convenience of using RFIDs and force themselves to live in a nightmare of passwords.

Another technique that could be used is to prevent the reader for understanding a tag's reply by means of *blocking*. In one instantiation of this scheme, a blocker tag could be used that would disrupt all communications from selected tags [23]. The proposed blocker tag could be carried by users inside purses or clothes and selectively control which of the users' tags might respond to scanning attempts. In some sense this would be similar to a Faraday cage that shields radio communications, the only difference being that tags outside the "privacy" zone of the blocker tag could still respond to scan requests.

We feel, however, that these solutions add a burden to consumers. What if a person neglects to kill a tag, or carry a blocker tag or forget to implement some security policy, especially if *additional* steps are required to make them effective? (A prototype killer kiosk requires users to load one item at a time, clearly a time-consuming and inconvenient process [24].) What would happen if users that decide to disable their tags do not get the same benefits as the rest? Thus killing or blocking tags might create two types of customers: those that care about privacy and those who do not (or not having the knowledge or time to do so). Belonging to either class could have important, negative ramifications [25].

Finally, it may be helpful to consider policy-based approaches in protecting user privacy from covert use of RFID technology. CASPIAN (Consumers Against Supermarket Privacy Invasion and Numbering), an organization for the defense of individual liberties, believes that RFID technology and its implementation should be guided by strong principles of fair information practices (in a similar vein, Simson Garfinkel proposed an "RFID Bill of Rights" in [26]). In general, the following set of minimum guidelines (adapted from [25]) must be enforced:

- *Openness or transparency.* Individuals have a right to know about readers and RFID tagged products as well as their technical specifications. There should be no hidden tags and no tag-reading in secret.

- *Purpose specification.* RFID users must be warned about the purposes for which tags and readers are used.
- *Collection limitation.* Collection of information should be limited to a well-specified purpose.
- *Accountability.* Retailers and users of this technology should be legally responsible for complying with the principles.
- *Security safeguards.* Security should prevail the RFID environment (communications, database and system access) and should be easily verifiable by third-parties.

3.3 Protocol Approaches

In this section, we are going to review some proposals that enhance user privacy without requiring any specific actions by the user. Unfortunately, this review cannot be comprehensive due to the immense amount of work in RFID security (an online repository is available at [27]). Instead, we will focus on the main constructions that will also help us highlight some of the protocol design issues in providing for secure and anonymous tag-to-reader transactions (additional information can be found in [28, 29]).

In general, all protocols rely on the existence of a *secret* shared between the tag and the reader (back-end database). This secret can be common to all tags, however, compromise of even a single tag leads to compromising the entire system. Another possibility is to have different secrets per tag. The disadvantage in this case is that a mechanism is required to allow the reader to determine which secret was used for which tag. Unfortunately, most obvious approaches either send an index to the system database (which opens the possibility of tracing) or require the database to *exhaustively search* over the collection of keys to identify the correct tag. However, these approaches do not scale well, especially when the database may contain billions of items.

A closely related problem to efficient identification is *authentication* between the tag and the reader. Tags must reveal their identities only to authorized readers but this should happen only if the reader has been authenticated to the tag. However, there is a chicken-and-egg problem here: In order to eliminate spoofing attacks, the reader can only authenticate itself if it knows the secret of the tag; but this requires knowledge of the tag's ID. However, the tag cannot reveal its ID unless the reader has already been authenticated to it. In the remaining section, we will see how existing solutions aim to address these problems and we will point to several issues that need to be taken care when designing a secure RFID protocol.

Finally, a point that needs to be remembered is that security and privacy is really a multilayered issue [30]. A cryptographic protocol, for example, aims to secure the application layer. At the lower layers there exist anticollision and other basic RF protocols. What if some information is leaked because of the singulation protocol used to disambiguate a tag from millions others? Also, since tags need to abide to standards, couldn't a mix of standards help identify the person carrying the corresponding tags? Finally, what about radio fingerprinting in the physical layer? Couldn't a tag be traced by its signal transmitting variations while interacting with a reader? Despite these threats, in the remainder of the section, we will focus on the upper layer since this is the starting point for traceability.

3.3.1 HashLock Scheme – Weis, Sarma, Rivest, and Engels [31]

In the work of [31] the authors propose the HashLock scheme: The tag carries a key K and a *metaID* which is simply the hash value of the key, $h(K)$. When a reader interrogates the tag, the tag responds with its *metaID* which is forwarded to the back-end database. The database recognizes the tag from its *metaID*, the secret key K is forwarded to the reader which eventually reports back this value to the tag. At this point the tag hashes the value received and if it matches the stored *metaID* it unlocks itself. Although this scheme is simple and reliable it suffers from many drawbacks. Perhaps the most serious one is that a tag can still be tracked by its *metaID*. Furthermore, the valid key K is sent in the clear so an adversary can easily capture the key and later spoof the tag to the reader.

In an attempt to improve upon this scheme, the authors presented a *randomized* variant in which tag responses *change* with every query made (Fig. 4). In particular, the tag sends a pair $\langle r, h(ID, r) \rangle$, where r is a *nonce* that is a randomly generated number used only once. Clearly, this scheme solves the tracking problem mentioned above as no fixed IDs are sent over time. However, a new problem is introduced. As the tag responds with different values every time, the database must exhaustively search through its list of known IDs until it finds one that matches $h(ID, r)$, for the given number r. Apart from this scalability problem, there is also a very simple protocol attack that can be applied: an adversary can query a tag and learn a valid pair $\langle r, h(ID, r) \rangle$, which then allows the attacker to impersonate (spoof or clone) the tag to a legitimate reader. This is a serious security flaw as the reader will identify the tag. In addition, the scheme allows the location history of the tag to be traced if the tag itself is compromised. Hence forward secrecy is not guaranteed.

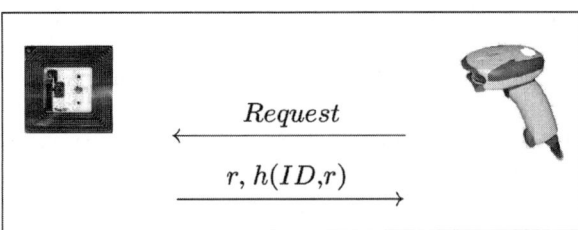

Fig. 4 A randomized variant presented in [31]

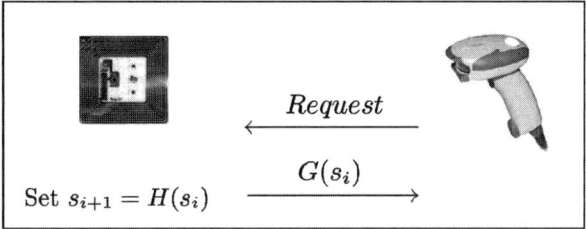

Fig. 5 A randomized and forward secure scheme presented in [32]

3.3.2 Ohkubo, Suzuki, and Kinoshita [32]

In [32], the authors devised a scheme that uses a low-cost *hash-chain* mechanism to defeat the problem of tracing and ensure forward security in tag transactions. The basic idea is to modify the tag ID each time the tag is queried so that the tag is recognized by authorized parties only. The scheme uses two hash functions H and G, one to refresh the secret in the tag, the other to make responses of the tag untraceable by eavesdroppers. Hence this scheme can be seen as an extension of the randomized version in [31] that also guarantees security of past transactions.

Initially, the tag is preloaded with a secret s_0. When the tag is queried by a reader it reports the value $G(s_i)$, where s_i is the tag secret during the ith transaction (Fig. 5). Then it updates its secret through the operation $s_{i+1} = H(s_i)$. However, just as in the previous scheme, scalability is problematic as it requires exhaustive search in the back-end database to locate the ID of the tag. Although in [33] a time–space memory tradeoff is presented, the scalability problem remains an issue. Furthermore, an attacker can still query a tag then replay the tag's response to authenticate itself to a valid reader.

3.3.3 Dimitriou [34]

In an attempt to solve the scalability problem, a protocol was proposed in [34], where the hash value of the secret ID of the tag is used by the back-end database to search and identify the tag. During system initialization the tag is loaded with an initial, secret identifier s_0 which is set to a random value. In a similar way, the back-end database contains the same data stored in the tag, together with a hash value of its ID, $h(s_0)$, that serves as the *main key* to look for any information related to this particular tag.

Upon query (Fig. 6), the tag answers back with a message of the form $\langle h(s_i), r, h_{s_i}(r)\rangle$, where s_i is the secret ID of the tag during the ith query and r a random number. The reader and eventually the back-end database upon receiving $h(s_i)$ uses this value to search and recover the identity s_i of the tag. Once the database has the s_i value, it can use the last part $h_{s_i}(r)$ to connect all the pieces together (this part essentially acts as a message authentication code) and verify the authenticity of the message. At this point the database accepts the tag as authentic and *renews*

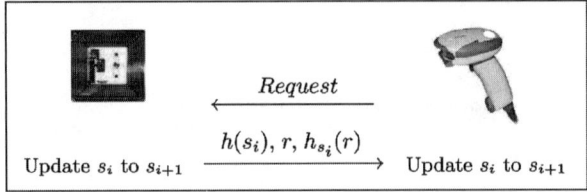

Fig. 6 A protocol presented in [34]

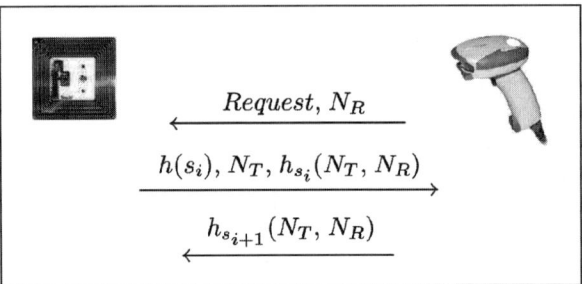

Fig. 7 An enhanced protocol presented in [34]

the secret identity from s_i to s_{i+1}. The tag does the same and erases any relevant information (r and s_i) from its memory. It should be clear by the description that this scheme is at least as good as the previous protocols since (1) it leads to faster identification, (2) it is forward secure, and (3) tracing is not possible. However, we should mention here that the scheme falls prey to a denial-of-service attack aiming at *desynchronizing* the tag from the database. If an invalid read attempt is made, the tag will update its secret from s_i to s_{i+1} resulting in desynchronization from the database. From then on, identification can only happen through exhaustive search.

To solve this desynchronization issue, another protocol was proposed that allows for mutual authentication between tag and reader (Fig. 7). Observe that this is just the simpler protocol enhanced with an extra message whose purpose is to authenticate the reader to the tag. This extra step essentially allows the tag to disregard any queries made by unauthorized readers. Thus the issue of desynchronization is eliminated entirely. There is, however, a penalty that is paid for this increase in efficiency.

While the scheme ensures that the secret identifier of the tag will change if the tag interacts with a valid reader, between valid sessions the tag ID remains the same. Tags are therefore subject to tracking during such intervals of inactivity. Any attempt to hide $h(\text{ID}_i)$ will incur a cost in searching the database, thus making the scheme not scalable as in [31, 32]. Privacy, however, can be regained if the user can make valid read requests, perhaps through the use of a proxy device [47].

3.3.4 Scalable Protocols

The scalability problem was addressed successfully in [35] and independently in [36]. These schemes work as follows. First, a tree is constructed whose leaves contain *all* possible tag identifiers $T_1, T_2, \ldots, T_n$ of interest. Then the edges of the tree are labeled with *secret keys* created during system setup. Each tag T_i is preloaded with the keys corresponding to the *path* from root to T_i, thus tags may also share secrets. If d is the length of such a path and $k_i^1, k_i^2, \ldots, k_i^d$ denote the secret values along the path to the ith tag, then the protocol for interacting with T_i is shown in Fig. 8.

To identify a particular tag the back-end database must distinguish it from the rest of the tags. This can be done as follows: recall that the database has obtained the values N_T, $\langle f^1, f^2, \ldots, f^d \rangle$, where each $f^j = F_{k_i^j}(N)$. Since it also knows the secrets of all nodes, it can try to infer the path that leads to tag T_i using the following method:

1. Consider the two keys k_l^1 and k_r^1, labeling the edges leaving the *root* of the tree. Thus k_l^1 is the first key preloaded to tags belonging to the left subtree and k_r^1 is the first key used by all tags in the right subtree. Compute $F_{k_l^1}(N)$ and $F_{k_r^1}(N)$ and compare with the received value f^1. If f^1 is equal to the first result, the tag belongs to the left subtree, otherwise it belongs to the right one.
2. Assume the path has been inferred up to a node at level j. Now consider the keys k_l^j and k_r^j, labeling the edges leaving that node. Again compute $F()$ using these keys on input N and compare with value f^j. Based on the output, continue to either or the right subtree.
3. Repeat Step 2 until a leaf node (tag) is reached. If at any point in the process a received value f^j does not match either of the two results, stop and reject the tag.

It should be clear at this point that a valid tag will eventually be identified by this process in time proportional to the *depth* of the tree (usually *logarithmic* in the number of tags). Hence the whole process is very efficient. It should also be clear that since no fixed identifiers are released, user privacy is enforced. Again, however, we see a recurring theme: increase in efficiency may result in some loss in security or privacy. This is because tags *share* secret keys, hence compromise of one tag may reveal information about others. This is illustrated in Fig. 9.

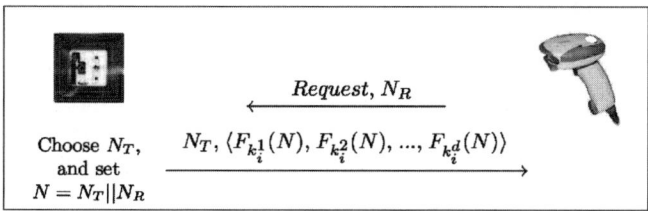

Fig. 8 A scalable tree-based protocol

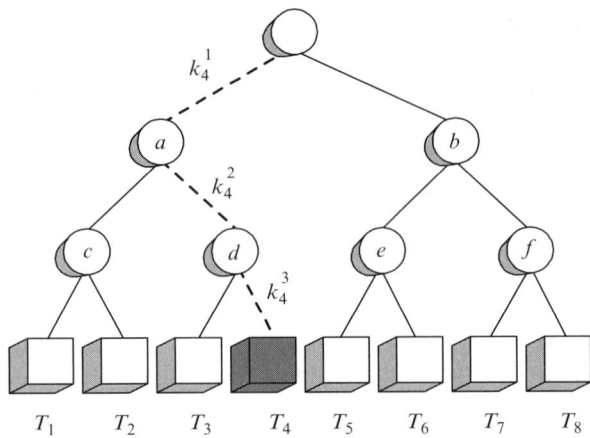

Fig. 9 Compromising a tag in the tree based protocol

Consider the case that T_4 is being compromised and an attacker gains access to the secret values stored in the tag. What does this mean about the privacy of the rest of the tags? Looking at the tree of Fig. 9 we can see that only the tags that share a common path are in danger of releasing private information. Nonetheless, some form of tracking is still possible. Consider a person carrying T_2 and T_7 with her all the time (say T_2 corresponds to an expensive watch and T_7 to a briefcase). An attacker, having compromised T_4, can find out that the first item is located in the subtree rooted at c, thus it can correspond to one of T_1 and T_2. This is so because when the attacker queries T_2 and applies the identification process described previously, she will succeed in computing the pseudorandom function (PRF) value using key k_4^1 but will fail when attempts to do so using k_4^2. This will be an indication that the tagged item belongs to the subtree rooted at node c. Similarly, the second item will be located under subtree b. Thus an attacker can use this information to track people not by the tags they carry but by the *subtrees* these tags fall into! And the more tags a person carries, the less probable it would be that this person is confused with some other one [36, 37].

The previous analysis clearly shows that tag secrets need to be *refreshed* occasionally. Forward secrecy ensures that messages seen in the past still remain secure even after the tag has been compromised. In the tree-based approach forward privacy is needed for another reason: As path keys are shared among tags, compromise of a path may lead to compromise of other tags' secrets. However, in the protocol this is exactly the reason that hinders forward privacy. If the secrets along a path get refreshed, tags belonging to either the left or the right of the specified path cannot longer be identified efficiently. Whether a tradeoff between efficient identification and forward privacy can be found remains an important research direction (see [38] for an attempt towards that direction).

3.3.5 Noncryptographic Solutions

The previous solutions assume that tags are capable of computing simple hash functions or pseudorandom ones in order to refresh tag IDs and make responses indistinguishable from random values. Although it is believed that cryptographic components are not suitable for low-cost tags, recent work [39,40] has led to more compact implementations of symmetric key primitives like the Advanced Encryption Standard (AES). This may lead to implementation of real tags capable of performing challenge–response protocols using strong cryptographic primitives.

There are, however, some solutions which do not use cryptographic operations [41, 42]. The authors in [41] proposed a set of challenge–response authentication protocols built from primitives that can be supported by low-cost RFID tags. The emphasis was on coming up with lower bounds on the abilities of the attacker and a trade-off between security and performance. However, these protocols can be easily broken as demonstrated in [43]. Furthermore, these protocols do not address the scalability problem and no attempt is being made to prevent tracking of the tags.

In [42], a protocol is proposed which is resistant in a rather weak adversary model: limited successive tag queries and limited successive eavesdropped interactions between the tag and legitimate readers. The key idea of the protocol is the use of pseudonyms to help enforce RFID privacy. Each time a tag is queried, it releases the next pseudonym from its list. In principle, then, only a valid verifier can tell when two different names belong to the same tag. Of course, an adversary could query a tag multiple times to harvest all names so as to defeat the scheme. So, the approach described involves some special enhancements to help prevent this attack. First, tags release their names only at a certain (suitably slow) prescribed rate. Second, pseudonyms can be refreshed by authorized readers. Although this scheme does not require the tags to perform any cryptographic operations (it uses only XOR operations) the protocol involves a large number of message exchanges and requires updating the keys and pads with new secrets, an operation which may be costly and difficult to realize.

3.3.6 Proxy Solutions

The protocols presented thus far aim to enhance user privacy against clandestine reading and scanning that can lead to tracking of users' activities by the tags they carry. *But tracking by whom*? All these solutions protect users against unauthorized scanning by third parties who want to find information about a user's tags. This protection is basically achieved by refreshing tag identifiers and making tag responses indistinguishable from random data. But there is still potential for violation of privacy. Recall, that in all these solutions the back-end database has knowledge of all the secrets associated with the company's tagged products. While this company may not be able to track the tags issued by other companies, it can still track the movements of tags issued by it. Furthermore, a coalition of such companies may create a federation network (similar to federated identity among enterprises) to exchange information about a user's movements, habits, and profile. Thus what is needed is the ability by the users themselves to have complete control of the tags they carry without losing any of the benefits that RFID technology has to offer [22]. The use of *proxy devices* serves exactly this purpose.

A proxy is a personal device such as a mobile phone (or any other similar device) enabled with reader capabilities [44] that can be used in acquiring and managing a set of tags on behalf of the user carrying RFID-tagged products. The proxy, being a more powerful device than an RFID tag, can specify a number of policies that readers must comply with. This can range from full release of tag data to more enhanced reader authentication protocols that are beyond the capabilities of today's simple tags.

So, in addition to the basic security requirements that were mentioned in Sect. 3.1, the following should also apply in a proxy environment:

- *Policy enforcement and access control.* We expect the proxy to act as mediator for tag access in order to minimize the privacy risks inherent in the use of RFID technology. It makes sense, therefore, to control which readers should have access to tag information. So, in certain cases it should be possible to release information about tags while in others to block such requests entirely (*not* by the use of jamming technology but by the use of proper cryptographic mechanisms). The user should be able to decide to the extend where access control will be used.
- *Transferability and tag release.* The main role of the proxy is to handle information about the tags a user owns and enhance user privacy. In many situations, however, it is necessary to bring a tag to its original state (tag release) or transfer it to a new user. In the first case, we require that the current owner should not be able to lie about the original ID of the tag, while in the second, the privacy of the new owner must be guaranteed.
- *Protection against impersonation and cloning attacks.* An adversary should not be able to impersonate either a tag or the mobile proxy. The first attack could lead to removal of a user's tagged items since a fake tag could still answer to proxy's challenges. The second attack could be useful in tracking a user's movements. If the tags the user carries respond to unauthenticated proxy requests, an adversary may learn valuable information about the user's habits and location. Additionally, an adversary should not be able to clone a tag or spoof the proxy by listening to the messages exchanged between the two.

The concept of having users protecting tags they carry by means of personal devices acting as *proxies* has been studied in [45–47]. The authors in [45], propose the use of RFID Guardian, a device that acts as an intermediary between tags and readers. Being a device with higher computational abilities, the Guardian can enforce various policies when interacting with readers. One drawback of this approach, however, is that the Guardian must always be alert in protecting tag responses from unauthorized read attempts. It has to either allow reader queries, appropriately reissuing queries in encrypted form, or *actively* block tag answers which may not always be feasible.

An improvement of the above scheme, is the RFID Enhancer Proxy or REP [46], which assumes the identities of tags and simulates them in the presence of reading devices by continuously relabelling the identifiers transmitted by tags. While the REP covers many of the security requirements mentioned here, it suffers from a few shortcomings that are attributed to the fact that tag identities need to be partially generated by the tag and match portions of its true ID.

Finally, in [47] a new framework was proposed that unifies the above approaches and proposes a complete solution for such issues as tag acquisition, proxy authentication, resistance to privacy attacks, ease of transfer and release, and so on. Once the proxy performs some initial transformations to the tags under its control, it can either *mediate* between tags and readers or let tags *directly* respond to scan requests. In the first case, the proxy can specify a number of policies that readers must comply with. This can range from full release of tag data to more enhanced reader authentication protocols that are beyond the capabilities of today's simple tags. In the latter case, it is ensured that tags do not emit any static identifiers thus providing ID anonymity and helping prevent tag tracing. Using the protocols described in the framework it is guaranteed that only authorized users can acquire and put tags under their control, user access to tags is authenticated, and when a user no longer needs the tagged product she can either make it readable to everybody or transfer it to another user in a way that guarantees the privacy of the new owner.

4 Conclusion

RFID is an important technology that can have remarkable consequences in every day life activities. In one instance, companies may benefit by improving manufacturing logistics and helping prevent product theft. In another, the use of this technology may help realize Mark Weiser's vision of *ubiquitous computing* in which small computers embedded in everyday objects could respond to people's presence and needs without being actively manipulated [48].

The use of this technology, however, may have important privacy ramifications. As highlighted in [25], "if used improperly, RFID has the potential to jeopardize consumer privacy, reduce or eliminate anonymity, and threaten civil liberties." We feel that consumer (as well as corporate) risk must be limited as much as possible through the use of technical solutions as well as education and policy efforts. Measures should be taken now, before widespread RFID adoption and deployment occurs, since it will be very difficult to integrate these measures later on. Thus, security should ship as "default" to help eliminate any potential threats to the personal data and privacy of millions of people worldwide.

References

1. WalMart (2003) Wal-Mart Details RFID Requirement. Article appears in http://www.rfidjournal.com/article/articleview/642/1/1/
2. DoD (2003) U.S. Military to Issue RFID Mandate. Article appears in http://www.rfidjournal.com/article/articleview/576/1/1/

3. S.E. Sarma, S.A. Weis, and D.W. Engels (2002) RFID systems, security and privacy implications. Technical Report MIT-AUTOID-WH-014, AutoID Center, MIT
4. About the EPCglobal network. http://www.epcglobalinc.org/about/
5. B. Fabian, O. Günther, and S. Spiekermann (2005) Security Analysis of the Object Name Service for RFID. In: Security, Privacy and Trust in Pervasive and Ubiquitous Computing
6. K.R. Foster and J. Jaeger (2007) RFID inside: The murky ethics of implanted chips. IEEE Spectrum, March 20–25. Available at http://www.spectrum.ieee.org/mar07/4939
7. Euro Bank Notes to Embed RFID Chips by 2005. Article appears in http://www.eetimes.com/story/OEG20011219S0016
8. ICAO (2004). Document 9303, Machine readable travel documents
9. A. Juels, D. Molnar, and D. Wagner (2005) Security and privacy issues in e-passports. In: D. Gollman, G. Li, and G. Tsudik, editors. IEEE/CreateNet SecureComm
10. Wired (2006) Hackers Clone E-Passports. Available at http://www.wired.com/science/discoveries/news/2006/08/71521
11. "Securing communications between mobile phones or other similar devices", SHA-1 fingerprint: 0x17503346d69b83f1cc9c2c4a43ee748e250b29c4, MD5 fingerprint: 0xae8e0db-474913e9162e058521cae30a4, Version 2, Manuscript 2007
12. M. Usami (2004) An ultra small RFID chip:μ-chip. In: IEEE Asia-Pacific Conference on Advanced System Integrated Circuits AP-ASIC 2004, Fukuoka, Japan, pp. 25
13. R. Stapleton-Gray (2005) Would Macys scan Gimbels? Competitive intelligence and RFID. In: S. Garfinkel and B. Rosenberg, editors, RFID: Applications, Security, and Privacy, Addison-Wesley, Reading, MA, pp. 283–290
14. S. Garfinkel, A. Juels, and R. Pappu (2005) RFID privacy: An overview of problems and proposed solutions. IEEE Security and Privacy, 3(3): 34-43
15. S. Garfinkel and B. Rosenberg, editors, Reading, MA, (2005) RFID: Applications, Security, and Privacy. Addison-Wesley
16. K. Albrecht and L. McIntyre (2005) Spychips: How Major Corporations and Government Plan to Track Your Every Move with RFID. Nelson Current
17. Sanjay E. Sarma, Towards the five-cent tag, Technical Report MIT-AUTOID-WH-006, MIT Auto ID Center, 2001. Available from http://www.autoidcenter.org
18. S.C. Bono, M. Green, A. Stubblefield, A. Juels, A. D. Rubin, and M. Szydlo (2005) Security Analysis of a Cryptographically-Enabled RFID Device. In: Fourteenth USENIX Security Symposium
19. J. Westhues (2005) Hacking the Prox Card. In: S. Garfinkel and B. Rosenberg, editors, RFID: Applications, Security, and Privacy, Addison-Wesley, Reading, MA, pp. 291–300
20. Z. Kfir and A. Wool (2005) Picking Virtual Pockets using Relay Attacks on Contactless Smartcard Systems. In: First IEEE/CreateNet International Conference on Security and Privacy for Emerging Areas in Communication Networks (SecureComm)
21. G. Hancke and M. Kuhn (2005) An RFID distance bounding protocol. In: First IEEE/CreateNet International Conference on Security and Privacy for Emerging Areas in Communication Networks (SecureComm)
22. R. Want (2004) RFID: A key to automating everything. Scientific American, 290(1): 56–65
23. A. Juels, R. Rivest, and M. Szydlo (2003) The blocker tag: Selective blocking of RFID tags for consumer privacy. In: Vijay Atluri, editor, ACM Conference on Computer and Communications Security CCS03, Washington, DC, USA, pp. 103–111
24. RFID Journal (2003) NCR prototype kiosk kills RFID tags. Available online at http://www.rfidjournal.com/article/articleview/585/1/1/
25. Consumers Against Supermarket Privacy Invasion and numbering-CASPIAN (2003) RFID Position paper. Available at http://www.privacyrights.org/ar/RFIDposition.htm
26. S. Garfinkel (2002) An RFID bill of rights. In: Technology Review, Available at http://www.technologyreview.com/articles/02/10/garfinkel1002.asp
27. G. Avoine Security and Privacy in RFID Systems. Online at http://lasecwww.epfl.ch/~gavoine/rfid/
28. A. Juels (2008) RFID security and privacy: A research survey. IEEE Journal on Selected Areas in Communication, Volume 24, Issue 2, Feb. 2006, Pages 381–394.

29. G. Avoine (2005) Cryptography in Radio Frequency Identification and Fair Exchange Protocols. PhD Thesis, EPFL

30. G. Avoine and P. Oechslin (2005) RFID traceability: A multilayer problem. In: Andrew Patrick and Moti Yung, editors, Financial Cryptography FC05, Volume 3570 of Lecture Notes in Computer Science, Springer, Berlin, pp. 125–140

31. S. Weis, S. Sarma, R. Rivest, and D. Engels (2003) Security and Privacy Aspects of Low-Cost Radio Frequency Identification Systems. In: First International Conference on Security in Pervasive Computing (SPC)

32. M. Ohkubo, K. Suzuki, and S. Kinoshita (2003) Cryptographic Approach to Privacy-friendly Tags. In: RFID Privacy Workshop, MIT, MA, USA

33. G. Avoine and P. Oechslin (2005) A Scalable and Provably Secure Hash Based RFID Protocol. In: The Second IEEE International Workshop on Pervasive Computing and Communication Security (PerSec), IEEE Computer Society Press, Washington, DC, pp. 110–114

34. T. Dimitriou (2005) A Lightweight RFID Protocol to protect against Traceability and Cloning attacks. In: First IEEE/CreateNet International Conference on Security and Privacy for Emerging Areas in Communication Networks (SecureComm)

35. D. Molnar, A. Soppera, and D. Wagner, A Scalable, delegatable pseudonym protocol enabling ownership transfer of RFID tags, Selected Areas in Cryptography, 2005

36. T. Dimitriou, A Secure and Efficient RFID Protocol That Could Make Big Brother (partially) Obsolete, in Fourth IEEE International Conference on Pervasive Computer and Communications (PerCom), 2006

37. K. Nohl and D. Evans (2006) Quantifying Information Leakage in Tree-Based Hash Protocols. In: Eighth International Conference on Information and Communications Security (ICICS), USA

38. L. Lu, Y. Liu, L. Hu, J. Han, and L. Ni (2007) A Dynamic Key-Updating Private Authentication Protocol for RFID Systems. In: Fifth IEEE Conference on Pervasive Computing and Communications (PerCom)

39. M. Feldhofer, S. Dominikus, and J. Wolkerstorfer, Strong authentication for RFID systems using the AES algorithm, Workshop on Cryptographic Hardware and Embedded Systems, 2004

40. M. Jung, H. Fiedler, and R. Lerch (2005) 8-bit microcontroller system with area efficient AES coprocessor for transponder applications. In: Ecrypt Workshop on RFID and Lightweight Crypto

41. I. Vajda and L. Buttyán (2003) Lightweight Authentication Protocols for Low-Cost RFID Tags In: Second Workshop on Security in Ubiquitous Computing

42. A. Juels (2004) Minimalist Cryptography for RFID Tags. In: C. Blundo, editor, Security of Communication Networks (SCN)

43. B. Defend, K. Fu, and A. Juels (2007) Cryptanalysis of Two Lightweight RFID Authentication Schemes. In: Fourth IEEE International Workshop on Pervasive Computing and Communication Security (PerSec)

44. Nokia unveils RFID phone reader. RFID Journal, 17 March 2004. Available at http://www.rfidjournal.com/article/view/834

45. M. Rieback, B. Crispo, and A. Tanenbaum (2005) RFID Guardian: A Battery-powered Mobile Device for RFID Privacy Management. In: Australasian Conference on Information Security and Privacy, vol. 3574 of LNCS, pp. 184–194

46. A. Juels, P. Syverson, and D. Bailey (2005) High-power proxies for enhancing RFID privacy and utility. In: Center for High Assurance Computer Systems – CHACS

47. T. Dimitriou (2008) Proxy Framework for Enhanced RFID Security and Privacy. 5th IEEE Consumer Communications and Networking Conference (CCNC 2008), Las Vegas, USA

48. M. Weiser (1991) The computer for the 21st century. Scientific American 265(3): 94–104

Part II
Security Protocols and Techniques

Design Trade-Offs for Realistic Privacy

Karsten Nohl* and **David Evans**

Abstract The integration of RFID technology into consumer products raises serious privacy concerns, but no privacy protection scheme that can be implemented on passive RFID tags is readily available. Existing proposals either sacrifice a core property of RFID systems, such as availability or scalability, or offer only limited privacy. The most promising approaches appear to be tree-based hash protocols, which sacrifice some privacy to maintain scalability. The amount of information that is leaked by these tree-based protocols depends on the tree setup, as well as the number and position of disclosed secrets. This leaked information is valued differently by different attackers. Some attackers aim to collect most information from many tags to build customer profiles; some need detailed information from a representative subset of tags to derive turnover rates of goods while others need very detailed information on selected tags to track individuals. Modifications of the tree protocol can improve privacy but need to be evaluated under the applicable attacker model. In this chapter, we first introduce privacy issues in RFID systems and techniques for measuring achieved privacy. Then, we describe protocols designed to enhance privacy and evaluate their effectiveness against different types of attackers. We find that some measures such as pseudonyms and periodic key updates improve privacy against some attackers, while hurting privacy against other attackers. Some measures such as restructuring the tree improve privacy against all attackers but incur additional computational cost for the legitimate reader. To find the best privacy protocol for a known attacker all available trade-offs should be considered.

1 Introduction

RFID tags provide easy access to information needed in logistics and are useful in a variety of other applications. Current RFID systems are designed without access control and therefore RFID tags can be read by anyone, which compromises the

K. Nohl
University of Virginia, Charlottesville, VA, USA
e-mail: nohl@cs.virginia.edu

P. Kitsos, Y. Zhang (eds.), *RFID Security: Techniques, Protocols and System-on-Chip Design,* © Springer Science+Business Media, LLC 2008

privacy of the tag bearer. Implementing access control as part of an RFID system increases the complexity of the system, incurs extra computational costs, and leads to shorter reading ranges. As with many new technologies, the design of first generation RFID systems focused mainly on creating a working, standardized system, without adequate consideration of security and privacy aspects. Similar experiences with other systems (such as credit cards and e-mail) teach us that after a technology has been established, adding security features is usually difficult due to legacy applications. Similarly, changing RFID standards to include privacy protection becomes increasingly harder as RFID tags are more widely deployed.

For the upcoming wide-scale deployment of RFID tags on the item-level, privacy measures are needed that are inexpensive to implement and nondisruptive to normal functionality. This pressing need for an RFID privacy solution disqualifies approaches based on public-key cryptography. Even though advances in cipher design and VLSI technology suggest that public-key arithmetic can eventually be implemented on a tag [1], no such implementation is expected in the next several years. Privacy protection for RFIDs, therefore, requires new protection schemes that solely rely on symmetric cryptography.

This chapter discusses different approaches to privacy in RFID systems and illustrates the trade-offs between privacy, scalability, and availability (Sect. 3). We then analyze the tree-based RFID privacy protocol that provides a good balance between scalability, availability, and privacy (Sect. 4). To maximize the privacy of this protocol in different scenarios, we consider the incentives of three attackers (Sect. 5) and propose modifications that can defeat them (Sect. 6).

2 Background

RFID tags with no access control leak information that can compromise individual's privacy and enable corporate espionage. In the simplest case of information leakage, the tag identifier (ID) can be read from the tag and decoded using public databases. An attacker can easily learn which items a person is carrying and can track individuals based on their unique collection of items. One system that would allow this kind of direct information leakage is the planed Object Name Service (ONS) system designed by EPCglobal, which uses the Internet DNS to locate product information about tagged items [2]. To locate the right ONS server, the system requires that the tag IDs be highly structured (e.g., the same products need to have the same ID prefix), thus leaking significant information even when only small parts of an ID are disclosed.

A first step toward privacy-friendly RFID tags is to abandon structured IDs and thereby eliminate the possibility for public information retrieval solely based on the tag identifier. For the remainder of this chapter, we make the standard assumption that the tag IDs are assigned randomly from some key space. Even with random IDs, however, information can be learnt from the tags. In particular, if a tag is read at different times or in different locations, its movement through the physical world is known to the attacker. These traces of RFID movement are similar to Internet traces that record which web pages a user has visited [3]. In the same way that Internet traces have a value and are actively traded [4], we hypothesize that RFID traces, too, have a value that makes them worth collecting. Applications in which RFID traces can be used include surveillance of individuals and corporate spying. Competitors can, for example, estimate a merchant's turnover by repeatedly browsing through the aisles and reading the tags. Just like Internet traces are increasingly used to derive information about a user's preferences, the movement of individuals, too, encodes information about their shopping behavior, price sensitivity, and taste (among many other things). This information can be used to implement profit maximizing schemes, including price discrimination and targeted advertisement. It should, thus, be desirable for corporations and individuals to control and restrict access to their RFID tags.

3 Privacy Protection Schemes

The information stored on RFID tags can be protected from unauthorized access in many ways. Early proposals suggest breaking or deactivating the tags that leave the domain of the owner, typically at check-out time [5]. These approaches sacrifice the possibility of postsales applications including smart homes and warranty management. Other RFID applications such as payment tokens and passports require the tags to remain active when given to individuals and therefore need more elaborate privacy protection schemes. Furthermore, deactivating tags does not provide any protection from corporate espionage that happens before checkout.

Finer-grained privacy schemes employ cryptography to protect the tags from rogue readers. Cryptographic protocols obfuscate the tag ID using random numbers and secret keys. These privacy protocols for RFIDs are limited to only those cryptographic primitives that can be implemented within the hardware constraints of the tags. In particular, asymmetric cryptography such as RSA and elliptic curves cannot be implemented on low-cost RFID tags.

Protocols alone cannot guarantee perfect privacy since extra information is leaked through other channels such as radio characteristics and manufacturing variances. From all we know, the magnitude of this extra information appears to be relatively small, and hence we focus on information leaked from the protocol layer. A protocol provides *strong privacy* if it leaks no information at the protocol layer. Strong privacy is achieved if an attacker, using only information available at the protocol layer, cannot tell any pair of tags apart [6]. In Sect. 5.1, we consider how an attacker can combine information acquired through other sources with information leaked at the protocol layer.

Strong privacy is generally not achievable in any RFID systems and costly on the protocol layer. Protocol-layer Privacy, scalability, and availability cannot be simultaneously achieved for large systems without using asymmetric cryptography [7]. Therefore, all proposed privacy schemes make weakening assumptions that limit their applicability to certain systems. One such assumption is that degraded levels of protection are sufficient as long as the expected value of traces is lower than the cost of collecting them. Next, we introduce some of the proposed protocols and identify the weakening assumptions under which they provide sufficient protection.

3.1 Basic Hash Protocol

The first cryptographic protocol proposed for RFID privacy hides the tag ID from rogue readers by hashing it under a secret key. In this basic hash protocol, each tag is assigned a unique secret that is shared with the legitimate reader. When queried, the tag responds with

$$N, H(k \parallel N),$$

where N is a random number (nonce) freshly generated at each query, k is the secret key, and H is a one-way function (e.g., a hash function). To identify the tag, the reader hashes the nonce with the keys of all tags and checks which of the outputs matches the tag response [8].

Assuming a strong one-way function, the basic hash protocol provides strong privacy because an adversary would see no advantage in linking tag responses over randomly guessing the secret key [6]. For the legitimate reader (who knows all keys in the system), the cost of identifying a single tag grows linearly with the number of tags in the system. For large systems, the cost of trying millions or billions of keys for every read becomes prohibitively expensive. Hence, the protocol is only suitable for small systems. RFID applications that grow to millions or more tags require schemes more scalable than the basic hash protocol.

3.2 Hash Chains

Hash chain protocols protect privacy by changing the tag identifier with every read. When queried for the ith time, the tag responds with $G(\mathrm{ID}_i)$ and updates the identity as

$$\mathrm{ID}_{i+1} = H(k \parallel \mathrm{ID}_i)$$

where G and H are different one-way functions and k is the secret key unique to the tag. The reader maintains a copy of the current tag ID and a chain of c future IDs. Since the tag update and response are deterministic, these values can be precomputed and stored in an efficiently accessible data structure. To find the matching tag, the reader simply looks up the response in this structure, removes obsolete values from the chain, and adds new ones to reflect the tag's new state [9].

Only those tags that have never been read by rogue readers more than c consecutive times can be identified. Given the fraction of rogue or accidental reads, p, a tag exceeds this threshold during n reads with a probability of at least

$$\sum_{l=0}^{n-c} p^l (1-p)^c = \frac{p^{n-c+1} - 1}{p-1} (1-p)^c \xrightarrow{\text{large } n} (1-p)^{c-1}.$$

For a hash chain with length 20, an acceptably low failure probability of one in a million is achieved when on average less than 48.4% of all reads are rogue reads. For hash chains of larger length, this bound rises quickly; chains with length 40 require no more than 60.2% to be rogue reads. These values state the lowest possible failure rate since a random mix of rogue and nonrogue reads is assumed. In every realistic application, however, rogue reads are clustered, leading to higher failure rates. The conditions under which the hash chain protocol can operate reasonably well can only be found in scenarios such as backend logistics, where physical access is strictly controlled (which already makes privacy less of a problem). In almost all retail and postsales environments, many parties have access to the tags – intentional and unintentional reads are frequent. The hash chain protocol causes an unacceptably high failure rate and therefore does not provide a practical privacy solution for these scenarios.

To avoid the loss of tags whose hash chain has been exceeded, Ohkubo et al. [10] suggested combining the hash chain and basic hash protocols. Once a reader is unable to identify a tag through the hash chain, the more expensive basic hash protocol is executed. In most realistic scenarios, however, where the degraded availability of the hash chain causes problems, readers frequently read tags that belong to other systems. With every such read, they would have to execute the expensive basis hash protocol so the combined protocol would become prohibitively expensive.

3.3 RFID Firewall

Privacy advocates have argued that the companies that deploy RFID technology are unlikely to implement privacy protection on the tags, either because of additional costs or because they themselves want to harvest consumer data. Following this argument, privacy protection can only be enforced by the individuals who carry the tags. A protection mechanism could be implemented on an external, active device such as an RFID-enabled cell phone that acts as a firewall. In fact, prototypes of devices able to enforce RFID firewall rules have already been built [11]. Equipped with an RFID firewall, a user has to maintain a set of rules defining which tags can be accessed by which readers. Describing privacy preferences, however, has been shown to be a hard problem in other fields and it seems unlikely that most privacy-concerned users of RFID technology are able and willing to maintain a rule set. Further research will need to investigate to what extent general policies can describe the privacy preferences of many users. A central protection policy as used in heuristic virus detection always faces a trade-off between overly strict policies and easy circumvention.

Like any other firewall, an RFID firewall can be combined with cryptographic protocols to provide several layers of protection. The extra cost of obtaining and administrating the firewall device and the potential impact on the availability of the RFID system, however, make portable firewall solutions attractive only to a small group of privacy-paranoid individuals. Retail environments, on the other hand, are a more natural scenario for a policy-based privacy protection. The policy can be developed centrally and active RFID devices will be readily available in the retail space. The extra layer of privacy might help fight off persistent attackers and incurs little extra cost.

3.4 Probabilistic Protocols

RFID protocols that provide strong privacy sacrifice either scalability (e.g., basic hash protocol) or availability (e.g., hash chain protocol) while alternative solutions such as RFID firewalls provide an additional layer of privacy but place additional burdens on the consumer. Other protocols offer a trade-off between scalability and availability but only achieve less than strong privacy. These weaker protocols might still provide sufficient protection under certain weakening assumptions about the attacker, which will be defined in Sect. 5. Many attackers can, for example, be assumed to act rational, and only attack a system when the expected revenue of the attack exceeds the cost.

The protocols described next provide *probabilistic privacy*, meaning that an attacker who has compromised the keys from some tags is able to distinguish between certain tags, but not others. The probabilities with which randomly chosen tags can be distinguished depend on which secrets keys are known to the attacker (Sect. 4.1).

Protocols that provide probabilistic privacy have been developed based on the basic hash scheme, altered to share some secrets among several tags. All proposed schemes with probabilistic privacy are special cases of a general *tree-based hash protocol* [12].

The general tree-based hash protocol structures all secret keys in a tree. All tags are assigned to different leafs of the tree and each tag knows all the secrets on the path from the root to its leaf. The tree shown in Fig. 1 has nine leaves at the lowest level and hence holds nine tags. Each tag in this example tree is assigned two secrets,

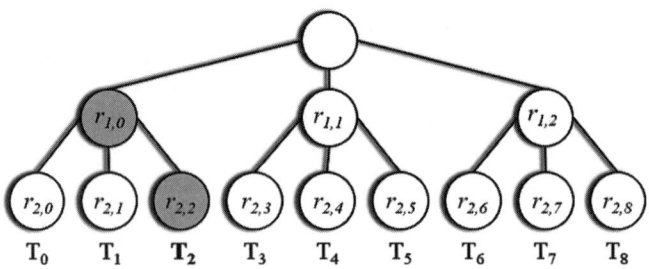

Fig. 1 Tree of secrets with nine tags

some of which are shared among several tags. The legitimate reader knows all the secrets and uses the basic hash protocol to identify a given tag. The basic protocol is executed on each tree level to find the branch that the tag resides in. For the example tree, the tag T_2 holds the two highlighted secrets and responds to a reader query with

$$N, H(r_{1,0} \parallel N), H(r_{2,2} \parallel N)$$

where N is a random nonce and H is a one-way function. The reader searches for a match of the first hash by hashing the nonce with all secrets on the first tree level. After identifying that the tag resides in the left third of the tree, the reader hashes the nonce with the three remaining possible secrets to find the tag.

In the most general case, the spreading factor (that is, the number of branches at a tree node) can vary for different tree nodes. Similarly, the depth of the tree can vary for different tags, which always increases cost and decreases privacy but might become necessary if a partly filled tree needs to be extended.

The tree protocol leads to degraded privacy, since an attacker who is able to extract secrets from some tags is able to identify groups of tags based on which of these secrets they share. Section 4 analyzes the effect of breaking tags on privacy.

Several special cases of the tree protocol have been proposed as RFID privacy protocols. One protocol uses a binary tree with a spreading factor of two at every node [12]. A binary tree with depth d holds $N = 2^d$ tags. To identify a tag, at most two hashing operations are needed on each tree level, totaling to $2d$ hashing operations (on average, $1.5d$ operations are needed). The cost in terms of hash operations grows logarithmically with the number of tags and is at most $2\log_2 N$.

A better size–cost ratio is achieved when the branching factor, k, grows with the number of tags. The minimum cost of $dN^{1/d}$ hash operations is realized for $k = N^{1/d}$. Besides minimizing cost, this setup also needs fewer secrets on the tags, where storage is particularly expensive, as well as on the reader. In Sect. 6.3, we will further demonstrate that this tree setup, besides minimizing cost, also provides better privacy than the binary tree, making it a strictly better design choice.

Another protocol that is related to the tree protocol structures the secrets in a matrix [13]. Each tag is assigned a unique set of secrets, one for each matrix row. As with the tree protocol, to identify the tag, the reader executes the basic hash scheme for each row. The matrix protocol can be viewed as a special case of the general tree protocol. The matrix is equal to a tree with some of the secrets replicated as shown in Fig. 2. When compared to a tree, the matrix provides the same computational cost

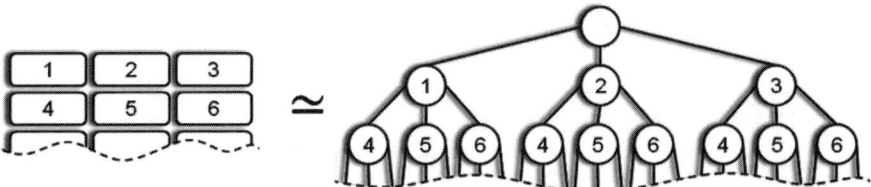

Fig. 2 Matrix tree equivalence

but much degraded privacy because an attacker learns a larger fraction of secrets from each tag compromise. The only apparent advantage of the matrix protocol is the smaller number of secrets stored on the reader. Since reader storage is not likely to be an important constraint, the tree is generally a better design choice than the matrix.

The tree and matrix protocols are just two examples from the large design space of tree-like protocols. Proposed variations of the tree protocol maximize privacy under different constraints, such as maximal reading time [14] or upper-bounded tree depth [7]. These variations achieve different, and apparently conflicting, goals. When maximizing privacy for a setup with upper-bounded reading time, the tree grows and has more secrets on the higher level; whereas maximizing privacy in a bounded tree puts more secrets on the lower levels and increases the reading time. The tree protocol offers similar trade-offs between privacy, reading time, tag cost, and scalability; but no general approach exists to date that identifies the best tree setup for a given scenario. Next, we introduce a framework in which the relative privacy of different tree setups can be compared, followed by a discussion of possible attack strategies.

4 Analyzing Probabilistic Privacy

The tree protocol provides only degraded privacy since secrets are shared among tags. An attacker can extract some tags' secrets and then distinguish other tags based on which of these secrets they use. The privacy of the protocol has been analyzed in several publications. The metrics used to quantify privacy are typically based on information theory and measure how well, on average, tags can be distinguished. Metrics include the anonymity set size (that is, the average number of tags that cannot be distinguished) and the entropy of these sets [7, 14–16]. All these analyses concluded that breaking one tag leaks little information, but breaking more tags severely hurt privacy. In these analyses, privacy is measured as a single scalar value that typically corresponds to the mean of some distribution. These metrics, however, fail to account for information about the type and variance of the distribution. We argue that privacy should be measured in form of a distribution to capture the different levels of privacy that different tags in the system exhibit. By preserving the extra information that the distribution encodes, we avoid making restrictive assumptions on the attacker a priori and can hence analyze the privacy of a protocol for different attackers.

4.1 Tree Grouping

In entropy-based privacy metrics, the number of tags a given tag cannot be distinguished from determines that tags privacy. This grouping of tags depends on which

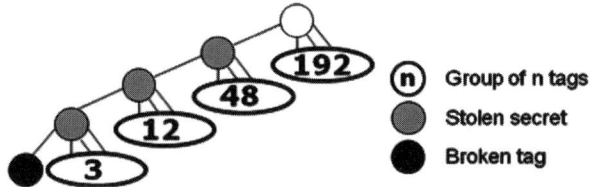

Fig. 3 Groups of tags that an attacker can distinguish after one tag compromise

secrets are disclosed to an attacker. The attacker can break a limited number of tags and extract all of their secrets.[1] Given the secrets from one tag, this tag can be distinguished from its $k - 1$ immediate neighbors in the tree, from their $k^2 - k$ neighbors, and so forth (where k is the branching factor). Figure 3 shows the groups of tags distinguishable within a tree of 256 tags of which one is broken.

The distribution of tags into groups of different sizes is upper-bounded by

$$\Pr(Z \leq z) \leq \frac{k}{k-1} \frac{z}{N}. \tag{1}$$

This function states the probability, $\Pr(Z \leq z)$, that a randomly chosen tag falls into a group of size z or smaller. As more secrets are disclosed, more groups of tags can be distinguished. In the worst case, each of the first k broken tags forms new groups of sizes $(k - 1)$, $(k^2 - k), \ldots, (N/k - 1/k^2)$. When expressing the distribution by the amount of encoded information, the *entropy* of the groups, the probability, $\Pr(X \geq x)$, that x nats (bits with base e) of information are leaked is upper-bounded by

$$\Pr(X \geq x) = b\frac{k}{k-1} \frac{1}{e^x}, \tag{2}$$

where b is the number of compromised tags. This upper bound along with the real distribution of entropy in the tree protocol is shown in Fig. 4a for the case where one tag is compromised. This plot shows the distribution of information for the case where an individual carries only one tag. In the scenarios we are considering, however, tags often appear in groups. An individual potentially carries one tag in each piece of clothing and in many other items. Assuming that the information from each tag follows the distribution in (2), the combined information from a collection of m tags is

$$\Pr(X = x) = \frac{x^{m-1}}{(m-1)!} b\frac{k}{k-1} \frac{1}{e^x}. \tag{3}$$

This distribution is shown in Fig. 4b for different numbers of tags that appear together.

---

[1] The compromised tags are no longer considered part of the system.

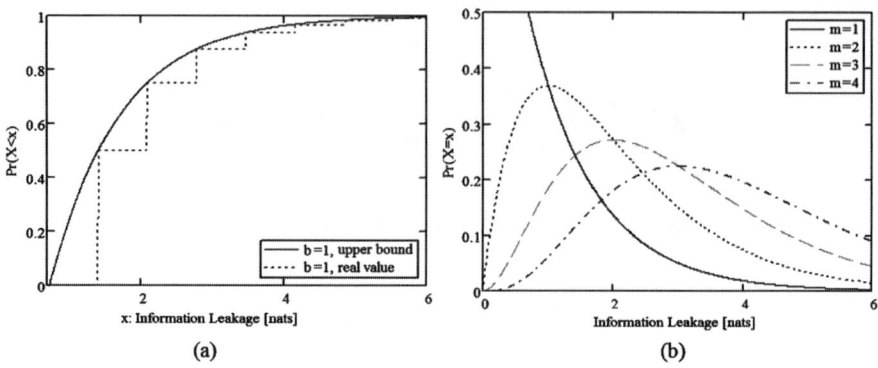

Fig. 4 (**a**) Probability that for a tree with b broken tags less than x nats of information are leaked by a randomly selected tag, real value vs. upper bound; k is large; one broken tag. (**b**) Distribution of information leaked by collection of m tags; one broken tag

5 Modeling Attackers

Measuring the level of privacy that consumers experience is a hard problem since the perception of privacy varies widely among different individuals. Furthermore, individuals typically lack a good understanding of how much protection is required to match their personal privacy preferences. Information that was disclosed voluntarily will often compromise privacy when combined with other information or when analyzed using data mining techniques. Because of these difficulties in modeling individual privacy perception, we model privacy as perceived by an attacker and argue that consumer privacy is achieved when no attacker has an incentive to compromise the privacy. We derive the incentives and capabilities of different attackers in this section and show in Sect. 6 how protocols can be designed to defeat particular types of attackers. Some protection measures improve privacy in the face of certain attackers, but hurt privacy when facing others. Next we identify characteristics common to all attackers, followed by a description of three distinct classes of attacker that represent three likely scenarios: corporate espionage, surveillance, and customer profiling.

5.1 Extracting Traces

All attackers have in common that they want to extract information from data collected by reading RFID tags. The data is composed of many intermixed traces from many tags. For most attackers, a trace only becomes valuable to the attacker when it can be separated from all other traces, thus identifying a set of readings from a single tag (or a conjoined group of tags). The likelihood with which a trace can be separated using data from the protocol level depends on the secrets known to the attacker. Section 4 explained how the expected sizes of tag groups can be computed for an attacker with a given number of compromised tags. To separate those traces that are indistinguishable at the protocol level, the attacker will further employ data mining techniques or use side channel information (e.g., place and time of read) to distinguish traces. To capture the success of the attacker in doing so, we introduce two functions: the *attacker strategy function*, which describes the sophistication of the attack, and the *binning function*, which captures the clustering of traces.

The attacker strategy function encodes the probability with which an attacker can tell traces apart that are indistinguishable at the protocol level. The function captures the side channel information and data mining techniques that are available to the attacker, and varies widely for different attackers. Even though the function is generally unknown (even to the attacker) we can derive an upper bound on it.

First, we define the average probability that a set of two traces can be distinguished to $p < 1$. This means that given a set of any number of readings known to come from two different tags, the attacker can separate the readings into two groups (based on the tag from which they were generated) with probability p. In reality, the actual probability also depends on the number of readings from the tags. The attacker, however, does not control the number of readings collected from a specific tag (in different locations), but only the total number of readings collected from all tags. Hence, our definition of p does not depend on the number of readings, just on the number of different tags which could have generated the readings. This accurately describes an attacker who tries to extract traces from a data set without having detailed prior knowledge of the composition of the data.

Given this definition of p for separating two traces, it follows that traces cannot be separated from groups of three intermingled traces with average probability better than p^2. If the chances of extracting a trace from a set of three traces were higher, an attacker could distinguish two traces with a probability higher than p by intermingling an additional trace. This extends to any number of intermingled traces. The attacker strategy function (that is, the probability that a trace can be extracted from a set of g traces) is upper-bounded by

$$S(t) = p^{g-1}. \tag{4}$$

Second, the binning function describes the distribution of traces into the different groups. These groups can be distinguished using protocol information leakage. Given a distribution of groups as derived in Sect. 4.1, a system with size N, and a data set with $t + 1$ intermingled traces, the probability that at most g traces from a group of size z are included in the data set is

$$B(z,g) \leq \binom{t}{g} \left(1 - \frac{z}{N}\right)^{t-g}. \tag{5}$$

Expressed in terms of encoded information, the probability that g indistinguishable traces with x nats of entropy (that is, traces from a group with size $z = N/e^x$) are found in the set is

$$B(x,g) = \binom{t}{g} \left(1 - \frac{1}{e^x}\right)^{t-g} \left(\frac{1}{e^x}\right)^g. \tag{6}$$

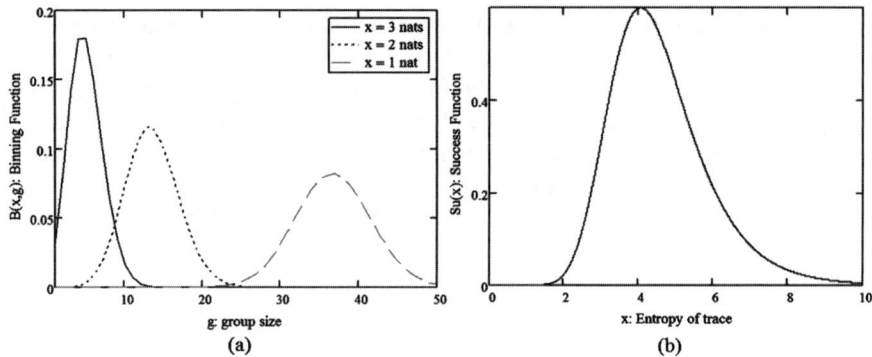

Fig. 5 (**a**) Binning function: probability that a trace with x nats of entropy is intermingled with g other traces; data set with $t = 100$ traces. (**b**) Success function: probability that a trace with x nats of entropy can be extracted; $t = 100$

The binning function is shown in Fig. 5a for traces with different entropies.

Last, the *success function* encodes the probability that a trace with given entropy can be extracted. The function is the probability that a tag is mixed with a certain number of tags, $B(x, g)$, multiplied with the probability that it can be extracted from that mix, $S(g)$, and summed over all possible mixes $(g = 0, 1, \ldots, t)$

$$Su(x) = \sum_{g=0}^{t} (B(x, g) \cdot S(g)). \tag{7}$$

The success function is shown in Fig. 5b. To determine whether an attack is successful, the success function needs to be interpreted in light of the attacker's requirements. Higher privacy generally requires a lower success probability. The specific requirements of three types of attackers are discussed next.

5.2 Attacker Profiles

Attackers can be divided into categories according to their capabilities and goals. We consider three likely classes of attacks distinguished by their goals: compromise of sensitive business information, tracking of customers for profiling, and surveillance of individuals.

5.2.1 Corporate Espionage Attacker

The incorporation of RFID tags into the logistics chain bears the risk of disclosing internal business information to competitors. Several years ago, RFID technology started infiltrating logistics and other corporate domains, where the tags help in automating processes. Soon, RFID tags will be further deployed into the retail space and become accessible by customers and competitors. Those retailers that use RFID technology without sufficient privacy protection risk leaking internal business data.

The *corporate espionage attacker* periodically reads tags in the retail space or eavesdrops on readings. For practical matters, access to the tags is restricted to small numbers of snapshots and may be limited to certain hours of the day. From the collected information, statistics about internal processes are derived (e.g., turn-over of different goods). The attacker is mostly interested in the number of items in some location and the rate with which these items get replaced. These statistics can be extrapolated from a representative subsample with sufficient accuracy which makes the espionage attacker relatively resilient to probabilistic privacy protection. The tree protocol, for example, protects tags only on average, potentially leaving too many tags unprotected. In Sect. 6.1, we show that the corporate attacker can likely be defeated by periodically updating the tag secrets thereby preventing an attacker from correlating different snapshots.

5.2.2 Rational Profiling Attacker

A second attacker on RFID systems uses RFID information to construct customer profiles. The profiles can be used to generate profit through, for example, directed advertisement and price discrimination. Businesses are eagerly building profiles in anticipation of future applications [17] and RFID data is an optimal source to be included in their information bouquet. Currently, there is little data available on the exact value of RFID traces, but the similarity of these traces to valuable Internet traces suggests that collecting RFID data is likely to become highly profitable.

The *profiling attacker* acts rationally and will only attack a system if the expected value of the traces exceeds the cost of the attack. A trace herein means several readings of the same tag at different times or locations. The profitability of different traces varies widely, but the attacker values all traces at the same average price because their true value only becomes evident after the attack (but not even then, necessarily). Decreasing this expected value will later prove to be key in defeating the profiling attacker.

The value that an attack has to the attacker depends on the likelihood that traces can be extracted from the collected data as derived in Sect. 5.1. Multiplying the success function (from Sect. 5.1) with the likelihood that a trace with a given entropy occurs (from Sect. 4.1) and integrating over all possible entropies (that is, the entropies from that largest to the smallest group) give us the expected value of the attack

$$\text{Value} = \int_{\text{ent(lrg)}}^{\text{ent(sm)}} \left(\text{Su}(x) \Pr(X = x) \right) dx. \tag{8}$$

As shown in Fig. 5(b), only some group sizes have a high likelihood to contain useable information. Hence, the majority of the attack value is generated by tags in these groups while all other tags contribute only very little value. In Sect. 6.3, we use this insight to design a protocol modification that decreases the attack value and defeats the profiling attacker.

5.2.3 Surveillance Attacker

Individuals carrying RFID tags can potentially become victims of surveillance. The surveillance attacker tries to track individuals through analyzing RFID readings. The attacker needs to learn significantly more information about an individual than the other two attackers. Anything short of distinguishing a surveillance subject from all other individuals will be considered to be a failure. Because the large amount of information required is not likely to be leaked from a single tag, the attacker has to pool information from several tags belonging to the same subject in order to collect enough information.

As with the profiling attacker, the abilities of the surveillance attacker can be expressed using the success function (Sect. 5.1). When compared to the profiling attacker, distinguishing several tags becomes more difficult as the attack will typically be limited to a smaller area, rendering the location meta-information less useful. On the other hand, when the subject's habits are known, the data can be filtered more efficiently.

Because of the difficulty of tracking individuals through tags that are not owned by the attacker, this will likely not be the preferred attack vector of a surveillance attacker. Although privacy advocates highlight the risks of surveillance posed by RFID systems, more realistic attackers determined to track a particular individual can employ other attacks that are probably easier than using information leaked from RFID systems. Such attacks include physical surveillance and tagging the individual with RFID tags that are owned by the attacker. Hence, most RFID systems should be designed with a focus on the corporate espionage and rational profiling attackers.

6 Strengthening the Tree Protocol

We propose three modifications to the tree protocol and identify which attackers they help defeat.

6.1 Update Secrets

The privacy of the tree protocol can be increased by periodically writing new secrets to the tags. An attacker who reads a tag before and after an update cannot link these two readings. The problem of changing the identifier of a tag has been considered in several ownership transfer protocols in which the old owner only knows one identity and the new owner only knows the other identity [18, 19]. These protocols can also be used to update the keys on the tag to improve privacy, but simpler protocols are sufficient as well. For example, this simple key update protocol can be used to update all the keys on a tag

$$\begin{array}{rc}
\text{Tag} \rightarrow \text{Reader} & N_0 \\
\leftarrow & N_1, \ H_{K_0}(N_0||N_1) \oplus K_1, \ H_{K_1}(N_0||N_1) \\
\leftarrow & N_2, \ H_{K_0}(N_0||N_2) \oplus K_2, \ H_{K_2}(N_0||N_2) \\
\leftarrow & \cdots
\end{array}$$

where K_0 is the unique secret of the tag and $K_1, K_2, \ldots$, are the other secrets on the tag that need to be updated. The tag initializes the protocol by sending a fresh nonce, N_0, to the reader. For each secret to be updated, the reader sends a reader nonce, N_i, and the new secret, K_i, XOR'd with the hash of the reader nonce, N_i, and the tag nonce, N_0. An attacker cannot learn the new secret without reproducing this hash, which requires knowledge of the secret K_0. The reader also includes a cryptographic integrity check which is the hash of the tag nonce and each new key. This integrity check ensures that an attacker cannot desynchronize reader and tag by altering the submitted key. If forward security is also required, the unique secret, K_0, should be updated as well.

The cryptographic primitives used by our key update protocol, namely a hash function and random number generator, are already available on tags that implement the tree protocol. The key update protocol, therefore, provides a cheap way to refresh the secrets on a tag. Writing to the tag memory requires significantly more power than reading the tag and can only be done over short distances. Should the tag exceed the maximal writing distance during the update, the protocol cannot be completed and all keys need to be reverted to their old values. Alternatively, the tag can acknowledge each update thereby keeping the set of secrets on reader and tag consistent.

The update scheme provides additional privacy only for sequences of readings that are interrupted by at least one update. If the attacker can learn information from short sequences of readings or when updates are rare, the measure is not effective. The scheme significantly increases the privacy in face of some attackers, but hurts privacy in face of others. An attacker can impersonate a compromised tag during the update process in order to learn more secrets. If the system has no way of knowing which tags are compromised, the attacker can learn a new set of secrets in each update round. Soon, all secrets are known to the attacker and privacy is lost. To avoid leaking new secrets to that attacker, the system must note the break of tags and exclude them from future updates. Some systems such as those designed to detect counterfeit tags already provide this property by checking whether readings are reasonable with respect to past readings [20]. To circumvent detection, an attacker would have to steal a tag, compromise it, and start impersonating it before the tag is reported as missing.

In summary, the update protocol improves security in an environment where an attacker has only limited access and where tag compromise can be detected. In scenarios where these assumptions hold (such as warehouses and retail spaces), periodic key updates can increase privacy at reasonable cost. Note, that for the same scenarios the hash chain protocol provides a good solution (but will potentially cause tag loss as described before). For other scenarios, key updates weaken privacy by disclosing new secrets to the attacker through the update protocol. As postulated before, the effectiveness of a measure must be seen in light of an attacker model and the update protocol is effective only against the corporate espionage attacker.

6.2 Pseudonyms

Another privacy measure that proves very effective against certain attackers is the use of pseudonyms. We propose that each tag be assigned several identities, one of which will be randomly chosen on each read. Unless the attacker knows the full set of pseudonyms of a tag, some readings of that tag are not linkable. When reading several tags from one person, the attacker cannot easily decide which identities belong to which tag. Since the attacker can only distinguish tags based on which groups of the tree they fall in, chances are high that the pseudonyms of all the tags carried by an individual occupy many of the groups and that the overlap of groups that two individuals' tags occupy is large. The probability of not being able to distinguish two individuals because of this effect grows with the number of tags included in the attack.

The surveillance attack is particularly thwarted by pseudonyms since the attacker has to pool information from several tags to uniquely identify individuals. Pseudonyms make it exponentially more difficult to combine information from several tags and still be able to distinguish individuals. The downside of pseudonyms is that each tag stores more secret keys and the attacker can learn more tree secrets when breaking a given number of tags. Having extracted more secrets, the attacker learns more information about every single tag. The use of pseudonyms is another example that emphasizes the need for considering the attacker when designing privacy measures.

6.3 Variable Branching Factor k

Key update and pseudonyms are not effective defenses against the profiling attacker. Both measures potentially increase the number of secrets leaked and hence increase the attack value derived in Sect. 5.2.2. To discourage the profiling attacker from attacking a system, the overall attack value must be decreased below the cost of the attack without leaking additional secrets. As shown in Fig. 5b, the amount that groups of different sizes contribute to the overall attack value varies widely. In fact, almost all value is generated by tags in groups smaller than a certain threshold. The attack value can hence significantly be decreased by restructuring the tree such that no tags fall in these small groups. The smallest group that an attacker can distinguish is determined by the branching factor at the lowest level of the tree. Increasing the branching factor on this level while structuring the rest of the tree for minimum cost provides an effective and reasonably cheap countermeasure against the profiling attacker.

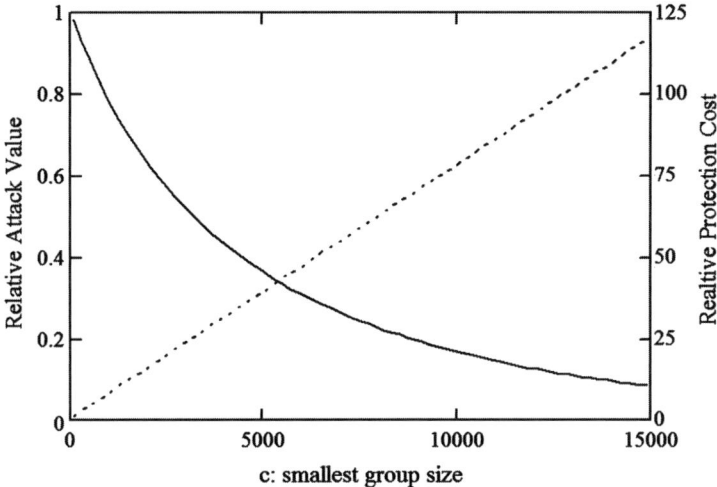

Fig. 6 Trade-off between attack value and computational cost. The attacker value function from Sect. 5.2.2 is used with $p = 0.5$ and $t = 100$. The analyzed system holds one million tags, of which 50 tags are broken; each individual carries 10 tags, and each tag stores four secrets

The modified tree incurs a higher cost for the legitimate reader. The trade-off between decreased attack value and number of hashing operations needed to authenticate a tag is shown in Fig. 6. The variable-k measure significantly decreases the attack value at a reasonable cost even for large systems. In the example, decreasing the attack value by 80% incurs a 67-fold increase in computational cost to 8,500 hashing operations, which can be computed in a split-second on any modern processor.

The variable-k measure is equally effective against the other two attackers, but more costly than the measures described before. Variable-k can be combined with key update or pseudonyms to build an even stronger protocol.

7 Conclusions

RFID privacy is a multifaceted problem and no general solution exists that offers protection in all scenarios where RFIDs will be used. Designers of RFID systems need to consider privacy requirements along with the goals and capabilities of potential attackers to balance privacy and cost. For their high scalability and availability, tree-based hash protocols provide the most versatile foundation for protection schemes against diverse attackers.

The design space for tree-based protocols is significantly larger than has been explored and there exist protocols that offer various trade-offs. For example, privacy can be traded off for increased scalability, shorter reading times, and lower per tag costs. The design space includes defenses against threats as diverse as corporate spies, consumer tracking, and surveillance of individuals. Certain protocol modifications guard against some attackers while degrading protection when facing others.

To find and evaluate effective defenses, the attacker's incentives for compromising privacy have to be taken into account. While no model can capture the often conflicting incentives of all attackers, their common goal of finding tag traces in large data sets can be modeled and analyzed. Based on this estimate on the probability of distinguishing different tags, the success of different attackers can be estimated. RFID systems should be designed to minimize this success probability at reasonable cost.

References

1. Wolkerstorfer, J. Is Elliptic-Curve Cryptography Suitable to Secure RFID Tags?. *Workshop on RFID and Lightweight Crypto*, 2005
2. Fabian, B., Guenther, O. and Spiekermann, S. Security Analysis of the Object Name Service for RFID. *International Workshop on Security, Privacy and Trust in Pervasive and Ubiquitous Computing*, 2005
3. Bauer, M., Fabian, B., Fischmann, M. and Gurses, S. Emerging Markets for RFID Traces. *arXiv.org*, 2006
4. Odlyzko, A. Privacy, Economics, and Price Discrimination on the Internet. *International Conference on Electronic Commerce*, 2003
5. Juels, A. RFID Security and Privacy: A research Survey. *Manuscript*, 2005
6. Juels, A. and Weis, S. Defining Strong Privacy for RFID. *Cryptology ePrint Archive*, 2006
7. Nohl, K. and Evans, D. Quantifying Information Leakage in Tree-Based Hash Protocols. *Conference on Information and Communications Security*, 2006
8. Weis, S., Sarma, S., Rivest, R. and Engels, D. Security and Privacy Aspects of Low-Cost Radio Frequency Identification Systems. *International Conference on Security in Pervasive Computing*, 2003
9. Ohkubo, M., Suzuki, K. and Kinoshita, S. Cryptographic Approach to "Privacy-Friendly" Tags. *RFID Privacy Workshop*, 2003
10. Zhai, J., Mok-Park, C. and Wang, G.-N. Hash-Based RFID Security Protocol Using Randomly Key-Changed Identification Procedure. *International Conference on Computational Science and its Applications*, 2006
11. Rieback, M., Crispo, B. and Tanenbaum, A. RFID Guardian: A Battery-Powered Mobile Device for RFID Privacy Management. *Australasian Conference on Information Security and Privacy*, 2005
12. Molnar, D. and Wagner, D. Privacy and Security in Library RFID: Issues, Practices, and Architectures. *ACM CCS*, 2004
13. Damgard, I. and Østergaard, M. RFID Security: Tradeoffs between Security and Efficiency. *Cryptology ePrint Archive*, 2006
14. Buttyan, L., Holczer, T. and Vajda, I. Optimal Key-Trees for Tree-Based Private Authentication. *Workshop on Privacy Enhancing Technologies*, 2006
15. Avoine, G. and Oechslin, P. RFID Traceability: A Multilayer Problem. *Financial Cryptography*, 2005
16. Nohara, Y., Inoue, S., Baba, K. and Yasuura, H. Quantitative Evaluation of Unlinkable ID Matching Schemes. *Workshop on Privacy in the Electronic Society*, 2006
17. Cate, F. and Staten, M. The Value of Information-Sharing. *Council of Better Business Bureau White Paper*, 2000

18. Lim, C.H. and Kwon, T. Strong and Robust RFID Authentication Enabling Perfect Ownership Transfer. *Conference on Information and Communications Security*, 2006
19. Molnar, D., Soppera, A. and Wagner, D. A Scalable, Delegatable Pseudonym Protocol Enabling Ownership Transfer of RFID Tags. *Selected Areas in Cryptography*, 2005
20. Staake, T., Thiesse, F. and Fleisch, E. Extending the EPC Network – The Potential of RFID in Anti-Counterfeiting. *Symposium on Applied Computing*, 2005

RFID Security: Cryptography and Physics Perspectives

Jorge Guajardo*, Pim Tuyls, Neil Bird, Claudine Conrado, Stefan Maubach, Geert-Jan Schrijen, Boris Skoric, Anton M.H. Tombeur, and Peter Thueringer

Abstract In this chapter, we provide an overview of mechanisms that are cheap to implement or integrate into RFID tags and that at the same time enhance their security and privacy properties. We emphasize solutions that make use of existing (or expected) functionality on the tag or that are inherently cheap and thus enhance the privacy friendliness of the technology "almost" for free. Technologies described include the use of environmental information (presence of light, temperature, humidity, etc.) to disable or enable the RFID tag, the use of delays to reveal parts of a secret key at different moments in time (this key is used to later establish a secure communication channel), and the idea of a "sticky tag," which can be used to re-enable a disabled (or killed) tag whenever the user considers it to be safe. We discuss the security and describe usage scenarios for all solutions. Finally, we summarize previous works that use physical principles to provide security and privacy in RFID systems and the security-related functionality in RFID standards.

1 Introduction

RFID as a technology is a rather old one, dating back to the Second World War when the Royal Air-Force used it to identify allied planes from their enemy counterparts [18,58]. However, it was not really until 1999, when RFID started to experiment a boom. This boom's main reason was (and continues to be) the envisioned ubiquity of RFID tags in everyday life. It is clear that the major advantages of tagging objects with RFID tags, as Juels [46] points out, are the ability to uniquely identify objects and the ability to automate tasks that previously had to be performed by a human. This will result in clear advantages to manufacturers of products or service providers. However, one may ask what is the general public's case for tagging

J. Guajardo
Philips Research Europe, Eindhoven, The Netherlands
e-mail: jorge.guajardo@philips.com

P. Kitsos, Y. Zhang (eds.), *RFID Security: Techniques, Protocols and System-on-Chip Design,* © Springer Science+Business Media, LLC 2008

everyday objects? RFID tags also have the potential to enable new applications (only limited by the reader's imagination) such as smart refrigerators that are able to tell when a product's life has expired or when you have run out of milk, washing machines that simply need to be started and know based on clothing information what wash cycle it should run, intelligent posters that allow a consumer to know in which cinemas and at what times a movie is playing, and finally, as an enabling technology in smart homes for the elderly and the cognitively impaired [46].

The pervasiveness of RFID tags, their ability to carry more information than bar codes, their expected low cost (below 10 US dollar cents), and their lack of need for line of sight communication also pose interesting challenges to those interested in their widespread adoption. Such challenges include both privacy and security concerns. On the privacy front, we can identify concerns on the part of consumers who will be carrying tagged objects. In particular, the wireless communication capabilities of RFID tags and their simple functionality (when queried they simply reply with their unique identifier) could make it easier to track people based on tag identifiers as well as to find out consumer preferences clandestinely. Similarly, companies and defense organizations will also be more vulnerable to espionage as it will be much easier to gather information on the competition or the enemy and much harder to detect such spying activities [46, 51]. We refer the reader to [46, 51] for a comprehensive survey of privacy issues in RFID.

On the security front, we have the authentication problem. In other words, how a legitimate party can assess whether an RFID tag embedded in an object (and thus the object itself) is authentic. The ability to authenticate legitimate tags has direct implications on industry's ability to decrease the counterfeit market, which in 2004 was estimated to surpass the 500 billion USD per year mark [31, 79]. The counterfeiting problem has been shown to be a threat as the following examples show: (1) in 2005 Bono et al. [9] showed how a popular transponder built by Texas Instruments and used by several auto makers in their ignition keys could be successfully cloned and (2) Carluccio et al. [11, 12] show how to build cheap RFID readers which could be used for tracing individuals via RFID chips embedded in passports. Thus, it is clear that solutions for authentication and privacy in RFID systems need to be developed. In fact, as we will see, both the academic and business communities have dedicated a lot of effort to these problems.

Based on the solutions that are known today and for purposes of presentation, we can divide security and privacy solutions for RFID into two groups: algorithmic solutions and solutions that either combine cryptography and physical principles, or that simply take advantage of a physical process. By algorithmic solutions, we mean those solutions based on traditional cryptographic mechanisms (e.g., public-key and symmetric-key primitives) or mechanisms which have been developed explicitly for the RFID environment but which make use of some type of cryptographic primitive (even if the primitive in question is not a standardized one such as the AES [61]). Examples of RFID security solutions based on algorithmic methods include basic access control through passwords as specified in standards [2, 23], minimalistic cryptography [45] and lightweight protocols [49, 64], solutions based on symmetric-key cryptography (e.g. [17, 26]), hash functions (e.g. [87]), and elliptic curve based solutions [5, 57, 73, 85]. Despite the proliferation of protocols including traditional cryptographic primitives, it is still not clear if such solutions will be viable for the cheapest of RFID tags. We note that the "low cost" requirement is an economic one. On the one hand, RFID tags are envisioned as more powerful substitutes for bar codes. On the other hand, if they are to be widely deployed (as bar codes are) then they also need to be in the same price range as a bar code, which only requires ink to be printed on a given item and thus, has cost close to zero. An additional consequence of the low cost of RFID tags is that in most papers when the term RFID tag is used what is implied is *passive* RFID tags. What is a passive tag will become clearer in Sect. 2, where we give an overview of general characteristics of RFID tags, types of tags according to their communication capabilities, their performance, costs, and security requirements as put forth in standards.

In the search for cheaper solutions, researchers have turned away from algorithmic approaches. Thus, ideas have been developed such as the `kill` command,[1] the blocker tag [47, 50] and similar blocking/proxy mechanisms [28, 71]. More engineering oriented approaches have also been introduced such as the IBM clipped tags [53] or distance bounding protocols [60]. Finally, we have begun to see the development of techniques that take advantage of noise in the communication channel between reader and tag to camouflage their communication [13, 14]. We will refer to such approaches as *algsics* methods[2] following [7]. Algsics solutions are based on combining the physical properties of RFID tags (or their environment) with traditional cryptographic primitives or simply making use of physics to enhance the privacy friendliness and security of tags. Section 3 provides an overview of such algsics solutions available in the literature. In Sects. 4–6, we take a look at three solutions recently proposed in [7] to enhance the security and privacy of the individual carrying the RFID tags. In particular, these mechanisms make use of environmental information to disable or enable the RFID tag. As with most algsics, we would like to point out that we do not claim that the solutions presented in this chapter will constitute stand-alone solutions to the privacy (or security) problems in RFID. Rather, we believe that these solutions will enhance other security and privacy solutions. It is possible that such methodology will in the end be the way toward securing RFID. Finally, we end with some conclusions in Sect. 7.

2 RFID Primer

RFID tags are made of a tiny chip, which implements the tag functionality, together with an antenna, which allows the chip to communicate with the external world. Tags come in three main varieties: passive, semiactive, and active tags. Passive tags do not carry a battery on them and use energy received from the reader's generated

---

[1] Although not application friendly, the `kill` command is a rather effective mechanism to safeguard the privacy of individuals.

[2] The first three letters of algorithmic and the last four of physics.

electromagnetic field to perform its operations and send data back to the reader. Semiactive tags use a battery for functions requiring to keep a state (for example if the tag needs to store sensor data) but its communication with the reader is still passive. Finally, active tags require a battery even for communications with the reader. This greatly extends the communication range (from a maximum of a few meters in the passive and semipassive case to 100 m or more in the active case). In this paper, we are mainly concerned with passive tags as these are the cheapest ones and the ones expected to be used for item-level tagging, thus making them also ubiquitous.

In general, the chip in an RFID tag has three main components (1) the analog front-end, which performs the modulation and demodulation of incoming signals as well as any voltage scaling if required (i.e., the reader provides electromagnetic energy which needs to be scaled down to the appropriate range); (2) the digital part, which implements the functionality of the chip (including security functionality); and (3) memory. The memory may be of several types, e.g., NV-RAM (nonvolatile RAM, only active when tag is powered), a ROM part (one-time burnable, unchangeable memory), and E-ROM (erasable ROM, which can have its contents permanently cleared). It is assumed that the tags carry a unique identifier (e.g., the EPC). In some cases, the tag may also carry product information (PI). This can be, for instance, information for distributors or shopkeepers such as the product type and price, but also contain information for end-users, such as washing or cooking instructions, expiration date, etc.

2.1 RFID Tag Classes

In [76], Sarma and Engels propose a framework to classify RFID tags according to their functionality. Recently, the RFID class structure has been slightly modified [19] to reflect the merging of Class 0 and Class 1 tags into a single lowest functionality Class 1. Table 1 summarizes the classes in both [19, 76] and their associated functionality. Note that already back in 2003, it was expected [76] that Class 0 and Class I RFID tags would eventually be phased out in favor of a Class I Generation 2 RFID tag [69]. A general idea of the framework is that each higher tag class will include the functionality of the lower level classes and incorporate new functionality. Thus, a Class III tag will support at a minimum all the functionality of a Class II tag and add other functionality on top of that. Note that this implies that even active tags, once their batteries are exhausted, can operate as passive Class I tags.

Table 2 summarizes the security functionality described in different RFID standards. We have also included information such as the size of the programmable memory available in the card according to the standard, the frequency band in which it operates, and the type of anticollision protocols specified.

Table 1 RFID tag class general characteristics

Class	Year	
	2003 (Source [70, 76])	2005 (Source [19, 22])
0	Read-only passive backscatter identity tag programmed at manufacturer time	
I	Read passive backscatter tag with one-time field programmable nonvolatile memory	Lowest functionality identity tag, passive backscatter, it contains user memory, system memory accessed either directly by the tag or by specific commands (not general *Read* or *Write* commands), EPC memory and tag identifier memory (TID), implements a "kill" function that permanently disables the tag, it provides optional password-protected access control, and optional user memory
II	Passive backscatter tag with up to 65 kbyte read–write memory and additional functionality like encryption	In addition to Class I functionality: an extended TID, extended user memory, authenticated access control and additional features to be defined
III	Semipassive backscatter tag with up to 65 kbytes read–write memory. It may support broadband communication	In addition to Class II functionality: an integral power source and an integrated sensing circuitry
IV	Active tag with microchip and active antenna to communicate with a reader. It may be capable of broadband peer-to-peer communication with other active tags in the same frequency band and with readers	In addition to Class III functionality: Tag-to-Tag communications, active communications, and Ad hoc networking capabilities
V	Active tag with microchip and active antenna that can power up other Class I, II, and III tags as well as communicate with class IV or V tags. Essentially a reader	In addition to the Class IV functionality a class V tag is capable of initiating passive communications and powering up passive tags. In addition, unlike other tag classes a Class V tag (reader) might have a connection to a back-end network

2.2 Costs and Performance Requirements of RFID Systems

From the beginning of the RFID boom back in 1999, the reduction in the cost of the tag (and consequently the chip) has been a major driver in the development and adoption of the technology. Security is directly affected by this, as the overall cost of the tag will also dictate the budget available for security functionality. This section summarizes some of the requirements on low-cost RFID systems available in the RFID security literature and their original sources.

It is generally accepted that a passive tag should cost in the range of US$0.05–US$0.10 for it to be successfully adopted by manufacturers and incorporated into

Table 2 RFID tag security related functionality according to different standards

Technology	Band	ID size (bits)	Security and performance related functionality
ISO/IEC 18000-2 [39, 66]	125 kHz or 135 kHz (LF)	Up to 1 kbyte of read–write memory	Reader talk first protocol, permanent manufacturer-set 64-bit ID, no protection on the read command, no encryption or authentication, optional lockable memory block capability, deterministic slotted anticollision algorithm, depending on frequency a read transaction takes 28 ms or 57–97 ms, 16-bit CRC on certain commands and mandatory on the tag side
ISO 11784, ISO/IEC 11785 [33, 44]	134.2 kHz (LF)	64	Reader talk first protocol, 16-bit CRC for error detection
ISO/IEC 10536-3 [37, 66]	4.9152 MHz (HF)	Read–write block	Reader talk first protocol, masked reader-to-tag communications, tag addressed by random number, quiet mode. The tag should draw no more than 150 mW of power from a single exciting field and a maximum of 200 mW, tag must respond within 40 ms of being powered up, probabilistic/slotted random anticollision algorithm, multiple tag modes are noninterfering
ISO/IEC 18000-3 [40]	13.56 MHz (HF)	Up to 8 kbytes of read–write memory	Reader talk first protocol, unique 64-bit manufacturer-set tag-ID programmed at manufacturing time and unchangeable thereafter, permanent memory-block lock via lock command, in mode 2 48-bit password field set by user to provide memory access control, 48-bit long password for read–write commands if required by the tag in mode 2, two tag modes defined (not interoperable and noninterfering), deterministic anticollision protocol with possible extension to probabilistic version, probabilistic anticollision in mode 2, random number generator equivalent to or better than a 32-bit maximal length LFSR, 16-bit CRC for command error detection in mode 1 (in mode 2, 16-bit CRC Interrogator-to-Tag, 32-bit CRC Tag-to-Interrogator)
EPC Class I [3]	13.56 MHz (HF)	64, 96, 256	kill command with 24-bit associated password, one-time memory programming safeguard by lock state, optional read–write memory safeguard by optional lock command. Passive tag, programmable with EPC and possibly other data, data can be individually destroyed, anticollision method with an identification rate of 200 tags per second, 8-bit CRC on commands (reader-to-tag link), 16-bit CRC on tag-to-reader link

Table 2 *Continued*

Technology	Band	ID size (bits)	Security and performance related functionality
ISO/IEC 15693-2, ISO/IEC 15693-3 [36,38,66]	13.56 MHz (HF)	64 unique ID, up to 8 kbytes of user programmable memory	Reader talk first protocol, lock block command to permanently lock (up to) 256-bit long memory block, optional protections on write command, optional password protection on lock command. 16-bit CRC on each request and response
ISO/IEC 14443-2, ISO/IEC 14443-3 [34,35]	13.56 MHz (HF)		Reader talk first protocol, 16-bit CRC for error detection, bit frame anticollision protocol with possible proprietary protocol extensions for type A cards and deterministic or probabilistic slotted anticollision protocol for type B cards
ISO/IEC 18000-7 [43]	433 MHz (UHF)	Up to 128 kbytes	Reader talk first protocol, 16-bit CRC, 32-bit tag ID assigned at manufacture time, 24-bit programmable owner ID, 32-bit password for access control with accompanying set password command, password protect bit to force use of password for all point to point commands, password protected unlock command, programmable 1–16 byte User ID, 24-bit owner ID for additional privacy (if owner ID is present in the tag and the owner ID supplied by the interrogator does not match it, the tag will not respond to the reader's request), probabilistic slotted anticollision protocol, slot is chosen via a pseudorandom number generator
EPC Class 0 [4]	900 MHz (UHF)	64, 96, 256	kill command with 24-bit associated password, use of two pseudonyms (ID0,ID1) in singulation ID1 is pseudorandom and fixed, ID0 is random and changed after every singulation, passive tag, 16-bit CRC on tag-to-reader link, reader-talk-first communication protocol, binary tree based scanning anticollision protocol, identification rate >1,000 tag per second, 8-bit commands plus parity bit
EPC Class I [1]	860 MHz – 930 MHz (UHF)		kill command with 8-bit associated password, 16-bit CRC computed on the EPC, A compliant tag will only change its internal state or perform backscatter modulation in response to commands defined in [1], all other commands will be interpreted as unknown commands and will be ignored, implements a lockID command, once the lockID is executed the EPC and kill command password cannot be changed. passive tag, commands have five parity bits, communications are half-duplex (reader talks first and tags listens or vice versa)

Table 2 *Continued*

Technology	Band	ID size (bits)	Security and performance related functionality
EPC Class I Gen 2 [22, 24]	860 MHz – 960 MHz (UHF)	96, 170, 195, 198, 202	32-bit kill command that permanently disables the tag, optional password-protected access control, 16-bit CRC used on certain commands, secured state in which all access commands can be executed, random or pseudorandom number generator mainly used during singulation, lock state to disable or enable the reading of a password or memory bank (permanently or temporarily). passive tag, communications are half-duplex (reader talks first and tags listens or vice versa)
ISO/IEC 18000-6 [42]	860 MHz – 960 MHz (UHF)	No minimum memory but must be multiple of 4 bytes, max. 8 kbytes	Reader talk first protocol, 64-bit unique ID, error detection in the forward link uses 5-bit CRC for all commands with additional 16-bit CRC for long commands for type A cards, all other links and type B cards use a 16-bit CRC, lock block command allowing to lock a 32-bit block permanently for type A cards, similar for type B cards but lock command implemented at byte level. Type A uses an Aloha probabilistic anti-collision protocol
ISO/IEC 18000-4 [41]	2.45 GHz (UHF)	Min. memory 64 bits and recommended min. 18 bytes for mode 1	Reader talk first protocol in mode 1, 64-bit unique tag ID in mode 1 16-bit CRC present in all communications to and from the tag in mode 1, each byte has a lock bit which can be set at manufacture time or during a lock command, passive backscatter RFID system probabilistic anticollision in mode 1

Notes:

- Both in EPC Class I [3] and EPC Class 0 [4] the 256-bit ID size is planned but not fully specified yet.
- Recommendation [1] does not specify any EPC sizes. However [20], which applies to all Generation I tags specifies 64-bit and 96-bit EPCs.
- In the EPC Class I Generation 2 Specification [22, 24], the ID size includes only the bits required for the EPC. Including all bits of storage and depending on the encoding standard, each tag will require either 224-bits or 336-bits of memory [24].
- In ISO/IEC 18000-4 [41] mode-2 assumes a battery assisted tag, thus, we have not included its functionality in Table 2.

Table 3 Gate density for different standard cell technologies

Technology	Gates/mm^2	Source
0.80 μm	1,500	[86]
0.50 μm	4,000	[86]
0.35 μm	10,000	[86]
0.25 μm	38,000	[86]
0.18 μm	60,000	[86]
0.13 μm	110,000	[16]
0.15 μm	182,000	[84]
0.13 μm	219,000	[83]
0.09 μm	436,000	[83]
0.065 μm	854,000	[83]

most packaging [86]. According to [86], to construct a US$0.05 tag, the IC cost should not exceed 2 cents. Weis [86] also states based on [74] (see also [75]) that the cost per mm^2 of silicon is roughly US$0.04. This implies that independent of the technology, we have a budget of 0.25–0.5 mm^2 for the whole RFID chip[3] if we want to attain the 5–10 cent tag. Despite the continued decrease in the cost of silicon, price pressure and competition is expected to keep these figures relatively stable. This amount of area can be translated into an approximate number of gates depending on the technology chosen. Table 3 shows the number of gates per mm^2 for different technologies available in 2006. Notice that in general, as we go down in technology and increase the gate density per mm^2, the cost of the technology also increases.

Based on these assumptions, Sarama et al. and Weis [77, 86] estimate that the number of gates that can be used for security functionality is between 250 and 2,000. Ohkubo et al. [62] estimate that this number can be increased to 5,000 gates. Ranasinghe et al. [68] from the Auto-ID Labs seem to agree with [62] and estimate that the number of gates destined for security should be between 400 and 4,000.

In terms of performance and power requirements, Ohkubo et al. [62] estimate that the transmission rate in the 13.56-MHz and 900-MHz bands available to the RFID tags is approximately 26 kbps per 50 tags at 13.56 MHz and 128 kbps per 200 tags at 900 MHz. Assuming that tag reading should not exceed 1 s, each tag can transmit about 500 bits. Feldhofer et al. [26] estimate that the power available to the digital part of the RFID tag is 20 μA based on an implementation compliant with the ISO 18000-3 standard. Finally, standards also dictate how long after a request a tag can wait to send a response to a reader. This effectively dictates the performance of an encryption operation. However, as Feldhofer et al. [26] have shown there are ways to get around these limitations.

---

[3] An example of an RFID chip in 2001 is the Hitachi's RFID μ-chip which was 0.06-mm thick and 0.4-mm long on each side (0.24 mm^2). It ran in the 2.45-GHz frequency band, it contained 128 bits of ROM memory and it was readable by a sensor within a 30-cm range [82].

3 Previous ALGSICS Solutions

In this section, we survey previous *algsics* methodologies found in the literature. They are organized according to the ideas in which they are based.

3.1 Killing and Sleeping

The "kill" command is a password-protected command available in the EPC Gen-1 [1, 3] and EPC Gen-2 standards [21, 22] that effectively renders a tag nonoperational. In general, a typical application of the kill command would be at a point of sale terminal. The user would get an article and upon checking out the tag associated with the bought item would be killed. This solves the privacy problem for the user but has the drawback of making many future applications of RFID in home environments (smart refrigerators, microwaves, washing machines, etc.) impossible. A similar idea is reported by Juels in [46] who explains how Marks and Spencer, a large clothing retailer in the UK, has begun to include RFID tags in the price tags of their garments, which are ultimately removed upon final purchase. In [52], Juels also suggests the idea of "putting tags to sleep" instead of completely killing them. Although interesting in nature, since the user would be able to wake the tag back up whenever he is in a safe environment, waking up tags implies that a wake-up command would have to be password-protected. This in turn causes two practical problems (1) the user will be in charge of managing the passwords for his tags and (2) the user will need to manage the sleep/wake patterns of tags. Juels suggests as a possible solution to have tags printed with their associated passwords and have the readers optically read the keys or wake a tag via physical contact as suggested in the *Resurrecting Duckling* approach from [80]. However, such approaches would do away with the benefits of a wireless system.

3.2 Blocker Tags

In [50], Juels introduces the blocker tag. This solution provides privacy by protecting consumers from the unwanted scanning of RFID tags that they may carry or wear. It protects against unauthorized readers but also to eavesdroppers who can listen to signals broadcasted by an authorized reader during the tree-walking or ALOHA singulation protocols. The blocker tag is an alternative to more simple solutions such as the kill command, the Faraday cage approach (has the drawback that no all objects can be shielded), and the active jamming approach (might be illegal in some countries depending on the radiated power and will create severe disruption to all RFID devices in its close proximity).

The approach is physical rather than cryptographic in nature, shielding tags from view of an unauthorized reader(s). A special tag, called "blocker" tag, blocks an RFID reader by simultaneously answering with 0 and 1 to every reader's request during the tree-walking singulation protocol. The reader is then incapable of singulating standard tags with k-bit identifiers. The blocker tag may block a reader universally (when it simulates the full set of 2^k possible RFID tag identifiers) or selectively (when it simulates tags with identifiers only in given ranges, e.g., under a subtree). In either case, the algorithm outputs all tags within the reader's range with identifiers in the targeted sets. In the selective case, the blocker tag might announce its policies to the reader to warn it to query only outside the blocked zones, where the blocker tag remains inactive. This is done to avoid that an honest reader stalls while performing a scan. Blocker tags can implement one or more privacy policies, which can be set on and off via the flipping of a bit on standard tags. Moreover, multiple blocker tags may cover multiple privacy zones. The blocker tag provides privacy protection and has a very low-cost implementation. Standard tags need not be modified or only little to support password-protected bit flipping. Blocker tags can be also made cheap with only very slight circuit changes on a standard tag. On the other hand, blocker tags can be also used to mount DoS attacks in which a malicious blocker tag universally blocks readers. The authors argue, however, that such an attack can be mounted, independent of the use of blocker tags for privacy protection. Another possible threat is that too many or very specific privacy zones or policies diminishes individual privacy as the policy can become a unique identifier. Reference [45] notes that the blocker tag has limited applicability as, for example, in industrial environments tags cannot be blocked as they must remain readable at all times. Thus, the blocker tag would not be an effective solution against the threat of industrial espionage.

In [47], the authors proposed a modification of the blocker tag which they named "soft-blocking." Soft-blockers simply express the privacy policy and preferences of the users carrying them to readers. Readers are expected to behave honestly and are audited, thus providing for a mechanism to check reader misbehavior. Soft-blocking involves the introduction of a software or a firmware module denoted "tag privacy agent" or TaPA in [47] whose main functionality is to filter out the data read from the tag depending on the specific tag's privacy policy or classification. As an example, [47] suggests that tags have three classifications (1) blocker, (2) private, and (3) public. Upon receiving the tag's classifications, the TaPA would output only data that is public if a blocker is detected and both public and private if a blocker is not found. Alternatively, one could imagine a system where everything is blocked by default and where a classification *unblocking* exists that makes (originally private) data public. This would protect individual's privacy by default. Although, soft-blocking does not provide the strict privacy guarantees of the blocker tag, it provides flexibility beyond what the blocker tag offers. Juels notes [46] that, in the simplest case, soft-blocking could rely only on reader auditing. This variant, although it has obvious technical deficiencies, might turn out to be the most practical form of blocking.

3.3 Privacy Sentinels

A generalization of the blocker tag is the privacy sentinel. This terminology was introduced by Sarma in [75]. In what follows, we will use the term privacy sentinel and watchdog tag interchangeably. Note that although the particular implementations might differ in specific features, the basic idea is the same: A proxy device that manages the communication of the RFID tag with the external world. The concept of the privacy sentinel was originally introduced in [28] with the name "watchdog tag." Similar approaches have also been introduced in [52, 71, 78]. The idea is to provide users with a more powerful trusted device (the privacy sentinel device) that takes care of their privacy, manages their privacy preferences, and could, for example, be integrated into a user's cell phone. The watchdog tag's main purpose is to manage the communication between the reader and the tags that the user is carrying. In addition, the watchdog tag could show warnings to the user, prompt him for authorization, and log all data transfers. Reference [71] extends the watchdog tag concept to include key management, authentication operations, and tag simulation (i.e., the privacy sentinel is able to mimic the operations of the less powerful tags that is managing). Juels et al. [52] consider the problems of tag relabeling, acquisition, and ownership transfer.

3.4 Channel Disturbances

Recently, Castelluccia and Avoine and Chabanne and Fumaroli [13, 14] have taken advantage of the noise present (or artificially generated) in the communication channel between reader and tag to enhance the security of their communication. Reference [14] takes advantage of the noise in the channel to allow readers and tags to share a secret without a *passive* adversary being able to learn it. Readers and tags perform a protocol where information reconciliation and privacy amplification take place through the use of universal hash functions. The scheme in [13] is somewhat different. It assumes the existence of *noisy tags* owned by the system which inject noise into the communication channel. The noisy tags also share a secret key with the reader, which is used to pseudorandomly generate noise. Whenever the tag sends its secret key to the reader, an eavesdropper will see a signal that is the sum of the signal corresponding to the tag's secret key and the noise injected by the noisy tags. On the other hand, the reader is able to replicate the noisy tags' noise and it is able to subtract the noise signal from the received signal, thus recovering the tag's secret key. A similar approach to [13] is presented in [30]. The difference is that the authors do not assume the presence of a noisy tag but rather assume that the reader and tag can synchronize their communications. Both tag and reader send a pseudorandom sequence to each other; whenever their bits are different an eavesdropper will not know which bit was sent by the tag and which bit by the reader. On the other hand, both the tag and the reader are able to obtain each other's keys.

3.5 Distance Bounding Protocols

Cryptographically secure distance bounding protocols date back to 1993 as introduced in [10]. However, Fishkin et al. [27] seems to be the first to suggest a protocol specifically suited to the RFID setting. Note that in the context of RFID protocols proximity implies trust. Fishkin et al. [27] find that looking at the signal noise (in particular the Fano factor, which is used to approximate signal noise) and at the actual signal strength received by an RFID tag correlates fairly well with the tag distance from the reader. They use this correlation to decide whether the energy received from the reader antenna can be considered to be in the far field or in the near field. Then, based on this decision, the RFID tag could have a policy of responding to the interrogating reader or not. This distance bounding protocol is combined in [27] with the idea of tiered revelation and authentication in which the tag reveals more and more information according to the level of authentication used by the reader. Reference [27] also notes that the tiered level can be associated with the amount of energy emitted by the reader. Thus, for example, a reader that requests more information will also be required to power the tag for a longer period of time while using a longer key size. The work in [29] proposes a new distance bounding protocol based on ultra-wideband pulse communication where the verifier is the reader and the prover the RFID tag. Thus, it considers the reverse problem, i.e., the reader wants to verify that it is talking to an honest tag. The protocol makes use of a keyed hash function or symmetric-key primitive to generate a sequence of pseudorandom bits which upon a challenge from the verifier are returned by the prover. Only an honest prover can generate the correct sequence as he also knows the secret key used to generate the sequence.

3.6 Changing-Tag Systems

By changing-tag systems, we mean systems in which the tag or tags change physically. Examples are the works presented in [32, 53] as well as in [8]. The work in [32] is interesting in that they suggest to physically split the IDs of RFID tags. In particular, their approach envisions splitting global RFID tag identifiers into a class ID (related to the class of objects) and a pure ID (which identifies the specific object, lot number, serial number, etc.). The idea is then for the user to be able to physically remove the class ID from the object and at a later stage attach a second tag with a different global ID, which might be unique in the user environment but not globally. The authors in [32] also note that the same effect (changing IDs) can be achieved by using rewritable memory in an RFID tag. Reference [8] considers systems in which an object is associated with multiple RFID tags. Then, chaffing and winnowing in the sense of [72] can be used to disguise the true identity of the object. Note that Weis [86] was the first to note that chaffing and winnowing can be used in the RFID context but he assumed that the readers would be the ones generating the chaff. In [53], the authors propose to physically disconnect the antenna and the chip in an RFID tag. In addition to allowing for visual confirmation (on the part of the consumer) that the tag communication capabilities have been disabled, it allows for this functionality to be "pasted" back on if the user desires to resurrect the RFID tag functionality once he/she is in a safe environment.

3.7 Tag Switches

The work in [88] explores the idea of physically deactivating a tag via a physical bit-dependent switch. If the bit is set to one, the RFID tag answers as usual to a reader query whereas if the bit is set to zero, then the tag is deactivated until the user activates it again. The idea is based on the assumption that only someone with physical access (or close proximity) to the tag can activate it again. Thus, consumer privacy is safeguarded and at the same time, tag functionality is preserved for privacy-friendly environments. The author describes three possible implementations of the physically changeable bit (PCB). The first implementation consists in physically (dis)connecting the antenna from the chip, much in the same way as the clipped tags in [53]. Other methods include electrically erasable ROM memory in the tag, writing or erasing the PCB depending on user wishes, and using "magnetic bits" in the tags to represent (and set or unset) the PCB bits. In this category, we also include the `kill` command, which works by completely disabling the tag if the tag is presented with the correct password. Although not application friendly, the `kill` command is a rather effective mechanism to safeguard individuals' privacy.

4 Environmental Context at the Service of Privacy

In this section, we describe solutions that enhance the privacy of users carrying objects with associated RFID tags. We assume that guidelines for RFID privacy have been followed, such as placing the tag on the outside of the object and that this position has been clearly identified. This also allows consumers to have the option of removing the tag if desired. We also assume the integration of sensors in the RFID tag functionality. This assumption gives rise to several questions. The first question we ask is if this approach is feasible at all from a technical point of view and if such a sensor-RFID tag can be implemented in a battery-free manner. The answer to these two questions is positive as [55,63,65] provide evidence of the feasibility of this approach. The second question regards price. How much such a sensor-RFID tag costs will in the end dictate whether such a solution will experience widespread adoption or not. To be successfully adopted at the item level, we require a price in the range of US$0.05 per tag [86]. The experience of [65] seems to indicate that today it is possible to build RFID tags including sensor functionality under a US$1 but far from the US$0.05 mark. In fact, some are already available, albeit only battery powered ones [81]. In the end, we expect that the continued decrease in silicon prices as well as consumer and customer requirements for additional functionality will enable the integration of sensor functionality into cheap RFID tags. In the following, we describe several scenarios which take advantage of embedded sensor functionality in an RFID tag to make the technology more privacy friendly. The basic idea in all the solutions is to use environmental information as an on/off switch. By environmental information, we mean data from temperature, light presence (or absence), or humidity readings of the environment surrounding the sensor-RFID tag. Depending on the setting and the application, a certain sensor might be more appropriate than another. Then, whenever the chosen environmental information attains a certain value (or range of values) or the user "creates" the right environmental conditions, the RFID tag is able to transmit data to an interrogating reader. Otherwise, the tag functions as if it was completely disabled. In the next sections, we describe usage scenarios for particular sensors and we discuss advantages and disadvantages of such solutions.

4.1 Tag Privacy Protection Via Light-Controlled Tag Activation

4.1.1 Idea

The idea is to control access to the powering circuit of the RFID tag via a fully integrated light-sensitive diode which can detect the presence of a laser beam, e.g., from a laser pointer. This allows for the presence of a secure light-controlled ON/OFF switch on the tag. When the tag is powered by a reader and a laser beam is pointed at the light sensor, a digital ON code is written into the RFID's nonvolatile memory. This ON code can, by means of an active switch (e.g., a MOS-transistor), be used to enable the power-supply voltage to parts of the RFID-chip, or enable other circuits to the rest of the chip, in such a way that the chip becomes fully functional. Even when the tag is taken out of the reader field, this ON state remains stored in memory. The tag can also be set in its OFF mode under similar conditions. When the tag is powered by a reader and a laser beam is pointed again to the light sensor, an OFF bit will be written in nonvolatile memory and the power supply voltage will be disabled from the rest of the tag. In that case, the tag is not functional anymore until it is switched ON again by means of the laser beam. Even though such a switch provides the desired functionality of access control to the tag, it suffers from the drawback that a laser beam needs to be pointed to the tag. Thus, this could be considered as undermining one of RFID's main advantages: no line of sight communication. As an alternative, it is also possible to make an RFID tag that will only function if enough environmental light is present. In this case, the user can protect his tags from being read by an unauthorized party simply by covering the tag such that no light can reach its photo detector or by keeping the tags in the dark. Note that in many situations, this would not be an unnatural thing to assume (just think of a grocery bag, a wallet, or a purse). Alternatively, an RFID tag could be part of a label that can be closed or opened (covered/uncovered) such that light to the tag is blocked or passed, respectively. This way the user is in control of the readout of his tags and can choose when and where his tags may be read. No special reader is required for reading out the RFID tag. The silicon-area required for the light-sensitive diode, including control circuits, can be very small [67]. This results in a cheap protection method that can be, if necessary, combined with other existing privacy enhancing technologies.

4.1.2 Discussion

A consumer carrying items with such a modified RFID tag disables the tag at the point of sale terminal and re-enables it again once he/she is in a safe environment, e.g., at home. Thus, future ambient intelligent applications are still supported and the user's privacy not affected. Another example application of such a solution is in the tagging of bank notes. By turning off the RFID interface in his/her bank notes via their light-enabled switch, a user very simply avoids tracking. Another attack that is prevented is that in which a thief targets passers-by who are carrying 500 Euro notes in their wallets [48] by simply reading their tags. On the other hand, any person or organization desiring to verify the authenticity of the bank note can do so upon obtaining the bank note as a form of payment for a service or product. Note that the light-enabled switch does not support all the properties put forward by Juels and Pappu in [48]. In particular, it would only allow law enforcement agencies (or any authorized entity) to trace bank notes after detaining a potential suspect and not in an unobtrusive manner as suggested in [48]. Finally, a potential attacker, intending to track someone via the RFID tags that his victim is carrying, would be required to point a light source at each consumer tag that needs to be enabled without this activity being detected by the victim.

4.2 Tag Privacy Protection Via Moisture-Dependent Contact and Other Sensors

4.2.1 Idea

Inclusion of RFID tags in clothing has been proposed as a means to support activities such as supply chain and retailer product management. However, including RFID tags in clothing raises privacy concerns to those that wear such garments (see for example [6]). To enhance the privacy of users in this situation, a modified tag is proposed. The tag operates normally prior to sale. At the point of sale, the tag is *disabled*, e.g., by burning a ROM component or wire, which can be done by applying a large amount of power to the tag at the point of sale reader/terminal. Note that we do not completely kill the tag but rather disable its RF interface. Once in the disabled state, the tag can still function but only if enough conducting moisture is present. This can be done by means of a switch (put in a strategic location such as the tag's antenna) that can only make electric contact if conducting liquid is present. Therefore, the tag is effectively disabled in the street (as long as it stays dry) and can be finally re-enabled when the washing machine pumps water onto the clothes. One may worry that tag readout is hampered by large volumes of water absorbing RF radiation. However, studies have shown that this is not a problem. In particular, it is well known that at low frequencies (in the 10–20 MHz range) water is transparent to an RF signal [54, pp. 2-6–2-7]. At higher frequencies, the attenuation is significant and it is highly frequency dependent. For example, the study in [15] shows that the attenuation of the signal traveling a distance of 6 cm varies between 7 and 23.5 dB for frequencies between 100 and 950 MHz. Note, however, that there are solutions starting to appear that can perform well in the presence of water and metals at high frequencies as shown in [56]. Finally, for the particular case of an RFID-tag operating in the 13.56-MHz band, a weakening of the signal by 10 dB is deemed acceptable. It can be shown experimentally that at frequencies around 10 MHz the RF signal penetrates 25 cm into salty liquid, which is more than sufficient for the washing machine example.

4.2.2 Discussion

In addition to supporting activities such as supply chain and retailer product management, RFID tags associated with clothing items could also support other applications such as smart washing machines. Smart washing machines could be equipped with an RFID reader, which allows the machine to access clothing information. Therefore, the machine could autonomously select a washing program based on that information or it could advise the user to remove an item that needs a different washing program via an alarm. A second example of a sensor used to enhance privacy is a temperature sensor for a smart refrigerator application. In this setting, RFID tags could be allowed to be read-only in certain temperature ranges. Thus, when the groceries are in the refrigerator at a certain temperature range, the RFID tags associated with the groceries would be readable and otherwise not. Such an RFID tag would enable applications as diverse as checking whether a product has been at the correct temperature during the whole supply chain or placing an automatic order when the user has run out of certain food items. On the other hand, one can argue that whenever the temperature outside was also in the range of the refrigerator temperature, the RFID tag would be allowed to transmit and thus, the user would be traceable. However, the ability that an attacker has to trace someone would be highly dependent on weather conditions and not on the attacker's choice. This diminishes the attacker's tracing abilities or forces him to change environmental conditions around his target. In this case, security is also highly dependent on how close the attacker can get to his target and stay there for extended periods of time. Clearly the closer the attacker is to his target, the easier it is for him to be discovered but also the more successful he will be in cheating the system. Finally, note that a single sensor will probably not be applicable to all scenarios, with the possible exception of the light sensor. For example, a humidity sensor might be suitable for clothing but not for electronic items, and similarly temperature sensors might work well with food but not with clothing. Light sensors, on the other hand, seem to allow a wide range of applications.

5 Sticky Tags and Privacy

Current privacy preserving solutions for RFID are such that they either add cost to the tag by including additional hardware to perform cryptographic functions or require the modification of current tag specifications to perform additional operations. On the other hand, the most widely available (standardized) solution for privacy concerns is the kill command that permanently disables the tag. This solves the privacy problem but it gives up the advantages that RFID tags can provide in other applications. Thus, the idea proposed in this section can be seen as middle ground between the two extremes of rendering tags completely useless with the kill command or having additional costs added to current RFID tags. It can also be seen as yet another instantiation (with different properties and characteristics) of a privacy sentinel [75] or watchdog tag [28].

5.1 Idea

The basic idea is to allow the kill command to completely disable the RF functionality of the RFID tag but to allow access to the information in the tag via a second interface, which requires proximity to the tag. This second interface could take different forms. The simplest instantiation of the second interface would be a contact-based interface. In this case, proximity means "as close as it is physically possible," i.e., touching the disabled tag. We emphasize that adding a contact interface to an RFID tag is not new. However, to the authors' knowledge the idea that a second interface can be used in combination with a second (more powerful) tag to "resurrect" the functionality of the killed tag and guarantee privacy (and security) for the user is novel. Note that the resurrecting functionality is different from the resurrecting duckling security policy of Stajano and Anderson [80], where a node in an ad hoc network establishes a secure channel after being "resurrected" by an adjacent node. A second possibility is a modified antenna system which upon receiving the kill command changes its configuration. For example, the read-range could be limited by the kill command to 1 mm. By a modified antenna system, we mean both an antenna which changes its range (for example, via clipped tags as in [53]) or simply a system consisting of two antennas. The first antenna has a normal range and it gets disabled upon the tag receiving the kill command whereas the second antenna has a very short range and it is not affected by the kill command. Note that this instantiation might succumb to relay attacks. The second interface can then

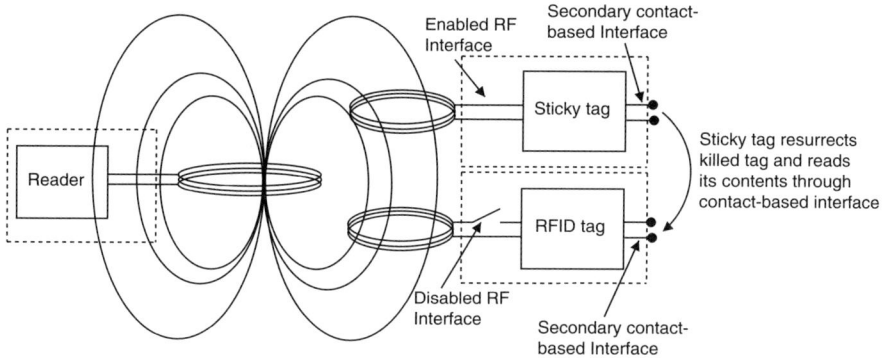

Enabled RF Interface

Secondary contact-based Interface

Sticky tag

Sticky tag resurrects killed tag and reads its contents through contact-based interface

Reader

RFID tag

Disabled RF Interface

Secondary contact-based Interface

Fig. 1 Sticky tag in the presence of a reader with a secondary contact-based interface

be used by another device, presumably a more powerful RFID tag both in terms of computational power and security, to access the data in the original RFID tag and communicate in a secure manner with RFID readers. We will refer to this device in what follows as a *sticky tag* to illustrate the fact that we expect such devices to be implemented as a sticky label that adheres to objects whose original RFID tags have been killed. "Sticking" our new more powerful tag on the less powerful tag has the effect of "resurrecting" the tag. Figure 1 depicts an illustration of the system. In particular, a standard reader powers up both antennas, the sticky tag's antenna and the original RFID tag's antenna. Since the RFID tag's antenna has been disabled, only if the sticky tag is present will the reader obtain a response from the RFID tag. Note that the sticky tag acts as a bridge between the disabled RFID tag and the RFID reader. As such, the sticky tag, when queried, forwards the information residing in the original RFID tag to the reader. Also the sticky tag must not have an identifier (e.g., EPC) of its own. In addition, the sticky tags do not necessarily have to be more powerful devices. A sticky tag could simply be a much cheaper device without memory or functionality other than reviving the killed RF interface of the original tag. This instantiation would have the advantage of extremely low cost. Finally, an added advantage of sticky tags is that they could be used to resurrect RFID tags with a defective RF interface.

5.2 Discussion

As usual, at check-out the RFID tag is disabled. However, by attaching a sticky tag to the killed tag now the user is able to take advantage of the information stored in the killed tag just as if the tag in the object had never been killed. This has the added advantage that the identifier is transmitted to the readers in a secure manner (if the sticky tag is equipped with cryptographic functionality) or in a more secure environment, since it is the user that decides where and when to resurrect the killed tag. The sticky tag is also envisioned to be reusable, i.e., users could have a bag of such sticky tags and attach them to objects whose RFID tags have been killed. Once the object's usable life has expired, the user could simply detach the tag and store it for future use after discarding the object. The manufacturer who would also like to check an object's information once the object is in the recycling phase, could similarly resurrect the originally embedded RFID tag by using a sticky tag as well. A final usage case is the scenario in which a user returns a product to the shop because of regular maintenance, repair, or malfunction. In this case, the shop can use a sticky tag to read the product information available in the original tag associated with the object. Admittedly, a main issue with the sticky tags is usability. Can we expect that users will tag their groceries so that they can make use of their smart refrigerator? Note that owning a smart appliance implies that the user has an interest in using the intelligence features in the refrigerator, otherwise he would not have bought it in the first place. In addition, attaching a sticky tag both at home and at the repair shop scenarios does not need to be a cumbersome activity. It could be similar to the customary practice of detaching antitheft tags at clothing stores once an item has been sold or to adding a pricing tag to an item as it had been done for years (and in some places it is still done) before the widespread adoption of bar codes. On the other hand, a main advantage of the sticky tags is that they are an opt-in solution. By default, we are safeguarding individual's privacy and if they desire they can regain many of the advantages that RFID offers. Sticky tags would be best suited to objects that are meant for home use once they have bought (e.g., groceries, TVs, DVD players, etc.). Similarly, using sticky tags for clothing for example, would imply that the user needs to remember to detach the sticky tag from his clothing before going out. Otherwise, he could risk traceability. This seems a burden not likely to be accepted by most people.

6 Time-Released Secrets and RFID

The idea of using a delay to enhance security is not new in cryptography. In particular, May [59] introduces timed-release cryptography as a new primitive. The solution that we present here can be seen as a timed-release system in a different timescale and with different granularity as the system of [59]. In the context of RFID security, Juels [45] seems to be the first to use delays to limit the ability of an attacker to perform successive queries to a tag by using a hardware-based throttling mechanism for his pseudonyms scheme. However, schemes such as the ones presented in this section and the ability to turn on and off the delays were not discussed.

6.1 Idea

This solution tries to hinder the ability of a reader to read or identify a tag when a person passes by. This is achieved by implementing an actual physical time-delay functionality in the RFID tag. This time delay forces the reading of sensitive data to require more time when the tag is in an unprotected environment than when it is in a protected setting. In this case, the tag itself acts as the agent that releases the secret at a given time in the future. The user or user's devices (e.g., smart home appliances) are the party requesting access to the secret-key information. The unprotected environment may be, for instance, the user's path from shop to home. In this case, the chances that an unauthorized reader is able to obtain any information from the tag are decreased thanks to the time delay between a reader requesting information (powering up the tag) and the time when the tag actually responds. On the other hand, when the tag is in a protected environment, e.g., the shop or the user's home, the tag responds without delay, thus not hindering trusted applications. Note that the delay can be used to send the tag identification number, product information stored on the tag, or a key used to encrypt the previously mentioned data. One can think of many different configurations for the delay. For example, the delay could occur before any actual data is transmitted from the tag to the reader (after which the message would be transmitted normally) or there could be a permanent delay introduced between the bits (bytes, or any other part) of a message being transmitted. In the latter case, a one-time switch can be used to permanently change a fast-readable tag into a slow-readable tag. In what follows, we describe a particular implementation of the above idea.

An RFID built to support these delays could contain three areas of ROM. The first area stores the EPC and product information PI in Erasable ROM (E-ROM), which is fast readable. The second area stores the symmetric encryption of the EPC and the PI, $\mathrm{Enc}_K(\mathrm{EPC}||\mathrm{PI})$, which is also fast readable, while the third area stores the encryption key K, which is slowly readable. Before purchase, the shop can quickly read the EPC and the PI from the E-ROM. When the product is sold, this fast reading path is destroyed or blocked, e.g., by erasing the E-ROM. Thus in an unprotected environment only the value $\mathrm{Enc}_K(\mathrm{EPC}||\mathrm{PIs})$ can be read fast by any reader. Not that this could potentially allow the tracking of the tag via the persistent identifier, $\mathrm{Enc}_K(\mathrm{EPC}||\mathrm{PI})$, but it does not reveal anything about the EPC or the PI, themselves. Finally, in the users home, a trusted device can slowly read the key K, quickly read the encrypted value $\mathrm{Enc}_K(\mathrm{EPC}||\mathrm{PI})$, and store the pairs $(\mathrm{Enc}_K(\mathrm{EPC}||\mathrm{PI}), K)$ in a product database. When product information is needed, the home devices can use the quickly sent value $\mathrm{Enc}_K(\mathrm{EPC}||\mathrm{PI})$ as an identifier to search the database for the key K which can in turn be used to decrypt $\mathrm{Enc}_K(\mathrm{EPC}||\mathrm{PI})$ to give the EPC and the PI. A variation of the above scheme that does not require a switch is shown in Fig. 2. The advantage here is that the EPC$||$PI value is never sent in the clear (even in the shop). In addition, there is no need for erasing or destroying the fast-reading path as in the previous system.

The tags' tracking problem can be solved if the tags are assumed to have more capabilities, namely, a random number generator and the capability to evaluate hash values. This protocol is depicted in Fig. 3. Following the protocol of Fig. 3, we guarantee that only a reader that has time to slowly read the key K (less likely for an attacker) is able to correctly respond to the challenge and learn the value $\mathrm{Enc}_K(\mathrm{EPC}, \mathrm{PI})$. Adding the reader authentication step to the protocol comes at the

1. **Common Input:** Dashed arrows indicate delayed transmission of value.
2. **Tag Input:** The tag has stored in memory $\text{Enc}_{KShop}(K)$, $\text{Enc}_K(EPC\|PI)$, and K.
3. **Honest Reader Input:** An honest shop reader knows the secret key $KShop$.
4. **Protocol in the shop:**

Tag   Reader

$\cdots\cdots\cdots\cdots\; K \;\cdots\cdots\cdots\cdots\blacktriangleright$ ignored

$\underline{\text{Enc}_{KShop}(K)}\!\!\blacktriangleright \;\; \text{Dec}_{KShop}(\text{Enc}_{KShop}(K))$

$\underline{\text{Enc}_K(EPC\|PI)}\!\!\blacktriangleright \;\; \text{Dec}_K(\text{Enc}_K(EPC\|PI))$

5. **Protocol in safe environment:**

Tag   Reader

$\cdots\cdots\cdots\cdots\; K \;\cdots\cdots\cdots\cdots\blacktriangleright$

$\underline{\text{Enc}_{KShop}(K)}\!\!\blacktriangleright$ ignored

$\underline{\text{Enc}_K(EPC\|PI)}\!\!\blacktriangleright \;\; \text{Dec}_K(\text{Enc}_K(EPC\|PI))$

Fig. 2 Delayed tag identification without physical switch

1. **Common Input:** Ability to compute hash functions. Dashed arrows indicate delayed transmission of value. The tag has the ability to generate cryptographically strong random numbers.
2. **Tag Input:** The tag's secret key K.
3. **Protocol:** The protocol involves the exchange of the following messages:

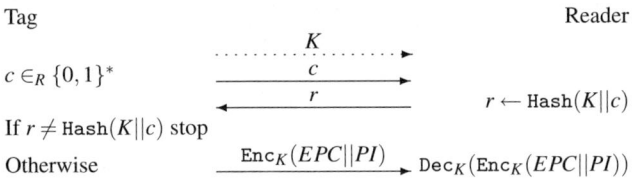

Fig. 3 Delayed tag identification with reader authentication

added cost of requiring hardware to compute hashes, which tends to be expensive as shown in [25]. Finally, another simple variant would have the tag send the EPC and/or the PI at normal speed at the shop and with a delay after the product is sold.

6.2 Discussion

The protocols presented here seem to be well suited for many applications. However, we would like to point out that in any version of the protocol, an attacker is successful if he is able to keep the attacked tags in his reader field long enough to obtain the secret key K. In particular, if the tag is stationary for long periods of time, then the attacker can seriously compromise the privacy of the user. Clearly, then security and usability can be traded off against each other. The longer it takes for the tag to release the next bit of its secret key, the longer the attacker will have to be present in the surrounding of the tag and thus, the less likely that he will obtain the whole secret information. On the other hand, the longer it takes for the tag to release the secret key, the longer that the legitimate user will have to wait when he wants to access the tag's encrypted information at home.[4] Given this limitation, delays appear to be well suited for objects that will not be carried outside the safe environment of the user very often (e.g., food, TVs, home electronics, etc.). On the other hand, tags incorporated into clothing would be less likely to be a privacy problem if using different privacy enhancing solutions such as those based on sensors. We end by noting that the assumption that the attacker tends to be stationary and thus unable to query tags for extended periods of time is not new in the RFID setting (see for example [45]).

7 Conclusions

In this paper, we have discussed and introduced solutions that show how the physics present in RFID systems can be leveraged to enhance security and privacy solutions at a low cost. We believe that this approach is promising in the sense that the cheapest RFID tags are constrained devices which will not allow (due to pricing requirements) the implementation of expensive cryptographic primitives. We point out, as it has been done also in previous works, that the security guarantees provided by *algsics* methods are not the same as those provided by crypto protocols using sophisticated primitives (for example, most algsics solutions provide security in a weak model against passive adversaries). However, it is also true that in many cases such guarantees might be enough. For example, it might not be feasible to implement an active attack without being discovered. Finally, the future might show that algsics solutions turn out to be effective additional countermeasures against attacks. In other words, when combined with other more sophisticated methods, the overall security (or privacy) guarantees of the system are enhanced.

References

1. Auto-ID Center, Massachusetts Institute of Technology, Cambridge, MA 02139-4307, USA. *860 MHz–930 MHz Class I Radio Frequency Identification Tag Radio Frequency & Logical Communication Interface Specification Candidate Recommendation, Version 1.0.1*, November 14th, 2002. Technical Report. Available at http://www.epcglobalinc.org/standards_technology/specifications.html

---

[4] This is only true the first time that the tag is queried at home.

2. Auto-ID Center, Massachusetts Institute of Technology, Cambridge, MA 02139-4307, USA. *13.56 MHz ISM Band Class 1 Radio Frequency Identification Tag Interface Specification: Candidate Recommendation, Version 1.0.0*, February 3rd, 2003. Technical Report. Available at http://www.epcglobalinc.org/standards_technology/specifications.html

3. Auto-ID Center, Massachusetts Institute of Technology, Cambridge, MA 02139-4307, USA. *13.56 MHz ISM Band Class 1 Radio Frequency Identification Tag Interface Specification: Candidate Recommendation, Version 1.0.0*, February 3rd, 2003. Technical Report. Available at http://www.epcglobalinc.org/standards_technology/specifications.html

4. Auto-ID Center, Massachusetts Institute of Technology, Cambridge, MA 02139-4307, USA. *Draft protocol specification for a 900 MHz Class 0 Radio Frequency Identification Tag*, February 23rd, 2003. Available at http://www.epcglobalinc.org/standards_technology/specifications.html

5. L. Batina, J. Guajardo, T. Kerins, N. Mentens, P. Tuyls, and I. Verbauwhede. Public key cryptography for RFID-tags. Printed handout of Workshop on RFID Security – RFIDSec 06, pp. 61–76. ECRYPT Network of Excellence, July 2006. Available at http://events.iaik.tugraz.at/RFIDSec06/Program/index.htm

6. E. Batista. 'Step Back' for Wireless ID Tech? Wired News, April 8th, 2003 Available at http://www.wired.com/news/wireless/0,1382,58385,00.html

7. N. Bird, C. Conrado, J. Guajardo, S. Maubach, G.-J. Schrijen, B. Skoric, A.M.H. Tombeur, P. Thueringer, and P. Tuyls. ALGSICS – Combining Physics and Cryptography to Enhance Security and Privacy in RFID Systems. In F. Stajano, C. Meadows, and S. Capkun, editors, *Security and Privacy in Ad-hoc and Sensor Networks – ESAS 2007*, number 4572 in LNCS, pp. 187–202, Springer, Berlin, 2007

8. L. Bolotnyy and G. Robins. Multi-tag radio frequency identification systems. In *Workshop on Automatic Identification Advanced Technologies – AutoID 2005*, pp. 83–88, 345 E. IEEE, 47th St, New York, NY 10017, USA, October, 2005

9. S. Bono, M. Green, A. Stubblefield, A. Juels, A. Rubin, and M. Szydlo. Security analysis of a cryptographically-enabled RFID device. In P. McDaniel, editor, *USENIX Security Symposium – Security'05*, pp. 1–16, 2005

10. S. Brands and D. Chaum. Distance-bounding protocols (extended abstract). In T. Helleseth, editor, *Advances in Cryptology – EUROCRYPT'93*, volume 765 of LNCS, pp. 344–359, Springer, Berlin, 1994

11. D. Carluccio, T. Kasper, and C. Paar. Implementation details of a multi purpose ISO 14443 RFID-tool. Printed handout of Workshop on RFID Security – RFIDSec 06, pp. 181–197. ECRYPT Network of Excellence, July 2006. Available at http://events.iaik.tugraz.at/RFIDSec06/Program/index.htm

12. D. Carluccio, K. Lemke, and C. Paar. E-passport: the global traceability or how to feel like an UPS package. Printed handout of Workshop on RFID Security – RFIDSec 06, pp. 167–180. ECRYPT Network of Excellence, July 2006. Available at http://events.iaik.tugraz.at/RFIDSec06/Program/index.htm

13. C. Castelluccia and G. Avoine. Noisy tags: A pretty good key exchange protocol for RFID tags. In J. Domingo-Ferrer, J. Posegga, and D. Schreckling, editors, *International Conference on Smart Card Research and Advanced Applications – CARDIS 2006*, volume 3928 of LNCS, pp. 289–299, Tarragona, Spain, April 2006. IFIP, Springer, Berlin

14. H. Chabanne and G. Fumaroli. Noisy cryptographic protocols for low-cost RFID tags. *IEEE Transactions on Information Theory*, 52(8): 3562–3566, August 2006

15. Y. Chan, M.Q.-H. Meng, K.-L. Wu, and X. Wang. Experimental study of radiation efficiency from an ingested source inside a human body model. In *IEEE Annual International Conference of the Engineering in Medicine and Bilogy Society – IEEE-EMBS 2005*, pp. 7754–7757, September 1–4, 2005

16. CS81 Series Standard Cell. 0.18 μm CMOS Technology. Available at http://www.fujitsu.com/downloads/MICRO/fma/pdf/cs81.pdf, 1999

17. S. Dominikus, E. Oswald, and M. Feldhofer. Symmetric authentication for RFID systems in practice. Printed handout of Workshop on RFID and Light-Weight Crypto, pp. 25–31. ECRYPT Network of Excellence, July 13–15, 2005

18. J. Eagle. RFID: The Early Years 1980–1990. Available at `http://members.surfbest.net/eaglesnest/rfidhist.htm`. Website. Updated September 27, 2002

19. D.W. Engels and S. Sarma. Standardization Requirements within the RFID Class Structure Framework. Technical report, Auto-ID Laboratories, Massachusetts Institute of Technology, Cambridge, MA 02139-4307, USA, January 2005. Available at `http://ken.mit.edu/web/`

20. EPCGlobal Inc., Princeton Pike Corporate Center, Suite 202 Lawrenceville, NJ 08648, USA. *EPC^{TM} Generation 1 Tag Data Standards Version 1.1 Rev. 1.27 – Standard Specification*, May 10, 2005. Available at `http://www.epcglobalinc.org/standards_technology/specifications.html`

21. EPCGlobal Inc., Princeton Pike Corporate Center, Suite 202 Lawrenceville, NJ 08648, USA. *EPC^{TM} Radio-Frequency Identity Protocols Class-1 Generation-2 UHF RFID Conformance Requirements – Version 1.0.2*, February 1, 2005. Available at `http://www.epcglobalinc.org/standards_technology/specifications.html`

22. EPCGlobal Inc., Princeton Pike Corporate Center, Suite 202 Lawrenceville, NJ 08648, USA. *EPC^{TM} Radio-Frequency Identity Protocols Class-1 Generation-2 UHF RFID Protocol for Communications at 860 MHz–960 MHz – Version 1.0.9*, January 31, 2005. Available at `http://www.epcglobalinc.org/standards_technology/specifications.html`

23. EPCGlobal Inc., Princeton Pike Corporate Center, Suite 202 Lawrenceville, NJ 08648, USA. *EPC^{TM} Radio-Frequency Identity Protocols Class-1 Generation-2 UHF RFID Protocol for Communications at 860 MHz-960 MHz – Version 1.0.9*, January 31, 2005. Available at `http://www.epcglobalinc.org/standards_technology/specifications.html`

24. EPCGlobal Inc., Princeton Pike Corporate Center, Suite 202 Lawrenceville, NJ 08648, USA. *EPCglobal tag Data Standards Version 1.3. Ratified Specification*, March 8, 2006. Available at `http://www.epcglobalinc.org/standards/EPCglobal_Tag_Data_Standard_TDS_Version_1.3.pdf`

25. M. Feldhofer and C. Rechberger. A case against currently used hash functions in RFID protocols. Printed handout of Workshop on RFID Security – RFIDSec 06, pp. 109–122. ECRYPT Network of Excellence, July 2006

26. M. Feldhofer, S. Dominikus, and J. Wolkerstorfer. Strong authentication for RFID systems using the AES algorithm. In M. Joye and J.-J. Quisquater, editors, *Cryptographic Hardware and Embedded Systems – CHES 2004*, volume 3156 of LNCS, pp. 357–370, Springer, Berlin, 2004

27. K.P. Fishkin, S. Roy, and B. Jiang. Some methods for privacy in RFID communication. In C. Castelluccia, H. Hartenstein, C. Paar, and D. Westhoff, editors, *Security in Ad-hoc and Sensor Networks – ESAS 2004*, volume 3313 of LNCS, pp. 42–53. Springer, Berlin, 2005

28. C. Floerkemeier, R. Schneider, and M. Langheinrich. Scanning with a purpose – supporting the fair information principles in RFID protocols. In H. Murakami, H. Nakashima, H. Tokuda, and M. Yasumura, editors, *International Symposium on Ubiquitous Computing Systems – UCS 2004*, volume 3598 of LNCS, pp. 214–231, Tokyo, Japan, Springer, Berlin, November 2004

29. G. Hancke and M. Kuhn. An RFID distance bounding protocol. In *Conference on Security and Privacy for Emerging Areas in Communication Networks – SecureComm 2005*, pp. 67–73. IEEE Computer Society, September 2005

30. E. Haselsteiner and K. Breitfuss. Security in near field communication (NFC). Printed handout of Workshop on RFID Security – RFIDSec 06, pp. 151–166. ECRYPT Network of Excellence, July 2006

31. ICC Policy Statement: The fight against piracy and counterfeiting of intellectual property. Submitted to the 35th World Congress, Marrakech, Document no 450/986, ICC, June 1, 2004

32. S. Inoue and H. Yasuura. RFID privacy using user-controllable uniqueness. RFID Privacy Workshop, November 2003

33. International Organization for Standardization, Geneva, Switzerland. *ISO/IEC 11785:1996 – Radio frequency identification of animals – Technical concept*, October 15, 1996

34. International Organization for Standardization, Geneva, Switzerland. *ISO/IEC 14443-2 – Identification cards–Contactless integrated circuit(s) cards–Proximity cards–Part 2: Radio frequency interface power and signal interface*, September 14, 2000. Final Draft

35. International Organization for Standardization, Geneva, Switzerland. *ISO/IEC 14443-3 – Identification cards – Contactless integrated circuit(s) cards–Proximity cards – Part 3: Initialization and anticollision*, January 13, 2000 Final Draft

36. International Organization for Standardization, Geneva, Switzerland. *ISO/IEC 15693-2:2000 – Identification cards – Contactless integrated circuit(s) cards – Vicinity cards – Part 2: Air interface and initialization*, May 1, 2000

37. International Organization for Standardization, Geneva, Switzerland. *ISO/IEC 10536-3:1996 – Identification cards – Contactless integrated circuit(s) cards – Part 3: Electronic signals and reset procedures*, August 13, 2001

38. International Organization for Standardization, Geneva, Switzerland. *ISO/IEC 15693-3:2001 – Identification cards – Contactless integrated circuit(s) cards – Vicinity cards – Part 3: Anticollision and transmission protocol*, April 1, 2001

39. International Organization for Standardization, Geneva, Switzerland. *ISO/IEC 18000-2:2003(E)-2 – Information technology – Radio frequency identification for item management – Part 2: Parameters for air interface communications below 135 kHz*, November 26, 2003

40. International Organization for Standardization, Geneva, Switzerland. *ISO/IEC 18000-3:2003(E) – Information technology – Radio frequency identification for item management – Part 3: Parameters for air interface communications at 13,56 MHz*, February 13, 2003

41. International Organization for Standardization, Geneva, Switzerland. *ISO/IEC 18000-4:2003(E) – Information technology – Radio frequency identification for item management – Part 4: Parameters for air interface communications at 2.45 GHz.*, March 25, 2003. Working document

42. International Organization for Standardization, Geneva, Switzerland. *ISO/IEC 18000-6:2003(E) – Information technology – Radio frequency identification for item management – Part 6: Parameters for air interface communications at 860 MHz to 960 MHz*, November 26, 2003

43. International Organization for Standardization, Geneva, Switzerland. *ISO/IEC 18000-7 – Information technology – Radio frequency identification for item management – Part 7: Parameters for active air interface communications at 433 MHz*, September 30, 2003. Working document

44. International Organization for Standardization, Geneva, Switzerland. *ISO/IEC 11784:1996 – Radio frequency identification of animals – Code structure*, August 15, 2004

45. A. Juels. Minimalist cryptography for low-cost RFID tags. In C. Blundo and S. Cimato, editors, *Security in Communication Networks – SCN 2004. Revised Selected Papers*, volume 3352 of LNCS, pp. 149–164. Springer, Berlin, September 8–10, 2004

46. A. Juels. RFID Security and privacy: A research survey. *IEEE Journal on Selected Areas in Communications*, 24(2): 381–394, February 2006. Extended version available from http://www.rsasecurity.com/rsalabs/node.asp?id=2029

47. A. Juels and J.G. Brainard. Soft blocking: flexible blocker tags on the cheap. In V. Atluri, P.F. Syverson, and S. De Capitani di Vimercati, editors, *ACM Workshop on Privacy in the Electronic Society – WPES 2004*, pp. 1–7, ACM Press, New York, NY, October 28, 2004

48. A. Juels and R. Pappu. Squealing Euros: Privacy Protection in RFID-Enabled Banknotes. In R.N. Wright, editor, *Financial Cryptography – FC'03*, volume 2742 of LNCS, pp. 103–121, IFCA, Springer, Berlin, January 2003

49. A. Juels and S.A. Weis. Authenticating pervasive devices with human protocols. In V. Shoup, editor, *Advances in Cryptology – CRYPTO 2005*, volume 3126 of LNCS, pp. 293–308, Springer, Berlin, August 2005

50. A. Juels, R.L. Rivest, and M. Szydlo. The blocker tag: selective blocking of RFID tags for consumer privacy. In S. Jajodia, V. Atluri, and T. Jaeger, editors, *ACM Conference on Computer and Communications Security – CCS 2003*, pp. 103–111, ACM Press, New York, NY October 27–30, 2003

51. A. Juels, R. Pappu, and S. Garfinkel. RFID privacy: An overview of problems and proposed solutions. *IEEE Security and Privacy*, 3(3): 34–43, May/June 2005. Extended version available from http://www.rsasecurity.com/rsalabs/node.asp?id=2029

52. A. Juels, P. Syverson, and D. Bailey. High-power proxies for enhancing RFID privacy and utility. In G. Danezis and D. Martin, editors, *Privacy Enhancing Technologies – PET 2005*, volume 3856 of LNCS, pp. 210–226, Springer, Berlin, 2005

53. G. Karjoth and P. Moskowitz. Disabling RFID tags with visible confirmation: Clipped tags are silenced. In *Workshop on Privacy in the Electronic Society – WPES*, Alexandria, Virginia, USA, ACM, ACM Press, New York, NY, November 2005

54. T. Karygiannis, B. Eydt, G. Barber, L. Bunn, and T. Phillips. *Draft Special Publication 800-98, Guidance for Securing Radio Frequency Identification (RFID) Systems*. National Institute for Standards and Technology, Gaithersburg, MD, USA, September 2006. Available for download at http://csrc.nist.gov/

55. H. Kitayoshi and K. Sawaya. Long range passive RFID-tag for sensor networks. In *IEEE 62nd Vehicular Technology Conference – VTC-2005*, pp. 2696–2700, IEEE Computer Society, Los Alamitos, CA, USA, 25–28 Sept, 2005

56. KU Information & Telecommunication Technology Center. The University of Kansas. UHF KU-RFID Tag, 2006. Available at http://www.rfidalliancelab.org/publications/ittc_press_release.shtml

57. S.S. Kumar and C. Paar. Are standards compliant elliptic curve cryptosystems feasible on RFID? Printed handout of Workshop on RFID Security – RFIDSec 06, pp. 41–60. ECRYPT Network of Excellence, July 2006. Available at http://events.iaik.tugraz.at/RFIDSec06/Program/index.htm

58. J. Landt. Shrouds of Time – The History of RFID. Whitepaper, AIM Inc., October 1, 2001. Available at http://www.transcore.com/pdf/AIM%20shrouds_of_time.pdf

59. T.C. May. Timed-release crypto. Posting to the Cypherpunks Mailing List, February 10, 1993. Available at http://cypherpunks.venona.com/date/1993/02/msg00129.html

60. J. Munilla, A. Ortiz, and A. Peinado. Distance bounding protocols with void-challenges for RFID. Printed handout of Workshop on RFID Security – RFIDSec 06, pp. 15–26. ECRYPT Network of Excellence, July 2006

61. National Institute for Standards and Technology, Gaithersburg, MD, USA. *FIPS 197: Advanced Encryption Standard (AES)*, November 2001. Available for download at http://csrc.nist.gov/encryption

62. M. Ohkubo, K. Suzuki, and S. Kinoshita. Cryptographic approach to "privacy-friendly" tags. In *RFID Privacy Workshop*, MIT, Cambridge, MA, USA, November 2003. Available at http://lasecwww.epfl.ch/~gavoine/rfid/

63. K. Opasjumruskit, T. Thanthipwan, O. Sathusen, P. Sirinamarattana, P. Gadmanee, E. Pootarapan, N. Wongkomet, A. Thanachayanont, and M. Thamsirianunt. Self-powered wireless temperature sensors exploit RFID technology. *IEEE Pervasive Computing*, 5(1): 54–61, Jan.–March 2006

64. P. Peris-Lopez, J.C. Hernandez-Castro, J. Estevez-Tapiador, and A. Ribagorda. LMAP: A real lightweight mutual authentication protocol for low-cost RFID tags. Printed handout of Workshop on RFID Security – RFIDSec 06, pp. 137–148. ECRYPT Network of Excellence, July 2006. Available at http://events.iaik.tugraz.at/RFIDSec06/Program/index.htm

65. M. Philipose, J.R. Smith, B. Jiang, A. Mamishev, R. Sumit, and K. Sundara-Rajan. Battery-free wireless identification and sensing. *IEEE Pervasive Computing*, 4(1): 37–45, Jan–March 2005

66. T. Phillips, T. Karygiannis, and R. Kuhn. Security standard for the rfid market. *IEEE Security and Privacy*, 3(6): 85–89, November–December 2005

67. S. Radovanovic, A.J. Annema, and B. Nauta. High-speed lateral polysilicon photodiode in standard CMOS technology. In *33rd European Solid-State Circuits Conference – ESSDERC'03*, pp. 521–524. IEEE Computer Society, 16–18 Sept. 2003

68. D.C. Ranasinghe, D.W. Engels, and P.H. Cole. Low-cost RFID systems: Confronting security and privacy. In *Auto-ID Labs Research Workshop*, Zurich, Switzerland, September 2004

69. RFID Journal. RFID Tag Market in Flux. Available at http://www.rfidjournal.com/article/articleview/971/1/1/, June 2004

70. RFID Journal. A Summary of RFID Standards. Available at http://www.rfidjournal.com/article/articleview/1335/1/129/, 2005

71. M. Rieback, B. Crispo, and A. Tanenbaum. RFID guardian: A battery-powered mobile device for RFID privacy management. In C. Boyd and J.M. González Nieto, editors, *Australasian Conference on Information Security and Privacy – ACISP'05*, volume 3574 of LNCS, pp. 184–194, Brisbane, Australia, Springer, Berlin, July 2005

72. R.L. Rivest. Chaffing and winnowing: Confidentiality without encryption. *CryptoBytes*, 4(1): 12–17, Summer 1998

73. K. Sakiyama, L. Batina, N. Mentens, B. Preneel, and I. Verbauwhede. Small-footprint ALU for public-key processors for pervasive security. Printed handout of Workshop on RFID Security – RFIDSec 06, pp. 77–88. ECRYPT Network of Excellence, July 2006. Available at http://events.iaik.tugraz.at/RFIDSec06/Program/index.htm

74. S. Sarma. Towards the 5c Tag. White paper mit-autoid-wh-006, Auto-ID Center, Massachusetts Institute of Technology, Cambridge, MA 02139-4307, USA, November 1, 2001. Distribution restricted to sponsors until February 1, 2002

75. S. Sarma. Some issues related to RFID and security. Introductory Talk – RFIDSec 06, July 2006. Available at http://events.iaik.tugraz.at/RFIDSec06/Program/index.htm

76. S. Sarma and D.W. Engels. On the Future of RFID Tags and Protocols. Technical report mit-autoid-tr-018, Auto-ID Center, Massachusetts Institute of Technology, Cambridge, MA 02139-4307, USA, June 1st, 2003. Early Released July 2003. Available at http://www.epcglobalinc.org/standards_technology/specifications.html

77. S. Sarma, S. Weis, and D. Engels. Radio-frequency identification: Security risks and challenges. *Cryptobytes*, 6(1): 2–9, Winter/Spring 2003. Available at http://www.rsasecurity.com/rsalabs/

78. A. Soppera and T. Burbridge. Off by default – RAT: RFID acceptor tag. Printed handout of Workshop on RFID Security – RFIDSec 06, pp. 151–166. ECRYPT Network of Excellence, July 2006

79. T. Staake, F. Thiesse, and E. Fleisch. Extending the EPC network – The potential of RFID in anti-counterfeiting. In A. Omicini H. Haddad, L.M. Liebrock and R.L. Wainwright, editors, *ACM Symposium on Applied Computing – SAC 2005*, pp. 1607–1612. ACM Press, New York, NY, March 13–17, 2005

80. F. Stajano and R.J. Anderson. The resurrecting duckling: Security issues for ad-hoc wireless networks. In B. Christianson, B. Crispo, J.A. Malcolm, and M. Roe, editors, *Security Protocols Workshop*, volume 1796 of LNCS. Springer, Berlin, April 19–21, 2000

81. C. Swedberg. DHL Expects to Launch "Sensor Tag" Service by Midyear. *RFID Journal*. Available at http://www.rfidjournal.com/article/articleprint/2986/-1/1/, January 19th, 2007

82. K. Takaragi, M. Usami, R. Imura, R. Itsuki, and T. Satoh. An ultra small individual recognition security chip. *IEEE Micro*, 21(6): 43–49, November–December 2001

83. TSMC Advanced Technology Overview. Available at http://www.tsmc.com/download/english/a05_literature/Advanced_Technology_Overview_Brochure_2006.pdf, May 2006

84. TSMC Standard Cell Libraries. Available at http://www.cadence.com/datasheets/4456_TSMC_SC_ds.pdf

85. P. Tuyls and L. Batina. RFID-tags for anti-counterfeiting. In D. Pointcheval, editor, *Topics in Cryptology–CT-RSA 2006*, volume 3860 of LNCS, pp. 115–131. Springer, Berlin, February 13–17 2006

86. S. Weis. Security and privacy in radio-frequency identification devices. Master Thesis, Massachusetts Institute of Technology (MIT), Massachusetts, USA, May 2003

87. S.A. Weis, S.E. Sarma, R.L. Rivest, and D.W. Engels. Security and privacy aspects of low-cost radio frequency identification systems. In D. Hutter, G. Müller, W. Stephan, and M. Ullmann, editors, *First International Conference on Security in Pervasive Computing – SPC 2003*, volume 2802 of LNCS, pp. 201–212. Springer, Berlin, March 2003

88. C.C. Zou. PCB: Physically Changeable Bit for Preserving Privacy in Low-End RFID Tags. RFID White Paper Library, *RFID Journal*, May 2006

RFID Anticounterfeiting: An Architectural Perspective

Tieyan Li* and Tong-Lee Lim

Abstract Counterfeit goods have always been an enormous threat to the world economy, but they could potentially be combatted by employing an emerging technology – Radio Frequency IDentification (RFID) – in the near future. In this chapter, we present an architectural perspective on RFID-based anticounterfeiting solutions. An overview of an RFID-enabled anticounterfeiting system is described and analyzed. In the end system, we emphasize on the importance of a secure binding between the target object and RFID tag, as well as the security of the mutual authentication protocol between the RFID tag and reader. As for the backend system, we describe the closed-loop systems that are deployed in the pharmaceutical industry, and the open-loop solution specified by the EPCglobal committee. On building practical and cost-effective anticounterfeiting solutions in realistic environments, we can learn and gain valuable experience from the current efforts put into RFID pilots. While we may potentially be facing more challenges ahead of us, we are optimistic that with advancements in RFID technology, better and more complete solutions toward anticounterfeiting can be provided.

1 Introduction

The International Chamber of Commerce estimates that 7% of the global world trade is in counterfeit goods, with the counterfeit market being worth approximately US$600 billion annually. Existing anticounterfeiting mechanisms, such as holograms, smart cards, biometric markers and inks, represent a flexible portfolio of

T. Li
Cryptography and Security Department, Institute for Infocomm Research (I^2R),
1 Fusionopolis Way, #21-01 Connexis, Singapore 138632
e-mail: litieyan@i2r.a-star.edu.sg

P. Kitsos, Y. Zhang (eds.), *RFID Security: Techniques, Protocols and System-on-Chip Design*, © Springer Science+Business Media, LLC 2008

solutions against counterfeiting. More recently, RFID was reportedly used in anti-counterfeiting solutions to achieve a higher degree of automation when checking the originality of a product. For example, Euro banknotes are attached with RFID chips to combat counterfeiting by European Central Bank. The United States Food and Drug Administration (US FDA) has issued a report that endorses RFID as a tool to combat counterfeiting of pharmaceuticals. So far, these RFID-based solutions seem pretty promising [20]. With wide adoption of RFID technology witnessed in various industries, the future of RFID in anticounterfeiting looks optimistic.

The main objective of an anticounterfeiting solution is to distinguish a genuine product from a fake one. The basic concept of applying RFID to anticounterfeiting lies in its original function of *identification*. Imagine a scenario in the future, in which every object will be attached with an RFID tag that contains a unique number belonging to the object. Once the tag is interrogated, the unique object number is emitted and interpreted by the back-end system to identify the object. If, for instance, all the unique object numbers are stored in a database, we can then check the database to verify the identity of the object. Unfortunately, identification alone is insufficient for solving the anticounterfeiting problem. Problems exist in such a straightforward solution. For example, the unique object number can be eavesdropped and copied onto blank tags to produce clones, and the database would not be able to distinguish a legitimate tag from a cloned tag containing the same object number. There are many other ways to attack such a simplified identification system. For example, a counterfeiter can remove a tag from an authentic product, perform reverse engineering on the tag to extract out key attributes, and replicate these attributes onto black tags.

In fact, anticounterfeiting has stronger requirements on security and needs a more complex system to implement. RFID-based anticounterfeiting solutions leverage on the benefits provided by the RFID tags and the back-end information system within the RFID-enabled production and distribution flow. RFID tags can have certain security functions implemented in them, which raises the barrier for counterfeiting them. Furthermore, a counterfeiter would now need to counterfeit both the product and the tag, which raises his costs for counterfeiting. The back-end information system assists in drawing and maintaining real-time profile over the movements and activities of goods, thereby facilitating fast tracking of the goods. Essentially, a simplified anticounterfeiting system could consist of the following components – the object that is to be protected, the RFID tag that is attached onto the object, the RFID reader and the back-end system. Figure 1 depicts the components in a generic RFID-enabled anticounterfeiting system.

Traditional anticounterfeiting methods rely on optical technologies such as watermarks, holograms, and microprinting to authenticate and verify goods. Other more advanced methods include the use of biological, chemical, or even nanotechnologies (e.g., using DNA markers, nanolevel material characteristics, etc.). RFID technology, with the use of RFID tags that are attached to goods, opens up a new way to authenticate products. Like optical solutions, RFID technology authenticates the information stored on an external object (the RFID tag) rather than the product itself. If the RFID tag is authenticated, we claim that the product is authenticated too.

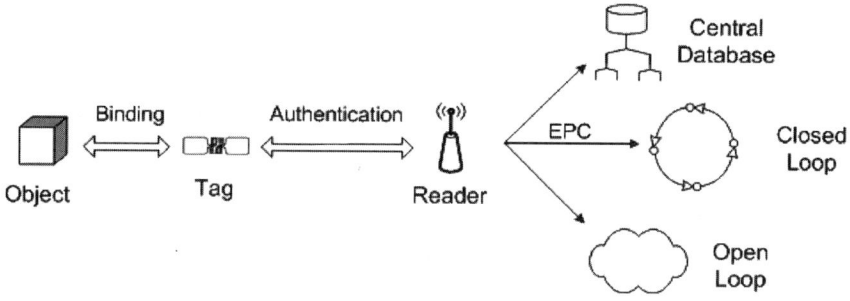

Fig. 1 Components in an RFID anticounterfeiting system

To ensure the effectiveness of such a solution, the RFID tag needs to be securely bound to the product. Some secure binding mechanisms that are used in RFID systems will be discussed in greater detail in Sect. 3.

The authentication of an RFID tag is carried out through interactions with an RFID reader. RFID tag-to-reader authentication protocols resemble much of the existing two party authentication protocols based on challenge–response. In fact, a large number of research works conform to this principle and rest on symmetric or public key cryptographic primitives. We summarize these solutions in Sect 4. Unfortunately, these solutions do not provide a practical solution in realistic anticounterfeiting scenarios. This is because most RFID tags (for example, those being used on fast moving consumer goods) are too cheap to incorporate even lightweight cryptographic primitives. Currently, there exists a gap between what needs to be implemented for a substantial level of security on the tag and what could be realistically supported on the tag. Achieving proper authentication with low-cost RFID tags is still very much a highly challenging task.

Besides the secure binding of an RFID tag to an object and the authentication between an RFID tag and a reader in the end system, another area that needs to be considered for a more complete anticounterfeiting solution is that of the back-end system. In a supply chain, as the goods are moved from one part of the world to another, many different activities can be taking place at each intermediate point. In fact, each intermediate point could potentially represent a point of vulnerability, where counterfeiting behavior might exist. Hence, in addition to checking at the end points, checks may need to be conducted at each intermediate point as well. This requires a systematic back-end support that connects itself to all the intermediate points. The simplest back-end system is a single standalone database that records up-to-date information on the goods by collecting data at each intermediate point. A verifier can then check the database for the details and/or status (e.g., ID, some stored secret, current location, history, etc.) of a particular product, and based on this knowledge, determine the authenticity of the product. With a powerful database, there is a high chance that even a perfectly cloned tag can be detected. However, collecting and collating all relevant information into one single database is rather ambitious and unlikely to be scalable. How to disseminate these information into decentralized locations is still very much a big challenge that implementors have to face. In this chapter, two possible classes of solutions – closed loop and open loop solutions – are examined.

Anticounterfeiting solutions may be customized for different application scenarios by considering hybrids involving the closed loop solution and the open loop solution. For example, an E-pedigree solution for combating counterfeit drugs is promoted and piloted as a major anticounterfeiting effort of the US FDA. The potential high risk of drug misuse and increasing market of counterfeit drugs are the main drivers of this countermeasure. In general, for an anticounterfeiting solution to be feasible, the cost of implementing the solution must be lower than the losses suffered due to counterfeiting activities. Moreover, the cost of breaking the system should be high in order to provide a substantial barrier against counterfeiting behavior. Hence, when customizing an anticounterfeiting solution, we need to consider the cost-effectiveness of the customizations. Challenges arise when we face dynamic and complex application environments, such that each of them requires a different security level. In such cases, it would be difficult to design an optimal solution that fits all the requirements.

The rest of this chapter is organized as follows: In Sect. 2, we discuss the ideal solutions that are expected by the end customers. Section 3 focuses on the secure binding of an RFID tag to the target object. In Sect. 4, we elaborate on relevant works that deal with RFID authentication protocols. Then, we present two network level solutions in Sects. 5 and 6. Finally, we conclude the chapter with some ending remarks.

2 Ideal Solutions

In this section, we discuss the two extreme scenarios for ideally identifying a counterfeit object. We study the pros and cons of these solutions and try to find a good balance between these two extremes. The factors that were considered when measuring the performance of each solution include the costs on the tag, the reader (on checking), and/or the back-end system vs. the cost of breaking the entire anticounterfeiting system.

2.1 Scenario One: Offline Object Authentication

We first consider a completely offline RFID object authentication scenario. An RFID-tagged product is dispatched by the manufacturer, distributed along the supply chain, and finally comes under possession by an end user. The end user would verify the product with an authentication device. The device is standalone and has no online network support. That means that the end user can only rely on this device to authenticate the product and make a judgment based on the authentication result reported by the device:

1. *Checking the physical binding.* The verifier visually checks whether the packaging over the product and tag has been tampered with. He does the same for the RFID tag and the product itself. Any traces of tampering witnessed on the packaging, tag, or product should cause the authentication to fail.
2. *Checking electronic binding (optional).* The unique features of the product could be stored on the authentication device, with the tag attached to the product storing a copy of the features as well. The verifier can then check whether the features stored in the tag match with those stored on the device. A mismatch implies a potential attack (e.g., a counterfeiter reapplying a legitimate tag on a fake product) could have occurred on the product. In this case, we note that a digital signature should be used to ensure integrity and authenticity of the stored information.
3. *Checking authenticity of tag.* The verifier scans the tag to obtain its ID and takes part in a challenge–response authentication protocol to prove that the tag owns some shared secret. As long as the secret on the tag is not disclosed, the authenticity of the tag is guaranteed. This can resist certain copycat attacks where all data except the secret of the tag is cloned on another tag.

In such a solution, the requirements on the authentication device are high. The device integrates a combination of functions including cryptographic algorithms, physical feature extraction functions, and huge memory to store the relevant information for all tags. The end system is expensive due to the cost of tag, the binding, and the authentication device. However, the system/network overhead is rather low in this ubiquitous setting.

2.2 Scenario Two: Centralized Database Checking

Next, we consider the online centralized solution, where the verifier relies on a back-end database to check the authenticity of the product. The tag on the product has no resources to support cryptographic calculations, thus it probably cannot be authenticated. A verifier uses a typical RFID reader to scan the EPC code of the tag and request the back-end system to perform a plausibility check:

1. *Checking the identity of tag.* The verifier scans the EPC code of the RFID tag. According to EPCglobal network, the verifier can retrieve the product information from the manufacturer site. If the retrieved information matches the printed information on the product, the tag can be identified successfully. Otherwise, the check fails.
2. *Checking the authenticity of tag.* Since the tag cannot authenticate itself, the verifier sends a request to the product track and trace service provided by the manufacturer site and expects a reply.
3. *A plausibility check.* Suppose the manufacturer maintains all the product's activities during its life cycle, it can check the history of the product to see whether the information in the new request is logically sound (e.g., a drug that is mandated to be sold only in US should not be available in South Africa). The result of this check is sent back to the verifier.

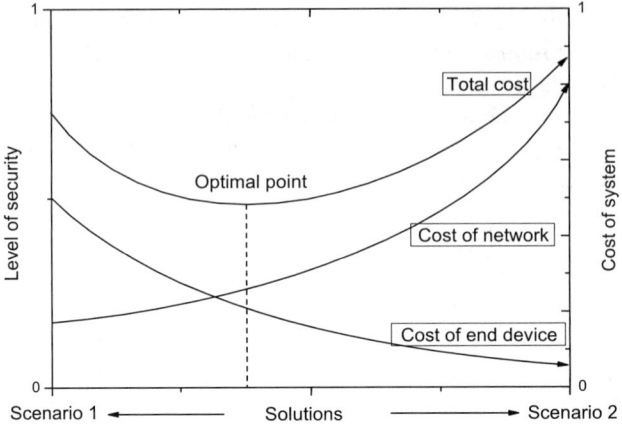

Fig. 2 Cost vs. security level of an RFID anticounterfeiting system

The cost of the end system is relatively low, while the cost of maintaining the centralized back-end database is extremely high. The collection and analysis of the status information of a tag is not likely to be easy. Moreover, defining the granularity of the information collected is also important. Other challenging issues include the sensitivity and/or privacy of the data that is to be shared and the requirement for protecting against a single point of failure. Figure 2 provides a diagram that roughly depicts the various costs incurred by an anticounterfeiting system. To achieve a certain level of security, tradeoffs involving higher costs may have to be considered. Referring to Fig. 2, the system designing process may involve trying to achieve an optimal point, where both the costs of the end devices (which include the tag, the reader and the binding) and the network are not too high, and the total cost of the entire system is minimized.

3 Secure Binding between Tag and Object

An RFID-enabled anticounterfeiting system typically authenticates the RFID tag attached to the product, instead of the product itself. Hence, the authenticity of the product can only be ensured if the RFID tag is securely bound to the product and is not tampered with. There are generally two categories of secure binding – physical binding and electronic binding.

Physical binding (packaging) refers to the use of physical means (which may involve the use of mechanical or chemical mechanisms) to pack the RFID tag with the product tightly so that the binding is either impossible to be tampered with (tamper-resistant) or leaves clear evidence when the it has been tampered with (tamper-evident). An example of such binding is the electronic seal used to guarantee the integrity of containers [13]. Secure physical binding is used to defend against attacks based on removal and reattachment of RFID tags.

Electronic binding refers to methods in which the unique fingerprint of a product is stored on the RFID tag. During authentication, an authentication device would be used to regenerate the fingerprint and compare it with the value stored on the RFID tag. The fingerprint is typically signed by the manufacturer of the product and can be verified by the authentication device. The digital signature guarantees the authenticity of the product, but not the authenticity of the tag, since the fingerprint, together with its signature, can be skimmed and copied onto other tags. It is possible that the cloned tag not only contains a part of authentic information, but also some other misleading information about this product. Thus, it is natural to bind the RFID tag with the product using methods proposed in [17] (the secure binding of object unique feature on tags) and [16] (the integration of tags on machine readable documents).

In [17], the authors proposed a method of secure binding that is achieved by signing on the unique features of the product, as well as that of the attached tag. For the tag, the Tag (or Transponder) IDentification number (TID) was used as the unique feature. The TID is essentially a globally assigned unique number that is programmed onto the tag by the chip manufacturer and set to a "locked" state. One cannot easily "unlock" the state and change the TID, although dedicated attackers might break it with some invasive attacks. The EPC is another globally assigned unique number for a specific product, but it is written by the product manufacturer and can be erased and overwritten with another EPC so that the tag can be reused. In short, it is easy to clone the EPC, but difficult to clone the TID [1]. Hence, we consider the TID to be a good authenticator of an RFID tag that can be used to tighten the binding proposed in [17].

Here, we stress that there is no such thing as absolute security. All security measures can very likely be broken given the time and resources. Nonetheless, for an anticounterfeiting solution to provide "good enough security", it should guarantee cost-effectiveness in preventing and detecting massive counterfeits in a timely manner. For the products that require very high level of security, strict security design techniques should be used and stringent tests and analyses should be carried out on those techniques before they can be put to deployment.

4 Tag-to-Reader Authentication Protocol

The RFID security research community has been paying a lot of attention on RFID authentication, with numerous works on privacy-enhanced authentication protocols having been proposed in existing literature. For the case of anticounterfeiting, privacy could be a counteract property, as it typically acts against tracking and tracing of products. We thus focus on the authenticate-only protocols, i.e., authentication without privacy enhancement. Basically, these protocols involve the use of some secret information, although not all RFID tags, for example EPC class 1 tags [6], can be authenticated this way due to their inability to store and protect the secret information. We focus our attention on tags that come with the capability to store some secret values, and we categorize these tags into three different classes based on the resources available on them – namely Crypto-tag, Light-tag, and Gen2-tag. Crypto-tags support classic cryptographic primitives and hence, traditional authentication schemes can be applied here. Light-tags cannot perform cryptographic functions, but can conduct bitwise operations such as XOR. Gen2-tags conforming to the EPC Class 1 Generation 2 specification, which can only perform 16-or 32-bits bitwise operations and are embedded with 16-bit PRNG and CRC functions.

4.1 Authentication with Classic Cryptographic Primitives

The objective of such an authentication protocol is for the RFID reader to verify whether a Crypto-tag knows some secret key that is shared between the reader and the tag. The reader first sends a challenge to the tag. The tag uses the challenge and its secret key as inputs to some cryptographic function and computes a result, which is returned to the reader as a response to the challenge. The response will then be checked by the reader for verification. If the reader needs to authenticate a lot of tags, it has to store the IDs and secrets of all these tags, which is not scalable.

With regards to Crypto-tags, one widely adopted assumption is that these tags can support a one-way hash function. The very first approach of using hash function was the "hash lock" scheme, proposed by Sarma et al. [19]. Following that, numerous RFID authentication protocols based on hash function have been proposed. Besides these hash-based solutions, there were other solutions that require a Pseudorandom Function (PRF) on a tag or make use of symmetric ciphers instead of hash functions. Another work [14] even assumed the use of public key cryptographic primitives, in which tags update their IDs with a re-encryption scheme. Although public key cryptography can reduce the key management overload, it is still too heavy to be implemented on medium-cost Crypto-tags. A fair comparison in terms of power consumption, chip area, and clock cycles on the implementations of some standardized cryptographic algorithms (e.g., SHA-256, SHA-1, MD5, AES-128, and ECC-192) on passive RFID tags is presented in [10].

In addition, there are some existing or on-going research efforts that lead to ultra-lightweight cipher designs. For example, the block cipher PRESENT-80 [2] features a compact implementation of only $1,570$ GEs. Comparable lightweight stream ciphers, like Grain, has about $1,300$ GEs [12]. More efficient hardware/software stream cipher designs are proposed and evaluated (currently within the ECRYPT project) for minimal footprint hardware implementation even in low-cost RFID tags.

4.2 Authentication Without Classic Cryptographic Primitives

Light-tags are restricted to a much lower gate count (less than hundreds of GEs) for the implementation of security features than Crypto-tags. Some authentication schemes that do not rely on assumptions on classic cryptographic primitives have been proposed so that they can be supported on such tags.

The HB family of RFID authentication protocols. In [23], Weis introduced the Hopper and Blum Protocol (HB) under the RFID setting. Subsequently, Juels and Weis proposed a lightweight authentication protocol (HB$^+$) in [15]. The security of both the HB and HB$^+$ protocols are based on the Learning Parity with Noise (LPN) problem, whose hardness over random instances remains as an open question. However, Gilbert et al. showed that HB$^+$ is not secure against a simple man-in-the-middle attack [11]. To defend against such active attacks, Bringer et al. extended the protocols to HB^{++} protocol [3].

The ultra-lightweight RFID authentication protocols. In [22], Vajda and Buttyan presented a set of extremely lightweight challenge–response authentication protocols that are suitable for authenticating tags, but their protocols can be broken by a powerful adversary as was shown in [4]. Besides this, there are a number of approaches employing existing or self-designed mathematical primitives to build ultra-lightweight mutual authentication protocol for low-cost RFID tags. Unfortunately, almost all such lightweight protocols are being attacked in one way or another, and their practical deployment could be at risk unless strict security analysis is conducted beforehand.

4.3 Authentication Based on Gen2 Functions

Some approaches, conforming to EPC Gen2 specifications [7] that rely solely on the specified functions like 16-bit CRC and PRNG, have also been proposed. In [5], Duc et al.'s authentication protocols used 16-bit PRNG, CRC and XOR operations to replace the 128-bit strong cryptographic PRNG and MAC functions. But the penalty of the replacement is the reduced (perhaps better than nothing) security. Thus far, all of the authentication protocols based on Gen2 functions are vulnerable even under a weak security model. Obviously, Gen2 tags provide almost no security at this moment, but the security issues are being investigated and improved in the next generation (Gen3) specification. With the fast development of lightweight cryptographic research and semiconductor technologies, we are optimistic on expecting lower cost and stronger security RFID tags being massively produced in the near future.

5 Legacy Closed-Loop Anticounterfeiting Solution

In Sect. 2, we described an extreme case of authentication involving an authentication device in the end system that can authenticate the tag and product without any online support. Such an ideal solution requires a secure binding between the RFID tag and the product (Sect. 3) and the tags must be capable of taking part in an authentication protocol (for example, the Crypto-tags in Sect. 4.1). The high cost of such an end system limits its application to supporting high-value products only. For ordinary products, a more economical anticounterfeiting solution would have to be used and the cost-effectiveness of the solution has to be weighed carefully. To support high-volume usage, the item-level tags for ordinary products would have to be extremely low-cost and thus, it is unlikely that there would be sufficient resources to support security features.

Even when Crypto-tags are used, these tags could still be compromised by side channel attacks [18]. Hence, under some circumstances, there might be a need to rely on a back-end system for stronger authentication. This gives rise to the other extreme case described in Scenario 2 of Sect. 2, where a central database dominantly grasps all product information. The database monitors all activities of a product and based on these information, make decisions over the authenticity of the product. This, however, is another imperfect solution since it does not scale well. Moreover, such a solution can potentially suffer from Distributed Denial of Service (DDoS) attacks and result in a single point of failure. In what is to follow, we shall study some distributed anticounterfeiting solutions that are practical, economical, and reasonably scalable.

One good example of such a distributed solution is the existing *E-pedigree* solution in pharmaceutical supply chain. Initially promoted by the US FDA, the E-pedigree specification was then ratified by EPCglobal in the beginning of year 2007 [8]. The purpose of the new standard is to provide the pharmaceutical supply chain partners with a common format on collecting pedigree information and building their pedigree software platforms. The standard comprises instructions on how supply chain partners can create an E-pedigree, update information on it and digitally sign it. Many companies are accelerating their initiatives toward integrating E-pedigree pilots into existing legacy supply chain systems to enhance product integrity and further protect patient safety.

An E-pedigree system consists of all partners involved in the distribution of a drug, including the drug manufacturer, the wholesaler, the retailer, and the pharmacy or any other entities administering or dispensing the drug. These partners form a limited distributed system and establish some business relationship between each other (Public Key Infrastructure, or PKI, is typically assumed in this scenario for establishing entity trust relationships). As the drug goes through the distribution path, it forms a growing certified chain of custody while each participant contributes to the E-pedigree. Figure 3 shows a sample form of the E-pedigree [8].

Here, we briefly describe how an E-pedigree system works:

1. A drug is produced by a manufacturer and attached with a unique RFID tag. The manufacturer starts to build the initial E-pedigree with the drug's serial number, transaction information and other product-related information. Then, the E-pedigree is digitally signed with the public key of the manufacturer. The E-pedigree, together with the digital certificate of the manufacturer, is ready to be sent to a downstream partner (typically before the real drug is shipped out).
2. On receiving the E-pedigree, the downstream partner first authenticates the E-pedigree by verifying the digital signature with the public key in the certificate. If the verification is successful, the partner continues to match the information

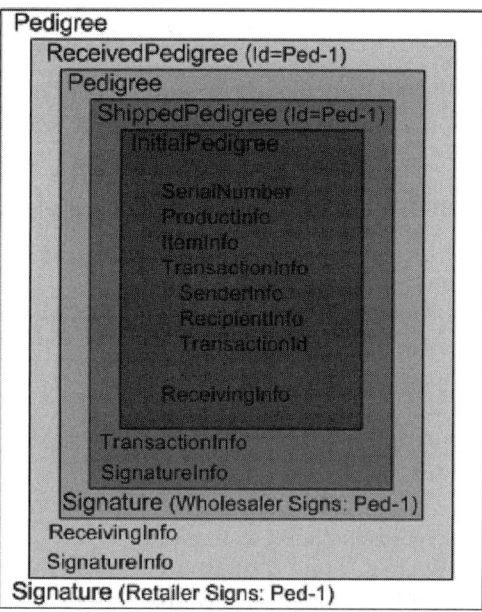

Fig. 3 A sample form of E-pedigree. An E-pedigree, which is originated from the wholesaler and distributed to the retailer, is signed by both the wholesaler and the retailer

on the E-pedigree with that on the product (assuming the drug has been shipped in at this moment). A successful match completes the verification procedure. If the drug is going to be shipped out, the partner needs to update the E-pedigree with its own information and signs on the renewed E-pedigree. Once again, the updated E-pedigree, together with the certificate of the partner, is ready to be transmitted to the next downstream player in advance.

3. The same procedure is repeated by every participant in the distribution path until the drug reaches its destination. The procedures described above actually represent a typical (aggregated) document authentication flow. It does not really need to involve an RFID tag, except that a tag's ID is recorded in the E-pedigree for an additional match. In fact, the tag can be made more useful by strengthening its binding with the E-pedigree. As in Sect. 3, we know that the tag could be a good authenticator due to its fabricated TID. A manufacturer can combine the tag's unique feature with the initial E-pedigree tightly by signing on them together and storing the signature on the tag (take for example TI's electronic marking scheme [21]). Then, the signature can be verified by the forthcoming partners. Under a secure access infrastructure, this piece of additional information can even be encrypted to ensure its confidentiality. The strong binding between the tag and the E-pedigree provides another layer of security.

The E-pedigree solution has been adopted rapidly in pharmaceutical supply chains since it is a natural extension of the legacy enterprise systems. Many of them already have existing internal, closed-loop RFID systems. Although the solution is promising, its success in real world applications will depend more on the nontechnical issues such as privacy protection and legal agreements among multiple partners.

6 Toward an Open-Loop Anticounterfeiting Solution

An E-pedigree revolves around a number of supply chain participants. However, it is more desirable that individual products can be tracked throughout the global supply chain to realize the greatest benefits of RFID technology. This inspires a globally available service – an **EPCglobal network** that offers another huge opportunity to obtain services from an open and standard interface (via Internet). As an essential part of the new supply chain management system, the emerging network enables real-time visibility of all products throughout the supply chain, improves efficiency in inventory control and reduces occurrences of product loss. EPCglobal network essentially resembles the Internet, but is an overlay of the Internet architecture. EPCglobal network architecture is shown in Fig. 4.

The EPCglobal Network [9] employs Electronic Product Code (EPC) to allow companies to track individual product through the global supply chain. The network provides real-time or near real-time tracking and product life cycle monitoring that make business processes more efficient. To realize these benefits, the EPCglobal committee specifies a standard framework to regulate the tracking, security, and collaboration between different supply chain partners. The EPCglobal Network manages RFID information through a number of core services: *Object Name Service (ONS), EPC Information Services (EPC-IS),* and *EPC Discovery Services*

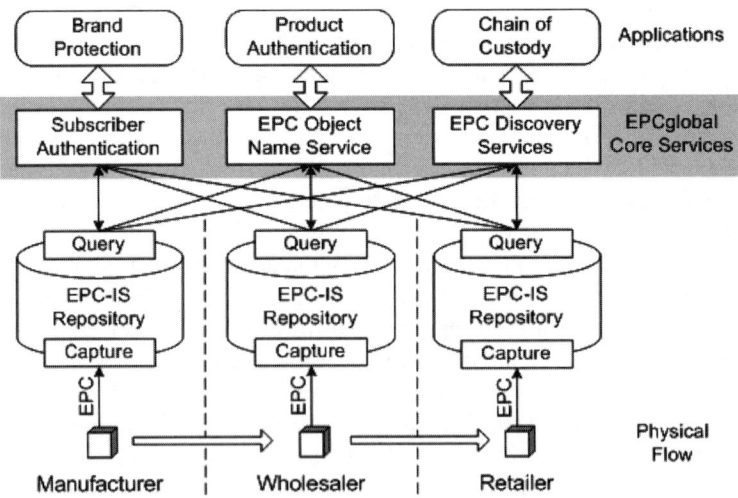

Fig. 4 EPCglobal network architecture

(EPC-DS). Of which, the EPCglobal Architecture Framework identifies three possible ways to locate the informative service according to a specific EPC of an object:

- A party may use the Object Name Service (ONS) to locate the EPC-IS service of the EPCglobal subscriber who commissioned the EPC of the object.
- A party may know in advance exactly where to find the information by means of being given the network address of the other party's EPC-IS service as part of a business agreement.
- A party may use "Discovery Service" (EPC-DS) to locate the EPC-IS services of trading partners that have information about the object, including partners other than the one who commissioned the EPC of the object.

These core services are briefly described below:

ONS. With an EPC that uniquely identifies a single product unit, one can query the ONS to look up the address of the product manufacturer's EPC-IS. Thus, ONS can be thought of as a lookup service that takes an EPC as input, and produces as output the address (in the form of a Uniform Resource Locator, or URL) of an EPC-IS repository designated and implemented by the EPC Manager of the EPC. This is similar to the Domain Name Service (DNS) on Internet, which matches the internet domain names to the IP addresses. From the EPC Manager's EPC-IS repository, one can then obtain detailed product information relating to the EPC.

EPC-IS. EPC-IS regulates the specification for supply chain partners to share EPC-related data. It controls the storage and retrieval of detailed product information on individual product units. It provides a standard data model to enable track and trace, product authentication, diversion detection, and other use cases involving supply chain partners across multiple industries. EPCIS defines a capture interface and a query interface to obtain and share business event information. The standard may be implemented by applications, but the applications themselves are developed by end users and solution providers. As such, EPCIS does not address issues such as purchasing, forecasts, bidding, and billing, which are typically exchanged via EDI in a business transaction between two parties.

EPCIS is the bridge between the physical world and information systems. Many businesses have completely internal business processes that involve the handling of goods, and EPCIS provides a standard way of managing visibility into those processes. In fact, EPC-IS is the foundation for increasing visibility, accuracy, and automation throughout the supply chain. Security is a vital property as the trading partners would only share their data they wish to share on an on-demand basis.

EPC-DS. The product information might be stored not only at the manufacturer's site, but also at different sites along the supply chain (for example the ship-in and ship-out information of a product might be stored at intermediate locations where the product transits). This raises the question of how a trading partner identifies and locates all of the other parties who may have relevant EPC-IS data. The EPC-DS provides the lookup service to all these fragmented sources of information. It serves as a search engine for the EPCglobal Network with restricted access, where subscribers can query it with an EPC to obtain a list of EPC-ISs that they can query directly for more detailed information.

EPC-DS provides visibility in the supply chain for all parties who have a right to know. The discovery of where data resides, the actual exchange of data, and the security policies governing these activities are all related. Of which, authentication and authorization are intimately connected with discovery. For example, merely discovering that one party in a supply chain has information about a particular EPC may or may not be privileged information subject to data authorization policies.

Beyond that, the EPCglobal committee is also working on some security services such as user authentication, authorization, and sensitive information control. The ongoing efforts of the committee also includes the establishment of some specific business cases such as brand protection, product authentication, and chain of custody. These use cases could utilize a combination of the core services described above. For example, the *EPC Product Authentication Service* (EPC-PAS), once regulated, might provide an all-in-one interface for the entities within a supply chain to authenticate a product.

While the EPC-PAS solution is very much desirable, it is not easy to regulate and could potentially encounter many obstacles when put under real operations. One of the major challenges in the design is the privacy of partners along the supply chain. There can be issues with regards to how much information a partner would want to keep with itself instead of sharing them with other partners and how to define the minimal level of authentication-relevant information that should be shared. If there is insufficient information available on product visibility, then one cannot make a good judgment on the authenticity of a product.

In addition, the solution provided by EPC-PAS faces other limitations. First, only authorized personnel can access the service, which is in conflict with our expectation toward a public service where everyone can authenticate a tagged product in hand. Second, even if the service is not provided to all, but to a group of subscribers, there could exist several desired service levels for different groups (e.g., for ordinary users or for supply chain partners). Under such circumstances, how to define the privacy levels for different groups in a dynamic deployment setting would be a big issue. Third, we need to think of how to prevent these services from abuse for malicious purposes, such as the tracking of a particular person. In addition, there is also a lack of practical experience on handling such a huge information system. Beyond that, there are also other issues like the likelihood of social acceptance and legislative support.

7 Conclusion

In this chapter, we presented an architectural view on RFID-based anti-counterfeiting solutions. Compared with existing anticounterfeiting solutions, RFID not only acts as an additional authenticator for product authentication, but also provides an easy way to share a product's information through the global supply chain. Although the solutions are not perfect at this moment (and is unlikely to be in the near future), they look promising with the potential to act against massive counterfeits. The heartening thing is that the anticounterfeiting solutions are being piloted and deployed at many companies. With these precious experiences gained, implementors should be equipped with better knowledge and be in a better position to design optimal security solutions in their fight against counterfeiters.

References

1. AIM Global Analysis: Counterfeit Tags, June 2005
2. A. Bogdanov, L.R. Knudsen, G. Leander, C. Paar, A. Poschmann, M.J.B. Robshaw, Y. Seurin, and C. Vikkelsoe. PRESENT: An ultra-lightweight block cipher. *Cryptographic Hardware and Embedded Systems – CHES 2007*, Vienna, Austria, Sept. 2007
3. J. Bringer, H. Chabanne, and E. Dottax. HB^{++}: A lightweight authentication protocol secure against some attacks. In: *Proc. of SecPerU'06*, pp. 28–33, IEEE Computer Society Press, Washington, DC, 2006
4. B. Defend, K. Fu, and A. Juels. Cryptanalysis of two lightweight RFID authentication schemes. In *Fourth IEEE International Workshop on Pervasive Computing and Communication Security (PerSec) Workshop*, March 2007
5. D.N. Duc, J. Park, H. Lee, and K. Kim, Enhancing security of EPCglobal GEN-2 RFID tag against traceability and cloning, In *The 2006 Symposium on Cryptography and Information Security*, 2006
6. EPCglobal, 13.56 MHz ISM Band Class 1 Radio Frequency (RF) Identification Tag Interface Specification
7. EPCglobal, EPC Radio-Frequency Identity Protocols Class-1 Generation-2 UHF RFID Protocol for Communications at 860MHz-960MHz Version 1.0.9
8. EPCglobal, Pedigree Standard v1.0 http://www.epcglobalinc.org/standards/pedigree/Pedigree_1_0-StandardRatified-20070105.pdf
9. EPCglobal, Architecture Framework Standard v1.0 http://www.epcglobalinc.org/standards/architecture/Architecture_1_0-StandardApproved-20050701.pdf
10. M. Feldhofer and J. Wolkerstorfer. Strong crypto for RFID TagsCa comparison of low-power hardware implementations, In: *IEEE International Symposium on Circuits and Systems (ISCAS 2007)*, pp.1839–1842, New Orleans, USA, May 27–30, 2007
11. H. Gilbert, M. Bobshaw, and H. Silbert, An active attack against HB^{+} – A probable secure lightweight authentication protocol, *Cryptology ePrint Archive, Report 2005/237*, 2007
12. T. Good, W. Chelton, and M. Benaissa. Hardware results for selected stream cipher candidates. In *SASC 2007*, February 2007
13. R. Johnston, Tamper-indicating seals, *American Scientist*, Nov–Dec 2005
14. A. Juels and R. Pappu. Squealing euros: Privacy protection in RFID-enabled banknotes. In: *Proc. of FC'03*, LNCS 2742, pp. 103–121, Springer, Berlin, 2003
15. A. Juels and S. Weis. Authenticating pervasive devices with human protocols. In: *Proc. of CRYPTO'05*, LNCS 3126, pp. 293–308, Springer, Berlin, 2005
16. M. Lehtonen, T. Staake, F. Michahelles, and E. Fleisch, Strengthening the security of machine readable documents by combining RFID and optical memory devices. In *Conference on Ambient Intelligence Developments – AmID,* Sophia-Antipolis, France, September 2006
17. Z. Nochta, T. Staake, and E. Fleisch, Product specific security features based on RFID technology. In *Proceedings of the International Symposium on Applications and the Internet* Workshops, IEEE Computer Society press, Washington, DC, 2006
18. Y. Oren and A. Shamir. Remote password extraction from RFID tags. In: *IEEE Transactions on Computers*, 56(9): 1292–1296, 2007

19. S. Sarma, S. Weis, and D. Engels. RFID systems and security and privacy implications. In: *Proc. of CHES'02*, LNCS 2523, pp. 454–469, Springer, Berlin, 2002

20. T. Staake, F. Thiesse, and E. Fleisch, Extending the EPC network – The potential of RFID in anti-counterfeiting. In *Proceedings of the 2005 ACM symposium on Applied computing*, pp. 1607–1612, ACM Press, New York, NY, 2005

21. Texas Instruments and VeriSign Inc.: Securing the pharmaceutical supply chain with RFID and public-key infrastructure technologies. *Whitepaper*, 2005

22. I. Vajda and L. Buttyan. Lightweight authentication protocols for low-cost RFID tags. In: *Proc. of UBICOMP'03*, 2003

23. S. Weis. Security parallels between people and pervasive devices. In: *Proc. of PERSEC'05*, pp. 105–109, IEEE Computer Society Press, Washington, DC, 2005

An Efficient and Secure RFID Security Method with Ownership Transfer

Kyosuke Osaka*, Tsuyoshi Takagi, Kenichi Yamazaki, and Osamu Takahashi

Abstract We are facing privacy and security problems and challenges to RFID systems. Recent papers have reported that RFID systems have to achieve the following requirements (1) indistinguishability, (2) forward security, (3) resistance against replay attack, (4) resistance against tag killing and (5) ownership transferability. We have to design the RFID system that achieves the above requirements. Existing RFID security schemes achieve some of them, but no one has been constructed that achieves all requirements. In this chapter, we analyze previously reported RFID security schemes, and propose an RFID security method that achieves all requirements, based on a hash function and a symmetric key cryptosystem. Our proposed method provides not only high security but also high efficiency.

1 Introduction

Generally, an RFID system consists of a large number of RFID tags, some RFID readers, and a back-end database. We are facing privacy and security problems and challenges to RFID systems. The privacy and security problems arise from RFID tag limitations such as small memory and low computing power. Due to such limitations, data transmission between RFID tags and readers are unencrypted; In addition RFID tags provide no tamper resistance. That is, RFID tags and readers communicate with each other using insecure wireless channels. Recent papers have reported that RFID systems have to achieve the following requirements:

- *Indistinguishability* [17]: No adversary can distinguish output from RFID tags.
- *Forward security* [17]: Even if present data on RFID tags are leaked to an adversary, past data still remain secure.

K. Osaka

Future University-Hakodate 116-2, Kamedanakano, Hakodate 041-8655, Japan

e-mail: g2107003@fun.ac.jp

P. Kitsos, Y. Zhang (eds.), *RFID Security: Techniques, Protocols and System-on-Chip Design,* © Springer Science+Business Media, LLC 2008

- *Resistance against replay attack* [18]: No adversary can succeed in a replay attack that spoofs a legitimate RFID tag.
- *Resistance against tag killing* [7]: RFID tags shall be resistant against a large number of incoming queries.
- *Ownership transferability* [21]: Ownership is transferable without violation of previous and present RFID tag owner's privacy.

Many RFID security schemes achieve some of the above requirements, but none achieves all of them [3, 9–11, 14, 17, 18, 21, 24].

In this chapter, we present the security requirements and analyze previously reported RFID security schemes. Moreover, we propose an RFID security method that achieves all of the requirements, based on a hash function and a symmetric key cryptosystem. Our proposed method provides not only high security but also high efficiency.

The rest of this chapter is organized as follows. Section 2 describes the RFID system and its security requirements. In Sect. 3, we analyze previously reported RFID security schemes. Our proposed method is described in Sect. 4. Finally, Sect. 5 offers conclusions.

2 RFID System

2.1 The Components of the RFID System

Generally, an RFID system consists of a large number of RFID tags, some RFID readers, and a back-end database. The details of each components are described in the following:

1. Tags

Tags are attached to products or objects, e.g., passports, books, or clothes, and might be used in place of bar code. Each tag is assigned a unique identification number (ID) which consists of a serial number, product code, or the like, depending on the standard-setting organization such as EPCglobal, uIDcenter [1, 5]. Each tag sends its ID to a reader by radio transmission when it receives a query from the reader.

2. Readers

Readers are wireless communication devices which read/write data on tags. A reader broadcasts a query to a tag, then it receives the tag's ID. The reader communicates with the tag in an insecure wireless channel. The reader requests detailed information of the tagged product from the database by sending the ID. In contrast with communication between the tag and the reader, the reader communicates with the database in a secure wired channel.

Database   **Reader**   **Tag**

Fig. 1 Basic protocol of the RFID system

3. Database

The database stores IDs and information associated with each tag, and also computes complex operation instead of tags or readers due to their low computing power. This information may consist of merchandise information, production information, price, personal data, and so on [1, 5]. The database gives the information of the tagged products to a reader when it receives a request from the reader.

2.2 Basic Protocol of the RFID System

Figure 1 shows the basic protocol of the RFID system, which works as follows:

1. A reader broadcasts a query to a tag.
2. The tag sends its ID to the reader.
3. The reader sends the ID to the database.
4. The database gives the information of the tagged product to the reader.

2.3 Security Requirements for RFID System

We describe the following five security requirements as the security goal of RFID systems.

2.3.1 Indistinguishability (IND) [17]

As mentioned in Sect. 2.1, tags may be attached to products or objects, such as clothes, accessories, wallets, bags, etc. which are carried by a person. In such cases, if an adversary could read their tags, they can distinguish and trace the tag carrier. In the case that IDs are encrypted, although the adversary cannot understand its meaning, the adversary can distinguish and trace the target encrypted ID because the encrypted ID is fixed. This privacy problem is called the violation of "location privacy" [22, 24]. In order to solve this, RFID systems are required to provide

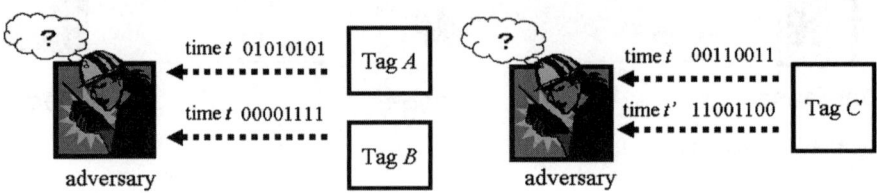

Fig. 2 Indistinguishability

indistinguishability [17]. This means that no adversary can distinguish output from tags. To this end, indistinguishability satisfies the following two conditions, which are described in Fig. 2.

Condition 1. Even if an adversary obtains IDs from different tags at the same time, the adversary cannot distinguish between them.

Condition 2. Even if an adversary obtains several IDs from same tag, the adversary cannot recognize this.

RFID systems can be said to provide indistinguishability if the adversary cannot win the following game [15]:

- Let input be two tags (tag A, B) which are randomly chosen from N tags.
- Let $A(t)$ be the output from tag A at time t, and let $B(t)$ be the output from tag B-at time t.
- R is a 1-bit random number generator, and outputs 0 or 1 as output b.
- Challenger changes its output depending on b.

  - If $b = 0$, challenger outputs $B(t_0)$, which corresponds to Condition 1.
  - If $b = 1$, challenger outputs $A(t_1)$.

- Let O_1 be $A(t_0)$, and let O_2 be an output of challenger, i.e., $B(t_0)$ or $A(t_1)$.
- The adversary X can choose an arbitrary tag as input. The adversary X can obtain both O_1 and O_2 without b.
- Based on the result of inputs and outputs, i.e., tag A, B, output O_1, O_2, the adversary X tries to guess value of b. Let b' be an output of X.

In the case of $b = 0$, X obtains $O_2 = B(t_0)$, which corresponds to Condition 1. In the case of $b = 1$, X obtains $O_2 = A(t_1)$, which corresponds to Condition 2.

Let $P(b' \leftarrow X, b = b')$ be the accuracy rate of the adversary X. The advantage of X is represented as the following equation:

$$\text{Advantage}(X) = \left| P(b' \leftarrow X, b = b') - \frac{1}{2} \right|.$$

If $\text{Advantage}(X)$ is negligible for every adversary X, then the RFID system can be said to provide indistinguishability.

2.3.2 Forward Security (FS) [17]

Tags contain valuable information, and this information must not be accessible to third parties. However, tags provide no tamper resistance and it is possible that information may be leaked to an adversary by physical methods such as a power analysis attack.

Consider the following situation. First, the adversary obtains a tagged product that the tag owner threw out. The adversary then obtains the present data on the tag by tampering. Next, the adversary uniquely distinguishes the target tag by using the tampered present data and past output of the tag which they collected in advance. Then, the movement history of the tag owner, who threw out the tagged product, is known by the adversary.

In order to solve this, RFID systems are required to provide forward security [17]. Forward security means even if the present data on a tag are leaked to the adversary, past data, which were on the tag, still remain secure. This means that the movement history of the tag owner is known by nobody. Forward security is shown in Fig. 3.

RFID systems can be said to provide forward security if the adversary cannot win the following game [17]:

- Let H be a hash function that updates the ith secret s_i on a tag by computing $s_{i+1} = H(s_i)$. If the hash function H is usually a one-way function, its inverse function H^{-1} is uncomputable.
- There exists an oracle that returns ith secret s_i as an answer when the adversary X inputs counter i s.t. $i > j$.

Let $P(s_j \leftarrow X, H^{i-j}(s_j) = s_i)$ be the probability that the adversary X guesses s_j from s_i s.t. $H^{i-j}(s_j) = s_i$. This probability equals the advantage of X which is represented as the following equation:

$$\text{Advantage}(X) = P(s_j \leftarrow X, H^{i-j}(s_j) = s_i).$$

If $\text{Advantage}(X)$ is negligible for every adversary X, then the RFID system can be said to provide forward security.

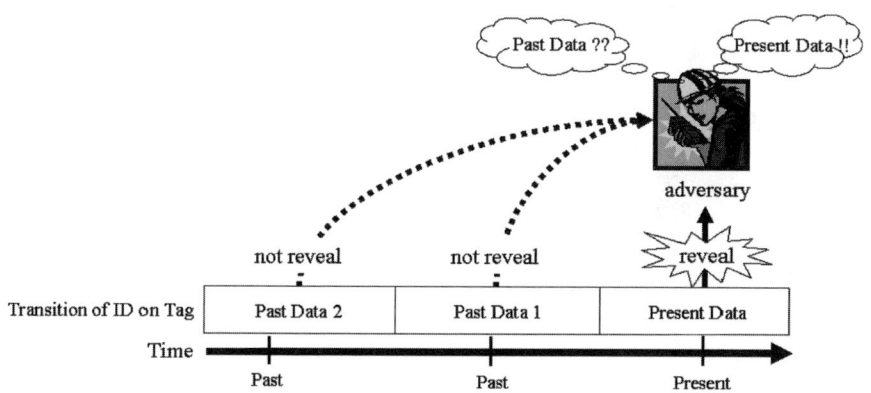

Fig. 3 Forward security

2.3.3 Resistance Against Replay Attack (RA) [18]

This security problem arises when the adversary can succeed in a replay attack that spoofs a legitimate tag. A replay attack may be carried out as follows. An adversary obtains the ID from tag A at time t, then the adversary's tag B spoofs the legitimate tag A at time $t' > t$ by sending the ID of tag A to a legitimate reader.

Consider the following situation. Let $100 be the price of a tie to which tag A is attached, and let $1,000 be the price of a suit to which tag B is attached. An adversary obtains the ID from tag A at time t, then writes the ID of tag A into tag B. Next, tag B, rewritten by the adversary, spoofs tag A at time $t' > t$ by sending its ID, namely ID of tag A, to a legitimate reader. The adversary then purchase the suit with the modified tag A at a price of $100, thus saving $900. We have already mentioned in Sect. 2.1 that all tag's IDs are stored in the database, and these link each tag with the information associated with it, e.g., price.

Currently, a popular method of countering replay attacks is the use of a one-time password [3, 9, 10, 14, 18]. We think that one-time passwords are very effective, and in this chapter will show how they are used.

Let H be a function, let R and $R'(\neq R)$ be random numbers, and let s_i be a secret (e.g., tag's ID) which is stored in a tag. The output of the tag should satisfy the following inequation:

$$H(s_i, R) \neq H(s_i, R').$$

If the function H is a collision-resistant hash function, then the above inequation holds. That is, if the function H is a collision-resistant hash function, then the RFID system can be said to provide a resistance against replay attack. This is shown in Fig. 4.

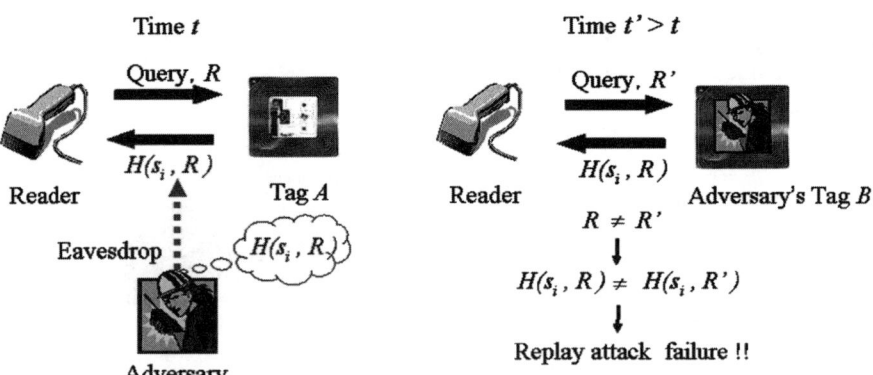

Fig. 4 Resistance against replay attack

2.3.4 Resistance Against Tag Killing (TK) [7]

Tag killing [7] was proposed by Han et al. in 2006. Tag killing is an attack whereby an adversary broadcasts a large number of queries to a tag and overloads the tag's memory. Tags need to be made cheply, so they have a limited range of operation.

Consider the following situation. If a tag is overloaded and no longer works due to tag killing, then the adversary might be able to easily shoplift the tagged product.

In the past, RFID security methods had two main types; challenge–response and Ohkubo [17]. Regardless of the type, most RFID security methods were vulnerable to tag killing. The reasons are as follows. In the case of methods based on the challenge–reponse, tags require additional memory for storing random numbers each reading a tag. Therefore if a large number of queries are broadcasted to a tag, then the tag's memory will be exhausted. On the other hand, although methods based on the Ohkubo method use hash chain technique [17], the hash chain has an upper limit of reading tags. Therefore if a large number of queries are broadcasted to a tag, then the tag goes beyond the limit and stops working.

In order to prevent tag killing, RFID systems require the following two conditions:

1. Tags require no additional memory for storing random numbers each reading a tag.
2. Tags have no upper limit of reading tags.

If they are satisfied, then the RFID system can be said to provide a resistance against tag killing. This is shown in Fig. 5.

2.3.5 Ownership Transferability (OT) [21]

RFID systems are expected to improve the efficiency of supply chain management i.e., the movement of products from the manufacturer to the wholesaler, retailer, and

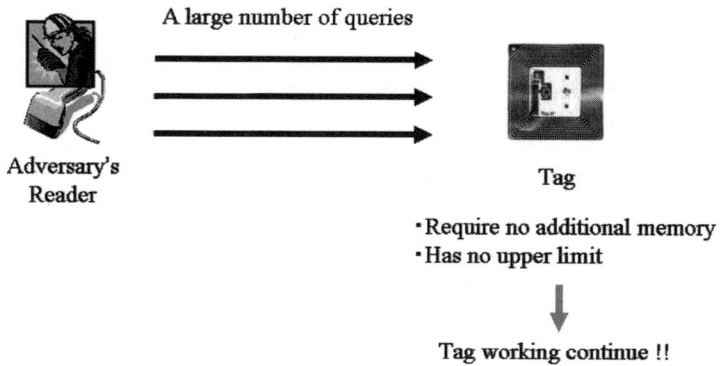

Fig. 5 Resistance against tag killing

consumer. However, it is hard to apply most RFID security methods to supply chain management because the methods do not account for the movement of products along the supply chain. Note that ownership transferability is considered in only some papers [13, 21].

Recall that tags are attached to products or objects. They are transferred, for example, to the wholesaler from the manufacturer at time t. At that point, the manufacturer should transfer the information of tagged products to the wholesaler. If the manufacturer, which is the previous owner, could read IDs freely at time $t' > t$ i.e., after ownership transfer, the manufacturer could trace the wholesaler which is the present owner. Then the privacy of the wholesaler might be violated. Ownership transferability [21] means that in such cases, the ownership is transferred without violation of the previous and present owner's privacy.

In this chapter, to achieve ownership transferability, we use the method that ownership is transferable by changing an owner's key which is used for encryption. RFID systems require the following two conditions:

1. A protocol to be able to securely change an owner's key used for encryption is built into the RFID system.
2. Owners can change their key at any time.

If they are satisfied, then the RFID system can be said to provide ownership transferability. This is shown in Fig. 6.

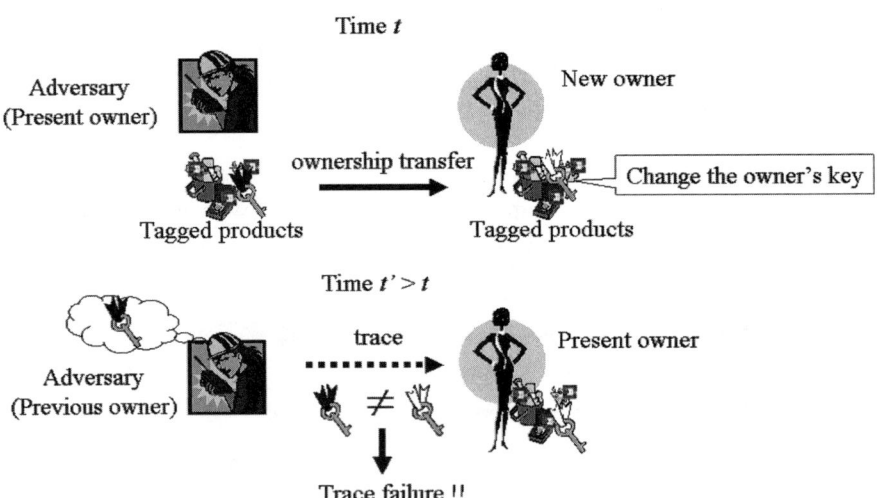

Fig. 6 Ownership transferability

3 Analysis of Previous Schemes

In this section, we describe previously reported RFID security schemes and point out their security flaws. The previous schemes are classified mainly into challenge–response type, Ohkubo type and other type. Some of the security analyses are beyond the scope of this section.

3.1 Challenge–Response Type Schemes

3.1.1 Hash Lock Scheme [23, 24]

The hash lock scheme [23,24] is a simple scheme based on a one-way hash function, which was proposed by Weis et al. in 2003. Tags have a one-way hash function h, and shift to either a "locked" or an "unlocked" state. Owners can freely choose the state. In the case of the locked state, the tag always sends the same $metaID$, which is a hashed key chosen randomly by the owner, to every reader. This scheme therefore can prevent unauthorized readings. When the owner wants to read the tag information, the owner changes the state to the unlocked state. The key and the $metaID$ are registered to the database by the owner. The reader authenticates the tag by comparing the received $metaID$ with the saved $metaID$. The tag hashes the key received from the reader, then authenticates the reader by comparing the hashed key with the saved $metaID$. The protocol of the hash lock scheme is as follows.

The Protocol of Hash Lock Scheme

This consists of two protocols for locking and unlocking, as shown below. Also, these are shown in Figs. 7 and 8:

1. Protocol for Locking:

(a). A reader randomly chooses a key and computes $metaID = h(key)$.
(b). The reader writes the $metaID$ into a tag.
(c). The tag enters the locked state.
(d). While the tag is locked, the tag always sends the same $metaID$ to every reader.

2. Protocol for Unlocking:

(a). A reader sends $metaID$ to the database.
(b). The database gives (key, ID) corresponding with the $metaID$ to the reader.
(c). The reader sends the key to the tag.

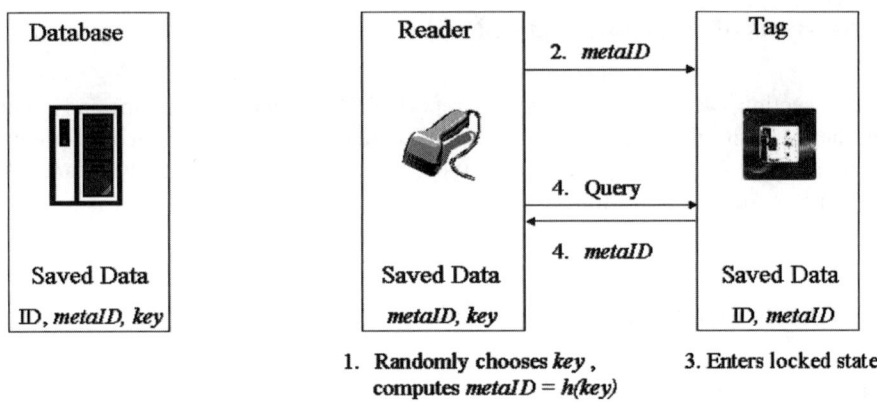

Fig. 7 Protocol for locking of hash lock scheme

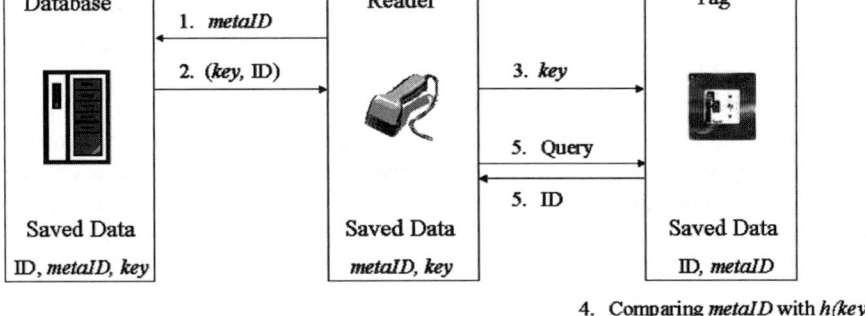

Fig. 8 Protocol for unlocking of hash lock scheme

(d). The tag compares value of hashed *key* with the *metaID*. If the $h(key)$ corresponds with the *metaID*, the tag enters the unlocked state.

(e). While the tag is unlocked, the tag always sends the ID to every reader.

Security of Hash Lock Scheme

In this scheme, tags require no additional memory for each query. Therefore, this scheme achieves resistance against tag killing. Consider the following attack. An adversary obtains a *metaID* by broadcasting a query. Then, the adversary can impersonate a legitimate tag by sending the *metaID* to a legitimate reader. This scheme therefore is vulnerable to the replay attack. The security of the hash lock scheme is shown in Table 1.

Table 1 Security of hash lock scheme

IND	FS	RA	TK	OT
×	×	×	○	×

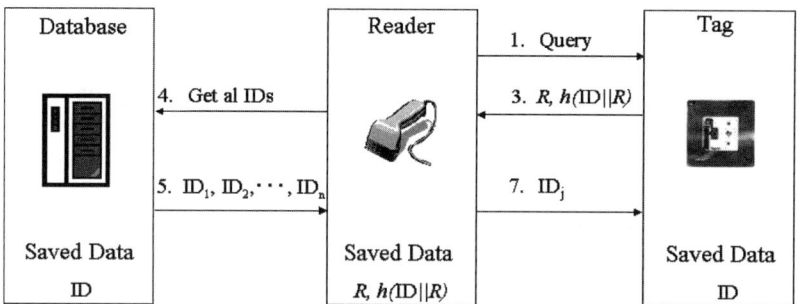

6. Comparing $h(\text{ID}_i||R)$ with $h(\text{ID}||R)$ 2. Generates R, computes $h(\text{ID}||R)$
8. If $\text{ID}_j = \text{ID}$, enters unlocked state

Fig. 9 Protocol of randomized hash lock scheme

3.1.2 Randomized Hash Lock Scheme [23, 24]

The randomized hash lock scheme [23, 24] is a modified hash lock scheme, which was proposed by Weis et al. in 2003. Tags have a one-way hash function h and a random number generator, and shift to either a "locked" or an "unlocked" state in the same way as the hash lock scheme. In this scheme also, owners can freely choose the state. The unlocked tag is locked with a simple instruction from the reader; no protocol is necessary.

Let R be a random number. The reader compares $h(\text{ID}||R)$ received from the tag with $h(\text{ID}_i||R)$ computed by the reader, then authenticates the tag. Then the reader sends the corresponding ID_j to the tag. The tag compares the ID_j received from the reader with the ID stored in the tag, then authenticates the reader. The protocol of the randomized hash lock scheme is as follows. Also, this is shown in Fig. 9.

Protocol of Randomized Hash Lock Scheme

1. A reader broadcasts a query to a tag.
2. The tag generates a random number R and computes $h(\text{ID}||R)$.
3. The tag sends R and $h(\text{ID}||R)$ to the reader.
4. The reader queries IDs of all tags to the database.
5. The database gives all IDs to the reader.
6. The reader compares $h(\text{ID}_i||R)$ with $h(\text{ID}||R)$ for $1 \leq i \leq n$, received from the database, then identifies the corresponding ID_j.

7. The reader sends the ID_j to the tag.
8. The tag compares the ID_j with the stored ID. If it does match, the tag enters an unlocked state.

Security of Randomized Hash Lock Scheme

In this scheme, the tag sends $h(ID||R)$, which is changed each query, to the reader. However, the adversary can distinguish the target tag due to ID_j sent by the reader. Therefore, this scheme achieves no indistinguishability. Consider the following attack. An adversary eavesdrops on R and $h(ID||R)$ which are output of a legitimate tag. Then the adversary can impersonate the legitimate tag by using the obtained R and $h(ID||R)$. This scheme therefore is vulnerable to the replay attack. In this scheme, tags require no additional memory for each query. Therefore, this scheme achieves resistance against tag killing. The security of the randomized hash lock scheme is shown in Table 2.

3.1.3 Hash-Based ID Variation Scheme [9]

The hash-based ID variation scheme [9] is a scheme based on a one-way hash function h and XOR operation, which was proposed by Henrici and Müller in 2004. In this scheme, there exists some database for tag owners. Parameters of this scheme is shown in Table 3.

Each tag needs to contain fields for DB-ID, ID, TID, and LST. In addtion, the database needs to contain a table with entries for ID, TID, LST, AE, HID, and DATA each record row. TID, LST, and AE have the purpose of counteracting replay attacks. The protocol of the hash-based ID variation scheme is as follows. Also, this is shown in Fig. 10.

Table 2 Security of randomized hash lock scheme

IND	FS	RA	TK	OT
×	×	×	○	×

Table 3 Parameters of hash-based ID variation scheme

DB-ID	Database-identifier which denotes the owner's database
ID	Tag-identifier
TID	Transaction number. Initial value is a random number
LST	Last successful transaction number. Initial value is TID
AE	Associated DB entry
HID	Hash value of current ID. Initial value is $h(ID)$
RND	Random number generated by the database
DATA	Tag/user data

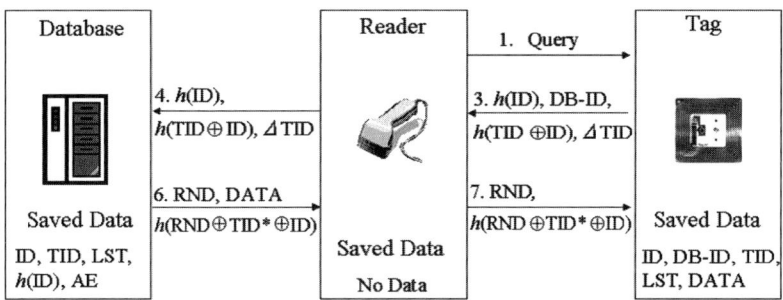

Database		Reader		Tag
			1. Query →	
	4. h(ID),		3. h(ID), DB-ID,	
	h(TID⊕ID), ΔTID		h(TID⊕ID), ΔTID	
	6. RND, DATA		7. RND,	
Saved Data	h(RND⊕TID*⊕ID)		h(RND⊕TID*⊕ID)	Saved Data
ID, TID, LST,		Saved Data		ID, DB-ID, TID,
h(ID), AE		No Data		LST, DATA

5. Identifies ID, computes TID* = ΔTID + LST.
   Comparing h(TID* ⊕ID) with h(TID ⊕ID),
   if TID*>TID,
   then updates TID ← TID*, ID* ← ID ⊕RND

2. Computes TID = TID+1, h(ID),
   h(TID⊕ID), ΔTID = TID – LST

8. If h(RND ⊕TID*⊕ID) = h(RND⊕ TID ⊕ ID),
   then updates ID ← ID*, LST ← TID

Fig. 10 protocol of hash-based ID variation scheme

Protocol of Hash-Based ID Variation Scheme

1. A reader broadcasts a query to a tag.
2. The tag increases its transaction number TID by one and computes h(ID), h(TID ⊕ ID) and ΔTID = TID − LST.
3. The tag sends h(ID), DB-ID, h(TID ⊕ ID), and ΔTID to the reader.
4. The reader identifies an owner's database from DB-ID, then sends h(ID), h(TID ⊕ ID), and ΔTID to the database.
5. The database identifies ID from h(ID), then computes TID* = ΔTID + LST, where TID* is the correct TID at the time. Next, the database computes h(TID* ⊕ ID), then compares it with h(TID ⊕ ID). If they do not match, the message is discarded. If they do match, the database compares TID* with TID stored in the database. If TID* ≤ TID, a replay attack is in progress and the message is discarded. If TID* > TID, the database updates TID to TID*. Then the database generates a random number RND and computes ID* = ID ⊕ RND, where ID* is a new ID.
   If an AE exists, the database updates this record row to a new record row (ID*, h(ID*), TID, LST, AE), where ID* = ID ⊕ RND, h(ID*) = h(ID), TID = TID*, LST = TID*, and AE = h(ID). If no AE exists, the database appends this record row (ID*, h(ID*), TID, LST, AE). Then the database computes h(RND ⊕ TID* ⊕ ID).
6. The database sends RND, h(RND ⊕ TID* ⊕ ID), and DATA to the reader.
7. The reader sends RND and h(RND ⊕ TID* ⊕ ID) to the tag.
8. The tag computes h(RND ⊕ TID ⊕ ID), then compares it with h(RND ⊕ TID* ⊕ ID) received from the reader. If they do not match, the message is discarded and no further action is taken. If they do match, the tag updates ID to ID*, and also LST to TID.

Table 4 Security of hash-based ID variation scheme

IND	FS	RA	TK	OT
✕	✕	○	○	✕

Security of Hash-Based ID Variation Scheme

The transaction number TID^* has the purpose of counteracting replay attacks. The database and the tag confirm whether a replay attack is in progress by comparing TID^* and TID. Therefore this scheme achieves resistance against replay attack. In this scheme, tags require no additional memory for each query. Therefore, this scheme achieves resistance against tag killing.

Consider the following attack. The value of ΔTID is always "1" if there is no error or attack. Even if authentication ends up in failure, TID is increased by one each query. The adversary broadcasts some queries to a tag. Then, although authentication ends up in failure, the value of ΔTID is not "1" but $\Delta TID > 1$ because of the increment of TID. If the adversary broadcasts a large number of queries to a target tag, the adversary could distinguish the tag because the value of ΔTID becomes a large value. Therefore, this scheme achieves no indistinguishability. The security of the hash-based ID variation scheme is shown in Table 4.

3.1.4 The RKKW Scheme [18]

The RKKW scheme [18], based on a one-way hash function h and random numbers, was proposed by Rhee et al. in 2005. A reader first broadcasts a query to a tag with a random number r_1. Then the tag responds using r_1 and another random number r_2 generated by itself. This scheme therefore provides resistance against replay attack. The protocol of the RKKW scheme is as follows. Also, this is shown in Fig. 11.

Protocol of the RKKW Scheme

1. A reader generates a random number r_1 and sends it together with a query to a tag.
2. The tag generates a random number r_2 and computes $\sigma = h(ID||r_1||r_2)$, where $||$ is a concatenate function.
3. The tag sends r_2 and σ to the reader.
4. The reader sends r_1, r_2 and σ to the database.
5. The database finds the ID s.t. $h(ID||r_1||r_2) = \sigma$. If the ID is found, the database computes $\tau = h(ID||r_2)$.
6. The database sends τ to the reader.
7. The reader sends τ to the tag.
8. The tag checks whether $h(ID||r_2)$ corresponds with τ. If $h(ID||r_2) = \tau$, the tag authenticates the reader.

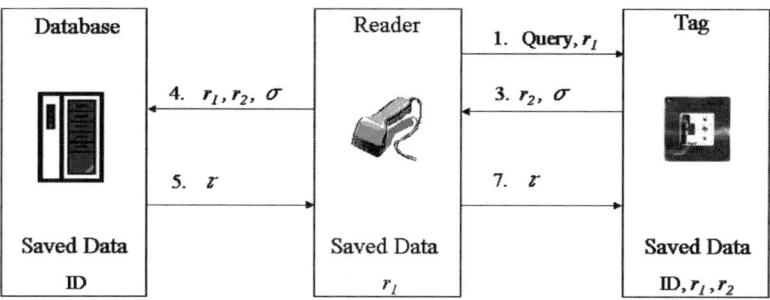

6. Finds ID s.t. $h(\text{ID}\| r_1 \| r_2) = \sigma$
   If received data are correct,
   authenticates Tag and computes $z = h(\text{ID} \| r_2)$

2. Generates r_2, computes $\sigma = h(\text{ID}\| r_1 \| r_2)$

8. If $z = h(\text{ID}\| r_2)$, authenticates Reader

Fig. 11 Protocol of the RKKW scheme

Table 5 Security of RKKW scheme

IND	FS	RA	TK	OT
○	×	○	×	×

Security of the RKKW Scheme

This scheme achieves indistinguishability and resistance against replay attack due to the different random numbers r_1 and r_2 which are changed by each query. However, tags require new additional memory for the random number r_2. Therefore if an adversary broadcasts a large number of queries to a tag, the tag's memory will be exhausted. This scheme therefore is vulnerable to tag killing. The security of the RKKW scheme is shown in Table 5.

3.1.5 Kang–Nyang's Scheme [10]

Kang–Nyang's scheme [10], which is the first scheme that achieves forward security among challenge–response type schemes, was proposed by Kang and Nyang in 2005. The nomenclature of this scheme is shown in Table 6. In this scheme, all possible values of $H(\text{ID}\|C)$ for each ID should be prepared by the database. For one tag, the database requires 2^m numbers of hash operation, where m is the bit length of C. Hence, for the N tags, the database requires $N * 2^m$ numbers of hash operation. Although the database seems to require much computation, hash operations can be performed after authentication in the idle time. The protocol of Kang–Nyang's scheme is as follows. Also, this is shown in Fig. 12.

Table 6 The nomenclature of Kang–Nyang's scheme

SFlag	A Session flag which indicates whether the tag is under a session. Initially it is set to *false* When a session starts, it is set to *true*, and when a session ends, it is set to *false*
R1	A random nonce which is changed each session. $R \in \{0,1\}^n$. If *SFlag* is *true*, it is replaced by saved *R1*. If *SFlag* is *false*, it is generated by the tag newly
C	A counter which is increased or randomly changed each session. $C \in \{0,1\}^m$. If *SFlag* is *true*, it is replaced by saved *R1*. If *SFlag* is *false*, it is changed by the tag
THR_COUNT	A threshold counter which indicates how many trials have occurred. When *THR_COUNT* reaches *THR_MAX*, the tag terminates current session and starts a new session. It can report tag killing
CWD	A confirm word used to confirms ID by the database. *CWD* $\in \{0,1\}^n$
ID	A tag identifier. ID $\in \{0,1\}^n$
H	A cryptographic hash function

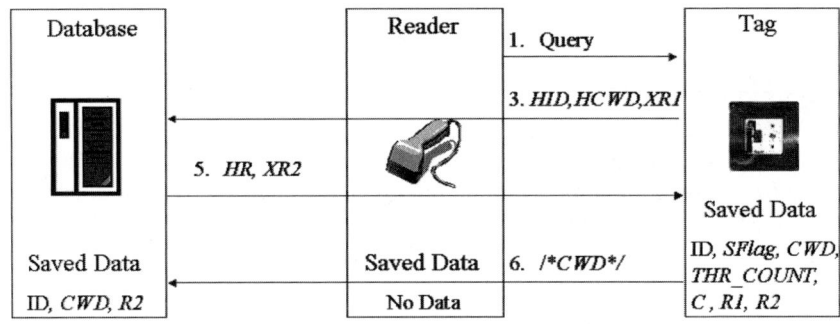

4. Search ID
   If found ID
      Choose R2 and computes
      $HR \leftarrow h(R1||R2)$, $XR2 \leftarrow ID \oplus R2$

7. If *CWD* is correct,
      $ID \leftarrow ID \oplus R1 \oplus R2$
      $CWD \leftarrow R1 + R2 \mod 2^n$

2. If *SFlag* = *false*
      Choose *R1* and *C* , *SFlag* $\leftarrow$ *true*
      $HID \leftarrow h(ID||C)$, $XR1$=ID$\oplus R1$,
      $HCWD \leftarrow h(R1||CWD)$

6. $THR\_COUNT \leftarrow THR\_COUNT+1$
      If $THR\_COUNT < THR\_MAX$
         $R2 \leftarrow \overline{X}R2 \oplus ID$
         If $HR = h(R1||R2)$
            /* sends *CWD*/
            $ID \leftarrow ID \oplus R1 \oplus R2$
            $CWD \leftarrow R1 + R2 \mod 2^n$
      Else
         SFlag $\leftarrow$ *false*

Fig. 12 Protocol of Kang–Nyang's scheme

Protocol of Kang–Nyang's Scheme

1. A reader broadcasts a query to a tag.
2. The tag checks *SFlag*. If *SFlag* is *true*, the tag uses $R1$ and C which are already in the memory. Otherwise, the tag chooses the random numbers $R1$ and C. Also the tag sets *SFlag* to *true*. Then, the tag computes $HID = H(ID||C)$, $HCWD = H(R1||CWD)$ and $XR1 = ID \oplus R1$.
3. The tag sends *HID*, *HCWD*, and *XR1* to the database through the reader.
4. The database saves all possible values of *HID*. The database identifies ID from *HID*, then gets $R1$ by computing $R1 = XR1 \oplus ID$. The database checks whether *HCWD* is the same as $H(R1||CWD)$, where $R1$ is the obtained candidates, *CWD* is the candidates stored in the database.
   If the database cannot find any satisfactory ID, the database regards this message as an attack and ignores it. If the database finds ID, the database chooses another random nonce $R2$, then computes $HR = H(R1||R2)$ and $XR2 = ID \oplus R2$.
5. The database sends HR and $XR2$ to the tag through the reader.
6. The tag increases *THR_COUNT* by one, then computes $R2 = XR2 \oplus ID$. If and only if $THR\_COUNT < THR\_MAX$, the tag checks whether HR received from the reader corresponds with $H(R1||R2)$ computed by itself. If it does match, the tag sends *CWD* to the database, then updates ID to step 7, (1) and *CWD* to step 7, (2). If it does not match, the tag should wait another messages while $THR\_COUNT < THR\_MAX$. If $THR\_COUNT = THR\_MAX$ (or expires), the tag sets *SFlag* to *false*, then ignores all further messages and waits until another reader opens a new session.
7. If the database receives the correct *CWD*, the database updates the ID to (1) and *CWD* to (2). Otherwise, the database halts the process.

$$ID_i = ID_{i-1} \oplus R1_{i-1} \oplus R2_{i-1}, \tag{1}$$

$$CWD_i = (R1_{i-1} + R2_{i-1}) \bmod 2^n, \tag{2}$$

where i denotes the current session and $i-1$ denotes the previous session.

Security of Kang–Nyang's Scheme

In this scheme, data eavesdropable by an adversary are $HID, HCWD, XR1$ in Step 3, $HR, XR2$ in Step 5, and CWD in Step 6. All of these data are converted at each query by performing the XOR operation. Therefore, this scheme achieves indistinguishability and resistance against replay attack. Moreover, this scheme achieves forward security due to the updating of ID on the tag at each query. In this scheme, when the tag receives another query in the authentication session, the tag generates neither the random number ($R1$) nor the counter (C), but uses the $R1$ and the C saved in the tag. Therefore even if the adversary broadcasts a large number of queries to a tag, the tag's memory will not be exhausted because the tag uses saved $R1$ and C. This scheme therefore achieves resistance against tag killing. The security of Kang–Nyang's scheme is shown in Table 7.

Table 7 Security of Kang–Nyang's scheme

IND	FS	RA	TK	OT
◯	◯	◯	◯	×

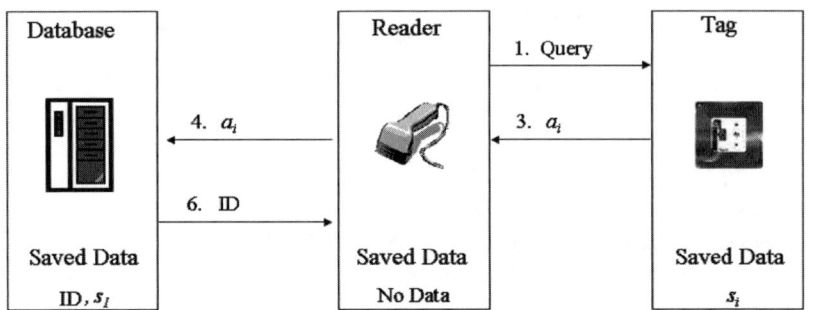

5. Finds s_1 s.t. $G(H^{i-1}(s_1))= a_i$ 2. Computes $a_i = G(s_i)$, $s_{i+1} = H(s_i)$

Fig. 13 Protocol of Ohkubo scheme

3.2 Ohkubo Type Schemes

3.2.1 Ohkubo Scheme [17]

The Ohkubo scheme [17], proposed by Ohkubo et al, is based on the hash chain technique. They use the hash chain technique to renew the secret information s_i stored in the tag. Initially tag has initial information s_1. In the ith session with the reader, the tag performs as follows:

1. Sends answer $a_i = G(s_i)$ to the reader
2. Renews secret $s_{i+1} = H(s_i)$ as determined from previous secret s_i

where H and G are hash functions. The database requires $O(nm)$ computation of hashing for searching of ID, where n is the number of the tag, m is the hash chian size (authentication upper limit), i.e., $1 \leq i \leq m$. The protocol of the Ohkubo scheme is as follows. Also, this is shown in Fig. 13.

Protocol of Ohkubo Scheme

1. A reader broadcasts a query to a tag.
2. The tag computes $a_i = G(s_i)$ from its secret s_i, then renews its secret to $s_{i+1} = H(s_i)$.
3. The tag sends a_i to the reader.
4. The reader sends a_i to the database.
5. The database finds s_1 s.t. $a_i = H^{i-1}(s_1)$, then identifies the corresponding ID.
6. The database sends the ID to the reader.

Table 8 Security of Ohkubo scheme

IND	FS	RA	TK	OT
◯	◯	✕	✕	✕

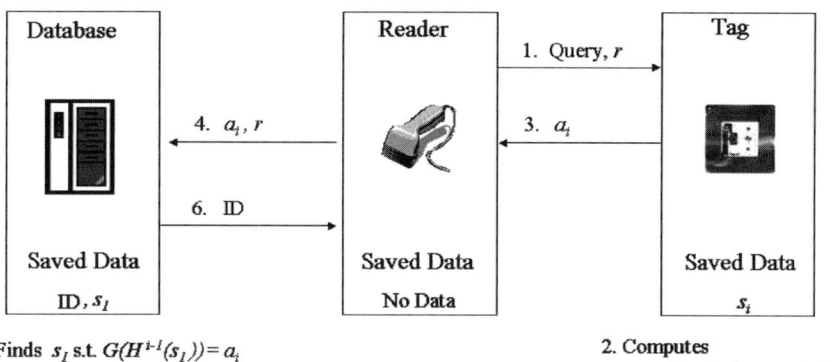

Fig. 14 Protocol of modified Ohkubo scheme

Security of Ohkubo Scheme

In this scheme, the tag outputs a different value and renews the secret information stored in it at each session by using H and G. This scheme therefore achieves indistinguishability and forward security. However, this scheme is vulnerable to a replay attack because one-time password is not used. This scheme is also vulnerable to tag killing due to the authentication upper limit, i.e., $i \leq m$. The security of the Ohkubo scheme is shown in Table 8.

3.2.2 Modified Ohkubo Scheme [3]

The modified Ohkubo scheme [3], based on the Ohkubo scheme, was proposed by Avoine et al. in 2005. They considered a time-memory trade-off to reduce the cost of searching for an ID in the database. When this scheme applies a time-memory trade-off, the searching costs could be reduced to $O(n)$ from $O(nm)$, where n is the number of the tag and m is the hash chain size. The protocol of the modified Ohkubo scheme is as follows. Also, this is shown in Fig. 14.

Protocol of Modified Ohkubo Scheme

1. A reader generates a random number r and sends it together with a query to a tag.
2. The tag computes $a_i = G(s_i \oplus r)$ from its secret s_i and the received r, then renews its secret to $s_{i+1} = H(s_i)$.

Table 9 Security of modified Ohkubo scheme

IND	FS	RA	TK	OT
○	○	○	×	×

3. The tag sends a_i to the reader.
4. The reader sends a_i and r to the database.
5. The database finds s_1 s.t. $a_i = H^{i-1}(s_1 \oplus r)$, then identifies the corresponding ID.
6. The database sends the ID to the reader.

Security of Modified Ohkubo Scheme

This scheme, in the same way as the Ohkubo scheme, achieves indistinguishability and forward security. Moreover, in this scheme, each tag outputs a different value by using random number at each session. Therefore this scheme achieves resistance against replay attack. The security of the modified Ohkubo scheme is shown in Table 9.

3.3 Other Type Schemes

3.3.1 Unidentifiable Anonymous ID Scheme [11]

The unidentifiable anonymous ID scheme [11], a scheme in which each tag stores anonymous ID converted by randomization, symmetric encryption, or public encryption, was proposed by Kinoshita et al. in 2004. The protocol of this scheme consists of a reading phase and an updating phase. In the reading phase, a reader reads an anonymous ID of a tag and gets the ID by sending it to the database. In the updating phase, an anonymous ID of a tag is updated in order to prevent longer term tracing. The protocol of the unidentifiable anonymous ID scheme is as follows. Also, this is shown in Fig. 15.

Protocol of Unidentifiable Anonymous ID Scheme

1. Reading Phase

(a). A reader broadcasts a query to a tag.
(b). The tag sends the anonymous ID to the reader.
(c). The reader sends the anonymous ID to the database.
(d). The database identifies the ID from the anonymous ID, then gives the ID to the reader.

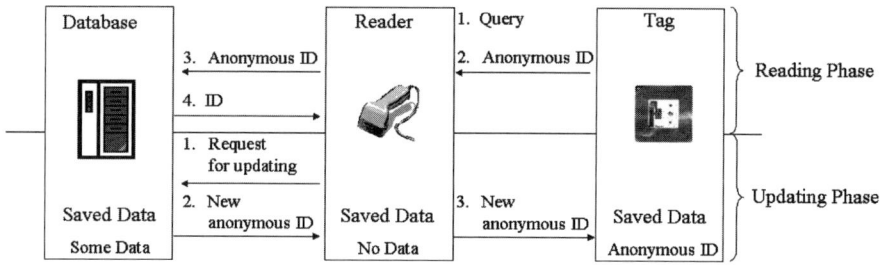

Fig. 15 Protocol of unidentifiable anonymous ID scheme

Table 10 Security of unidentifiable anonymous ID scheme

IND	FS	RA	TK	OT
×	○	×	○	×

2. Updating Phase

(a). A reader sends a request for updating to the database.
(b). The database obtains the key ID included in the anonymous ID, then generates a new anonymous ID and sends it to the reader.
(c). The reader writes the new anonymous ID into the tag.

Security of Unidentifiable Anonymous ID Scheme

In this scheme, the reader writes a new anonymous ID into the tag in the updating phase when the reader wants to update the anonymous ID of the tag. However, the anonymous ID is not updated in each session but regularly. This scheme prevents longer-term tracing but achieves no indistinguishability. On the other hand, even if the anonymous ID is leaked to an adversary, neither the ID nor the past anonymous IDs are available to the adversary unless they have the past symmetric/secret keys. Therefore, this scheme achieves forward security. When the tag receives a query, all tags have to do is to send the anonymous ID to the reader. Therefore this scheme achieves resistance against tag killing. The security of the unidentifiable anonymous ID scheme is shown in Table 10.

3.3.2 Owner Change Scheme [19, 21]

The owner change scheme [19, 21], in which ownership is transferable securely, was proposed by Saito et al. in 2005. Each tag saves the ID, the symmetric key k and the counter C, and achieves ownership transferability by changing the symmetric key k, This scheme consists of only the tag and the reader. The protocol of this scheme

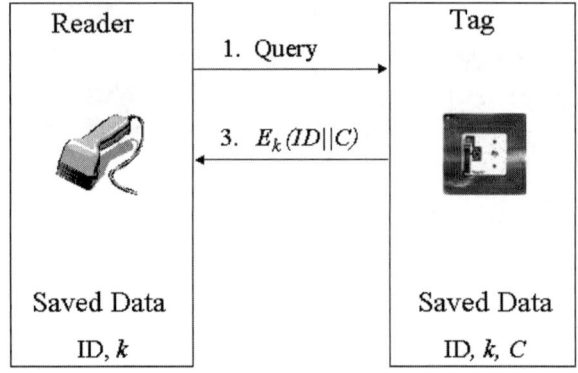

2. Computes $E_k(ID||C)$, $C \leftarrow C+1$

Fig. 16 Authentication protocol of owner change scheme

Table 11 Security of owner change scheme

IND	FS	RA	TK	OT
○	×	×	○	○

consists of a key change protocol and an authentication protocol. For more detail on the key change protocol, refer to [19, 21]. The authentication protocol of the owner change scheme is as follows. Also, this is shown in Fig. 16.

Authentication Protocol of Owner Change Scheme

1. A reader broadcasts a query to a tag.
2. The tag computes $E_k(ID||C)$, where E_k is a symmetric encryption by using symmetric key k and increases C by one.
3. The tag sends $E_k(ID||C)$ to the reader.

Only the legitimate tag owner can decrypt $E_k(ID||C)$ by using the symmetric key k.

Security of Owner Change Scheme

In this scheme, the tag outputs $E_k(ID||C)$ by using the counter C which is increased by one at each session. This scheme therefore achieves indistinguishability. Moreover, this scheme achieves ownership transferability by changing the symmetric key k. The security of the owner change scheme is shown in Table 11.

4 Proposed Method

In this section, we propose an RFID security method which satisfies the security requirements mentioned in Sect. 2.3. It is based on a hash function and a symmetric key cryptosystem. The nomenclature of the proposed method in Table 12.

4.1 Protocol of Proposed Method

The protocol of the proposed method consists of:

- A writing process
- An authentication process
- An ownership transfer process

In the writing process, a manufacturer generates a symmetric key k and writes $E_k(\text{ID})$ into the tag. In the authentication process, the database authenticates the transmitted data from the reader and gives $Info(\text{ID})$ to the reader. In the ownership transfer process, ownership is transferred without violation of previous and present owner's privacy by changing the symmetric key k. In this process, some data (e.g., symmetric key, ID, $Info(\text{ID})$, etc.) are transmitted from the present owner to the new owner.

We prepare some notations in the following. $A \xrightarrow{x} B$ is a transmitting map of data x from entity A to B. $x * y \mapsto z$ is a converting map from data x and y to z by operation $*$. $C(x) \mapsto y$ is a converting map from data x to y by function C. The three processes of the proposed method is as follows. Also, the authentication process of the proposed method is shown in Fig. 17.

4.1.1 Writing Process

A manufacturer generates a symmetric key k, then encrypts the ID of a tag by the encryption E_k. The manufacturer then writes $E_k(\text{ID})$ into the tag by using their reader.

$$\text{Reader} \xrightarrow{E_k(\text{ID})} \text{Tag}.$$

Table 12 The nomenclature of proposed method

ID	Unique number used to identify a product. S bit
a	Converted ID which is outputted by the tag at each session. S bit
r	Random number. R bit
$H(\cdot)$	Collision-resistant random oracle hash function. $H : \{0,1\}^* \to \{0,1\}^S$
k, k'	Symmetric key used for encryption/decryption
$E_k(\cdot)$	Symmetric encryption function by key k
$D_k(\cdot)$	Symmetric decryption function s.t. $\text{ID} = D_k(E_k(\text{ID}))$
e	New symmetric key information s.t. $e = E_k(\text{ID}) \oplus E_{k'}(\text{ID})$
$Info(\text{ID})$	Detailed information of the tagged product which is assigned an ID

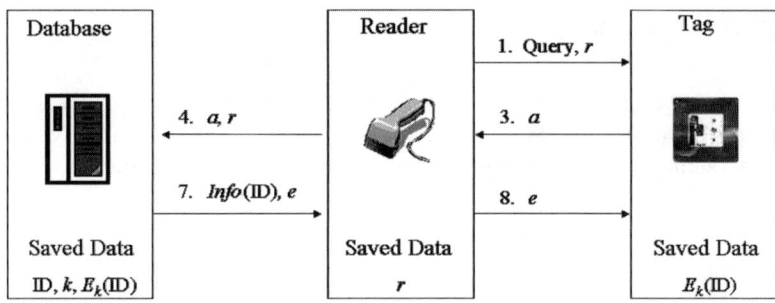

Fig. 17 Authentication process of the proposed method

4.1.2 Authentication Process

1. A reader generates a random number r and sends it together with a query to a tag.

   Reader $\xrightarrow{\text{Query},r}$ Tag.

2. The tag computes a hash value a by the hash function H.

   $$H(E_k(\text{ID}) \oplus r) \mapsto a.$$

3. The tag sends the hash value a to the reader.

   Tag $\xrightarrow{a}$ Reader.

4. The reader sends the hash value a and r to the database.

   Reader $\xrightarrow{a,r}$ Database.

(I) Phase without changing a symmetric key

5. (1) First, the database tries to find an $E_k(\text{ID})$ which satisfies $H(E_k(\text{ID}) \oplus r) = a$ for both a and r received from the reader.

   (2) Second, the database obtains the ID by performing decryption D_k.

   $$D_k(E_k(\text{ID})) \mapsto \text{ID}.$$

   (3) Third, the database finds the $Info(\text{ID})$ from the ID.

6. Finally, the database gives the $Info(\text{ID})$ to the reader.

   Database $\xrightarrow{Info(\text{ID})}$ Reader.

(II) Phase with changing a symmetric key

5. The owner generates a new symmetric key k' and transmits it into the database via a secure wired channel, by using their reader.

$$\text{Reader} \xrightarrow{k'} \text{Database.}$$

6. (1) First, the database performs above phase (I), Step 5, ((1)–(3)).

(2) Second, the database encrypts the ID by new symmetric key k'.

(3) Third, the database computes the new symmetric key information e.

$$E_k(\text{ID}) \oplus E_{k'}(\text{ID}) \mapsto e.$$

(4) Fourth, the database updates the saved data to k' from k and to $E_{k'}(\text{ID})$ from $E_k(\text{ID})$.

7. Finally, the database sends the $Info(\text{ID})$ and e to the reader.

$$\text{Database} \xrightarrow{Info(\text{ID}),e} \text{Reader.}$$

8. The reader sends e to the tag.

$$\text{Reader} \xrightarrow{e} \text{Tag.}$$

9. The tag computes $E_{k'}(\text{ID})$, and updates the saved data to $E_{k'}(\text{ID})$ from $E_k(\text{ID})$.

$$E_k(\text{ID}) \oplus e \mapsto E_{k'}(\text{ID}).$$

4.1.3 Ownership Transfer Process

1. The present owner A changes the symmetric key k to a new symmetric key k' in order to protect their privacy.

2. In order to transfer ownership, the present owner A gives some data (k', ID, $Info(\text{ID})$) to the new owner B in a secure wired channel.

$$\text{Present owner } A \xrightarrow{k',\ \text{ID},\ Info(\text{ID})} \text{New owner } B.$$

3. The new owner B changes the symmetric key k' to a new symmetric key k'' in order to protect their privacy.

4. Then the new owner B will use k'' as their own symmetric key.

In the ownership transfer process, ownership is transferable without violation of the owner's privacy. Although the present owner A has to give their symmetric key to the new owner B, the privacy of the present owner A is protected by changing the symmetric key of A to k'. In the case that the previous owner A can still access tag's secret information after the ownership transfer, the privacy of the present owner B is violated. In the ownership transfer process, the present owner B prevents their violation by changing their symmetric key k' to k'' after the ownership transfer.

4.2 Security of Proposed Method

In this section, we show that our proposed method achieves all of the security requirements mentioned in Sect. 2.3. The security of the proposed method is shown in Table 13.

4.2.1 Indistinguishability (IND)

The tag outputs the hash value a at each session, and it looks random to the adversary due to the randomness of hash functions. The tag's output satisfies the following:

Condition 1. $H(E_k(\text{ID}) \oplus r) \neq H(E_k(\text{ID}') \oplus r)$ for $\text{ID} \neq \text{ID}$

Condition 2. $H(E_k(\text{ID}) \oplus r) \neq H(E_k(\text{ID}) \oplus r')$ for $r \neq r'$

The tag receives the new symmetric key information $e = E_k(\text{ID}) \oplus E_{k'}(\text{ID})$ from the reader. e satisfies the following:

Condition 1. $(E_k(\text{ID}) \oplus E_{k'}(\text{ID})) \neq (E_k(\text{ID}') \oplus E_{k'}(\text{ID}'))$ for $\text{ID} \neq \text{ID}'$

Condition 2. $(E_k(\text{ID}) \oplus E_{k'}(\text{ID})) \neq (E_{k'}(\text{ID}) \oplus E_{k''}(\text{ID}))$ for k, k', and k'' are pairwisely different

If above conditions are all satisfied then Advantage(X) in indistinguishability is negligible, which means that the adversary cannot distinguish tags at all. Therefore, the proposed method achieves indistinguishability.

4.2.2 Forward Security (FS)

The tag saves $E_k(\text{ID})$ which is encrypted ID by the symmetric key k. The saved encrypted ID is updated by changing symmetric key. Even if $E_{k_i}(\text{ID})$ is leaked to the adversary, neither the ID nor $E_{k_j}(\text{ID})$ leaks to the adversary, where k_i represents ith symmetric key s.t. $i > j$. Unless the ID is leaked, the encryption function E satisfies the one-wayness. This means that the adversary cannot compute $E_{k_j}(\text{ID})$ from $E_{k_i}(\text{ID})$ unless the ID is leaked. Therefore Advantage(X) in forward security is negligible, and the proposed method achieves forward security.

Table 13 Security of the proposed method

IND	FS	RA	TK	OT
○	○	○	○	○

4.2.3 Resistance Against Replay Attack (RA)

The tag outputs the hash value a which is different in each session due to the one-time random number. Each random number is used only one time, therefore, the random number r_i in ith session and the random number r_j in jth session are different s.t. $i < j$. Hence, $H(E_k(\mathrm{ID}) \oplus r_i)$, which is eavesdropped by the adversary, and $H(E_k(\mathrm{ID}) \oplus r_j)$, which is generated by the legitimate tag, are also different due to the collision resistant property of hash functions. The tag's output satisfies the following:

$$H(E_k(\mathrm{ID}) \oplus r) \neq H(E_k(\mathrm{ID}) \oplus r') \text{ for } r \neq r'.$$

The adversary therefore cannot spoof a legitimate tag, and the proposed method achieves resistance against replay attack.

4.2.4 Resistance Against Tag Killing (TK)

First, tags in the proposed method do not have to store the random number used as one-time password unlike some challenge–response schemes [10, 14, 18]. Second, the proposed method is not based on hash chain technique [3, 17]. That is, the proposed method satisfies the following:

1. Tags require no additional memory for storing random numbers each reading a tag.
2. Tags have no upper limit of reading tags.

Therefore, the proposed method achieves resistance against tag killing.

4.2.5 Ownership Transferability (OT)

Due to the ownership transfer process, the proposed method satisfies the following:

1. A protocol to be able to securely change an owner's key used for encryption is built into the RFID system.
2. Owners can change their key at any time.

That is, the ownership transfer process enables ownership transfer without violation of the owner's privacy. Therefore, the proposed method achieves ownership transferability.

Remark 1. We considered another approach based on the hash chain technique to achieve the above security requirements. The hash chain technique [3, 17] have two demerits;

1. The tag can only update its secret $(m - 1)$ times, where m is the hash chain size.
2. m cannot be increased, as this would also increase the cost of searching for an ID in the database.

Ohkubo type schemes using hash chain technique are therefore vulnerable to tag killing. In our approach, the cost of searching for an ID in the database does not depend on m because j, the session number, is encrypted and sent to the reader. We can therefore use a large m and prevent tag killing. Moreover, this method is able to transfer ownership by changing the symmetric key which encrypts j.

4.3 Efficiency of Proposed Method

In this section, we compare the efficiency of the proposed method with those of the previous security schemes mentioned in Sect. 3.

Table 14 shows the efficiency using the hash function H, the random number generation RNG, and the encryption/decryption by a symmetric key cryptosystem SKC. Those parameters for estimating the efficiency have been used in previous papers. Let n be the number of tag, and let m be the hash chain size [3, 17].

From Table 14, we can see that the required algorithm and the required memory are not limiting factors because those of the proposed method are comparable with previous security methods. The computation time of the proposed method is faster than all other security methods listed except the Unidentifiable Anonymous ID Scheme.

Table 14 Comparison with some previous security methods

Security methods	Component	Reqd. Algo	Computation time	Reqd. memory
Kang–Nyang's scheme [10]	Tag	1H, 1RNG	3H+1RNG	7
	Reader		No cost	
	Database	1H, 1RNG	$O(n)$H+1RNG	$3*n$
Ohkubo scheme [17]	Tag	2H	2H	1
	Reader		No cost	
	Database	2H	$O(n*m)$H	$1*n$
Modified Ohkubo scheme [3]	Tag	2H	2H	1
	Reader	1RNG	1RNG	1
	Database	2H	$O(n*m)$H	$2*n$
Unidentifiable anonymous ID scheme [11]	Tag		No cost	1
	Reader		No cost	
	Database	1SKC	1SKC	No description
Ownership transfer scheme [21]	Tag	1H, 1SKC	1H+1SKC	3
	Reader		No description	
	Database		No description	
Proposed method	Tag	1H	1H	1
	Reader	1RNG	1RNG	1
	Database	1H, 1SKC	$O(n)$H+1SKC	$3*n$

5 Conclusion

In this chapter, we presented several security requirements for RFID systems. Moreover, we also analyzed previously reported RFID security schemes, and proposed an RFID security method that achieves our security requirements: indistinguishability, forward security, resistance against replay attack, resistance against tag killing, and ownership transferability. None of the previous security schemes has achieved all above requirements. Finally the proposed method is reasonably efficient compared with the previous security methods, e.g., the searching cost of ID in Database is the order of the number of tag.

References

1. Auto-ID Center (2003). 860MHz–930MHz Class 0 Radio Frequency Identification Tag Protocol Specification Candidate Recommendation, Version 1.0.0
2. G. Avoine and P. Oechslin (2005). A Scalable and Provably Secure Hash-Based RFID Protocol. In PerSec 2005, IEEE Computer Society Press, Washington, DC pp. 110–114
3. G. Avoine, E. Dysli, and P. Oechslin (2005). Reducing time Complexity in RFID Systems. In SAC 2005, LNCS 3897, pp. 291–306
4. D.N. Duc, J. Park, H. Lee, and K. Kim (2006). Enhancing Security of EPCglobal Gen-2 RFID Tag against Traceability and Cloning. In SCIS 2006, Proceedings of SCIS 2006, p. 97
5. EPCglobal (2004). EPC Tag Data Standards Version 1.1 Rev. 1.24
6. P. Golle, M. Jakobsson, A. Juels, and P. Syverson (2004). Universal Re-Encryption for Mixnets. In CT-RSA 2004, LNCS 2964, pp.163–178
7. D.G. Han, T. Takagi, H.W. Kim, and K.I. Chung (2006). New Security Problem in RFID Systems "Tag Killing". In ACIS 2006, LNCS 3982, pp. 375–384
8. M.E. Hellman (1980). A cryptanalytic time-memory trade-off. IEEE Transactions on Information Theory, IT-26(4) 401–406
9. D. Henrici and P. Müller (2004). Hash-Based Enhancement of Location Privacy for Radio-Frequency Identification Devices using Varying Identifiers. In PerSec 2004, IEEE Computer Society press, washington, DC, pp. 149–153
10. J. Kang and D. Nyang (2005). RFID Authentication Protocol with Strong Resistance Against Traceability and Denial of Service Attacks. In ESAS 2005, LNCS 3813, pp. 164–175
11. S. Kinoshita, F. Hoshino, T. Komuro, A. Fujimura, and M. Ohkubo (2004). Low-cost RFID privacy protection scheme. IPSJ Journal, 45(8) 2007–2021 (In Japanese)
12. J. Kwak, K. Rhee, S. Oh, S. Kim, and D. Won (2005). RFID System with Fairness within the Framework of Security and Privacy. In ESAS 2005, LNCS 3813, pp. 142–152
13. C.H. Lim and T. Kwon (2006). Strong and Robust RFID Authentication Enabling Perfect Ownership Transfer. In ICICS 2006, LNCS 4307, pp. 1–20
14. D. Molnar and D. Wagner (2004). Privacy and Security in Library RFID: Issues, Practices, and Architectures. In ACM CCS, ACM Press, New York, NY, pp. 210–219
15. Y. Nohara, S. Inoue, K. Baba, and H. Yasuura (2005). Quantitative Evaluation of Unlinkable ID Matching Schemes. In WPES 2005, ACM press, New York, NY, pp. 55–60
16. P. Oechslin (2003). Making a Faster Cryptanalytic Time-Memory Trade-Off. Crypto 2003, LNCS 2729, pp. 617–630
17. M. Ohkubo, K. Suzuki, and S. Kinoshita (2003). Cryptographic Approach to "Privacy-Friendly" Tags. RFID Privacy Workshop
18. K. Rhee, J. Kwak, S. Kim, and D. Won (2005). Challenge–Response based RFID Authentication Protocol for Distributed Database Environment. In SPC 2005, LNCS 3450, pp. 70–84

19. J. Saito and K. Sakurai (2005). Owner Transferable Privacy Protection Scheme for RFID Tags. In CSS 2005, Proceedings of CSS 2005, vol. 1, pp. 283–288 (in japanese)
20. J. Saito, J.C. Ryou, and K. Sakurai (2004). Enhancing Privacy of Universal Re-encryption Scheme for RFID Tags. In EUC 2004, LNCS 3207, pp. 879–890
21. J. Saito, K. Imamoto, and K. Sakurai (2005). Reassignment Scheme of an RFID Tag's Key for Owner Transfer. In ECU 2005 Workshops, LNCS 3823, pp. 1303–1312
22. S.E. Sarma, S.A. Weis, and D.W. Engels (2003). Radio-frequency identification: Security risks and challenges. Cryptobytes (RSA Laboratories), 6(1) 2–9
23. S.A. Weis (2003). Security and Privacy in Radio-Frequency Identification Devices. Master Thesis, University of California Berkeley
24. S.A. Weis, S.E. Sarma, R.L. Rivest, and D.W. Engels (2003). Security and Privacy Aspects of Low-Cost Radio Frequency Identification Systems. In SPC 2003, LNCS 2802, pp. 201–212
25. S.S. Yeo and S.K. Kim (2005). Scalable and Flexible Privacy Protection Scheme for RFID Systems. In ESA 2005, LNCS 3813, pp. 153–163

Digital Signature Transponder

Ulrich Kaiser

Abstract While the first RFID car immobilizer systems were based on read-only transponders, in 1995 followed the Digital Signature Transponder – the first RFID device containing a real encryption module and using the state-of-the-art challenge–response protocol.

After an introduction about the immobilizer operation and system properties the development of the Digital Signature Transponder is described, also discussing different trade-offs. Section 3 covers attack scenarios and the re-engineering work that led to a machine for exhaustive key search. Section 4 concludes with important lessons learnt and future development directions.

1 Introduction

Due to worldwide increases in automobile theft, there is a rapidly growing demand for less obtrusive, yet highly secure "key-based immobilizers." After the introduction of electronic immobilizers in 1993 the number of reported thefts is decreasing again [13]. In addition, a second segment requiring secure transactions, for example the Automatic Recognition of Customers, has been successfully introduced at gas stations (later at fast-food shops) in the USA [39].

Both systems are based on RFID technology [18, 27, 30]. They consist of a transponder and a reader (control unit, transceiver, and antenna). The transponder is a tiny, battery-less, electronic device containing (at least) a small antenna and an IC of mixed-signal technology that controls the transfer of energy and data as well as containing electrical erasable programmable memory (EEPROM) for storing, e.g., for the unique secret code, unique serial number, and configuration bits. The physical characteristics of the tags are related to the frequency range the tags are operating. 13.56-MHz transponders are flat and have mainly printed antennas;

U. Kaiser
Texas Instruments Deutschland GmbH, 85350 Freising, Germany

P. Kitsos, Y. Zhang (eds.), *RFID Security: Techniques, Protocols and System-on-Chip Design*, © Springer Science+Business Media, LLC 2008

whereas transponders operating at 125 or 134.2 kHz use mainly ferrite or air coils with copper wires for the antenna.

In the immobilizer application, the driver places the ignition key in the ignition lock cylinder. The controller detects this and sends a command to the transceiver, which in turn sends out an RF energy burst to activate the transponder followed by a command. After a few milliseconds, the transponder sends back its coded response (modulated RF telegram). The transceiver then demodulates this telegram and passes the code to the controller for verification. If the code is correct, the engine management computer is enabled; without a valid signature the car will not start.

The related system advantages are:

1. Security is based on a passive, contact-less device (no wear-out, no batteries)
2. Unobtrusive to the user (no user actions required, transaction duration less than 200 ms)
3. Confined RF signals (no "broadcast" to thieves (code grabbing))

1.1 Immobilizer Operating Principle

The immobilizer system presented here uses a battery-less transponder working at 134.2 kHz (Fig. 1). Data transmission is achieved by means of pulse width modulation (PWM) for the downlink (line of sight from transceiver to the transponder) and frequency-shift keying (FSK) for the uplink.

The system works in half-duplex (HDX) mode [30]: First, energy and data are transmitted to and stored in the transponder. Second, the data-telegram is sent back to the listening reader unit (see also Chap. 1 for details).

The Immobilizer-IC block diagram is shown in Fig. 2. The external LC tank L_R, C_R, serves to receive energy from the reader unit and for the data exchange. The

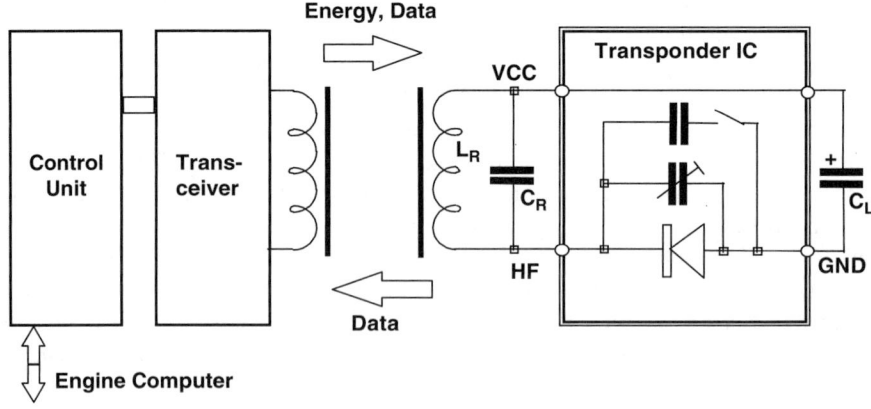

Fig. 1 RFID system block diagram

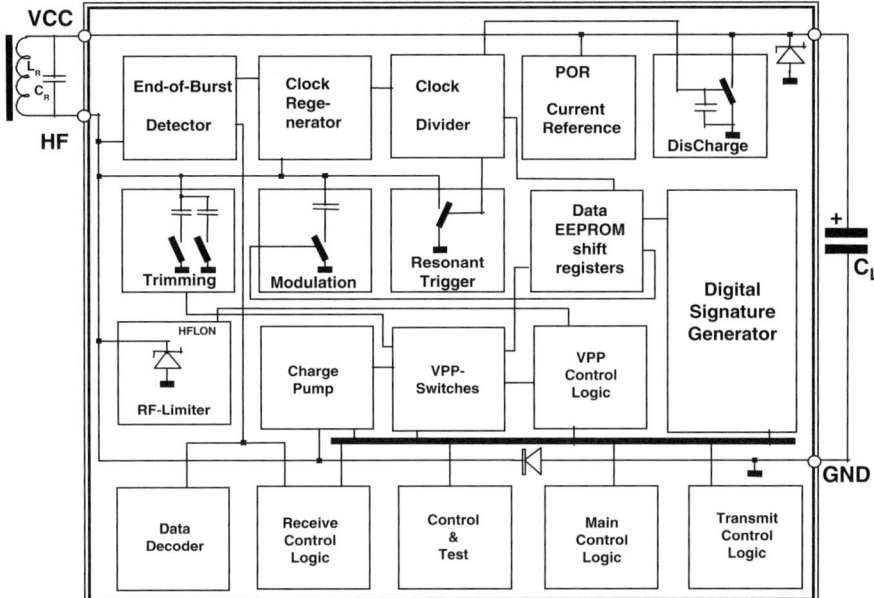

Fig. 2 Block diagram of the integrated circuit

external supply capacitor C_L stores energy for use between bursts and during the uplink phase. The specialized low-power blocks for burst detection, clock recovery, trimming, maintenance of oscillation, charge pump, modulation, power-on-reset, discharge, and limitation are described in [16, 17, 32] in detail.

1.2 Digital Signature Generator Transponder System

Various security levels are possible for immobilizers [6, 8, 13]:

(a) *Fixed code transponders* are true read-only transponders with *unique* numbers.
(b) *Rolling code transponders* modify the code in their EEPROM for each transaction [6].
(c) *Password protected transponders* require a code known only by the microcontroller of the vehicle to trigger the response.
(d) *Digital signature generator transponders* (DSG) use a technique referred to as authentication, encryption, or challenge–response [6, 8, 12, 13, 22] to create the most advanced level of security for immobilizers.
(e) *Mutual authentication transponders* add an encryption cycle to the process to validate the reader (vehicle) before responding [22].

The challenge–response operating principle is based upon the controller sending a random number (of predetermined length) to the transponder, which then

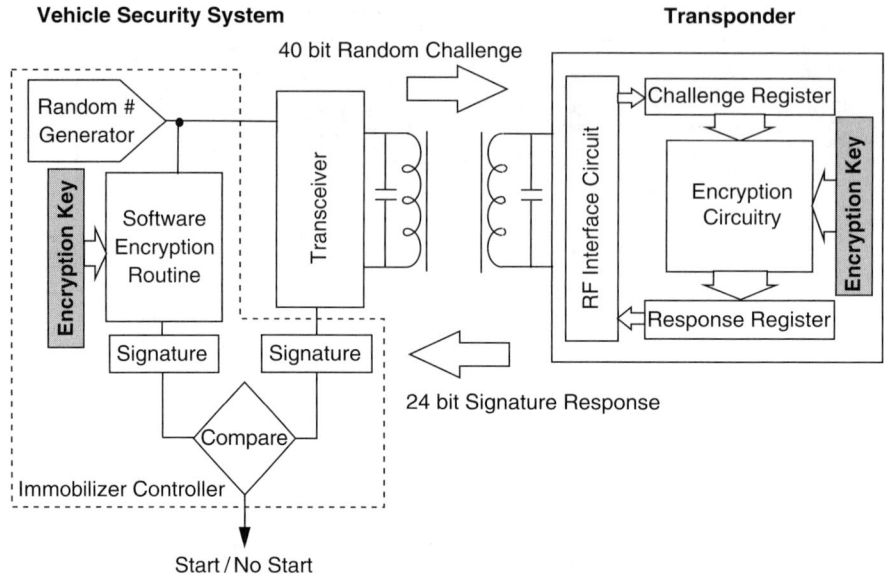

Fig. 3 Digital signature generator transponder system

scrambles it in a unique way such that the controller can authenticate the response as being only from a valid transponder. Each transponder would scramble the challenge in a unique way, based on a hidden encryption key known only by the controller and transponder (Fig. 3).

There are many advantages to this configuration [6].

No portion of the critical encryption key code is ever transmitted (except once at installation and initialization in a secure environment). The encryption key code cannot be read-out or easily determined from the key alone, or a combination of the key and vehicle. Only the immobilizer controller knows what the next proper response will be. Transaction times are fast, i.e., below 150 ms.

Microprocessor software is simplified because no re-synchronization or password routines are needed. This device provides a 24-bit encrypted telegram (signature) in response to a 40-bit random number challenge. Sixteen bits of the 40-bit calculated result are intentionally kept hidden in order to strengthen the encryption system against dictionary attacks [6, 22, 29].

Figure 4 shows the characteristic supply voltage variation of the DSG during the authentication process. After charge-up of the supply capacitor, a command and the random challenge is written [32] into the transponder, and the DSG is initialized. The reader unit provides further energy, so that the response (signature) is calculated in about 3 ms. Then, the response is sent back as documented in [17] in about 15 ms, and at the end the supply capacitor is discharged and the chip is internally reset in order to ensure that the device is accessible again quickly. During the cryptographic processing the total power used by the tag is typically less than 50 μW.

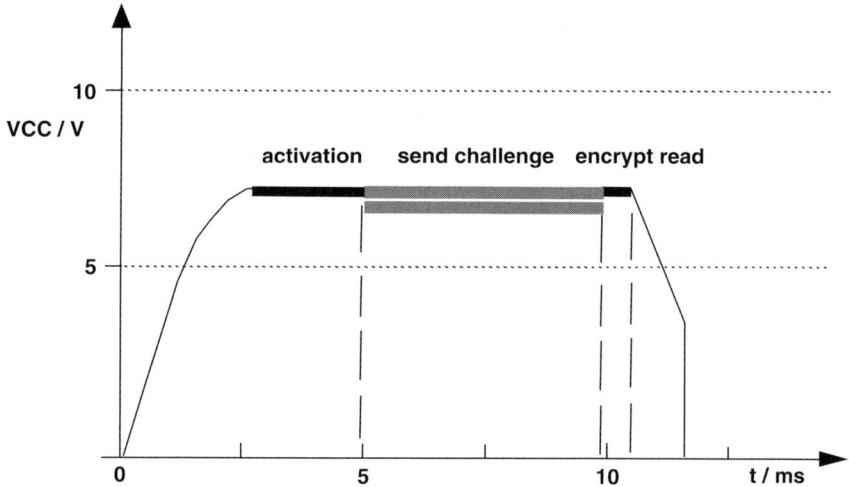

Fig. 4 Transponder charge voltage pattern

2 Development of the DST

2.1 Introduction

The very first electronic car immobilizers were based on the read-only principle, with contact-based or wireless interface. Texas Instruments' read-only transponders were first used in Great Britain to protect against car theft and indiscriminate triggering of airbag systems by juveniles. Insurance companies, in particular, were driving the adoption of electronic immobilizers not only in new cars but also in form of extension kits for older cars, and cars with an immobilizer were assigned a lower insurance rate. It did not take very long before all new cars got electronic immobilizers as standard equipment.

The move from read-only transponders to systems with encryption came in 1995, this being the next logical step [6]. The DST algorithm developed in the year 1995 was the first cryptographic primitive embedded in an RFID device.

In the following, the design objectives during development of the DST algorithm are discussed.

2.2 Commercial Trade-Offs

All cryptographic algorithms can be broken. Security requirements are satisfied if the amount of effort to break the algorithm is uneconomic in relation to the possible gains. These gains are typically financial, but might also involve peer recognition, e.g., within the scientific community.

The broad categories of attack considered are Scanning, Dictionary Attack, and Cryptanalysis [6, 22, 29, 33]. The Cryptanalysis methods considered include, e.g., Exhaustive Key Search, Key Homing, Correlation Attacks, Linearity Attacks, Linear Cryptanalysis, and Differential Cryptanalysis [6, 9–11, 22, 29, 33]. The DST algorithm was designed to frustrate these methods [7, 23, 26, 28].

The designers of the DST algorithm assumed that an attacker:

1. Has a knowledge of cryptanalytic techniques
2. Has a detailed knowledge of the algorithm
3. Has access for 24 h to the targeted transponder prior to an attempted violation
4. May use a computer for 10 days at a private location to help break the system
5. Would not invest in building a specialized hardware based key cracker machine
6. Is restricted to about 10 min at the scene of the attempted violation
7. Is not restricted to experiments with the transponder at the scene of the attempted violation

Hardware, such as an RSA processor with 74 kgates [38], is too expensive for automotive applications. To fulfill the requirements of high speed, low power, and small die size (to keep costs acceptably low) as well as compatibility to the HDX principle, a proprietary cryptographic algorithm was designed. Encryption with a 40-bit key was chosen as a sound compromise between chip size, i.e., cost and the required security level.

Assuming that one encryption takes $400\,\mu s$ on the fastest personal computer of 1995, the exhaustive search takes $2^{40} \times 400\,\mu s = 5,090$ days. Thanks to Moore's law [21] it was known that computer performance would approximately double every 18 months. The consequence of this is a security loss of two key bits every 3 years, resulting in estimated 20 days for an exhaustive search in 2007.

The length of the challenge must be long enough to withstand dictionary attacks but also short enough to fulfill the customer's requirements regarding downlink data communication speed. It was set to 40 bits so that the downlink bit transmission takes less than 40 ms. However, 40 ms is quite long for an attacker and a dictionary of $2^{40} \times 3\,\text{bytes} = 3,300\,\text{GB}$ is quite large.

A more realistic scenario for an attack is the case when a car is in a repair shop for one day and the "bad guy" can collect a large number of challenge–response pairs to build a small dictionary. Assuming a complete protocol takes 125 ms the attacker can store eight pairs per second resulting in 691,200 pairs for the day. Later he can perform 4,800 trials in maximum 10 min with his transponder emulator. Now, the probability of success is $4,800 \times 691,200/2^{40} = 3.017 \times 10^{-3} = 0.3\%$, which is quite low for the effort spent.

The response (signature) of the transponder should be shorter than the result of the cryptographic algorithm in order to hide some bits, provoke collisions, and increase the security strength against roll back attacks. Alternately, a very short response allows guessing, i.e., the attacker ignores the challenge and transmits a random or even a constant number to the reader unit. For the DST40 system a signature of 24 bits was defined resulting in 16 hidden bits and a guessing probability of $2^{-24} = 59.6 \times 10^{-9}$.

2.3 Design Trade-Offs

The first Digital Signature Transponder (DST1) was designed in 1.6-μm CMOS with EEPROM capability, i.e., it carried double-level polysilicon. For signal connections and power routing only one metal level and polybridges were available. The EEPROM was composed of (a) two pages of 8 bits each, (b) one page of 32 bits for the serial number, and (c) one page of 40 bits for the encryption key.

2.4 Layout Considerations

The Boolean functions in the first logic level of the encryption core had to be connected to the shift register outputs of the challenge and the key. There were two possibilities: Either make the logic compact and route the 80 wires, or use a matrix of wires and distribute the logic. The better solution with respect to the layout area was the matrix approach. This kind of ROM-array was built with metal layer and poly layer in perpendicular style. The poly wires were connected to the drivers, i.e., the register outputs, and to related local inverters. The metal wires were connected to the inputs of the next Boolean logic level. Then many N-channel transistors were distributed in the matrix in order to build the AND-terms for the functions in the first logic level [12].

The result of this layout style can be seen in Fig. 5 with the EEPROM key shift register on the left side and the challenge register on the right side of the large Boolean function array. The metal wires in the array are built in vertical direction. The second and third level functions are at the top and bottom of the array, the rest is filled with control logic. This device TMS3791 (DST1) is fabricated in a 1.6-μm CMOS process with EEPROM capability (SLM, DLP, n-well), carries about 14,400 devices, and measures 5.46 mm × 2.1 mm.

After the successful development of the DST1, the next version was designed in order to obtain a smaller chip for cost reduction. This DST2 chip is fabricated in a

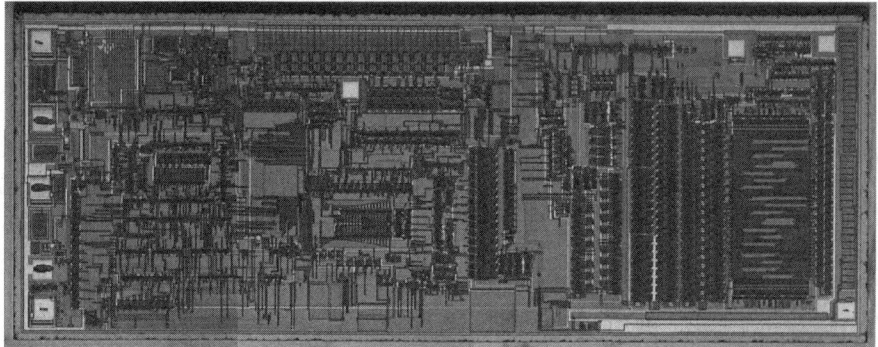

Fig. 5 Chip photograph of the TMS3791 (DST1)

1.2-μm CMOS process with EEPROM capability (SLM, DLP, n-well), carries about 14,500 devices, and measures 2.75 mm × 2.0 mm. The DSG in the DST2 has been optimized with respect to low-power consumption, and small layout area. Although including some control logic and 40 bits of EEPROM, it measures only $1,000 \mu m \times 750 \mu m$.

2.5 Testing

The DST incorporates an autonomous built-in self-test (BIST) block. A special challenge–key pair is chosen so that only 105 clock cycles are required to ensure that every function's output in the Boolean function is toggled at least once; even during this high node activity it takes only 9.3 μA at 5 V operation.

3 Attacks

The DST40 algorithm, a trade secret of Texas Instruments, was made available to customers under individual nondisclosure agreements (NDA). The customers themselves have no interest that the algorithm goes to the public. However, there are always a few people who try to gain something by discovering secrets – whether it be financial reward or peer recognition.

Regarding the discovering work one can identify different scenarios:

1. Reverse-Engineering the integrated circuit (IC)
2. Taking the algorithm as block box and analyzing the input/output dependencies

In 2004, a team of three students under professor Rubin at the Johns Hopkins University (JHU), supported by two scientists from RSA Laboratories, re-engineered the DST40 algorithm using the black box method. They were attracted by the controversy about RFID privacy and the huge number of car immobilizers based on the DST. Their work was published for the first time on January 28 2005 in a paper of 19 pages [1].

The JHU team bought an inexpensive low-frequency design kit and some unprogrammed DST devices. They learned how the RFID system works, the details of the protocols, the programming of the device, and reading distance limitations. Since they could program encryption keys into the devices they could collect key–challenge–response triples at will. Using this information the team first programmed devices with the constant 0×0000000000 encryption key in order to bring the key scheduling process to a stuck state. The Boolean core logic was re-engineered by means of hypotheses (e.g., oracles, models), proofs and counterevidences guided by the challenge–response pairs under the constant key. This reverse-engineering method is described in detail in [1, 34]. However, this process was hampered by inaccuracies in the schematic sketch.

To overcome such a lack of information one can guess more alternatives leading to even more experiments, analyze the power profile [19], or peek into the reassembled software. After the Boolean functions and their signal routing were found the key scheduling was left for black box analysis. The taps of the 40-bit LFSR were found and there were only three different alternatives for the clock gating counter.

The next step of the JHU team was to implement the algorithm in software. They found that the code was too slow to build a key-cracker; even after hand-optimizations the computation rate was less than 200,000 encryptions per second on a 3.4-GHz workstation leading to over 20 weeks of exhaustive search for a single key.

A cryptanalytic inspection of the algorithm did not reveal any weakness that could be exploited to shorten the key search process. The team therefore built a hardware key cracker using a FPGA board. The FPGA had 32 algorithm cores implemented in parallel. With its 100-MHz clock the board was able to test nearly 16 million keys per second against a chosen challenge–response pair. Due to the collisions provoked by the shortened 24-bit response, a second challenge–response pair had to be tested, too. Expanding the scope, the team built a larger version of the key cracker consisting of 16 FPGA boards, with each board covering a 16th of the whole key space resulting in a crack time of less than 1 h.

In order to demonstrate the successful discovery of the DST40 algorithm, the JHU team built up a set of special hardware working as a transponder emulator [1, p. 16ff] and they started their own Ford Escape 2005 model. Additionally, they presented how one can collect challenge–response pairs of an external RFID device with help of a bag containing a battery-powered reader system and a notebook while the other person sitting nearby is unaware that data is being "stolen" and their privacy being violated.

Although these demonstrations seem very dramatic it is quite difficult to build small clone devices.

4 Conclusion

This chapter contains two parts because on one hand it is important to give a summary about the gathered experiences and lessons learnt, on the other hand future directions need to be depicted because the developments in the field of cryptography will never stop.

4.1 Lessons Learned

Based on the experiences described in Sects. 1–3 several conclusions and lessons can be established:

1. The DST transponder design has been a commercial success.

2. Respecting Kerckhoff's principle [22] is important, but in some cases hiding algorithm details buys time and reduces costs.
3. Security requirements should be reviewed every year with respect to the changing environment. Even if only small clouds appear on the horizon one should plan ahead for the next security concept.
4. Build a cryptographic module that is easily extendable for wider key sizes, e.g., doubled key size.
5. Do not sell unprogrammed cryptographic devices freely.

4.2 Current and Future Developments

Regarding software implementation, the DST40 algorithm is slow. In contrast, during the design of Rijndael [4] the designers considered both hardware and software applications. The same holds for the design of UICE [14, 36], which should be fast in software, even on small microprocessors having only an 8-bit data path.

Figure 6 shows the data flow of the UICE core. One sub-round consists of the following operations:

(a) Get the Accu content and one challenge-byte, addressed by pointer p1
(b) Perform the first EXOR operation
(c) Get one key-byte, addressed by pointer p2
(d) Perform the second EXOR operation
(e) Apply the 8×8 S-Box function
(f) Store the result in the Accu and overwrite the challenge-byte
(g) Update pointer p1 to the challenge register bank and p2 to the key register bank.

Although the content of the final S-Box [31, 35, 37] does not need to be defined in detail, the relative run times can be determined, here (as presented in Table 1) by means of a notebook ASUS 7400 with Pentium 2, 400 MHz, Linux SuSE 8.0, gcc

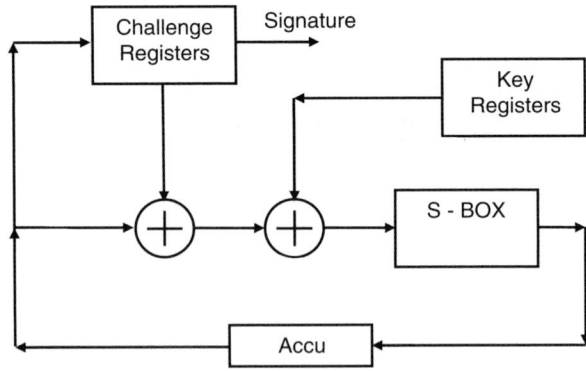

Fig. 6 UICE algorithm subround principle

Table 1 Run times of different cryptographic algorithms

Algorithm	Time (μs)	Challenge (bytes)	Key (bytes)	Cipher Rounds	Notes
DST40	480	5	5	10	(1)
UICE40	5.6	5	5	10	(2)
UICE64	8.6	8	8	10	(2)
UICE128	8.6	8	16	10	(2)
AES128	163	8	16	10	(3, 4) [14]

Note (1) DST40 is based on many shift register operations, which are slow to implement in software

Note (2) UICE is based on byte-operations, which can be executed quickly in software

Note (3) AES (i.e., Rijndael) was chosen as an appropriate compromise between hardware and software applications

Note (4) AES128 is based on rijndael-alg-ref.c and needs three tables of 256 bytes each and also a table of 30 bytes [pulled from http://www.esat.kuleuven.ac.be/~rijmen/rijndael/index.html], but with the code for key length 192 and 256 eliminated. The 16 bytes per block are filled with eight bytes of the challenge; the rest is filled with zeros. As result only eight bytes are taken

2.95.3, and ANSI-C. It is impressive that UICE40 is 86 times faster than DST40, and that UICE128 is 19 times faster than AES128. The details about UICE can be found elsewhere [15].

Despite new proprietary developments, there is also AES128 [4], an attractive alternative for RFID, if the design-to-cost model fits. High-speed variants such as [20, 25] are too costly with respect to IC area, while a slow variant such as [5] needs only 3,400 gates but as many as 1,032 clocks. If AES is not suffering from algebraic attacks [2, 3] in the near future it is probably suitable for immobilizers and other parts of the car that are linked by means of cryptographic protocols for theft prevention and communication security since it is a FIPS standard [24].

References

1. Bono et al. 2005, Security Analysis of a Cryptographically-Enabled RFID Device, Johns Hopkins University and RSA Laboratories, Baltimore, MD, 28 January 2005, http://www.rsa.com/rsalabs/staff/bios/ajuels/publications/pdfs/DSTbreak.pdf
2. Courtois 2004, General Principles of Algebraic Attacks and New Design Criteria for Components of Symmetric Ciphers. Invited talk, AES 4 Conference, Bonn, May 10–12 2004, LNCS 3373, pp. 67–83, Springer, Berlin
3. Courtois 2006, How Fast can be Algebraic Attacks on Block Ciphers? http://eprint.iacr.org/2006/168
4. Daemen and Rijmen 1999, AES Proposal: Rijndael, Version 2, 03/09/99, 45 pages and related reference code in C http://www.esat.kuleuven.ac.be/~rijmen/rijndael/rijndaelref.zip
5. Feldhofer et al. 2004, Strong Authentication for RFID Systems Using the AES Algorithm, CHES 2004, LNCS 3156, Springer, Berlin
6. Gordon 1994, Designing Codes for Vehicle Remote Security Systems, Concept Laboratories Ltd. and Police Scientific Development Branch, Hertfordshire, GB, pp. 1–22

7. Gordon and Retkin 1981, Are Big S-Boxes Best?, IEEE Workshop on Communication Security, Santa Barbara, pp. 1–6
8. Gordon et al. 1996, A Low Cost Transponder for High Security Vehicle Immobilizers, Proceedings of ISATA, Florence, Italy, 3–6 June 1996, Automotive Electronics, 96AE001
9. Heys 2001, A Tutorial on Linear and Differential Cryptanalysis, Technical Report CORR 2001-17, March 2001, http://www.engr.mun.ca/~howard/PAPERS/ldc_tutorial.ps
10. Heys and Tavares 1994, On the Design of Secure Block Ciphers, Queen's 17th Biennial Symposium on Communications, Kingston, Ontario, Canada, May 1994, 6 pages
11. Heys and Tavares 1996, Substitution–Permutation Network Resistant to Differential and Linear Cryptanalysis, Journal of Cryptology, 9(1): 1–19
12. Kaiser 1999, A Low-Power Digital Signature Transponder IC for High Performance RFID Authentication, Proceedings of European Conference on Circuit Theory and Design, ECCTD'99, Stresa, Italy, Aug. 29–Sep. 02 1999, pp. 45–48
13. Kaiser 2003, Theft Protection by means of Embedded Encryption in RFID-Transponders (Immobilizer), ESCAR – Embedded IT-Security in Cars, Bochum, Germany, 18–19 Nov. 2003
14. Kaiser 2004, UICE Universal Immobilizer Crypto Engine – The Little Brother of AES, AES4, May 2004, Bonn, http://www.aes4.org/english/events/aes4/downloads/AES4_UICE_slides.pdf
15. Kaiser 2007, UICE: A High-Performance Cryptographic Module for SoC and RFID Applications, http://eprint.iacr.org/2007/258
16. Kaiser and Steinhagen 1994, A Low Power Transponder IC for High Performance Identification Systems, Proceedings of CICC'94, San Diego, CA, USA, May 1–4 1994, pp. 14.4.1–14.4.4
17. Kaiser and Steinhagen 1995, A low power transponder IC for high performance identification systems, IEEE Journal of Solid-State Circuits, 30: 306–310
18. Kern 1999, RFID-Technology – Recent Development and Future Requirements, Proceedings of European Conference on Circuit Theory and Design, ECCTD'99, Stresa, Italy, Aug. 29–Sep. 02 1999, pp. 25–28
19. Kocher et al. 1999, Differential Power Analysis, Advances in Cryptology, CRYPTO'99, LNCS 1666, 10 pages, Springer, Berlin
20. Kuo and Verbauwhede 2001, Architectural Optimization for a 1.82 Gbits/sec VLSI Implementation of the AES Rijndael Algorithm, CHES 2001, LNCS 2162, pp. 51–64, Springer, Berlin
21. Loney 2003, Moore's Law is the biggest thread to privacy, according to Phil Zimmermann, news.zdnet.co.uk and www.silicon.com, 29 April 2003
22. Menezes, Oorshot and Vanstone 1997, Handbook of Applied Cryptography, CRC, Boca Raton, FL
23. NIST 2001, FIPS 140-2, Security Requirements for Cryptographic Modules, May 25, 2001, http://csrc.nist.gov/cryptval and http://csrc.nist.gov/publications/fips/fips140-2/fips1402.pdf
24. NIST 2001, National Institute of Standards and Technology, FIPS PUB 197, 26 May 2002, http://csrc.nist.gov/publications/fips/fips197/fips-197.pdf
25. Rejeb et al. 2003, Hardware Implementation of the Rijndael Algorithm for High-Speed Networks, ISPC 2003, Dallas, TX, 6 pages
26. Ritter 1997, S-Box design: A Literature Survey, Research Comments, http://www.ciphersbyritter.com/RES/SBOXDESN.HTM
27. Sarma et al. 2002, RFID and Security and Privacy Implications, CHES 2002, LNCS 2523, pp. 454–469, Springer, Berlin
28. Scheerhorn 1997, DSG Algorithm Evaluation, EVAL11.DOC, CCI, Meppen, 2 Dec. 1997
29. Schneier 1995, Applied Cryptography, Addison Wesley, Reading, MA
30. Schuermann and Meier 1993, TIRIS – Leader in Radio Frequency Identification Technology, Texas Instruments Technical Journal, 10(6): 2–14
31. Seberry et al. 1994, Pitfalls in Designing Substitution Boxes, S-Box, Crypto'94, p. 383ff
32. Steinhagen and Kaiser 1994, A Low Power Read/Write Transponder IC for High Performance Identification Systems, Proceedings of ESSCIRC'94, Ulm, Germany, September 20–22, 1994, pp. 256–259

33. Stinson 1995, Cryptography, Theory and Practice, CRC, Boca Raton, FL
34. Vahlis 2005, Security Analysis of a Cryptographically Enabled RFID, University of Toronto, http://www.eecg.toronto.edu/~lie/Courses/ECE1776-2005/Presentations/RFID-Eugene.pdf
35. Webster and Tavares 1986, On the Design of S-Boxes, Proc. CRYPTO 1985, LNCS 218, pp.523–534, Springer, Berlin
36. Wollinger et al. 2000, How Well are High-End DSPs Suited for the AES Algorithms? – AES Algorithms on the TMS320C6x DSP, The Third Advance Encryption Standard (AES3) Candidate Conference, New York, April 2000, 11 pages
37. Xu and Heys 1997, A New Criterion for the Design of 8×8 S-Boxes in Private-Key Ciphers, IEEE Canadian Conference on Electrical and Computer Engineering (CCECE'97), May 1997
38. Yang et al. 1998, A new RSA cryptosystem hardware design based on Montgomery's algorithm, IEEE Transactions on Circuits and Systems II, 45(7): 908–913
39. Speedpass, http://en.wikipedia.org/wiki/Speedpass

Scalability Issues in Privacy-Compliant RFID Protocols

Gildas Avoine

Abstract Like all growing technologies, radio frequency identification brings along its share of security-related problems. Such problems are impersonation of tags, denial of service attacks, leakage or theft of information, malicious traceability, etc. to name a few.

To carry out her attack, an adversary can try to penetrate into the back-end database, to tamper with some tags, or she can try to eavesdrop or even modify the information exchanged between the tags and the readers. The latter approach is the one we focus on in this chapter: We address the conception of tag–reader protocols that avoid malicious traceability. Finding such a protocol is far from being an easy task, due to the weak resources available on tags. Indeed, we consider that tags are not able to use public-key cryptography. With such an assumption, protocols that resist to malicious traceability do not scale well, and so cannot be used in most of the current applications.

In what follows, we recall the basic knowledges about RFID protocols and malicious traceability. Then, we present protocols that scale well but which are not secure. We so exhibit common design-related mistakes one can encounter when analyzing RFID protocols. Next, we introduce protocols based on the well-known challenge–response scheme. We explain why they are secure, but also why they do not scale well. In the last part of this chapter, we present techniques that have been suggested to reduce the computation complexity of challenge–response-based protocols.

G. Avoine
UCL, Louvain-la-Neuve, Belgium
e-mail: gildas.avoine@uclouvain.be

P. Kitsos, Y. Zhang (eds.), *RFID Security: Techniques, Protocols and System-on-Chip Design,* © Springer Science+Business Media, LLC 2008

1 Introduction

1.1 Goal of the Protocol

A major issue when designing a protocol is defining its purpose. This issue, as trivial as it may seem, does not have any immediate solution when we speak of RFID. Determining if it is an identification or an authentication protocol that is required, and knowing what each term signifies remains a confusing issue and has provoked several misunderstandings.

Indeed, when we analyze RFID applications, it is possible to think of two large application categories: those whose goal is to provide security to a system, and those whose goal is to provide functionality, in other term users' convenience, without any security concerns. In the first category, we find applications such as access control (badge to access a restricted area, car ignition key, etc.) or prevention against counterfeits. So, the objective is to ensure the tag authentication. In the second category, we find RFID being used to improve supply chains, to facilitate location and stocking of books in libraries, to localize people in amusement parks, to facilitate sorting of recyclable material, to improve consumer services in shops, to count cattle, etc. Clearly, the security aspect is not the driving force in this category. Here, the objective is simply to obtain the tag's identity, without any proof being required.

We define below the concepts of *authentication, mutual authentication,* and *identification.*

Definition 1 (Authentication Protocol). An authentication protocol allows a reader to be convinced of the identity of a queried tag. Conversely, it can allow a tag to be convinced of the identity of a querying reader. If both properties are ensured, we speak of *mutual* authentication.

Definition 2 (Identification Protocol). An identification protocol allows a reader to obtain the identity of a queried tag, but no proof is required.

The basic *identification* protocol that is used in today's applications (e.g., pet identification) is depicted in Fig. 1. It consists in a request from the reader to the tag and an answer from the tag to the reader, containing the identifier ID of the tag. If the tag is a legitimate one, then the system's database also contains ID. This protocol is simple, sound, but we will see in Sect. 1.5 that it brings out privacy issues.

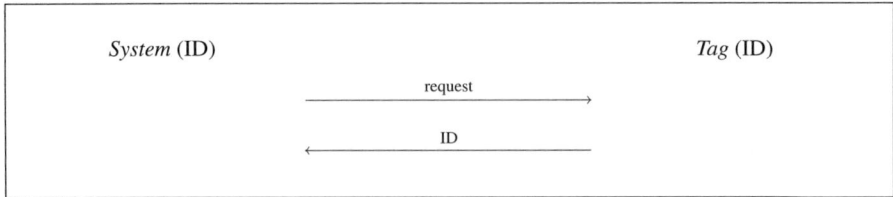

Fig. 1 Basic identification scheme

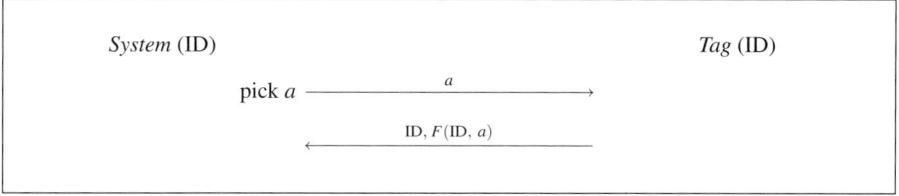

System (ID)   *Tag* (ID)

$$\text{pick } a \xrightarrow{\quad\quad a \quad\quad}$$

$$\xleftarrow{\quad \text{ID}, F(\text{ID}, a) \quad}$$

Fig. 2 Basic challenge–response authentication scheme

The basic *authentication* protocol that is used in today's applications (e.g., car ignition key) is depicted in Fig. 2. It consists in a common challenge–response protocol: the reader sends a nonce a to the tag and the latter answers ID and $F(\text{ID}, a)$ where F is a given function, e.g., an encryption function. Again, this protocol endangers the privacy of the tag's bearer as explained in Sect. 1.5.

1.2 Toward a Definition of Untraceability

Designing and analyzing RFID protocols, either for identification or authentication, is still a real challenge because no universal adversary model has been defined yet: up until now designs and attacks have been made in an ad hoc way.

Even though we are aware of concepts such as *chosen plaintext attacks* (CPA), *nonadaptive chosen ciphertext attacks* (CCA1), or *adaptive chosen ciphertext attacks* (CCA2) for confidentiality, and concepts such as *known-message attacks* (KMA) and *adaptive chosen message attacks* (CMA) for signature schemes, in radio frequency identification, the adversary's resources are defined in an ad hoc manner. Depending on the publications, the adversary is passive, active but limited in the number of successive queries to the tag [26], active but cannot modify the exchange between a legitimate reader and a tag [29], active but cannot tamper with the tags, or finally the only restriction of the adversary can be the reader's tamper-resistance [38].

Note that tags' tamper-resistance is a controversial issue. The question to be asked is whether it would be possible to obtain data from a tag by carrying out a physical attack on it [2, 44]. Tampering with a tag is possible but expensive. So, the adversary's benefit from tampering with a tag must be less than the cost of the attack. Consequently, the common assumption is that tags should not share secret information because such an approach increase the benefit of an attack: Tampering with one tag would reveal secret information about the other tags.

The goal of the adversary is also not clearly defined today. While a public key encryption scheme verifies, for example, the properties of *indistinguishability* (IND) or of *nonmalleability* (NM), or that a signature scheme is resistant to *forgery* or to *total break*, the concept of *untraceability* is not yet formally defined, although some works have been done in that direction [4, 30].

In this chapter, we informally define *untraceability* as follows: Given a set of readings between tags and readers, an adversary must not be able to find any relation between any readings of a same tag or set of tags. Since tags are not tamper-resistant, a priori, an adversary may even obtain the data stored in the tags' memory in addition to the readings from the readers/tags. Thus, she might become capable of tracing the tags' past events, given their content. Therefore, we define *forward untraceability*: Given a set of readings between tags and readers and given the fact that all information stored in the involved tags has been revealed at time t, the adversary must not be able to find any relation between any readings of a same tag or set of tags that occurred at time $t' \leq t$. *Privacy* also includes the fact that a tag must not reveal any information about the kind of item it is attached to. These rather informal definitions are needed to analyze the protocols given in Sects. 2 and 3. In Sect. 4, we provide a more formal but still ad hoc definition of untraceability in order to analyze Molnar and Wagner's protocol [35].

1.3 Why Dealing with Malicious Traceability is Needed

Tags are not cut out to contain or transmit large quantities of information. When a database is present in the system, the tag may only send a simple identifier, which only people having access to the database can link up to the corresponding object. However, even if an identifier does not allow obtaining information about the object itself, it allows us to trace it, that is to say, to recognize the object in different places and/or at different times. Thus we can know when a person passed through a given place, for example to work out his time of arrival or departure from his place of work. We could also piece together, from several readers, the path taken by a person, for example in a store or shopping mall.

Other technologies also permit the tracking of people, e.g., video surveillance, GSM, Bluetooth. However, RFID tags permit everybody to track people using low cost equipment. This is strengthened by the fact that tags cannot be switched off, they can easily be hidden, their lifespan is not limited, and analyzing the collected data can be efficiently automated.

Advocates of this technology arduously refute the argument that RFID puts respect for privacy in peril. A maximum reading distance reduced to a few centimeters is the principal defense argument. From the opposition's point of view, the short reading distance is not a relevant security argument. Indeed, by using a more efficient antenna and a stronger power, it is possible to go beyond the presupposed limit. Moreover, there are many cases where an adversary can get close enough to her victim to read his tags: on public transport, waiting in line, etc. Finally, the last argument of any substance is that the trend is not toward HF systems that have a short communication distance, but toward UHF systems, where the communication distance is a few meters.

Many voices have spoken out through several boycott campaigns, aimed at reversing the trend for omnipresent tags in everyday life. CASPIAN (Consumers Against Supermarket Privacy Invasion and Numbering), an organization for the defense of individual liberties led by Albrecht, have called for a moratorium on the use of tags for individual items. Wal-Mart, Procter & Gamble, Gillette and Tesco, who are not in favor of the moratorium, have had to face particularly virulent boycott campaigns. In Germany, it is the Metro group who has suffered such attacks at the hands of FoeBud, a militant human rights organization. They issued a "Big Brother Award" to Metro for their experimental shop in Rheinberg, where customers' loyalty cards unknowingly contained tags.

Consequently, even if the majority of people are not worried about the threat that RFID poses to privacy, the weight of organizations such as CASPIAN and FoeBud is sufficiently great to slow down the deployment of the RFID market. The problem of malicious traceability of individuals is thus an aspect of RFID that has to be taken into account and treated seriously.

1.4 Palliative Measures Protecting Privacy

The first technique that can be used to protect the tag's holders is to kill the tags. It is well suited for example to supply chains: when the tag reaches the end of the chain, e.g., during the shop checkout, it is killed. The technique is effective but has several drawbacks: The management of the keys is complex in case of keyed kill-command, the tag can no longer be used afterward, and there is no confirmation of successful disablement.

To avoid these drawbacks, Karjoth and Moskowitz [32] suggest the *clipped tags*. These devices are tags whose antenna can be physically separated from the chip. The bearer can carry out this operation by himself, and reactivation of the tag can only be done intentionally.

A less radical method consists of preventing the tag from hearing the request by enclosing it in a Faraday cage. This solution is only suitable for a few precise applications, e.g., money wallets or passports, but is not for general use: Animal identification is an example of an application that could not benefit from this technique. Today, US passports – which include an RFID tag – contain a thin radio shield in their cover, preventing the tags from being read when the passports are closed.

The third technique consists of preventing the reader from understanding the reply. The best illustration of this technique is surely the *blocker tag* [27,31] that aims at preventing a reader from determining which tags are present in its environment. Roughly speaking, the blocker tag relies on the tree walking anticollision protocol and simulates the full spectrum of possible identifiers.

Another technical approach consists in requiring an optical contact with the object prior to querying its tag remotely. Such an approach is very restrictive but is suited to certain applications. It is an approach suggested by Juels and Pappu [28] to protect banknotes. It is also the approach that is recommended by the International Civil Aviation Organization (ICAO) to protect the privacy of the passport bearers [24]: The officer must swipe the passport through an optical reader to get the key that gives access to the electronic data inside the tag.

Finally, a rather different and complementary approach is that of Garfinkel, who elaborates the so-called "RFID Bill of Rights" [15, 16], which outlines the fundamental rights of the tag's bearers. In the same vein, Albrecht proposed the "RFID Right to Know Act of 2003", which requires that commodities containing RFID tags bear labels stating that fact.

Even though the methods presented here are efficient, they have many constraints that render them inconvenient. Thus the objective is to conceive protocols that protect the holders' privacy without imposing any constraints on them.

1.5 Protocols Resistant to Malicious Traceability

In Sect. 1.1 we have seen RFID identification and authentication schemes. Unfortunately, the two presented sketches did not protect privacy at all since *the tag identifier is sent in clear to the reader*.

The naive idea in which the tag encrypts its identifier before sending it, with a key it shares with the reader creates more problems than it solves.

Indeed, if the same key is used by all tags then obvious security problems arise if the adversary tampers with one tag: she so obtains the key used by all the tags.

If a different key is used for each tag, then we have to consider two cases. In the first case, the encryption is deterministic, that is, the tag always sends back the same value, but with this, the traceability problem is not solved, as the ciphertext becomes the "marker" that can be used to trace the tag. In the second case, the encryption is randomized, which poses a complexity problem, as the reader has to test all the keys present in its database to find the one that matches the queried tag. This method, which is the only current one that ensures privacy with a realistic adversary model, is in fact a challenge–response where the verifier does not know the prover's identity. This is a fundamental point that differs from traditional cryptographic assumptions, where we generally suppose that each entity participating in the communication protocol knows the identity of the other entities participating in the same protocol.

Of course, a public-key *encryption* scheme could easily solve this problem: the prover encrypts his identity with the public key of the verifier. Thus, no eavesdropper is able to identify the prover. We stress that a public-key *authentication* scheme is not better than a symmetric authentication scheme in the following sense: in both cases, the system must test all the keys present in its database in order to identify the tag. A public-key authentication scheme may possibly reduce the computation on the tag's side (e.g., [17]) but computation on the system's side may become unmanageable when all keys must be tested. Anyway, we assume from the beginning of this chapter that public-key cryptography is too heavy to be implemented within low cost tags. Thus, we only deal with tags capable of using symmetric cryptography, in the best case.

To ensure untraceability, the key-point is that the adversary should not be able to distinguish the information sent by the tag from a random value. In other words, each time the tag is queried, it should send a new random pseudonym to the reader, and only the latter knows that all these pseudonyms belong to the same person. The pseudonym sent by the tag is either the tag identifier, which is then refreshed at each new identification, or is the encrypted version of the identifier; In the latter case, the identifier contained in the tag can remain unchanged, but the encryption must be randomized. What differentiates the existing protocols is the way in which information sent by the tag, i.e., the pseudonyms, is refreshed. We can classify the protocols into two categories, those where the information refreshment is reader-aided and those where the information is refreshed by the tag itself, without the reader's help.

1.5.1 Protocols Based on Reader-Aided ID-Refreshment

In the case where the reader participates in refreshing information sent by the tag, the RFID protocol is usually a 3-moves protocol. As shown in Fig. 3, the reader first sends a request to the tag; then the tag replies by sending an information that allows its identification, that is, its pseudonyms; and finally the reader sends data that allows the tag to refresh the information it would send during the next identification.

In this type of protocol, the difficulty lies in ensuring that the reader correctly carries out its work, that is, the sent data should allow proper refreshing of the information sent by the tag. If the adversary is active and not constrained on the number of successive requests that she can send to the tag, then she will eventually be able to trace the tag or ensure that the system itself can no longer identify it.

Thus from a traceability point of view, these protocols, for example, [18, 23, 26, 28, 43], are only secure when considering a weak adversary model. When designing

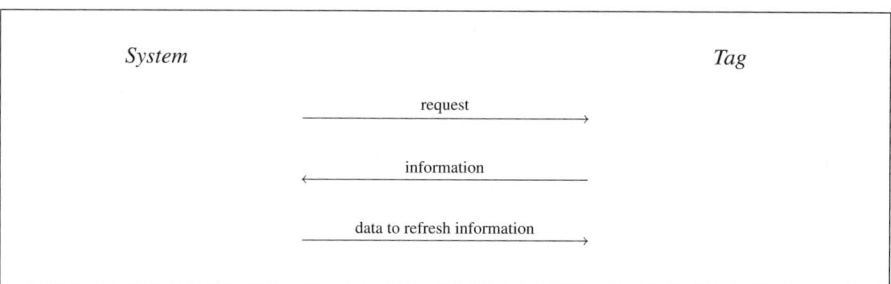

Fig. 3 Sketch of an identification protocol

such protocols, the challenge is thus to obtain the strongest possible adversary
model, that is, to render the adversary's life hard. Section 2 introduces some protocols based on this approach.

1.5.2 Protocols Based on Self-Refreshment

Protocols that rely on the fact that tags refresh themselves their identifiers, for example, [14, 35], are usually 2-moves or possibly 3-moves protocols when there is a
mutual authentication between the reader and the tag. These protocols usually use
a hash or an encryption function on the tag's side and thus are not suitable for very
low cost tags. Such protocols are given in Sect. 3.

2 Protocols Based on Reader-Aided ID-Refreshment

To avoid a tag being traced, one manner is modifying its identifier (ID) such that
only an authorized party is able to link the successive ID modifications.

A basic way is storing inside the tag a list of identifiers called pseudonyms, which
are used sequentially and cyclically. This simple idea, however, requires a large
amount of memory to store the pseudonyms, otherwise, the tag becomes traceable
as soon as the pseudonyms have been used once.

A more sophisticated approach consists in refreshing the tag identifier using a
deterministic or randomized process. We provide here a few examples of such protocols based on reader-aided ID-refreshment, that is protocols where the tags identifiers are refreshed by the reader, avoiding heavy computations on the tags. For each
protocol, we exhibit weaknesses or attacks that endanger the privacy of the tags. In
some cases, the threat appears because the authors focused on a too restricted adversary model that is not realistic. In some other cases the threat is due to weaknesses
in the protocol design.

Whatever the protocol, they all suffer from a common weakness, inherent to
the protocols that involve the reader in the refreshment process: The tag is always
traceable between two legitimate identifications. Indeed, the identifier is not refreshed between two successful identifications of the tag by the system that owns
or manages it. Other attacks rely on the fact that the tag sends values that are distinguishable from random values, e.g., an identification session counter. Another
important problem that we explain below is the desynchronization between the
tag and the system: Some protocols assume a kind of synchronization between the
system and the tags it manages. For example, it can be the last used pseudonym or
the number of the last successful identification. Finding a way to desynchronize the
system and the tags is therefore a very efficient way to trace the tags because the
identifiers can usually not be refreshed subsequently. We illustrate these recurrent
weaknesses in the examples given below.

2.1 Henrici and Müller

2.1.1 Description

In Henrici and Müller's protocol [22], the tag needs to store a (nonstatic) identifier ID and two variables k and k_{last}. When the system is launched, the tag contains its current identifier ID, the current session number k (both are set up with random values), and k_{last} that is equal to k. On the reader's side, a database contains a similar 3-tuple per tag it manages. Data in the database and in the tags are initially synchronized. The tag identification works as follows (see Fig. 4):

1. The reader sends a request to the tag.
2. The tag increases its current session number k by one and sends back $h(\text{ID})$, $h(k \oplus \text{ID})$, and $\Delta k := k - k_{\text{last}}$. $h(\text{ID})$ allows the database to recover the tag's identity; Δk allows the database to recover k and thus to compute $h(k \oplus \text{ID})$, whose goal is to thwart replay attacks.
3. The database checks the validity of these values according to its recorded data. If it matches, it sends a random number r and $h(r \oplus k \oplus \text{ID})$ to the tag and stores the new values.[1] Since the tag knows k and ID and receives r, it can check whether or not $h(r \oplus k \oplus \text{ID})$ is correct. If this is the case, it replaces its identifier by $r \oplus \text{ID}$ and k_{last} by k. Otherwise it does not refresh its identifier.

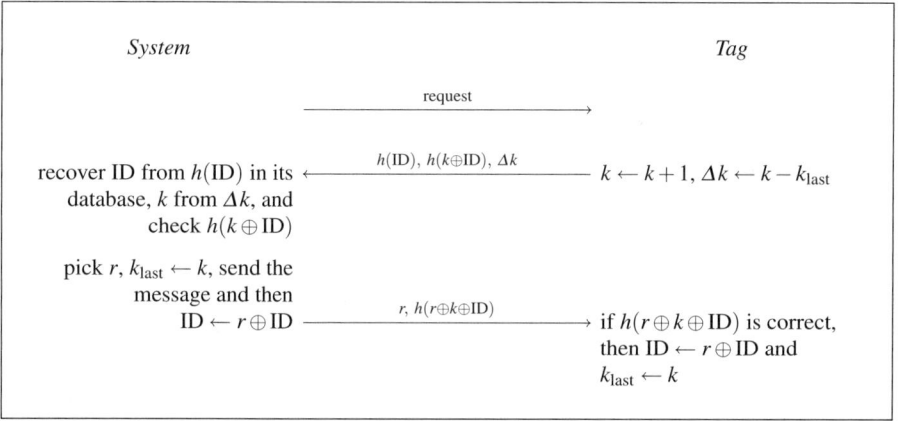

Fig. 4 Henrici and Müller's protocol

---

[1] Note that due to resiliency considerations, the entry $(\text{ID}, k, k_{\text{last}})$ in the database is not erased when the database has replied to the tag, but a copy is kept until the next correct session: If the third step fails, the database will still be able to identify the tag the next time with the "old" entry. Thus two entries per tag are used in turn.

2.1.2 Attack Based on Nonrandom Information

This attack consists of tracking a tag, taking advantage of the information supplied by Δk. Indeed, the tag increases its value k every time it receives a request (Step 2) even when the identification fails, but it updates k_{last} only when the identification succeeds (Step 3). Thus, an adversary may query the tag several times to abnormally increase k and in turn Δk. Because this value is sent in clear in the second message, the adversary is then able to later recognize its target according to this value: If the tag sends an abnormally high Δi, the adversary concludes that this is her target, even in a passive way.

2.1.3 Attack Based on Refreshment Avoidance

Another attack consists of corrupting the hash value sent by the reader. When this value is not correct, "the message is discarded and no further action is taken" [22], so the tag does not refresh its identifier. Note, however, that it is easier to inject than to modify a message into a wireless channel. We therefore propose a practical variant of this attack: When a reader queries a tag, the adversary queries this tag as well before the reader carries out the third step. Receiving the request from the adversary, the tag increases k. Consequently, the hash value sent by the reader seems to be incorrect since k has now changed. More generally, an adversary can always trace a tag between two correct identifications. Combined with a relay attack [5, 19, 20, 33], this attack is even easier to put into practice.

2.1.4 Attack Based on Database Desynchronization

A more subtle and definitive attack consists of desynchronizing the tag and the database (as a kind of denial of service attack). For that, the adversary performs the identification so that the random value r she sends is the neutral element of $\oplus$: the adversary replaces r by the null bit-string and replaces $h(r \oplus k \oplus \text{ID})$ by $h(k \oplus \text{ID})$ obtained by eavesdropping the second message of the current identification. We have trivially $h(\mathbf{0} \oplus k \oplus \text{ID}) = h(k \oplus \text{ID})$. Hence, the tag cannot detect the attack. Then the tag replaces its identifier by $\mathbf{0} \oplus \text{ID}$ (that is equal to its previous identifier) and it updates k_{last}. In the next identification, the tag and the database will be desynchronized since the tag computes the hash value using the previous ID and the fresh k_{last} whereas the database checks the hash value with the previous ID and the previous k_{last}: the test fails and the received message is discarded. Consequently, the database will never send the third message to refresh the tag's identifier and the tag is definitively traceable.

The above attack can be thwarted just by checking that $r \neq \mathbf{0}$. However, we show below that a desynchronization-based attack is still possible when $r \neq \mathbf{0}$. First of all, the adversary eavesdrops an interaction between her targeted tag and a legitimate reader. Let $h(k_i \oplus \text{ID})$ and $\Delta k_i = k_i - k_{i-1}$ be the collected information. Later, the adversary queries the tag again, obtaining thus $h(k_j \oplus \text{ID})$ and $\Delta k_j = k_j - k_{j-1}$.

Given that

$$k_i - k_j = \sum_{\ell=i}^{j-1} \Delta k_\ell,$$

she guesses $k_i \oplus k_j$. For example, if $k_i - k_j = 1$ (which is the common case in close environments) then $k_i \oplus k_j = 00\ldots01$ with probability $1/2$. While generating the third message, she takes $r = k_i \oplus k_j$ and $h(k_i \oplus \mathrm{ID})$. Upon reception of the third message of the exchange, the tag checks whether the received hash value is valid, which is true since $h(r \oplus k \oplus \mathrm{ID}) = h(k_i \oplus k_j \oplus k_j \oplus \mathrm{ID}) = h(k_i \oplus \mathrm{ID})$. As in the case $r = \mathbf{0}$, the attack desynchronizes the database and the tag, which definitively becomes traceable.

2.2 Golle, Jakobsson, Juels, and Syverson

2.2.1 Description

Golle et al.'s protocol [18] relies on the concept of *universal re-encryption*, i.e., a scheme where re-encryptions of a message m are performed neither requiring nor yielding knowledge of the public key under which m had been initially encrypted. The scheme consists of encrypting a plaintext m by appending two ciphertexts: the first one is the Elgamal encryption of m while the second one is the Elgamal encryption of the neutral element of $\mathcal{G}$, where $\mathcal{G}$ is the underlying group for the cryptosystem. We recall that encryption with the Elgamal scheme [13] of a message m under the public key y and a random number r is (my^r, g^r), where g is a generator of $\mathcal{G}$.

Below, we expand Golle et al.'s protocol. Let E be the Elgamal encryption scheme, and U be the corresponding re-encryption scheme, we have $U(m) := [E(m); E(1_\mathcal{G})]$. Let q be the order of $\mathcal{G}$, and g a generator. The universal re-encryption scheme is defined by the following four algorithms:

- *Key generation.* Output the private key $x \in \mathbf{Z}$ and the public Elgamal key $y = g^x$.
- *Encryption.* Let (r_0, r_1) be a random element picked in $(\mathbf{Z}/q\mathbf{Z})^2$. The encrypted value of a message m is

$$U(m) = [(\alpha_0, \beta_0); (\alpha_1, \beta_1)] = [(my^{r_0}, g^{r_0}); (y^{r_1}, g^{r_1})].$$

- *Decryption.* Given the ciphertext $[(\alpha_0, \beta_0); (\alpha_1, \beta_1)]$, if $\alpha_0, \beta_0, \alpha_1, \beta_1 \in \mathcal{G}$ and $\alpha_1/\beta_1^x = 1$, then the plaintext is α_0/β_0^x.
- *Re-encryption.* Let (r_0', r_1') be a random element picked in $(\mathbf{Z}/q\mathbf{Z})^2$. The re-encrypted value of a ciphertext $[(\alpha_0, \beta_0); (\alpha_1, \beta_1)]$ is

$$[(\alpha_0 \alpha_1^{r_0'}, \beta_0 \beta_1^{r_0'}); (\alpha_1^{r_1'}, \beta_1^{r_1'})].$$

We now describe the RFID protocol suggested by Golle et al. based on their universal re-encryption scheme. During the tag initialization, an encrypted identifier is stored in the tag. This encrypted identifier as well as the secret key corresponding to the tag are stored in the database. As depicted in Fig. 5, an execution is carried

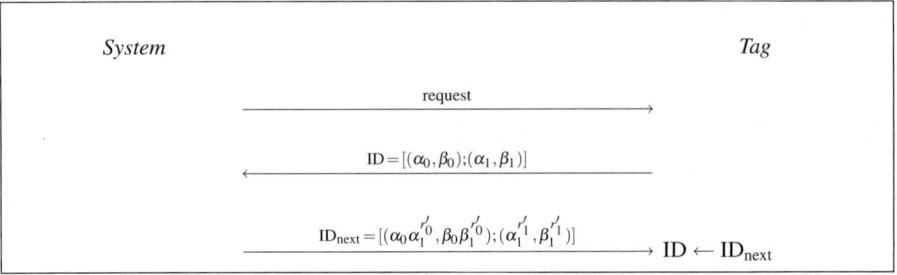

Fig. 5 Golle, Jakobsson, Juels, and Syverson's protocol

out as follows (1) the reader sends a request to the tag; (2) the tag sends back its encrypted identifier; (3) the reader re-encrypts the tag identifier using the universal re-encryption scheme described above and sends the new value to the tag (Fig. 5).

As noted in [18], if an adversary sends a fake re-encrypted identifier to the tag, the database will not be able to identify the tag in the future. The authors say that this attack does not allow the tag to be traced, it will only harm the normal functioning of the system. The authors do, however, reveal an exception: when an adversary replaces the value (α_1, β_1) by $(1_{\mathcal{G}}, 1_{\mathcal{G}})$ where $1_{\mathcal{G}}$ represents the neutral element of $\mathcal{G}$, the future re-encryptions will no longer change the identifier. The tag can protect itself from this attack by verifying that (α_1, β_1) is not equal to $(1_{\mathcal{G}}, 1_{\mathcal{G}})$ before changing its value. However, Golle et al.'s protocol also suffers from other weaknesses: Saito, Ryou, and Sakurai [43] stress that an adversary can replace the identifier of the tag by a value she has encrypted with her own public key. Thus, she is able afterward to decrypt the content of the tag and trace it. Recently, Ateniese, Camenisch, and de Medeiros [3] suggested an approach to thwart this attack. Their solution is based on a new cryptographic primitive, called *insubvertible encryption*.

2.2.2 Attack Based on Eavesdropping

The first thing to notice is that the protocol [18] does not resist to simple eavesdropping attacks. Indeed, since the tag sends in the second message what it received in the third message of the previous execution, an adversary is able to trace the tag by eavesdropping the communication.

2.2.3 Attack Based on Invariants

The weakness described below results from the fact that the ciphertext sent by the tag is not random. Taken independently, every element of the ciphertext $[(\alpha_0, \beta_0); (\alpha_1, \beta_1)]$ follows a uniform distribution assuming that the discrete logarithm is a random function, but these elements are not independent.

We denote by $[(\alpha_0^{(i)}, \beta_0^{(i)}); (\alpha_1^{(i)}, \beta_1^{(i)})]$ the message sent by the tag during the ith identification. If $[(\alpha_0^{(i)}, \beta_0^{(i)}); (\alpha_1^{(i)}, \beta_1^{(i)})]$ verifies a property $\mathcal{P}$ that is invariant by re-encryption, i.e., $\mathcal{P}$ remains verified after re-encryption, then the adversary is (almost certainly) able to trace the tag. We describe the attack below. Let us define $\mathcal{P}$ such as $[(\alpha_0^{(i)}, \beta_0^{(i)}); (\alpha_1^{(i)}, \beta_1^{(i)})]$ verifies $\mathcal{P}$ if and only if $\alpha_1 = \beta_1$. If $[(\alpha_0^{(i)}, \beta_0^{(i)}); (\alpha_1^{(i)}, \beta_1^{(i)})]$ verifies $\mathcal{P}$, then $[(\alpha_0^{(j)}, \beta_0^{(j)}); (\alpha_1^{(j)}, \beta_1^{(j)})]$ verifies it as well for any $j \geq i$. Indeed, the same (deterministic) operation is applied to both α_1 and β_1 during a re-encryption, that is both α_1 and β_1 are raised to a given power r_1'.

In order to trace a tag, the adversary queries it and sends the (third) message

$$\mathrm{ID}_{\mathrm{next}} = [(a, b); (c, c)],$$

where a, b, and c can be any values. When she queries the tag next and receives the message $[(\alpha_0^{(i)}, \beta_0^{(i)}); (\alpha_1^{(i)}, \beta_1^{(i)})]$, she verifies whether $\alpha_1^{(i)} = \beta_1^{(i)}$. In this case, the queried tag is her target with high probability. While the tag could detect such an attack by testing that $\mathrm{ID}_{\mathrm{next}}$ does not verify $\mathcal{P}$, there are other invariant properties, e.g., the property $\mathcal{P}'$ such that $[(\alpha_0^{(i)}, \beta_0^{(i)}); (\alpha_1^{(i)}, \beta_1^{(i)})]$ verifies $\mathcal{P}'$ if and only if $\alpha_1^{(i)} \cdot \beta_1^{(i)} = 1$ in $\mathcal{G}$.

2.3 Saito, Ryou, and Sakurai

Saito et al. also pointed out an attack [43] against Golle et al.'s protocol [18]. They subsequently suggested two RFID protocols based on [18]. The two protocols, described below, are, respectively, called "with a check" and "with one-time pad."

2.3.1 With a Check

The first protocol is an improvement of [18] where operations carried out by the tag have been modified: The tag checks the new value re-encrypted by the reader before accepting it as the new identifier. The aim is to detect an adversary who would send a wrong re-encrypted identifier. Therefore, when a tag is queried, it sends its current identifier, $[(\alpha_0, \beta_0); (\alpha_1, \beta_1)]$, and receives the new value $[(\alpha_0', \beta_0'); (\alpha_1', \beta_1')]$. If $|\alpha_0'|, |\beta_0'| \neq 1$ and if $\alpha_0'/\beta_0'^x = 1$, where x is the private key of the tag, then $[(\alpha_0', \beta_0'); (\alpha_1', \beta_1')]$ becomes the new current identifier. If not, the tag does not renew its content.

2.3.2 With a One-Time Pad

The second protocol suggested by Saito et al. is also based on the universal re-encryption scheme introduced in [18]. The primary difference compared to [18] is

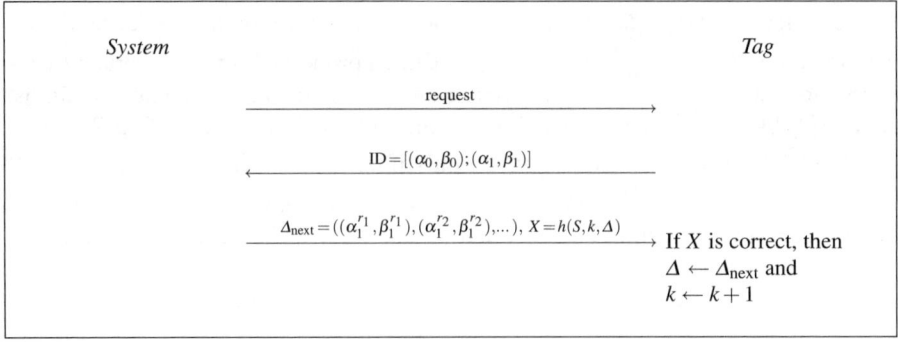

Fig. 6 Saito, Ryou, and Sakurai's protocol

that the re-encryptions are carried out by the tag itself and no longer by the reader. Since the tag is not able to carry out the exponentiations itself, precalculations are carried out by the database and sent to the tag from time to time. Below, we expand the protocol.

To begin with, the tag contains an identifier $\text{ID} = [(\alpha_0, \beta_0); (\alpha_1, \beta_1)]$. It also has a finite list of pairs of random values $\Delta = ((\alpha_1^{r_1}, \beta_1^{r_1}), (\alpha_1^{r_2}, \beta_1^{r_2}), \dots)$ that will allow it to re-encrypt its identifier. The tag also contains a variable k that is the session number, as well as a secret S. All these data are shared with the database. We must consider two distinct operations in this protocol: the reading of the tag and the update of its list of random values, which does not occur at every identification. The procedure unfolds in the following way (see Fig. 6):

1. The reader sends a request to the tag.
2. The tag sends back ID and replaces its identifier by

$$\text{ID}_{\text{next}} := [(\alpha_0 \alpha_1^{r_k}, \beta_0 \beta_1^{r_k}); (\alpha_1 \alpha_1^{r_{k+1}}, \beta_1 \beta_1^{r_{k+1}})]$$

$$\text{where } (\alpha_1^{r_k}, \beta_1^{r_k}), (\alpha_1^{r_{k+1}}, \beta_1^{r_{k+1}}) \in \Delta.$$

3. If an update of Δ is needed, the reader sends a new list Δ_{next} of random values and the key $X = h(S, k, \Delta)$ to the tag, where h is a hash function. If the key is correct, then the tag replaces Δ by Δ_{next} and increments the session number k. If not, the tag does nothing.

2.3.3 Attack Based on the Private Key

In Saito, Ryou, and Sakurai's first protocol [43], the fact that the tag carries out a test based on its public/private key transforms it into an oracle that responds whether or not this value has been encrypted with its public key. In other words, the oracle responds whether or not we are dealing with the traced tag. Let us, however, note that this response from the oracle is internal to the tag. The adversary therefore still has to recover this response. This is rather straightforward because the tag changes its identifier if and only if the test succeeds. So the adversary proceeds as follows. She requests its targeted tag for the first time thus obtaining a reference identifier $[(\alpha_0, \beta_0); (\alpha_1, \beta_1)]$. Subsequently, when the adversary wants to know if a tag corresponds to her target, she queries it: she receives (message 2) a value $[(\alpha'_0, \beta'_0); (\alpha'_1, \beta'_1)]$ and resends (message 3) the value $[(\alpha_0, \beta_0); (\alpha_1, \beta_1)]$ to the tag instead of resending the value $[(\alpha'_0, \beta'_0); (\alpha'_1, \beta'_1)]$ re-encrypted. She queries the tag once again. If she again receives $[(\alpha'_0, \beta'_0); (\alpha'_1, \beta'_1)]$, this means that the tag has not renewed its identifier and she is not dealing with the traced tag. The traced tag would have recognized $[(\alpha_0, \beta_0); (\alpha_1, \beta_1)]$ as a valid value, meaning encrypted with its public key, and would have used it to refresh its identifier.

2.3.4 Attack Based on the Random Values

In Saito, Ryou, and Sakurai's second protocol [43], knowing the list of the random values contained in the tag allows an adversary to easily trace a tag as she can calculate all the identifiers that will be used by it. So, eavesdropping the communication between the reader and the tag during an update is sufficient to subsequently trace the tag. Since the adversary has to be present during the update (which is only carried out from time to time), she can force the update using a man-in-the-middle attack. No authentication is used in the protocol. Thus the tag knows that Δ_{next} has been created by the database but it does not know who is sending it this value. On the other hand, the database does not know that it is sending Δ_{next} to the adversary instead of sending it to the tag. The session number prevents a replay-attack, not a man-in-the-middle attack.

2.3.5 Attack Based on Database Desynchronization

The danger that exists for protocols using synchronized values between the tag and the database (the session number k in Saito, Ryou, and Sakurai's second protocol [43]) is that an adversary can cause a desynchronization between the two parties. We have already pointed out such an attack against Henrici and Müller's protocol. Here, if an adversary causes the database to send the update message while the tag cannot receive it, then the session number stored by the database will be higher than that stored by the tag. Consequently, all the subsequent updates will fail as the calculation of the key X, which authorizes the update, takes into account the current session number.

2.4 Yang, Park, Lee, Ren, and Kim

2.4.1 Description

In order to complete our list of possible attacks against protocols based on reader-aided ID-refreshment, we put forward a rather straightforward attack against Yang et al.'s [25, 46] protocol.

As in all the previously presented protocols, an adversary can trace the tag between two legitimate identifications. In other words, if the adversary is able to *query* the tag at time t_1, then she is able to trace the tag at time $t_2 \neq t_1$ if and only if there is no legitimate identification between t_1 and t_2. We give here the following variant: if the adversary is able to *eavesdrop* a legitimate identification at time t_1, then she is able to trace the tag at time $t_2 > t_1$ if and only if there is no legitimate identification between t_1 and t_2. Below, we explain our attack.

The authors of [25,46] distinguish the reader from the back-end database whereas up until now, we have considered them as a unique entity. In Fig. 7, we only illustrate the exchanges between the reader and the tag because the adversary does not need to exploit the channel between the reader and the back-end database in order to carry out her attack. At the beginning, the tag and the system share three random values k_1, k_2, and C. k_1 and k_2 are refreshed during each legitimate identification while C is a static value. Three exchanges are required in this protocol:

1. The system queries the tag with a value S. The precise content of S is not relevant in the attack. More details are available in [25, 46].
2. The tag answers with $ID = h(k_1 \oplus S \oplus C)$ where h is a hash function. If the message is the expected one, the system computes $ID' := h(k_2)$.
3. Finally, the system sends ID' to the tag and the latter replaces k_1 by $k_1 \oplus ID'$ and k_2 by $k_2 \oplus ID$ if the received ID' is valid.

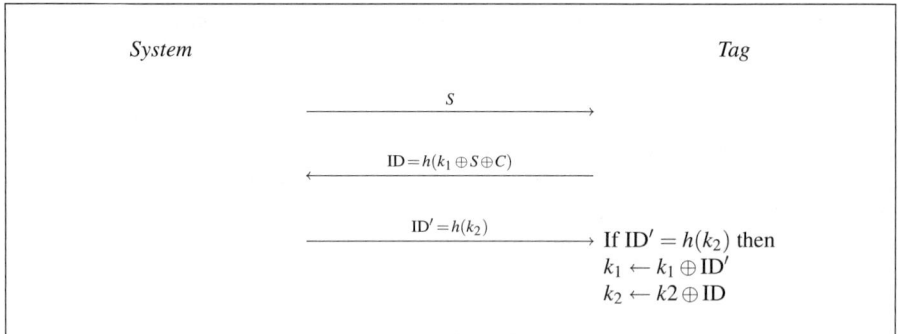

Fig. 7 Yang, Park, Lee, Ren, and Kim's protocol

2.4.2 Attack

Our very simple attack works as follows. If the adversary is able to eavesdrop a legitimate identification, she obtains the current exchanged values S_{cur}, ID_{cur}, and ID'_{cur}. Later, she queries the tag with $S_{next} = S_{cur} \oplus ID'_{cur}$. Subsequently, the tag computes $ID_{next} := h((k_1 \oplus ID'_{cur}) \oplus (S_{cur} \oplus ID'_{cur}) \oplus C)$, which is clearly equal to ID_{cur}. Thus, if $ID_{cur} = ID_{next}$, the adversary is certain that it is the same tag in both identifications.

More generally, using the operation $\oplus$ to design RFID protocol must be done very carefully, because it is usually (always?) easy to find a way to track the tags with such an approach. As an illustration, the protocols proposed by Peris-Lopez et al. [40–42], which are based on simple logic operations, have all been proven to be weak in terms of untraceability (see for example [34]).

3 Protocols Based on Self-Refreshment

In this section, we present radio frequency identification protocols based on self-refreshment, that is, protocols in which the tag refreshes its identifier by itself, without the help of the reader. These protocols rely on a challenge–response exchange and assure either identification or authentication, as defined previously. Such an approach avoids the attacks presented in Sect. 2 against protocols based on reader-aided ID-refreshment. Proof of security can even be supplied under certain assumptions. Thus, protocols based on self-refreshment seem to be precisely what we need both in theory and in practice. However, we will show in Sect. 4 that these protocols require a linear computation complexity on the system's side in order to identify only a single tag. Moreover, they require computations on the tag's side that are usually based on a pseudorandom function. Such a function is still rather heavy to implement in very low cost tags [1, 12, 14] and can only deal with tags that are a bit more expensive.

Below, we present RFID (identification) protocols that require two moves [38,45] and an RFID (authentication) protocol that require three moves [14]. For each of them, we describe the protocol by defining three phases (a) the **Setup** phase where the database of the system and the tags are initialized; (b) the **Interaction** phase where the system and the tag interact; and (c) the **Search** phase where the system looks for the tag identity in its database. Sometimes, **Interaction** and **Search** phases are interleaved.

3.1 Weis, Sarma, Rivest, and Engels

We describe Weis, Sarma, Rivest, and Engels's protocol [45] with "Randomized Access Control". In this protocol (see Fig. 8), the information sent by the tag each time it is queried consists of a random value a and a randomized hash value $\sigma = h(ID\|a)$

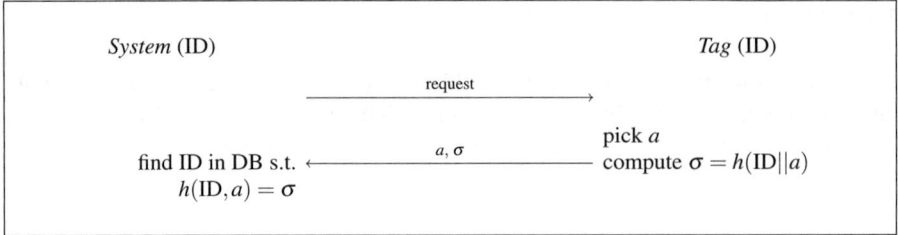

Fig. 8 Weis, Sarma, Rivest, and Engels's protocol using a hash function

where ID is the static identifier of the tag. In order to compute this information, the tag needs a pseudorandom number generator and an embedded one-way hash function but only stores its identifier. Below we give the three phases of the protocol.

3.1.1 Setup

Each tag is initialized with a randomly chosen identifier ID. For each tag it manages, the system stores an entry in its database that contains its identifier.

3.1.2 Interaction

The system sends a request to the tag. Upon reception of this message, the tag picks a random a and computes $\sigma = h(\text{ID}||a)$. It sends both a and σ to the system.

3.1.3 Search

Upon reception of σ and a, the system performs an exhaustive search in its database: for each entry ID, it computes $h(\text{ID}||a)$ until it finds σ.

Weis et al. emphasize that from a theoretical point of view, the hash functions guarantee by definition irreversibility, but not secrecy: input bits may be revealed. Consequently, they suggest another construction that relies on pseudorandom functions. In that variant, the tag shares a secret s with the database and, instead of sending a and $h(\text{ID}||a)$, the tag sends a and $\text{ID} \oplus f_s(a)$ where f_s is a pseudorandom function chosen in a set $\mathcal{F} = \{f_s\}_{s \in \mathbb{N}}$. The exchanges are depicted in Fig. 9.

3.2 Feldhofer, Dominikus, and Wolkerstorfer

Feldhofer, Dominikus, and Wolkerstorfer's protocols [14] differ from [45] in the fact that the pseudorandom function is replaced by AES. However, the general outline

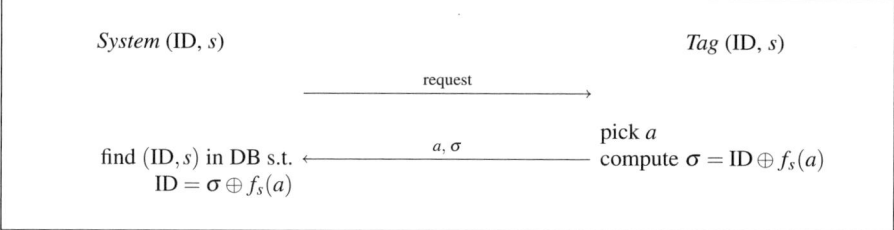

Fig. 9 Weis, Sarma, Rivest, and Engels's protocol using a pseudorandom function

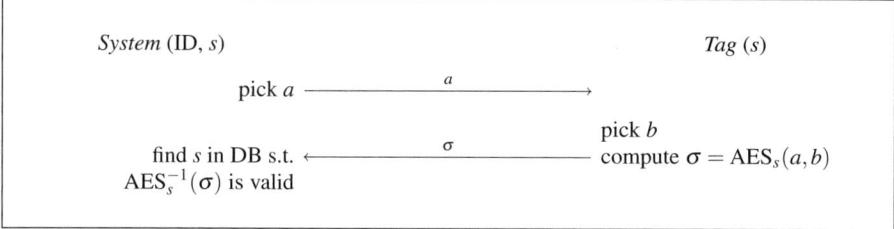

Fig. 10 Feldhofer, Dominikus, and Wolkerstorfer's authentication protocol

remains the same. Two variants are proposed: only the authentication of the tag is provided in the first variant, detailed below and in Fig. 10, while mutual authentication is ensured in the second one.

3.2.1 Setup

Each tag is initialized with a randomly chosen secret key s, which is also stored by the system with the corresponding tag's identifier.

3.2.2 Interaction

The system picks a random number a and sends a request including a to the tag. Upon reception of this message, the tag picks a random number b and computes $\sigma = \text{AES}_s(a,b)$, which is sent to the system.

3.2.3 Search

Upon reception of σ, the system performs an exhaustive search in its database: for each entry ID, it computes $\text{AES}_s^{-1}(\tau)$ until it finds a valid decryption.

Note that [14] does not precise whether or not the same secret s is used for all tags. If it is the case, the tag must stores its identifier ID, which must be encrypted together with a and b. The paper is a bit ambiguous regarding this point.

3.3 Ohkubo, Suzuki, and Kinoshita

The protocol put forward by Ohkubo, Suzuki, and Kinoshita [38] is fairly similar to the hash-version of [45]. The difference is that the hash is deterministic rather than randomized. In order to avoid traceability by an adversary, the identifier contained in the tag is consequently refreshed at each new identification, by using a second hash function, as depicted in Fig. 11. This brings an additional interesting property, which is the forward privacy as defined in Sect. 1.2. Thus, an adversary tampering with the tag is not able to trace it in the past. Clearly, this carries an extra cost in terms of computation as the tag has to calculate two hashes at each identification.

3.3.1 Setup

The personalization of a tag T_i consists of storing in its memory a random identifier s_i^1, which is also recorded in the database of the system. Thus, the database initially contains the set of random values $\{s_i^1 \mid 1 \le i \le n\}$. Two hash functions G and H are chosen. One hash function is enough if a one-bit parameter is added to the function.

3.3.2 Interaction

When the system queries T_i, it sends an identification request to the tag and receives back $r_i^k := G(s_i^k)$ where s_i^k is the current identifier of T_i. While T_i is powered, it replaces s_i^k by $s_i^{k+1} := H(s_i^k)$. The exchanges between the system and the tag can be represented as follows.

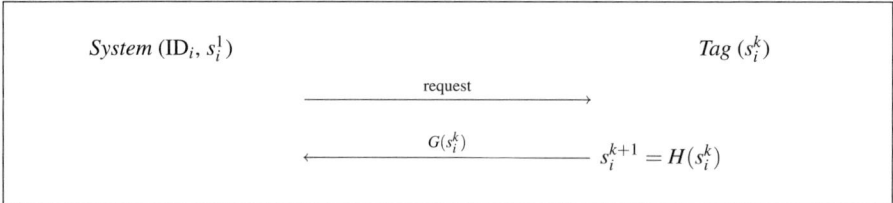

Fig. 11 Ohkubo, Suzuki, and Kinoshita's protocol

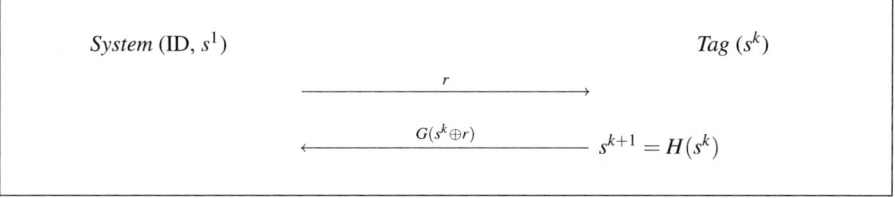

Fig. 12 Ohkubo, Suzuki, and Kinoshita's modified protocol

3.3.3 Search

From r_i^k, the system has to identify the corresponding tag. In order to do this, it constructs the hash chains from each n initial value s_i^1 until it finds the expected r_i^k or until it reaches a given maximum limit m on the chain length.

The lifetime of the tag is a priori limited to m identifications. However, when a tag is scanned by a legitimate reader, its field in the database can be refreshed. The threshold m is therefore the number of read operations on a single tag between two legitimate identification.

The main advantage of this protocol compared to the previous challenge–response protocols is that it also assures forward secrecy. However, it does not prevent replay attacks. Common techniques to avoid replay attacks are usually incremental sequence number, clock synchronization, or a fresh challenge sent by the verifier. This latter option is the one used by Avoine, Dysli, and Oechslin in [8] to modify Ohkubo, Suzuki, and Kinoshita's identification protocol into an authentication protocol, as depicted in Fig. 12.

4 Reducing Complexity on the System's Side

We have presented in Sect. 3 several protocols based on self-refreshment. They are based on classical challenge–response protocols (denoted CR in what follows), and so are secure in terms of untraceability, assuming that the hash functions and ciphers behave like pseudorandom functions. However, they all suffer from a complexity issue: Since the tag no longer sends its identifier "in clear," the system does not know which key it should use to treat the tag's answer. To identify or authenticate one tag, it must carry out an exhaustive search over the set of all the identifiers it manages. So, CR-based protocols are quite expensive in terms of computation: in a system which n tags belong to, existing protocols require $O(n)$ cryptographic operations to identify one tag and $O(n^2)$ in order to identify the whole system.

Molnar and Wagner [35] have suggested a method to reduce the complexity to $O(\log n)$. We denote this technique by CR/MW and present it below. We then show that it degrades privacy if the adversary has the possibility to tamper with at least one tag [8, 11, 36]. In Sect. 4.3, we go thoroughly into the Ohkubo, Suzuki, and

Kinoshita's protocol [38], denoted OSK, already introduced in Sect. 3.3. In Sect. 4.4, we present an approach proposed by Avoine and Oechslin whose goal is to improve OSK using a time-memory trade-off. This variant, called OSK/AO, is as efficient as CR/MW in practice but does not degrade privacy [7, 8]. Contrary to CR/MW, only the method used by the system to store its data is modified while CR/MW adapts the data stored by the tag and the data that are exchanged between the system and the tag. Thus the security of OSK/AO is equivalent to the security of OSK.

In order to illustrate our comparison between CR, CR/MW, OSK, and OSK/AO, we consider a real life scenario in a library, where tags are used to identify books. Several libraries already use this technology, for example the libraries of Santa Clara (USA), Heiloo (Netherlands), Richmond Hill (Canada), and K.U. Leuven (Belgium). Inside the library, tags make it possible to scan shelves for misfiled books and to identify books that have not been placed on the shelves. These tags also make the check-out and check-in of books much easier. When a user takes a book home, an adversary should not be able to find out what he is reading nor track him using the tags. Nor should the adversary be able to track him a posteriori, when the book has been brought back to the library. Indeed, the adversary could borrow the book and tamper with its tag to track the past events of the tag. In other words, the protocol should assure *forward privacy*. In a library scenario, it is realistic to assume that the tags can contain a secret-key cipher or a hash function because they are not disposable. Thus, a slightly higher cost is conceivable.

In the next sections, we assume that the system relies on a single computer which takes $\theta = 2^{-23}$ s to carry out a cryptographic operation. This value is rather arbitrary since it depends on the cryptographic building block itself, either an encryption function or a hash function. However, our goal is to choose a rather realistic value just to compare the protocols in a fair way, disregarding the underlying building blocks. We assume that inputs and outputs of the cryptographic functions are 128-bit long. The library manages 2^{20} tags. Current implementations allow a single reader to read several hundreds of tags per second, meaning that the system should spend at the most a few milliseconds to identify one tag. In the following sections, t_P denotes the average time to identify one tag using a protocol P. Because certain applications (in libraries, in amusement parks, etc.) may use numerous readers, the system should not become a bottleneck in terms of computation. Thus, the system should be capable of identifying the whole set of tags it manages in only a few seconds (e.g., for real-time inventories).

4.1 Molnar and Wagner's Protocol

4.1.1 Challenge–Response Building Block

The Molnar and Wagner's challenge–response building block (CR), depicted in Fig. 13, provides mutual authentication of the reader and the tag in a private way. It prevents an adversary from impersonating, tracing, or identifying tags.

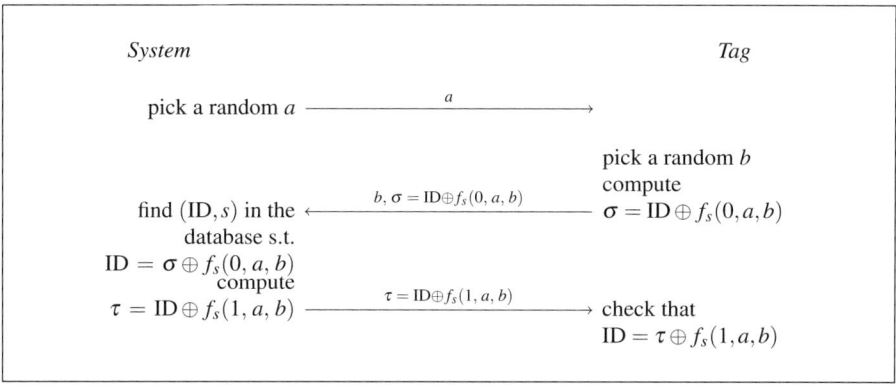

Fig. 13 Molnar and Wagner's challenge–response protocol

Let ID be the tag's identifier that is stored in both the system's database and the tag. They also share a secret key s. To initiate the authentication, the reader sends a nonce a to the tag. Subsequently, the tag picks a random b and answers $\sigma :=$ $\text{ID} \oplus f_s(0,a,b)$, where f_s is a pseudorandom function. The system retrieves the tag's identity by finding the pair (ID,s) in its database such that $\text{ID} = \sigma \oplus f_s(0,a,b)$. This completes the tag's authentication. Now, in order to achieve mutual authentication, the system sends back $\tau := \text{ID} \oplus f_s(1,a,b)$ to the tag. The tag can thus verify that the reader is authorized to read it by checking that $\text{ID} = \tau \oplus f_s(1,a,b)$.

4.1.2 Efficiency

In order to identify a tag, the system must carry out an exhaustive search on the n secrets stored in its database. Therefore the system's workload is linear in the number of tags. More precisely, the average number of cryptographic operations required to identify one tag is $n/2$ and therefore we have $t_{CR} = (n\theta)/2$. With $n = 2^{20}$ and $\theta = 2^{-23}$, we have $t_{CR} \approx 62$ ms which is too high in practice. Since CR does not scale well in a system with many tags, we now examine Molnar and Wagner's tree-based technique [35], whose main strength lies in the reduction of the system's workload from $O(n)$ to $O(\log n)$.

4.1.3 Tree-Based Technique

The technique suggested by Molnar and Wagner, called CR/MW, relies on a tree structure to reduce identification complexity. Instead of searching a flat space of secrets, let us arrange them in a balanced tree with branching factor δ. The tags are the leaves of this tree and each edge is associated with a value. Each tag has to store the values along the path from the root of the tree to itself. This sequence makes up its *secret*, and each value is called a *block of secret*. On the other side, the reader knows all the secrets. We describe the protocol below.

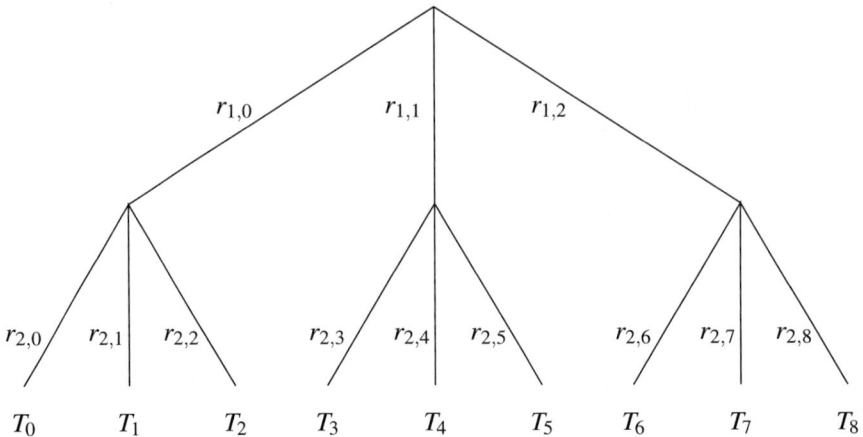

Fig. 14 Tags' secrets tree

4.1.4 Setup

Let n be the number of tags managed by the system and $\ell := \lceil \log_\delta n \rceil$ be the depth of the tree with a branching factor δ. Each edge in the tree is valued with a randomly chosen secret $r_{i,j}$ where i is the level in the tree and j is the branch index. Figure 14 represents such a tree with parameters $n = 9$ and $\delta = 3$. The secret of a given tag is the list of the values $r_{i,j}$ from the root to the leaf. For example, the secret of T_5 in Fig. 14 is $[r_{1,1}, r_{2,5}]$.

4.1.5 Interaction

The tag is queried level by level from the root to the leaves. At each level i, CR/MW runs CR for each secret of the explored subtree, that is, the reader tries every edge in turn to find out on which one the tag is. If CR fails for all current level's secrets, the tag rejects the reader and the protocol stops. If the reader has been successfully authenticated at each level, the protocol succeeds. Note that CR inevitably does not need to be performed δ times per level in practice. One run is enough if the reader checks the tag's answer with all current level's secrets as described below.

4.1.6 Search

At each level i, the system has to search in a set of δ secrets for the one matching the tag's secret. Given that $[s_1, \ldots, s_\ell]$ denotes a secret, at level i, on an average, the system has to compute $f_{s_i}(0, a, b)$ $\delta/2$ times, implying that $\frac{\delta}{2}\ell$ operations are required to identify one tag. Thus we have

$$t_{\text{CR/MW}} = \frac{\delta\theta}{2}\log_\delta n.$$

The identification of one tag is far below the threshold of a few milliseconds. Identifying the whole system would take more than 2 min when $\delta = 2^{10}$ and decreases to 2 s when $\delta = 2$. However, we will see below that having a small branching factor enables tracing the tags.

4.1.7 Privacy-Weakening Attacks

Tampering With Only One Tag

Avoine, Dysli, and Oechslin show in [8] how the tree technique suggested by Molnar and Wagner allows tracing a tag when the adversary is able to tamper with some tag. The attack consists of three phases:

1. The adversary has one tag T_0 (e.g., her own) she can tamper with and thus obtain its complete secret. For the sake of calculation simplicity, we assume that T_0 is put back into circulation. When the number of tags in the system is large, this does not significantly affect the results.
2. She then chooses a target tag T. She can query it as much as she wants but she cannot tamper with it.
3. Given two tags T_1 and T_2 such that $T \in \{T_1, T_2\}$, the adversary can query T_1 and T_2 as many times as she wants but cannot tamper with them. We say that the adversary succeeds if she definitely knows which of T_1 and T_2 is T. We define the probability to trace T as being the probability that the adversary succeeds.

Avoine et al. assume that the underlying challenge–response protocol assures privacy when all the blocks of secrets are chosen according to a uniform distribution. They consequently assume that the adversary cannot carry out an exhaustive search over the secret space. Hence, the only way for an adversary to guess a block of secret of a given tag is to query it with the blocks of secret she obtained by tampering with some other tags. When she tampers with only one tag, she obtains only one block of secret per level in the tree. Thus, she queries T, and then T_1, and T_2 with this block. If either T_1 or T_2 (but not both) has the same block as T_0, she is able to determine which of them is T. If neither T_1 nor T_2 has the same block as T_0, she cannot answer. Finally, if both T_1 and T_2 have the same block as T_0, she cannot answer, but she can move onto the next level of the tree because the reader's authentication succeeded. We formalize the analysis below. We denote the secrets of T, T_0, T_1, and T_2 by $[s_1, \ldots, s_\ell]$, $[s_1^0, \ldots, s_\ell^0]$, $[s_1^1, \ldots, s_\ell^1]$, and $[s_1^2, \ldots, s_\ell^2]$, respectively. We consider a given level i where s_i^1 and s_i^2 are in the same subtree. Four cases must be considered:

- $C_i^1 = ((s_i^0 = s_i^1) \wedge (s_i^0 \neq s_i^2))$ then the attack succeeds,
- $C_i^2 = ((s_i^0 \neq s_i^1) \wedge (s_i^0 = s_i^2))$ then the attack succeeds,
- $C_i^3 = ((s_i^0 \neq s_i^1) \wedge (s_i^0 \neq s_i^2))$ then the attacks definitively fails,
- $C_i^4 = (s_i^0 = s_i^1 = s_i^2)$ then the attacks fails at level i but can move onto level $i+1$.

When the number of tags in the system is large, we can assume that

$$\Pr\left(C_i^1\right) = \Pr\left(s_i^0 = s_i^1\right) \times \Pr\left(s_i^0 \neq s_i^2\right).$$

The same assumption also applies to C_i^2, C_i^3, and C_i^4. Thus we have

$$\Pr\left(C_i^1 \vee C_i^2\right) = \frac{2(\delta - 1)}{\delta^2} \quad (1 \leq i \leq \ell)$$

and

$$\Pr\left(C_i^4\right) = \frac{1}{\delta^2}.$$

The overall probability P that the attack succeeds is therefore

$$P = \Pr\left(C_1^1 \vee C_1^2\right) + \sum_{i=2}^{\ell}\left(\Pr\left(C_i^1 \vee C_i^2\right) \times \prod_{j=1}^{i-1}\Pr\left(C_j^4\right)\right)$$

$$= \frac{2(\delta - 1)}{\delta^2} + \sum_{i=2}^{\ell}\left(\frac{2(\delta - 1)}{\delta^2}\left(\frac{1}{\delta^2}\right)^{i-1}\right)$$

$$= \sum_{i=1}^{\ell}\left(\frac{2(\delta - 1)}{\delta^{2i}}\right)$$

$$= 2(\delta - 1)\frac{1 - \left(\frac{1}{\delta^2}\right)^{\ell}}{1 - \frac{1}{\delta^2}}\frac{1}{\delta^2}.$$

Given that $\delta^\ell = n$ we obtain

$$P = \frac{2}{\delta + 1}\left(1 - \frac{1}{n^2}\right).$$

The curve of P when $n = 2^{20}$ is plotted in Fig. 15.

Tampering With Several Tags

Avoine et al. then consider the case where the adversary can tamper with several tags, e.g., she borrows several books in the library in order to tamper with their tags. They examine the influence of the number of opened tags on the probability of tracing the target tag. The probability that the attack succeeds, according to the branching factor δ and given that k_0 tags have been opened among 2^{20} tags in the system is reported in Table 1 and plotted in Fig. 16 (details of the calculation can be found in [8]).

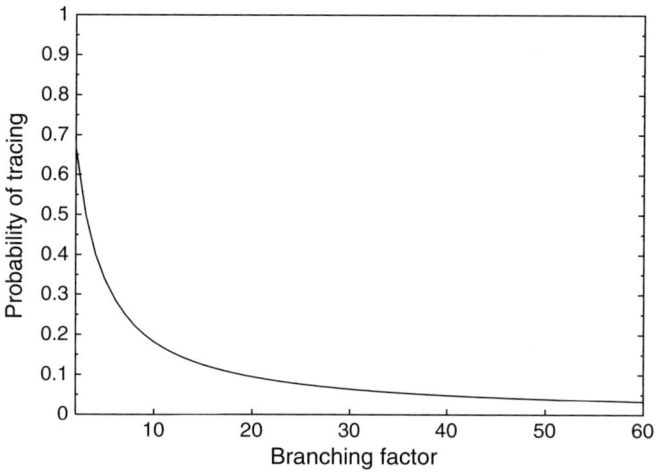

Fig. 15 Probability of tracing a tag when the adversary tampered with one tag

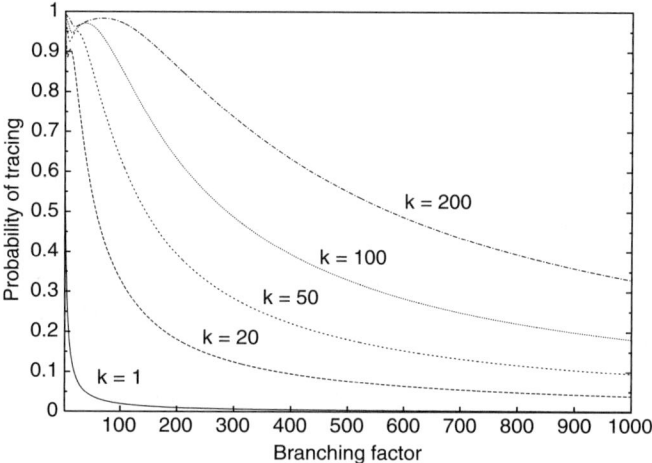

Fig. 16 Probability of tracing a tag when the adversary tampered with k_0 tags

Table 1 Probability that the attack succeeds

k_0 \ δ	2	20	100	500	1,000
1	66.6%	9.5%	1.9%	0.3%	0.1%
20	95.5%	83.9%	32.9%	7.6%	3.9%
50	98.2%	94.9%	63.0%	18.1%	9.5%
100	99.1%	95.4%	85.0%	32.9%	18.1%
200	99.5%	96.2%	97.3%	55.0%	32.9%

4.2 Avoine, Buttyán, Holczer, and Vajda's Protocol

4.2.1 Description

Avoine, Buttyán, Holczer, and Vajda's protocol [10] is an alternative to Molnar and Wagner's protocol, where the key-tree is replaced by groups of keys. Unfortunately, this approach, as Molnar and Wagner's protocol, degrades privacy if the adversary has the possibility to tamper with at least one tag. However, the authors show that for a given computation complexity (on the side of the back-end system), the group-based protocol provides a better level of privacy and a better complexity on the side of the tag than CR/MW. The protocol is depicted on Fig. 17.

4.2.2 Setup

The tags of the system are divided into groups of equal size. Each tag stores a key k_g that is shared with all the other members of the group, and a personal key k_t that is used to uniquely authenticate the tag within the group. Finally, each tag has a unique identifier that can be public.

4.2.3 Interaction

In order to authenticate a tag, the reader sends a challenge to it. The tag answers with two values: the first value is the encryption using the cipher E with the group key of the identifier ID of the tag, the challenge, and a nonce picked by the tag itself; the second value is the encryption using the cipher E with the personal key of the challenge and the nonce. Consequently, the first value authenticates the group while the second value authenticates the tag itself.

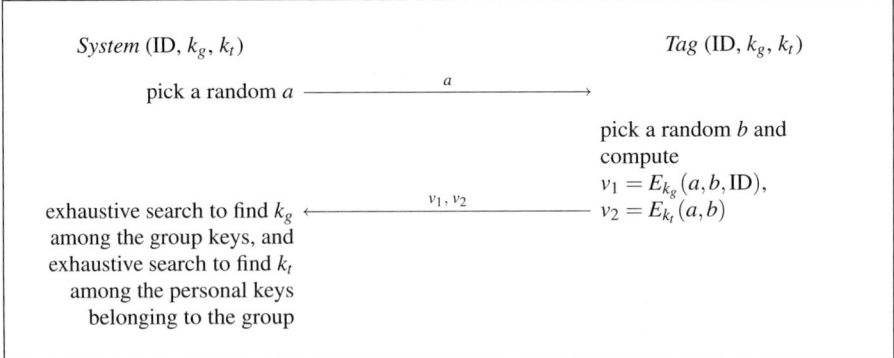

Fig. 17 Avoine, Buttyán, Holczer, and Vajda's protocol

4.2.4 Search

Upon reception of the two values, the system tries to decrypt the first value with all the group keys it manages. When it succeeds, this means that it authenticated the group the queried tag belongs to. Then he tries to decrypt the second value with all the personal keys belonging to the identified group.

4.2.5 Efficiency

The complexity of ABHV depends obviously on the number of groups, denoted γ. To obtain the same complexity as in CR/MW, γ must be chosen such that $\gamma = (\delta \log_\delta n) - 1$, where n is the total number of tags in the system and δ is the branching factor used in CR/MW. The dual property to complexity is privacy. In [10], the authors prove that ABHV is slightly better than CR/MW in terms of malicious traceability.

Other advantage of ABHV over CR/MW is that it reduces the complexity on the tag's side: storage complexity is constant (two keys and one identifier) instead of $\log_\delta n$ keys in CR/MW, and two encryptions performed by the tag instead of $\log_\delta n$ as well.

4.3 Ohkubo, Suzuki, and Kinoshita's Protocol

OSK is already defined in Sect. 3.3. We provide here the related efficiency analysis.

4.3.1 Efficiency

We describe here the efficiency analysis given in [7] of OSK described in Sect. 3.3. Outside the library, tags can be queried by foreign readers. This avoids maintaining synchronization between the tag and the system. Therefore, the complexity in terms of hash operations in order to identify one tag is $t_{OSK} = mn\theta$ on average (two hash operations are carried out $mn/2$ times). With the parameters $n = 2^{20}$, $\theta = 2^{-23}$, and chains of length $m = 128$, we have $t_{OSK} \approx 16$ s. Note that if we had considered that readers of the library may read foreign tags (held by people in the library), then the complexity would tend toward to $2mn$ because the system would have to explore the whole database to determine whether or not a tag is owned by the system. Note that even if tags and readers were able to stay synchronized, for example when the RFID system is deployed in a closed environment, the complexity of OSK cannot be better than CR if no additional memory is used.

4.3.2 Improvements

Note that Ohkubo, Suzuki, and Kinoshita modified their scheme in [39]. To reduce the complexity on both the system's side and tag's side, they propose not to apply the function H each time the tag is queried. Instead, they use a counter c and apply H only when the counter reaches its upper bound. Unfortunately, this technique degrades forward privacy because an adversary can trace the c last events of the tag (in the worst case) if she is able to tamper with it. Still worst, the value of the counter is sent to the reader each time the tag is queried and therefore it may be traced according to this value, which is not random.

4.4 Avoine and Oechslin's Complexity Improvement Technique

Because OSK suffers from a high computation complexity, Avoine and Oechslin suggest in [7] a technique to reduce OSK's complexity using a time-memory trade-off.

4.4.1 Time-Memory Trade-Off

To do so, they propose improving how secrets are managed by the system, without modifying the exchanges between tags and reader. For that, they suggest using a specific time-memory trade-off based on Hellman's original work [21] and Oechslin's optimizations [37]. This type of trade-off reduces the amount of work T needed to invert any given value in a set of N outputs of a one-way function E with the help of M units of memory. The efficiency follows the rule $T = N^2\gamma/M^2$, where γ is a small factor depending on the probability of success and the particular type of trade-off being used [9]. Compared to a brute-force attack, the trade-off can typically reduce the amount of work from N to $N^{2/3}$ using $N^{2/3}$ units of memory.

The basic idea of time-memory trade-off techniques consists in chaining (almost) all the possible outputs of E using a *reduction function* R that generates an arbitrary input of E from one of its outputs. By alternating E and R on a chosen initial value, a chain of inputs and outputs of E can be built. If enough chains of a given length are generated, most outputs of E will appear at least once in any chain. The trade-off comes from the fact that only the first and the last element of each chain is stored. Thus, a substantial memory space is saved, but computations will be required on-the-fly to invert a given element. Given one output r of E that should be inverted, a chain starting at r is generated. If r was part of any stored chain, the last element of a chain in the table will eventually be reached. Looking up the corresponding start of the chain, we can regenerate the complete chain and find the input of E that yields the given output r. To assure a high success rate, several tables have to be generated with different reduction functions. The exact way of doing this is what differentiates existing trade-off schemes.

In what follows, *perfect rainbow tables* are used because they have been shown to perform better than other types of tables [37]. The characteristic of the rainbow tables is each column of a table having a different reduction function. So, when two chains collide, they do not merge (except if they collide at the same position in the chain). When the residual merged chains are removed during the precomputation step, the tables are said to be perfect.

4.4.2 Adapting the Trade-Off to OSK

The time-memory trade-off technique described above cannot be directly applied to OSK. Indeed, the input of E must cover all the identifiers but no more. Otherwise, the system would have no advantage over the adversary. Consequently, it is important to choose E such that its input space is as small as possible. The function E is chosen as follows:

$$E : (i,k) \mapsto r_i^k = G(H^{k-1}(s_i^1)),$$

where $1 \leq i \leq n$ and $1 \leq k \leq m$. Thus, given the tag number and the identification number, E outputs the value which will be sent by the tag. The reduction function R is such that

$$R : r_i^k \mapsto (i',k'),$$

where $1 \leq i' \leq n, 1 \leq k' \leq m$. Avoine et al. suggest:

$$R(r) = \left(1 + (r \bmod n), 1 + \left(\left\lfloor \frac{r}{n} \right\rfloor \bmod m\right)\right).$$

There are still two important points that distinguish common time-memory trade-off from those to be applied to OSK.

First, the brute force method of OSK needs $n|s|$ units of memory to store the n values s_i^1 while usual brute-force methods do not require any memory. Thus, it makes sense to measure the amount of memory needed by the trade-off in multiples of $n|s|$. We call c the ratio between the memory used by the trade-off and the memory used by the brute-force. The memory used to store the tables is a multiple of the size of a chain while it is a multiple of s in the case of the brute-force. A stored chain is represented by its start and end point that can either be the output of E or its input. In the present case the input is smaller. Consequently, the stored pairs are the (i,k), thus requiring $2(|n| + |m|)$ bits of memory. The conversion factor from units of brute-force to units of trade-off is $\mu = |s|/(2|n| + 2|m|)$. In the scenarios we are interested in, μ is typically between 2 and 4.

Second, when used in the trade-off, E is more complex than when used in the brute-force. Indeed, in the brute-force, the hash chains are calculated sequentially, thus needing just one H and one G calculation at each step. In the trade-off, i and k are arbitrary results from R and have no incremental relation with previous calculations. Thus, on average, each step computes the function E $(m-1)/2+1$ times and the function G once. We can now rewrite the trade-off relation as:

$$T = \frac{N^2}{M^2}\gamma$$

$$= \frac{n^2 m^2}{(c-1)^2 \mu^2 n^2}\left(\frac{m-1}{2}+1\right)\gamma$$

$$\approx \frac{m^3 \gamma}{2(c-1)^2 \mu^2}.$$

We now show how this issue is mitigated by Avoine and Oechslin. So far, among the c shares of memory, $(c-1)$ shares are used to store the chains, and 1 share is used to store the n values s_i^1. If we not only store the first element of the chains, but also store the element at the middle of the chain, we sacrifice even more memory but we reduce the average complexity of E. We will have only $(c-2)$ shares of the memory available for the tables, but E will have a complexity of $\left(\frac{m-1}{4}+1\right)$ (we need to generate only a quarter of a chain on average). We therefore have a trade-off between the memory sacrificed to store the intermediary points and the complexity of E. In general, if we store x values per chain, sacrificing x shares of memory, the trade-off complexity becomes

$$T = \frac{n^2 m^2}{(c-x)^2 \mu^2 n^2}\left(\frac{m}{2x}+1\right)\gamma$$

$$\approx \frac{m^3 \gamma}{2x(c-x)^2 \mu^2}.$$

The optimal complexity is achieved when $x = c/3$. So we have

$$T_{\text{optimal}} \approx \frac{3^3}{2^3}\frac{m^3 \gamma}{c^3 \mu^2}. \tag{1}$$

Since a pair of (i,k) is 27-bits large (20-bits for i and 7-bits for k) we need at most 54-bits to store one chain. Thus, in the same amount of memory we would require to store one s ($\mu \geq 2$), we could actually store more than two chains. From (1), we can compute the time required to identify one tag. Assuming that all calculations are carried out on a single back-end equipped with $\frac{c(n|s|)}{8} = 2^{24}c$ bytes of memory, and that we choose a success rate of 99.9% ($\gamma = 8$), the time to read a tag is

$$t_{\text{OSK/AO}} \approx \frac{6^9 \theta}{c^3}\ (\text{s}).$$

Figure 18 shows the identification time according to the available memory. For example, with 1 GB of RAM (i.e., $c = 64$), we have $t_{\text{OSK/AO}} \approx 0.004$ milliseconds.

Note that the AO technique cannot be applied directly to the *modified* OSK. This is due to the randomization of the tag's answer. To apply the technique on the modified version of OSK, the tag must answer with both $G(s_i^k)$ and $G(s_i^k \oplus r)$. The former value enables the reader to identify the tag and the latter one allows to detect replay attacks.

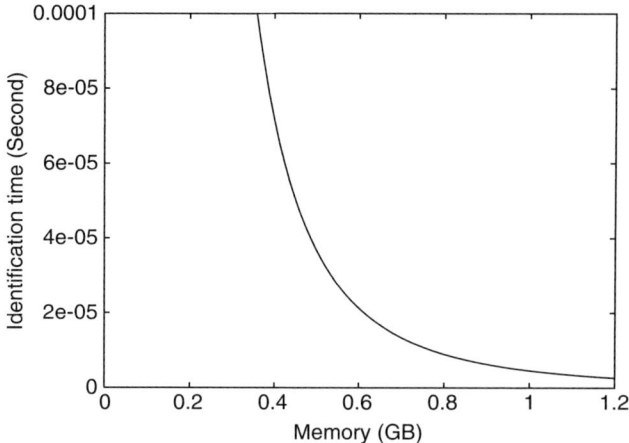

Fig. 18 OSK/AO time complexity

4.4.3 (P)recomputation of the Tables

Before the trade-off can be used to read the tags, the trade-off chains must be generated and their start and end stored in the memory. Since the chains contain arbitrary hashes, we need to generate slightly more than nm hashes to ensure that each hash appears at least once in the tables with a high probability. Again, the hashes are not created sequentially and each calculation of E incurs about $\frac{m}{2} + 1$ hash calculations. The effort to create the tables is thus $T_{\mathrm{precalc}} \approx nm^2/2$. This complexity is reduced by the fact that we store intermediate elements of the chains in some part of the memory.

If the set of tags in the system stays the same, the tables only need to be calculated once. If new tags must be added, the tables must be recalculated. Extra tags can be included in the tables, so that they need not be recalculated for every single new tag. Every time the tables are recalculated we can also remove tags that are no longer in use. Typically the tables could be recalculated off-line every night, week, or month.

Keeping m low increases the advantage of the trade-off over the brute-force method. The following procedure can be applied to keep m small. In the database that contains the s_i^1 we can keep track of how many times each tag was read. We know that the next time we read the tag, the result will be further down the hash chain. If tag i has been read k times, we can thus replace s_i^1 by s_i^k in the database when the next recalculation of the tables occurs. Thus m is no longer the number of times a tag is read in its lifetime but the maximum number of times it is read between two recalculations of the tables or the maximum number of times it is read by a foreign reader. Note that the adjustment of s_i^1 makes both the trade-off and the brute-force method faster but increases the speed-up factor between the two.

Time-memory trade-offs are probabilistic, thus there is an arbitrarily small chance that a tag may not be found in the tables because a particular s_i^k is not part of any chain that was generated. A pragmatic approach to this problem is simply to read the tag a second time in such a case (hoping that s_i^{k+1} will be found). A more deterministic approach would be to keep score of the hash values that are generated when the tables are calculated and to eliminate the s_i^1 for which not all hash values have appeared.

Given that $n = 2^{20}$, $m = 2^7$, and $\theta = 2^{-23}$, in our library example, the precomputation takes about 17 min.

5 Conclusion

This chapter tacked the computation complexity issue when dealing with RFID protocols that avoid malicious traceability. We presented several examples of exotic protocols where this complexity problem does not occur. In such protocols, the reader is involved in the refreshment of the identifier of the tag. Unfortunately, all the reader-aided ID-refreshment protocols are weak in terms of untraceability. The exhibited examples illustrate the most common weaknesses one can find in RFID protocols. Then, we introduce the protocols based on self-refreshment. We show that they are secure in terms of untraceability but they all suffer from a high computation complexity that avoids their use in large-scale applications. Finally, we presented and analyzed techniques to reduce the computation complexity of Challenge/Response protocols. We detailed CR/MW, ABHV, OSK, and OSK/AO. We summarize below their main advantages and drawbacks.

First, we consider the storage aspect. On the tag side, the storage of the identifiers becomes a real problem with CR/MW when the branching factor δ is small. Having a large δ is therefore preferable. Storage in the back-end system is the main drawback of OSK/AO because precomputation and storage of tables is required. In the considered example, 1 GB of RAM is used. Today, such a memory is available on Joe Blow's computer.

Next, we address the complexity question. CR/MW, ABHV, and OSK/AO are all parameterizable. CR/MW depends on δ which can be chosen between 2 and n. ABHV depends on γ that is the number of groups. Obviously, the cases $\delta = n$ and $\gamma = n$ leads to CR. Having $\delta > \sqrt{n}$ is possible but in this case the tree is no longer complete. In fact, a typical value could be $\delta = \sqrt{n}$. On the other hand, OSK/AO depends on the available memory. Considering the example of the library, one can show that the required time to authenticate one tag is about 62 s with the classical CR and 16,000 s with OSK. These values show that none of these two protocols can be used in large-scale applications. Since CR/MW, ABHV, and OSK/AO are parameterizable, they can reduce this time complexity to practical values. The cost for OSK/AO is in terms of memory consumption, while the cost for CR/MW and ABHV is in terms of privacy.

Indeed, while CR is secure, CR/MW degrades the privacy because, when an adversary is able to tamper with at least one tag, she is also able to trace other tags in a probabilistic way. The same problem occurs with ABHV even though it is mitigated in this case. We have shown that the probability to trace tags decreases when the computation complexity grows. Thus, CR/MW and ABHV can be seen as a trade-off between privacy and complexity. The probability to trace tags is far from being negligible. For example, when the branching factor is $\delta = 2^{10}$ in CR/MW, the probability to trace a tag is about 0.1% when only one tag has been opened, but it is about 32.9% when 200 tags have been opened. OSK/AO inherits from the security proofs of OSK, in particular the fact that OSK assures forward privacy, because it modifies neither the information exchanged, nor the tag's content: It only improves the way the system manages and stores the data.

Consequently, OSK/AO is the only protocol available today that can assure untraceability in large-scale applications.

Designing protocols that assure untraceability is one of the main topics of today's RFID security research [6]. However, RFID manufacturers do not seem affected by this problem. The only deployed application we know that addresses the malicious traceability problem is the e-passport. In this application, the complexity issue is overcame by printing the material needed to generate the key on the passport itself. Perhaps, today, the hopeless problem is finally not to design protocols that resist to malicious traceability, but to deploy these protocols in commercial applications.

Acknowledgments The main parts of this chapter follow from works achieved at the EPFL and MIT, while I was in the teams of Serge Vaudenay and Ron Rivest, respectively. I am so quite grateful to both of them. I would also like to thank the researchers I worked with on the topic addressed in this chapter, especially Philippe Oechslin, Levente Buttyán, and Tamás Holczer.

References

1. M. Aigner and M. Feldhofer. Secure symmetric authentication for RFID tags. In *Telecommunication and Mobile Computing – TCMC 2005*, Graz, Austria, March 2005
2. R. Anderson and M. Kuhn. Low cost attacks on tamper resistant devices. In B. Christianson, B. Crispo, M. Lomas, and M. Roe, editors, *International Workshop on Security Protocols – IWSP'97*, volume 1361 of *Lecture Notes in Computer Science*, pp. 125–136, Paris, France, April 1997, Springer, Berlin
3. G. Ateniese, J. Camenisch, and B. de Medeiros. Untraceable RFID tags via insubvertible encryption. In *Conference on Computer and Communications Security – CCS'05*, Alexandria, Virginia, USA, November 2005, ACM, ACM Press, New York, NY
4. G. Avoine. Adversary model for radio frequency identification. Technical Report LASEC-REPORT-2005-001, Swiss Federal Institute of Technology (EPFL), Security and Cryptography Laboratory (LASEC), Lausanne, Switzerland, September 2005
5. G. Avoine. *Cryptography in Radio Frequency Identification and Fair Exchange Protocols*. PhD Thesis, EPFL, Lausanne, Switzerland, December 2005
6. G. Avoine. *Bibliography on Security and Privacy in RFID Systems.* Available Online, 2007
7. G. Avoine and P. Oechslin. A scalable and provably secure hash based RFID protocol. In *International Workshop on Pervasive Computing and Communication Security – PerSec 2005*, pp. 110–114, Kauai Island, Hawaii, USA, March 2005, IEEE, IEEE Computer Society Press, Washington, DC
8. G. Avoine, E. Dysli, and P. Oechslin. Reducing time complexity in RFID systems. In B. Preneel and S. Tavares, editors, *Selected Areas in Cryptography – SAC 2005*, volume 3897 of *Lecture Notes in Computer Science*, pp. 291–306, Kingston, Canada, August 2005, Springer, Berlin

9. G. Avoine, P. Junod, and P. Oechslin. Time-memory trade-offs: False alarm detection using checkpoints. In *Progress in Cryptology – Indocrypt 2005*, volume 3797 of *Lecture Notes in Computer Science*, pp. 183–196, Bangalore, India, December 2005, Cryptology Research Society of India, Springer, Berlin

10. G. Avoine, L. Buttyán, T. Holczer, and I. Vajda. Group-based private authentication. In *IEEE International Workshop on Trust, Security, and Privacy for Ubiquitous Computing – TSPUC*, Helsinki, Finland, June 2007, IEEE, IEEE Computer Society Press,Washington, DC

11. L. Buttyán, T. Holczer, and I. Vajda. Optimal key-trees for tree-based private authentication. In *Workshop on Privacy Enhancing Technologies - PET 2006*, Cambridge, United Kingdom, June 2006

12. S. Dominikus, E. Oswald, and M. Feldhofer. Symmetric authentication for RFID systems in practice. *Handout of the Ecrypt Workshop on RFID and Lightweight Crypto*, July 2005

13. T. Elgamal. A public key cryptosystem and a signature scheme based on discrete logarithms. *IEEE Transactions on Information Theory*, 31(4): 469–472, July 1985

14. M. Feldhofer, S. Dominikus, and J. Wolkerstorfer. Strong authentication for RFID systems using the AES algorithm. In M. Joye and J.-J. Quisquater, editors, *Workshop on Cryptographic Hardware and Embedded Systems – CHES 2004*, volume 3156 of *Lecture Notes in Computer Science*, pp. 357–370, Boston, Massachusetts, USA, August 2004, IACR, Springer, Berlin

15. S. Garfinkel. Adopting fair information practices to low cost RFID systems. *Ubicomp 2002 – Workshop on Socially-Informed Design of Privacy-Enhancing Solutions in Ubiquitous Computing*, September 2002

16. S. Garfinkel. An RFID bill of rights. *Technology Review*, October 2002

17. M. Girault and D. Lefranc. Public key authentication with one (online) single addition. In M. Joye and J.-J. Quisquater, editors, *Workshop on Cryptographic Hardware and Embedded Systems – CHES 2004*, volume 3156 of *Lecture Notes in Computer Science*, pp. 413–427, Boston, Massachusetts, USA, August 2004, IACR, Springer, Berlin

18. P. Golle, M. Jakobsson, A. Juels, and P. Syverson. Universal re-encryption for mixnets. In T. Okamoto, editor, *The Cryptographers' Track at the RSA Conference – CT-RSA*, volume 2964 of *Lecture Notes in Computer Science*, pp. 163–178, San Francisco, California, USA, February 2004, Springer, Berlin

19. G. Hancke. A practical relay attack on ISO 14443 proximity cards. Manuscript, February 2005

20. G. Hancke and M. Kuhn. An RFID distance bounding protocol. In *Conference on Security and Privacy for Emerging Areas in Communication Networks – SecureComm 2005*, Athens, Greece, September 2005, IEEE, New York, NY

21. M. Hellman. A cryptanalytic time-memory trade off. *IEEE Transactions on Information Theory*, IT-26(4): 401–406, July 1980

22. D. Henrici and P. Müller. Hash-based enhancement of location privacy for radio-frequency identification devices using varying identifiers. In R. Sandhu and R. Thomas, editors, *International Workshop on Pervasive Computing and Communication Security – PerSec 2004*, pp. 149–153, Orlando, Florida, USA, March 2004, IEEE, IEEE Computer Society Press, Washington, DC

23. D. Henrici and P. Müller. Tackling security and privacy issues in radio frequency identification devices. In A. Ferscha and F. Mattern, editors, *Pervasive Computing*, volume 3001 of *Lecture Notes in Computer Science*, pp. 219–224, Vienna, Austria, April 2004, Springer, Berlin

24. ICAO DOC–9303. *Machine Readable Travel Documents*, Part 1, Volume 2, November 2004

25. Y. Jeongkyu. Security and privacy on authentication protocol for low-cost radio frequency identification. Master Thesis, Information and Communications University, Daejeon, Korea, December 2004

26. A. Juels. Minimalist cryptography for low-cost RFID tags. In C. Blundo and S. Cimato, editors, *International Conference on Security in Communication Networks – SCN 2004*, volume 3352 of *Lecture Notes in Computer Science*, pp. 149–164, Amalfi, Italia, September 2004, Springer, Berlin

27. A. Juels and J. Brainard. Soft blocking: Flexible blocker tags on the cheap. In S. De Capitani di Vimercati and P. Syverson, editors, *Workshop on Privacy in the Electronic Society – WPES*, pp. 1–7, Washington, DC, USA, October 2004, ACM, ACM Press, New York, NY

28. A. Juels and R. Pappu. Squealing euros: Privacy protection in RFID-enabled banknotes. In R. Wright, editor, *Financial Cryptography – FC'03*, volume 2742 of *Lecture Notes in Computer Science*, pp. 103–121, Le Gosier, Guadeloupe, French West Indies, January 2003, IFCA, Springer, Berlin

29. A. Juels and S. Weis. Authenticating pervasive devices with human protocols. In V. Shoup, editor, *Advances in Cryptology – CRYPTO'05*, volume 3621 of *Lecture Notes in Computer Science*, pp. 293–308, Santa Barbara, California, USA, August 2005, IACR, Springer, New York, NY

30. A. Juels and S. Weis. Defining strong privacy for RFID. Cryptology ePrint Archive, Report 2006/137, 2006

31. A. Juels, R. Rivest, and M. Szydlo. The blocker tag: Selective blocking of RFID tags for consumer privacy. In V. Atluri, editor, *Conference on Computer and Communications Security – CCS'03*, pp. 103–111, Washington, DC, USA, October 2003, ACM, ACM Press, New York, NY

32. G. Karjoth and P. Moskowitz. Disabling RFID tags with visible confirmation: Clipped tags are silenced. In *Workshop on Privacy in the Electronic Society – WPES*, Alexandria, Virginia, USA, November 2005, ACM, ACM Press, New York, NY

33. Z. Kfir and A. Wool. Picking virtual pockets using relay attacks on contactless smartcard systems. In *Conference on Security and Privacy for Emerging Areas in Communication Networks – SecureComm 2005*, Athens, Greece, September 2005, IEEE, New York, NY

34. T. Li and R. H. Deng. Vulnerability analysis of EMAP – an efficient RFID mutual authentication protocol. In *Second International Conference on Availability, Reliability and Security – AReS 2007*, Vienna, Austria, April 2007

35. D. Molnar and D. Wagner. Privacy and security in library RFID: Issues, practices, and architectures. In B. Pfitzmann and P. Liu, editors, *Conference on Computer and Communications Security – CCS'04*, pp. 210–219, Washington, DC, USA, October 2004, ACM, ACM Press, New York, NY

36. K. Nohl and D. Evans. Quantifying information leakage in tree-based hash protocols. In *Conference on Information and Communications Security – ICICS'06*, volume 4307 of *Lecture Notes in Computer Science*, pp. 228–237, Raleigh, North Carolina, USA, December 2006, Springer, Berlin

37. P. Oechslin. Making a faster cryptanalytic time-memory trade-off. In D. Boneh, editor, *Advances in Cryptology – CRYPTO'03*, volume 2729 of *Lecture Notes in Computer Science*, pp. 617–630, Santa Barbara, California, USA, August 2003, IACR, Springer, Berlin

38. M. Ohkubo, K. Suzuki, and S. Kinoshita. Cryptographic approach to "privacy-friendly" tags. In *RFID Privacy Workshop*, November 2003, MIT, Cambridge, MA, USA

39. M. Ohkubo, K. Suzuki, and S. Kinoshita. Efficient hash-chain based RFID privacy protection scheme. In *International Conference on Ubiquitous Computing – Ubicomp, Workshop Privacy: Current Status and Future Directions*, Nottingham, England, September 2004

40. P. Peris-Lopez, J. C. Hernandez-Castro, J. Estevez-Tapiador, and A. Ribagorda. LMAP: A real lightweight mutual authentication protocol for low-cost RFID tags. *Printed Handout of Workshop on RFID Security – RFIDSec 06*, July 2006

41. P. Peris-Lopez, J. C. Hernandez-Castro, J. Estevez-Tapiador, and A. Ribagorda. M2AP: A minimalist mutual-authentication protocol for low-cost RFID tags. In *International Conference on Ubiquitous Intelligence and Computing – UIC̆201906*, volume 4159 of *Lecture Notes in Computer Science*, pp. 912–923, September 2006, Springer, Berlin

42. P. Peris-Lopez, J. C. Hernandez-Castro, J. M. Estevez-Tapiador, and A. Ribagorda. EMAP: An efficient mutual authentication protocol for low-cost RFID tags. In *OTM Federated Conferences and Workshop: IS Workshop – IS'06*, volume 4277 of *Lecture Notes in Computer Science*, pp. 352–361. November 2006, Springer, Berlin

43. J. Saito, J.-C. Ryou, and K. Sakurai. Enhancing privacy of universal re-encryption scheme for RFID tags. In L. Jang, M. Guo, G. Gao, and N. Jha, editors, *Embedded and Ubiquitous Computing – EUC 2004*, volume 3207 of *Lecture Notes in Computer Science*, pp. 879–890, Aizu-Wakamatsu City, Japan, August 2004, Springer, Berlin

44. S. Weingart. Physical security devices for computer subsystems: A survey of attacks and de-fenses. In c. K. Koç and C. Paar, editors, *Workshop on Cryptographic Hardware and Embedded Systems – CHES 2000*, volume 1965 of *Lecture Notes in Computer Science*, pp. 302–317, Worcester, Massachusetts, USA, August 2000, Springer, Berlin
45. S. Weis, S. Sarma, R. Rivest, and D. Engels. Security and privacy aspects of low-cost radio frequency identification systems. In D. Hutter, G. Müller, W. Stephan, and M. Ullmann, editors, *International Conference on Security in Pervasive Computing – SPC 2003*, volume 2802 of *Lecture Notes in Computer Science*, pp. 454–469, Boppard, Germany, March 2003, Springer, Berlin
46. J. Yang, J. Park, H. Lee, K. Ren, and K. Kim. Mutual authentication protocol for low-cost RFID. *Handout of the Ecrypt Workshop on RFID and Lightweight Crypto*, July 2005

Dynamic Privacy Protection for Mobile RFID Service

Namje Park* and Dongho Won

Abstract Recently, mobile RFID has been studied actively as a primary technology in computing environments. The mobile RFID service is defined as a special type of mobile service using RFID tag packaging objects and RFID readers attached to mobile RFID terminals. While the mobile RFID system has many advantages, it may make new intrusions to the user's privacy. We propose the policy-based dynamic privacy protection framework leveraging globally mobile RFIDs. In this paper, we describe privacy infringements for the mobile RFID service environment and requirements for personal privacy protection, and develop privacy protection service based on a user privacy policy. The proposed framework provides a means for securing the stability of mobile RFID services by suggesting personal privacy-policy-based access control for personalized tags. This means a technical solution to privacy protection for the mobile RFID service system.

1 Introduction

Though the Radio Frequency Identification (RFID) technology is being actively developed with a great deal of effort to generate a global market, it is also raising fears of its role as a "Big Brother." So, it is necessary to develop technologies for information and privacy protection as well as promotion of markets (e.g., technologies of tag, reader, middleware, etc.) The current excessive limitations to RFID tags and readers make it impossible to apply present codes and protocols. The technology for information and privacy protection should be developed in terms of general interconnection among elements and their characteristics of RFID to such technology that meets the RFID circumstances [33–35, 37].

N. Park
Electronics and Telecommunications Research Institute (ETRI), 161 Gajeong-dong, Yuseong-gu, Daejeon 305-350, Korea
e-mail: namjepark@etri.re.kr, namjepark@gmail.com

P. Kitsos, Y. Zhang (eds.), *RFID Security: Techniques, Protocols*
and System-on-Chip Design, © Springer Science+Business Media, LLC 2008

While common RFID technologies are used in Business to Business (B2B) models like supply channels, distribution, logistics management, mobile RFID technologies are used in the RFID reader attached to an individual owner's cellular phone through which the owner can collect and use information of objects by reading their RFID tags; in case of corporations, it has been applied mainly for Business to Customer (B2C) models for marketing. Though most current RFID application services are used in fields like the search of movie posters and provision of information in galleries where less security is required, they will be expanded to and used more frequently in such fields as purchase, medical care, electrical drafts, and so on where security and privacy protection are indispensable [1,7]. Therefore, in this paper, we described a privacy preserving enhanced trust building mechanism that extends the extant to which trust building service mechanisms for mobile RFID networks can gain many advantages from its privacy control and dynamic capabilities. This is new technology to mobile RFID and will provide a solution for protecting absolute confidentiality from basic tags to user's privacy.

2 Background on Mobile RFID Technology

2.1 Mobile RFID Primer

Networked RFID comprises an expanded RFID network and communication scope to communicate with a series of networks, inter-networks, and globally distributed application systems, engendering global communication relationships triggered by RFID, for such applications as B2B, B2C, B2B2C, Government to Customer (G2C), etc. The mobile RFID loads a compact RFID reader into a cellular phone, thereby providing diverse services through mobile telecommunications networks when reading RFID tags through a cellular phone. Since the provision of these services was first attempted in Korea in 2005, their standardization has been ongoing. Korea's mobile RFID technology is focusing on the Ultra High Frequency (UHF) range [3, 11,27,42]. Thus, as a kind of handheld RFID reader, in the selected service domain the UHF RFID phone device can be used to provide object information directly to the end-user using the same UHF RFID tags that have been widely distributed.

The mobile RFID refers to a system whereby 900-MHz passive RFID reader can be built within a mobile phone. Therefore, the access target of mobile RFID should be 900-MHz passive RFID tag. Figure 1 shows the structure of how a mobile phone communicates with an RFID tag.

2.2 Network Architecture Component

The mobile RFID service has been defined as the provision, through the wireless Internet network, of personalized secure services – such as searching for product

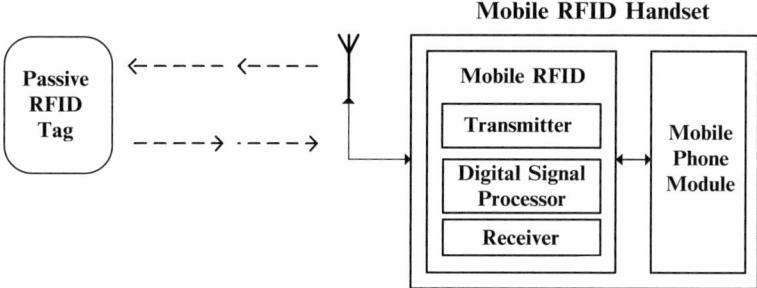

Fig. 1 Mobile RFID diagram

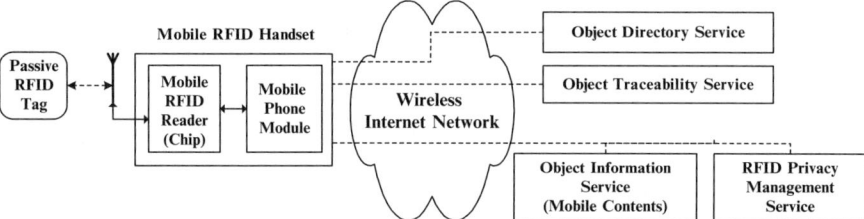

Fig. 2 Mobile RFID service's network architecture

information, purchasing, verifying, and paying for products – while on the move, by building the RFID reader chip into the mobile terminal [36, 39]. The service infrastructure required for providing such an RFID-based mobile service is composed of an RFID reader, handset, communication network, network protocol, information protection, application server, RFID code interpretation, and contents development; the configuration map is as follows.

The mobile RFID service structure is defined to support ISO/IEC 18000-6 A/B/C through wireless access communication between the tag and the reader; however, as yet there is no RFID reader chip capable of supporting all three wireless connection access specifications so that the communication specification for the mobile phone will be determined by the mobile communication companies [3, 9, 31]. It will also be possible to mount the RF wireless communication function on the reader chip using software-defined radio (SDR) technology and develop an ISO/IEC 18000-6 A/B/C communication protocol in software to choose from the protocols when needed [3, 12, 28]. Figure 2 shows the interface structure for the mobile RFID service's communication infrastructure and the types of relevant standards. RFID wireless access communication takes place between the RFID tag and a cellular phone; Code division multiple access (CDMA) mobile communication takes place between a cellular phone and base transceiver station (BTS)/access network transceiver subsystem (ANTS); and wire communication takes place between BTS/ANTS and a networked RFID application server.

Figure 2 represents the entities of the mobile RFID service network architecture. The Object Directory Service (ODS) server functions as a Directory Name

System (DNS) server which informs the mobile RFID phone of the contents/service server's location, as explained earlier [11, 31, 42]. While DNS returns the "IP address" corresponding to a "domain name" or vice versa, ODS returns Uniform Resource Identifier (URI) information that enables access to the information resource corresponding to the identification code of on-line or off-line objects. The ODS communication protocol is the same as the existing DNS protocol. Therefore, an ODS query message substitutes for a DNS query message and an ODS response message for a DNS response message. URI information contains details needed to gain resources from a content server such as the address of the content server, directory and file name of the resource, communication protocol name, and port number of the server.

The Object Traceability Service (OTS) server keeps a record of the tag readings in the RFID readers throughout the lifecycle of the objects. This refers to an information system that reads a tag and records its history to trace information on when the tag-attached product was rolled out, how it was distributed, and what processes it went through. Its main purpose is to track objects in the supply chain management (SCM).

The Object Information Service (OIS) system server retains details about a certain object, and they could be any information. This server records the reading of the RFID tag event in the OTS server and may provide additional detailed information on an object – such as manufacturing time, manufacturer's name, expiration time, etc. In this regard, OIS is similar to the Wireless Application Protocol (WAP)/Web content server that offer common contents. The WAP server provides WAP-based contents through the WAP protocol and is not concerned about the Information Schema of contents. The Web server also offers a variety of contents through the HTTP protocol, and does not rely on a particular information scheme. Since OIS holds details on objects and provides this information through the HTTP protocol, it can be classified into a Web or WAP server. However, the mechanism to access the OIS is different from that of other servers. When an ODS server saves Naming Authority Pointer (NAPTR) records for codes, if the code is allocated for every product item, it would not be difficult to register NAPTR records because a producer has limited product items. But if NAPTR records should be registered for individual products, it would cause ODS operation to be less efficient because the ODS server has to register NAPTR records for millions of products. Therefore, if the OIS server stores details about individual products and the ODS server keeps NAPTR records for the OIS server information on product items, the Client could gain OIS server information on product items from the ODS server and details about each product from the OIS. While the existing Web server directly accesses the requested content, using Uniform Resource Locator (URL), an OIS server provides detailed information on a particular product only after receiving its serial number, even though the server has the same content. That is, the Web server and OIS server are different in accessing even the same content. So, a unique protocol needs to be defined to gain access to an OIS server.

The RFID Privacy Management Service (RPS) refers to a policy-based privacy management service that enables decision making on and adoption of privacy protection policy so that tag information can be applied to the policy within the mobile RFID network infrastructure. This controls access to the information on the object in accordance with the privacy profile put together by the owner of the object.

The WAP and Web servers are contents servers that provide wireless Internet contents such as news, games, music, videos, stock trading, lotteries, images, and so forth.

The mobile RFID middleware is composed by extending the Wireless Internet Platform for Interoperability (WIPI) software platform to provide RF code-related information obtained from an RFID tag through an RFID reader installed in the mobile phone. The networked terminal's function is concerned with the recognition distance to the RFID reader chip built into the cellular phone, transmission power, frequency, interface, technological standard, personal identification number (PIN) specification, universal asynchronous receiver and transmitter (UART) communication interface, and WIPI application program interface (API) extended specification to control the reader chip. RFID reader chip middleware functions are provided to the application program in the form of the mobile platform's API. Here, the mobile RFID device driver is the device driver software provided by the reader chip manufacturer [19, 20, 23–25, 29].

The mobile RFID network function is concerned with communication protocols such as the ODS communication for code interpretation, the message transportation for the transmission and reception of contents between the mobile phone terminal and the application server, contents negotiation that supports the mobile RFID service environment and ensures optimum contents transfer between the mobile phone terminal and the application server, and session management that enables the application to create and manage the required status information while transmitting the message and the WIPI extended specifications that support these communication services [5, 11, 22, 27, 36].

A cellular phone requires a common control interface between the various RFID readers and the application or the middleware; to that end, EPCglobal, Inc. and International Standards Organization (ISO) are defining the functions that an RFID reader should commonly support, as well as various common command and standardizing message types. The mobile RFID functions will be extended continuously into standard cellular phone RFID readers, and the RFID supported WIPI extension model using WIPI will define the API required in using the reader suitable for the mobile environment as the API extension of WIPI, while maintaining compatibility among the various devices.

2.3 Basic Service Communication Model

The basic communication scenario for mobile RFID service is as follows.

First, a mobile RFID phone reads the RFID tags on an object and fetches the code stored in it [27, 28, 42]. Second, a mobile RFID phone should execute the code resolution with which the mobile RFID phone obtains the location of the remote server

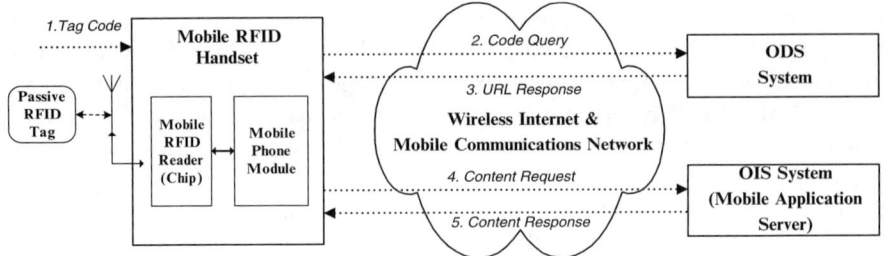

Fig. 3 Basic communication model for mobile RFID

that provides information on the product or an adequate mobile service. The code resolution protocol is identical with the DNS protocol. The ODS server in Fig. 3 functions as a DNS server and is similar to EPCglobal's Object Name Service (ONS) server. The mobile RFID phone directs queries on the location of the server with a code to the ODS server, then the ODS server replies by giving the location of the server. Finally, the mobile RFID phone requests contents or a service from the designated server whose location has been acquired from the ODS server.

Code resolution refers to the process based on the DNS protocol that obtains NAPTR record information on a certain code from an ODS server [12, 15, 21, 42]. The ODS server should register the corresponding NAPTR record for every code allocated for the information service. Code resolution procedure is the same as in "Reverse DNS lookup," but it sends NAPTR query instead of PTR. In the code resolution, code and URI information replace IP address and domain name address of the Reverse DNS lookup, respectively.

An ODS resolver is functionally different from a local ODS, but they are the same in terms of the software system. A local ODS server returns the NAPTR record for a certain code through an ODS query, but if it does not have the corresponding NAPTR record, it implements a code resolution itself involving the Root ODS at first and then other ODS servers according to the authorization hierarchy until, as a client, it can finally convey the NAPTR record information to the Stub ODS resolver. That is, the ODS resolver performs its function as a client by carrying out code resolution and also its local ODS function as a server by offering NAPTR record information from itself.

The DNS server address is basically set within a mobile phone, and so is the ODS server address, usually the address of the local ODS server of a user's mobile telecom company. In fact, such a local ODS server address represents the ODS resolver address because the local ODS server executes code resolution to return NAPTR record information to the stub ODS resolver within a mobile phone.

ODS resolver asks the ODS server for the URI information on a specific code upon the stub ODS resolver's request for the code resolution and sends back the response to the stub ODS resolver. The code resolution process follows the DNS operation system; whereby, it starts with the root ODS and then other servers in an iterative way.

The service model consists of tag, reader, middleware system, and information server. From the point of view of information protection, a serious problem for the RFID service is the threat of privacy [6, 32, 38]. Here, the damage of privacy is of exposing the information stored in tags and the leakage of information includes all data of the person possessing the tag, tagged products, and location. The privacy protection of the RFID system can be considered from two points of view. One is the privacy protection between the tag and the reader, which takes advantage of ID encryption, prevention of location tracking, and the countermeasure of the tag being forged. The other is of the exposure of what the information server contains along with tagged items [26, 40]. First of all, we will examine the exposure of information caused between tag and reader, and then discuss the solution proposed in this paper.

3 Mobile RFID-Oriented Vulnerability and Requirements

3.1 Some Mobile RFID-Oriented Privacy Vulnerability

The security vulnerability of the mobile RFID is the infringement of owners' privacy and the physical attack in cyber space [4, 8, 10, 13, 16, 18]. Typical exposures include threats to individual privacy due to the approval of unlimited access to an RFID tag owned by a person. Access to the information must be limited to those who need it for an application. An individual RFID tag also may become a means to track and locate its owner. The infringement of privacy in the Internet world results from the collection, storage, and use of customers by companies, but it has grown more serious in the mobile RFID world in that anyone with an RFID reader can read any information on anyone who keeps a tag-attached object. It is also possible to hack tags, prevent the normal use of tags, or get incorrect information from them by altering tag information or using a tag-kill function. Information can be collected illegally by hiding a system in a commodity or object for remote communication to wiretap, track, catch information, or profile. Security is also vulnerable to jamming, replay attack, and covert reading.

3.2 Security Requirements

The mobile RFID service structure provides its services by associating the mobile communication network and the RFID application service network based on the RFID tag. The areas to be considered with regard to security are essentially the RFID tag, reader terminal area, mobile communication network area, RFID application service network area, while other security issues such as confidentiality/integrity/authentication/permission/nonrepudiation shall be considered in each network area. Especially, as the mobile RFID service is the end-user service, the issue of privacy protection must inevitably become a serious issue to consider, and as content accessibility increases due to the off-line hypertext property of RFID, the authentication for adult services is also highly likely to become another important issue for consideration.

3.3 Personal Privacy Protection Requirements

The mobile RFID service is a B2C service for end users. It uses RFID tags, which make the privacy protection issue all the more important. With an RFID tag, a party with proper equipment can know where the owner is located, which device the owner is using, and which information the owner is reading through interpreting the tag codes. To secure an owner's location information, a location information protection function shall be implemented because it is easy to access the tag code in the current passive tag (ISO/IEC 18000-6 Type C: Gen2.) Moreover, to protect device information of the mobile RFID terminal, a mechanism shall be provided to provide application and contents information and the tag owner's information (such as which information the tag owner read, using which tag, and in which place) based on the privacy policy.

3.3.1 Location Privacy Protection

The above privacy issue in relation to the owner's location is raised, because it is possible to locate an individual who possesses an object with an installed tag through the tag code. This privacy issue can be handled through control of the tag code that can be used as an identifier. However, passive tags used in mobile RFID services do not support the access control function to the tag code.

We can protect the owner's location information either by allowing accesses to the tag code or by removing the identifier in the tag code. Therefore, in the mobile RFID application services that shall ensure privacy protection, the service operator shall provide a plan to protect an owner's location information.

3.3.2 Privacy Protection of Mobile Device

It is natural that the service should protect information of a person who possesses an object to which a tag is attached. The mobile RFID service is provided through the user's individual portable device so that privacy protection of the user using open tags shall be guaranteed. In other words, when an open tag is read, the reader can immediately transmit information of the mobile RFID terminal to the application service or contents provider through the network of a mobile telecommunication company. Not only the location data of the person who has a mobile RFID terminal but also how often the person collects that information in which way and which information the person tends to access can be collected by the application service or contents provider. These kinds of information can be collected when a reader is detecting the mobile RFID tag, regardless of the owner's purchasing activities. The application service or contents provider shall collect users' information in compliance with the privacy protection level that the users have agreed upon.

3.3.3 Privacy Protection of RFID Information

In most cases, an individual should be responsible for the protection of his/her privacy in the process of using and processing personal information. However, in mobile RFID services, privacy issues are raised because a party with proper equipment can identify the owner of an object to which a tag is attached and access the owner's personal information in the RFID network. Therefore, privacy protection shall be secured for the owners of a tag-attached object who need to keep using RFID services. For example, privacy protection may be provided through the RPS system. Through an interface with the RPS service, the mobile RFID service can allow individual users to select and modify their privacy protection levels. The RPS system shall be able to transmit each user's privacy protection policy to the application server so that the tag connection contents that the application server provides can reflect the user's privacy protection policy.

When privacy protection is enabled by the RPS system, the user data contained in the RFID tag shall include the default privacy level that was given at the time of privacy impact evaluation. The mobile RFID reader shall provide a default privacy level when negotiating the contents to allow access by the application server.

4 Multilateral Approaches with Improved Privacy

This technology is aimed at RFID application services like authentication of tag, reader, and owner, privacy protection, and nontraceable payment system where stricter security is needed:

- Approach of platform level
  This technology for information portal service security in offering various mobile RFID applications consists of application portal gateway, information service server, terminal security application, payment server, and privacy protection server and provides a combined environment to build a mobile RFID security application service easily.
- Approach of protocol level
  It assists write and kill passwords provided by Electronic Product Code (EPC) Class1 Gen2 for mobile RFID tag/reader and uses a recording technology preventing tag tracking. Information technology solves security vulnerability in mobile RFID terminals that accept WIPI as middleware in the mobile RFID reader/application part and provides End-to-End (E2E) security solutions from the RFID reader to its applications through WIPI-based mobile RFID terminal security/code treatment modules.

– Approach of privacy level

This technology is intended to solve the infringement of privacy or random acquisition of personal information by those with RFID readers from those with RFID attached objects in the mobile RFID circumstance except when taking place in companies or retail shops that try to collect personal information. The main assumptions are privacy in the mobile RFID circumstance when a person holds a tag attached object and both information on his/her personal identity (reference number, name, etc.) and the tag's information of the commodity are connected. Owners have the option to allow access to any personal information on the object's tag by authorized persons like a pharmacist or doctor but limit or completely restrict access to unauthorized persons.

5 Key Technology and Solution: Security Framework

The mobile RFID is a technology to install the RFID reader in a cellular phone and to provide users of mobile RFID readers with various application services. There are many ways to interfere with RFID circumstances, issues which are not only approved theoretically but also possible practically. Besides security vulnerabilities in RFID security like passive signal interception attack on RFID tags and readers, reading of RFID tags by unauthorized readers, falsifying tag or reader identity, use of attack tools against RFID tags, neutralization of RFID tags, and elaborate attack on RFID tags with cryptographic hacking methods, there are also similar vulnerabilities and possible infringement of privacy in mobile RFID circumstances. This circumstance requires proper security technologies. Furthermore, some information protection service models are needed that ensure security and privacy protection and management for service providers in practical compliance with present RFID specifications and mobile RFID standards, even when tags do not use code algorithms. This session suggests and analyzes these mobile RFID information protection service models considering the situations mentioned above.

For the provision of secure mobile RFID services, the author prepared the mobile RFID protection policy named MRF-Sec631 (Mobile RFID-Security 6.3.1) and proposed the mobile RFID information protection service model. Specifically speaking, MRF-Sec631 means the development of six typical standard security functions out of mobile RFID terminal platforms, applies three main security service mechanisms on the basis of such development, and the execution of secure mobile RFID services through application portal services. The six standards security functions are mobile RFID data encryption API function, mobile RFID secure communication API function, mobile RFID password management API function, EPC C1G2 security command API function, adult certification API function, and privacy protection API function. The three security service mechanisms are authentication service mechanism, privacy protection service mechanism, and secure location tracking service mechanism. The one secure application service is secure mobile RFID application portal service.

The provision of secure mobile RFID services needs a combined security framework resolving many security issues like security among domains, personal privacy profile, authentication, end-to-end security, and track prevention. The following is the configuration of the service framework based on the MRF-Sec631 development.

The main functions of the proposed mobile RFID information protection service model are the provision of WIPI-based mobile security middleware, tag authentication, tag tracking prevention, reader authentication, message security, and protection of profile-based privacy.

6 Key Technology and Solution: Privacy Protection

6.1 Policy-Based RFID Privacy Protection Primer

The policy-based mobile RFID privacy protection technology can protect personal privacy contained in information connected to customized tags in the mobile RFID circumstance. User of the mobile RFID can directly control all information connected to tags through the security service upon his/her purchase of any tag-attached items. With this technology based on profiles, users buying commodities with RFID tags can decide whether information on such commodities and other related information would be disclosed to whom and/or to what extent, and the mechanism to control access to the information. The greatest strength of this mechanism is to promote mobile RFID application services by not preventing access to the information in any case but allowing limited access for authorized persons. It also informs the user through an Internet message of the compliance with duties in connection with the execution of customer-defined profiles, minimizes the number of infringements of privacy by the analysis of system logs through privacy reinforcing inspection function, and allows the decision of a fault source when a problem is reported. Basic main service functions provided for such purposes include the preparation and management of privacy protection policy, prevention of access to information connected to customized tags according to the privacy policy set by the owner, notice of result of executed necessary operations as specified by the owner, and monitoring of privacy through monitoring log management. Figure 4 shows the conceptual security framework architecture of mobile RFID service including RPS.

Privacy protection in mobile RFID services refers to technological measures against unauthorized access of personal information. Access to platform resources can be controlled based on each user's privacy protection level. Privacy protection in mobile RFID services is based on protection systems that guarantee confidentiality and the integrity of private information on the network and ensure authorization of entities. There must be a means to provide detailed access control mechanisms that can manage object information, log data, and personal information by user group and communicate with RPS systems through secure communication paths. These systems must provide auditing functions with stronger security based on the privacy

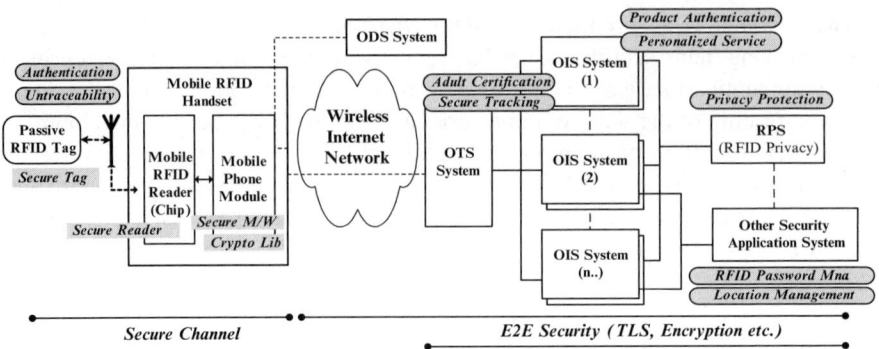

Fig. 4 Conceptual architecture for secure mobile RFID service framework

protection policy that each individual user defined in the RPS system and there must also be a mechanism to negotiate privacy policies with mobile RFID terminals to prevent them from gathering personal information.

6.2 General Service Scenario

The RFID user privacy management service provides mobile RFID users with information privacy protection service for personalized tag under the mobile RFID environment [2–5, 41, 43]. When a mobile RFID user possesses an RFID tagged product, RPS enables the owner to control his backend information connected with the tag such as product information, distribution information, owner's personal information, and so on.

The service scenario for profile-based privacy protection service generally arises from the tag personalizing process such as the tagged product purchase. Figure 5 illustrates a general profile-based privacy protection service scenario of a networked RFID application.

In this system, a consumer reads the ID code from the tagged product with his/her mobile terminal equipped with an RFID reader. The consumer then browses the product-related information from the application service network and purchases the product using one of the various payment methods. At this moment, the consumer becomes the tag owner. The networked RFID application then requests the owner-defined privacy profile from the RPS server, and the RPS server responds with the owner-defined privacy profile to the application, if the owner's privacy protection policy exists. The RPS server receives the owner's privacy protection policy for this application service. Anyone requesting information associated with this tag ID can browse all information provided by the application service if the requestor is the owner, but otherwise, the requestor only has access to a limited amount or no information according to the limits established by the owner. This generic service scenario is able to be modified according to each specific-networked RFID application.

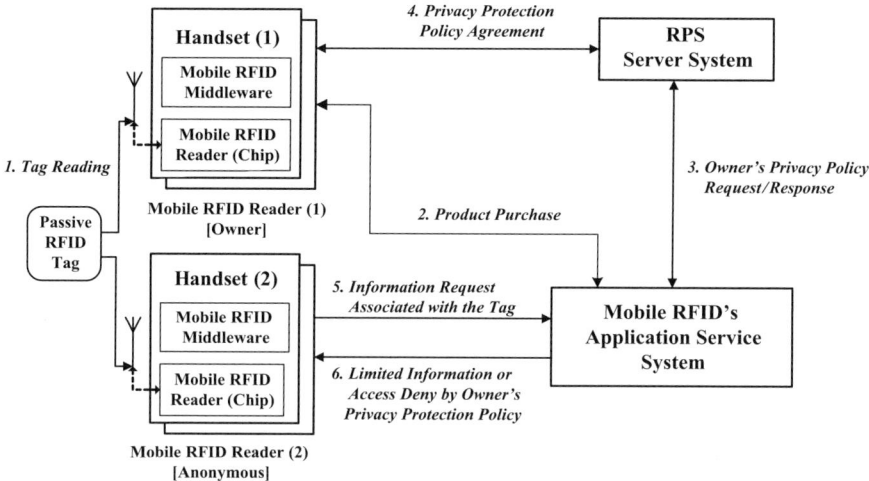

Fig. 5 Service scenario of policy-based RPS system

For example, in a healthcare-related service triggered by an RFID patient card, the product purchase process of step (2) can be considered the patient card issuing step.

6.3 Service Entities for Privacy Protection Service

Profile-based privacy protection service has the following three service entities (see Fig. 6). The privacy protection system consists of the RPS system, the application and contents information system on the service provider side, the user system, and the tag. The user systems include mobile terminals, PDA, and PC that can change the user's privacy protection policy. The privacy protection level is set in the tag, and users can set the privacy policy using the user terminal. The RPS manages privacy policies, sends privacy policies to the application and contents information server of the service provider, and checks the service provider's compliance with the rules. The application and contents information server provides privacy protected information for the mobile RFID terminals and informs RPS of the result. RPS sends privacy policy history to the user terminals to enhance the credibility of privacy protection.

As shown in Fig. 4, the privacy protection system in mobile RFID services consists of the user terminal, the tag, the application and contents information server of the service provider, and RPS. The functions and structure of user terminals include mobile terminals, PDAs, and PCs. It is an entity that has wireless (or wired) network access function and RFID reader function, if it is necessary. It can be a mobile terminal with RFID reader device. The owner – tag owner or/and device owner – can access the service-side and RPS system via this user-side system. The application and contents information server of the service provider delivers extended access control

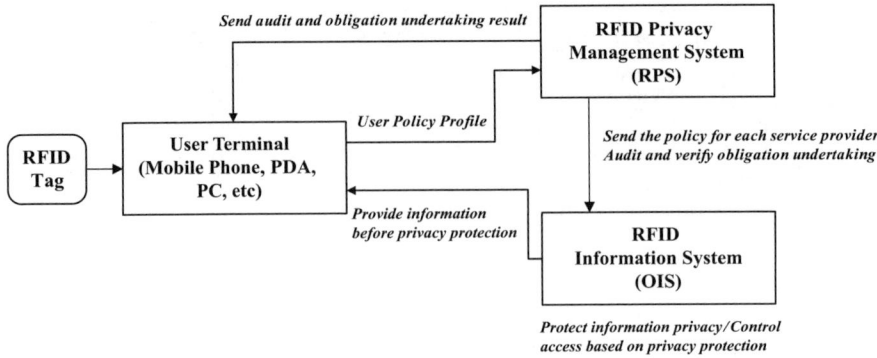

Fig. 6 Structure of mobile RFID privacy protection system

for greater stability. Depending on the tag owner's policy transmitted from RPS, the application and contents information server manages privacy contents, checks who are accessing information, and controls access based on the privacy protection level established by the object holder. RFID privacy protection management handles the owner's privacy policy by creating the owner's privacy profile for service-side system and manages the event logs from service-side or RPS system for auditing.

6.4 Major Functions of Policy-Based Privacy Protection Service

In order to satisfy the privacy protection requirements of users of the mobile RFID service, the profile-based privacy protection service incorporates the following functions: access control, registration, privacy profile management, privacy enhanced log management, obligation notification, and tag data refreshment [4, 5, 14, 17, 30, 40]:

- Access control
  An RPS system access control function may be used to authenticate the identity of the owner (or application service provider) and authorize the access of the owner's information resources, which are mainly the owner's privacy protection policies.
  On the other hand, it would seem to be a more essential component of the profile-based privacy protection service on the service-side system, since application service provider's server should control the access to all information resources and provided application services by the owner-defined privacy profile. The service-side system must be able to deduce where a requestor has access to a certain service or information from the owner-defined privacy profile which is a formatted set of the privacy protection rules and policies that are defined by the owner.
- Registration
  Application service provider (service-side system) and the user (user-side system) have a registration process with the RPS system. The RPS system provides the default privacy profile, which is a formatted set of privacy protection rules

and policies of the application service, to the service-side system. The default privacy profile can be created by the privacy profile management functionality.
- Privacy profile management
  Privacy profile management is a core function of the profile-based privacy protection service. It performs the establishment and management of the owner's (or default) privacy profile. The RPS system should create and manage the default privacy profile for each application service and the owner-defined privacy profile from user's privacy protection policy of the registration process. And hence, this privacy profile is able to be sent to the service-side system when it is requested by the application service provider.
- Privacy enhanced log management
  This function performs secure event log collection and limitation of event log collection by the owner's privacy profile. Furthermore it may be concerned with detection and analysis of privacy violation elements for collected event logs.
- Obligation notification
  The results of the obligation that should be performed by the service-side system can be notified to the owner via email, mobile phone text message, and so on. The service-side system should notify the RPS system of the obligation result, and the RPS system should notify the owner of the obligation result.
- ID code refreshment
  This function provides a mechanism that can refresh the ID code of the owner's tag, and the user-side system should write this refreshed ID code to his/her own tag. Furthermore, the service-side system should reestablish its relationship between the information and the tag's ID code. And hence, the ID code refreshment function would seem to satisfy the ID code requirement for being untraceable.

6.5 Service Level for Privacy Protection

The privacy protection system for the mobile RFID services shall have the privacy protection level that is defined by the RFID tag owner. The privacy protection level refers to the grade of information to be protected depending on the user who is accessing the tag. In other words, if information is of high privacy protection level, a third party may not be allowed to access this information or a third party will only be given access to low level information.

6.5.1 Basic Guideline for Classification of Privacy Levels

The following figure is an example of how the privacy levels are classified and how each level is applied. The privacy level is from 0 to 10. In level 0, virtually no privacy protection is provided, and in level 10, tags are killed or the levels are not in use. As shown in Fig. 7, levels actually used for privacy protection are from 1 to 9. These

Object Infomation	Low Level			Medium Level			High Level		
Privacy Level	1	2	3	4	5	6	7	8	9
Object Category	O	O	O	O	O	O	O	O	X
Object Name	O	O	O	O	O	O	O	O	X
Object Code	O	O	O	O	O	O	O	X	X
Object History	O	O	O	O	O	O	X	X	X
Price	O	O	O	O	O	X	X	X	X
Distribution Information	O	O	O	O	X	X	X	X	X
Object Description	O	O	O	X	X	X	X	X	X
Owner ID	O	O	X	X	X	X	X	X	X
Owner Account	O	X	X	X	X	X	X	X	X
Owner Personal	O	X	X	X	X	X	X	X	X

(X: Not to be disclosed, O: To be disclosed)

Fig. 7 Default privacy protection level (upon access by a third party in the logistics industry)

levels are again classified into low level (1–3), medium level (4–6), and high level (7–9.) Each level is for the privacy protection in each application service. However, it is recommended that the privacy protection system should support the following levels to ensure compatibility with the RPS system. In other words, privacy platforms have three groups of privacy protection levels that are from 1 to 10. Three groups of privacy protection levels include low level (where most information is disclosed), medium level (where object information and history are disclosed), and high level (where only part of the object information and object category are disclosed). The default privacy level is applied to the tag and the RPS system.

– Low level (open)
  Low levels refer to levels where privacy is least protected among all privacy protection levels. When the privacy level is a low level, most mobile RFID terminals can access the system and related information including parts of personal information. Low levels are allowed only for those who are reliable.
– Medium level (object information and history)
  When mobile RFID accessing individuals are reliable or information carried on the tag does not infringe on a users' privacy, medium levels are applied. In medium levels, parts of information are not protected because some security keys are disclosed and disclosed information does not affect security.
– High level (part of object information and object category)
  Access to high level information is not reliable, and all access is controlled. Only limited parts of information such as object names or object categories are allowed in high levels. For example, in a high level, mobile RFID service is sensitive to privacy and the object owner allows the least information to be exposed to third parties.

6.5.2 Privacy Level Data Structure in RFID Tag

When the mobile RFID application service provides a personal privacy protection function, the application data of the privacy protection level shall be stored in the user data area of the tag.

The privacy protection level stored in the user data area is defined in the "EPC Gen2 Data Format" standard. The privacy protection level has TYPE, LENGTH, and VALUE in the TLV (TYPE–LENGTH–VALUE) structure as shown below (Figs. 8–10).

– **TYPE**, Ex.) TYPE CODE 13 = Privacy Grade

Fig. 8 Privacy protection TYPE CODE Field

	TYPE CODE							
Bit	7	6	5	4	3	2	1	0
	0	0	0	0	1	1	0	1

– **LENGTH**

The privacy protection level stored in the user data area of the RFID is an integer, and the highest level is 10. The LENGTH field is 8 bits. The LENGTH field has been set as 1_{10} and is of 8 bits as shown below.

Fig. 9 Privacy protection LENGTH Field

	LENGTH							
Bit	7	6	5	4	3	2	1	0
	0	0	0	0	0	0	0	1
				1_{10}				

– **VALUE**, VALUE = INTEGER (VALUE TYPE = INTEGER)

VALUE = 0: Product or service that does not require privacy protection
VALUE = 0 or higher: Product or service that requires privacy protection of level 1 or higher level.

Fig. 10 Privacy protection VALUE Field

	VALUE							
Bit	7	6	5	4	3	2	1	0
	0	0	0	0	0	0	0	0

The maximum privacy level depends on the application services and the privacy level is from 1 to 9. The privacy level follows the privacy-applied standard of each application service; and if there is no standard, the privacy level is determined based on the privacy impact assessment result.

6.6 Service Procedure of RFID Privacy Protection

There are three privacy protection scenarios for mobile RFD services:

1. Storing of default privacy protection level
   In this process, the default privacy protection level is determined based on the privacy impact assessment result undertaken by the government and the default privacy protection level is stored as the default policy of the tag and the RPS server.
2. Privacy policy-setting stage

   - Subscribing to RPS and setting the privacy policy

   Figure 11 shows the procedure for subscribing to RPS. To use the privacy protection service, the user shall subscribe to RPS and define his/her privacy protection policy. In the same way, the service providers that intend to provide privacy-secured services shall also subscribe to RPS and comply with the default privacy policy of the corresponding service or the privacy protection policy set by the user.

   - Personalization of tag-attached object (privacy information combining phase)

   In Fig. 12, the privacy policy is applied when the tag-attached object is personalized or when privacy information is combined. The procedure involves the mobile RFID terminal reading the tag. Depending on the user's decision, the RFID terminal starts the application program to personalize the tag-attached object (such as purchase) The RFID terminal finds the mobile RFID application server in the ODS server and sends the request. At this time, the mobile RFID application server receives information requests and checks whether there is any privacy policy for the owner's tag-attached objects. When the application server does not have a privacy policy for the tag-attached object owner, the RFID terminal requests the policy from the RPS server. Then, RPS checks whether there is a privacy policy for the tag-attached object.
3. Provision of privacy protected information

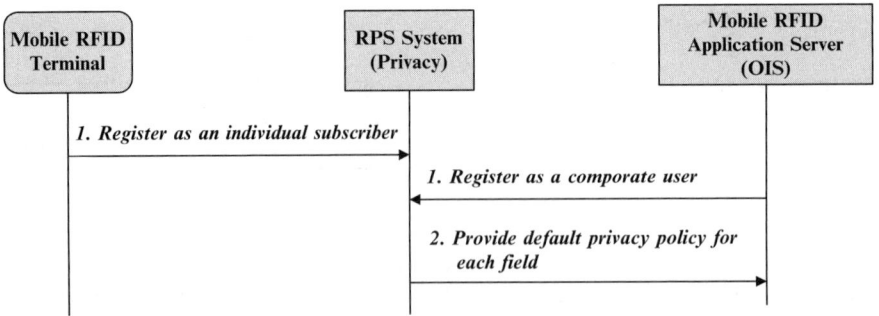

Fig. 11 Subscribing to privacy protection service

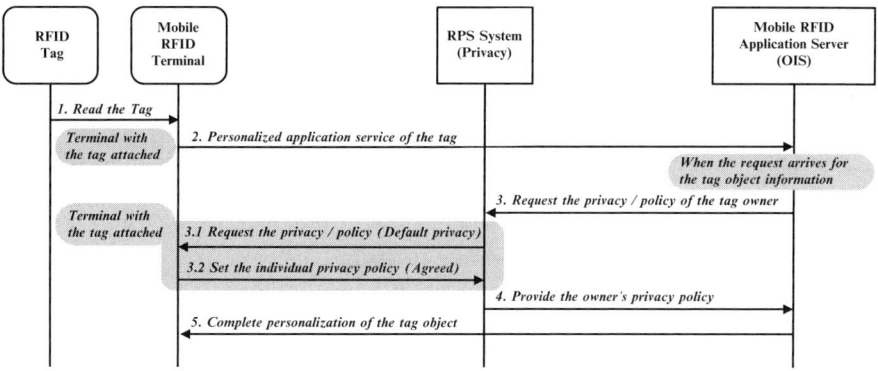

Fig. 12 Personalization of tag-attached object

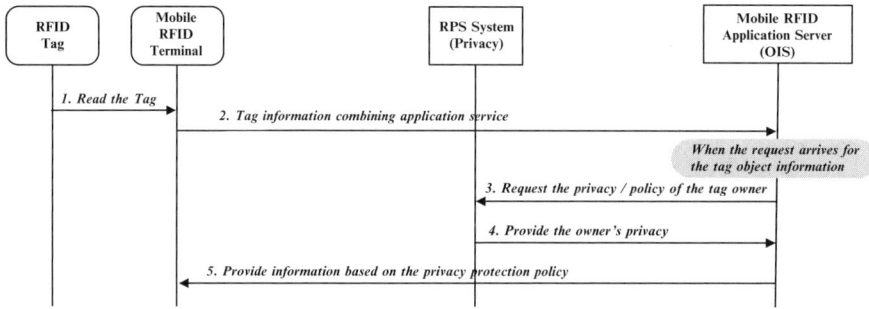

Fig. 13 Mobile RFID privacy protection scenario

In the above procedure, privacy protection information is provided in three ways as shown in Fig. 13. First, the mobile terminal reads the tag's information. At this time, the privacy policy stored in the tag is also sent to the mobile RFID terminal. The mobile RFID terminal is then coupled with the user's terminal application service and finds the mobile RFID application server in the ODS server. At this time, the mobile RFID application server receives the request and checks whether it has the privacy policy of the tag-attached object owner. When the mobile RFID application server does not have the privacy policy of the tag-attached object owner, the mobile RFID application server will request the RPS server to send the policy. Then, RPS checks whether it has the privacy policy of the tag-attached object owner. The object's personalized privacy policy stored in the RPS is then sent to the application server. When RPS does not have a policy, the owner will send the privacy information request in a short text message and will specify who is requesting the privacy information. To avoid delays, the default privacy protection policy determined based on the privacy impact assessment result is sent to the application server. The mobile RFID application server sends information that is at a lower privacy protection level than the one defined by the user. In other words, only privacy protected information is sent to the person or entity requesting the information.

6.7 RFID Privacy Protection Management System

The proposed RFID privacy protection management system architecture for mobile RFID service with privacy is shown in Fig. 14.

The main features of this service mechanism are owner's privacy protection policy establishment and management, access control for information associated with personalized tag by owner's privacy policy, obligation result notification service, and privacy audit service by audit log management. The brief personal privacy protection process using the above functions of RPS is as follows:

1. Information of the service system consists of Privacy Reference List (PRL) and Privacy Reference Profile (PRP) as follows

   - *PRL.* Information item list (like personal information, product information, distribution information) treated by each service field (like finance, trade, etc.)
   - *PRP.* Privacy level allotted profile by PRL items, which is decided through its effect estimation by the concerned authority (e.g., financial profile, trade profile, medical care profile, etc.)

2. The RPS registration and the generation of Default Privacy Reference Profile (DPRP) for a company are as follows

   - Each company may register RPS on RPS registration Web site and, upon its registration, has to provide RPS with its OIS schema
   - RPS generates DPRP by matching a privacy level to each attribute of schema of OIS, which means an item list treated by the company included in a certain service group (Fig. 15)

3. Owner's profile in the RPS system is generated as follows

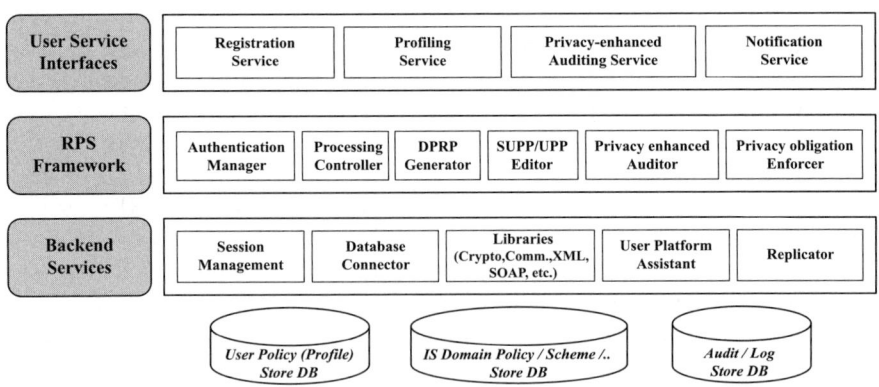

Fig. 14 RPS system architecture for mobile RFID service with privacy

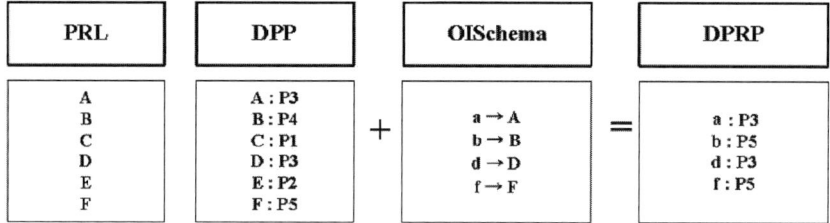

Fig. 15 DPRP's generation mechanism

- Get access to RPS Web server through cellular phone to set an access group profile for generating owner profile and allowing service information access by category groups
- Each profile is generated in eXtensible Markup Language (XML) form by RPS and sent to every service provider
- When the owner sets a privacy level (L1–L9) for each item in the service group, Owner-defined Privacy Policy (OPP) for information will be generated, which in turn results in individual Owner-defined Privacy Reference Profile (OPRP) for service providers according to their OIS schema included in the service group (Fig. 16)
- The owner must set a single profile input for each service (it is intended to reduce the owner's burden)
- Appendix 1 is an example of owner privacy reference profile

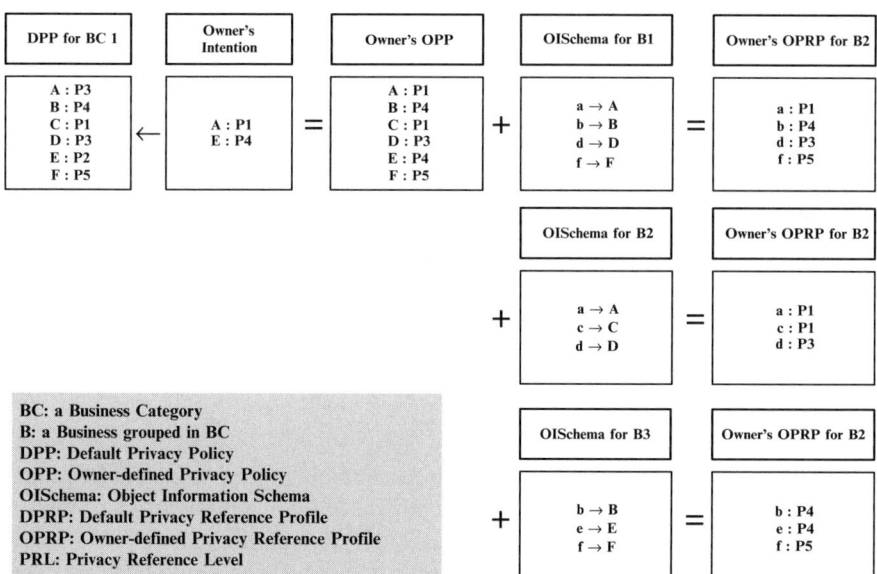

Fig. 16 OPRP's generation mechanism

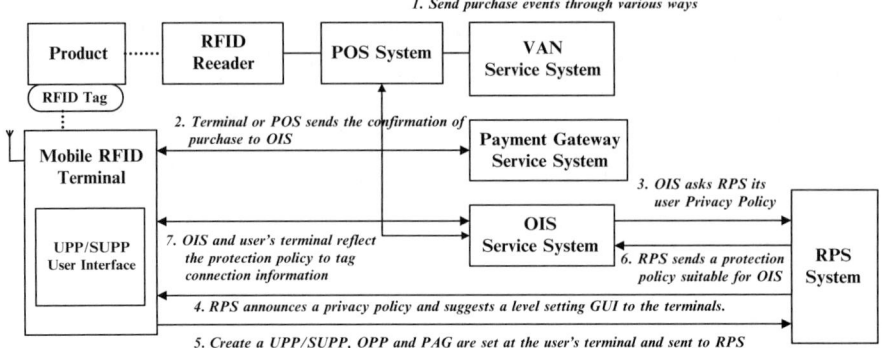

Fig. 17 Function process of RPS system

- As the owner registers cellular phone numbers of those who are allowed access for every access approval level (L1–L5), RPS will generate a security token indicating the approval of access and then the Profile for Access Group (PAG), including this token, will be sent to every service provider
- Appendix 2 is an example of access group profile by service information access groups

4. The following are the mechanism order in an RPS applied service network (Fig. 17)

- Send purchase events through various ways (for purchase via Point Of Sales management (POS) system, his/her owner information (cellular phone number) is sent to OIS)
- Either terminal or POS sends the confirmation of purchase to OIS, which in turn changes information in connection with the information on user
- OIS asks RPS its user privacy policy
- RPS announces a privacy policy and suggests a level setting Graphical User Interface (GUI) to the terminals
- OPP and PAG are set at the user's terminal and sent to RPS
- RPS sends a protection policy suitable for OIS
- OIS and user's terminal reflect the protection policy to tag connection information and receive only information satisfying the policy

7 Conclusion

RFID technology will evolve into an intellectual ubiquitous environment by mounting an RFID tag to every object, automatically detecting the information of the surrounding environment, and interconnecting them through the network. Especially, RFID technology requires technological security measures as it is vulnerable to privacy infringement like counterfeiting, falsification, camouflage, tapping, and global positioning through ill-intentioned attack. Therefore, it is necessary to enact laws and regulations that can satisfy all consumer protection organizations that are sensitive to individual privacy infringement, and develop and apply the security technology that can apply such laws and regulations to all products.

The mobile RFID technology is being actively researched and developed throughout the world and more efforts are underway for the development of related service technologies. Though legal and institutional systems endeavor to protect privacy and encourage protection technologies for the facilitation of services, the science and engineering world also has to develop proper technologies. Seemingly, there are and will be no perfect security/privacy protection technology. Technologies proposed in this paper, however, would contribute to the development of secure and reliable network RFID circumstances and the promotion of the mobile RFID market.

Acknowledgments This work was partly supported by the IT R&D program of MKE/IITA [2005-S088-04, Development of Security Technology for Secure RFID/USN Service] and the MKE (Ministry of Knowledge Economy), Korea, under the ITRC (Information Technology Research Center) support program supervised by the IITA (Institute of Information Technology Advancement) [IITA-2008-C1090-0801-0028]. The authors thank the members of the RFID/USN research team at ETRI for formulating many of the concepts discussed in this chapter. The authors also thank Dr. Seungjoo Kim (Sungkyunkwan University), Dr. Byunggil Lee (ETRI), and Dr. Dooho Choi (ETRI) for their encouragement and support. Due to the introductory character of this chapter, some of the material that provides detailed information relevant to the topic of the chapter is not included. Readers interested to know more about matter described here are directed to conference papers and more technical information on mobile RFID's security or Namje Park's Web site: http://njpark.name.

References

1. An Y., Oh S. (2005) RFID System for User's Privacy Protection. 2005 Asia-Pacific Conference on Communications, Perth, Western Australia, 3–5 October 2005
2. Avoine G., Oechslin P. (2005). RFID traceability: A multilayer problem. In Andrew Patrick and Moti Yung, editors, Financial Cryptography – FC'05. Lecture Notes in Computer Science, vol. 3570, Springer, Berlin, pp. 125–140
3. Chae J., Oh S. (2005) Information Report on Mobile RFID in Korea. ISO/IEC JTC1/SC 31/WG4 N0922, Information Paper, ISO/IEC JTC1 SC31 WG4 SG 5
4. Choi D., Kim H., Chung K. (2007) Proposed Draft of X.rfidsec-1: Privacy Protection Framework for Networked RFID Services. ITU-T, COM17C107E, Q9/17, Contribution 107, Geneva
5. Chug B. et al. (2005) Proposal for the Study on a Security Framework for Mobile RFID Applications as a New Work Item on Mobile Security. ITU-T, COM17D116E, Q9/17, Contribution 116, Geneva
6. Finkenzeller K. (2003) RFID Handbook: Fundamentals and Applications in Contactless Smart Cards and Identification, Wiley, New York
7. Garfinkel S., Rosenberg B. (2005) RFID: Applications, Security, and Privacy, Addison-Wesley, Reading, MA
8. Garfinkel S., Juels A., Pappu R. (2005) RFID Privacy: An Overview of Problems and Proposed Solutions. IEEE Security and Privacy 3(3): 34–43
9. ITU-T TSAG (2005) A Proposed New Work Item on Object/ID Associations
10. ITU-T TSAG RFID CG Deliverable (2006) Review Report of Identification Based Business Models and Service Scenarios

11. Kim Y., Koshizuka N. (2006) Review report of Standardization Issues on Network Aspects of Identification Including RFID. ITU-T, Paper TD315
12. Kim Y., Lee J., Yoo S., Kim H. (2006) A Network Reference Model for B2C RFID Applications. Proceedings of ICACT 2006
13. Konidala D.M., Kim K. (2006) Mobile RFID Security Issues. Proceeding of Symposium on Cryptography and Information Security
14. Kwak J., Rhee K., Oh S., Kim S., Won D. (2005) RFID System with Fairness within the Framework of Security and Privacy. Lecture Notes in Computer Science, vol. 3813, Springer, Berlin, pp. 142–152
15. Lee J., Kim H. (2006) RFID Code Structure and Tag Data Structure for Mobile RFID Services in Korea. Proceedings of ICACT 2006
16. Lee H., Kim J. (2006) Privacy Threats and Issues in Mobile RFID, Proceedings of the First International Conference on Availability, Reliability and Security, vol. 1
17. Lee B., Kim H., Chung K. (2006) The design of dynamic authorization model for user centric service in mobile environment, Proceedings of ICACT 2006, vol. 3, pp. 20–22
18. MIC (Ministry of Information and Communication) of Korea (2005) RFID Privacy Protection Guideline. MIC Report Paper 2005
19. Mobile RFID Forum of Korea (2005) WIPI C API Standard for Mobile RFID Reader. Standard Paper
20. Mobile RFID Forum of Korea (2005) WIPI Network APIs for Mobile RFID Services. Standard Paper
21. Mobile RFID Forum of Korea (2005) Mobile RFID Code Structure and Tag Data Structure for Mobile RFID Services. Standard Paper, http://www.mrf.or.kr
22. Mobile RFID Forum of Korea (2005) Access Right Management API Standard for Secure Mobile RFID Reader, MRFS-4-03. Standard Paper. http://www.mrf.or.kr
23. Mobile RFID Forum of Korea (2005) HAL API Standard for RFID Reader of Mobile Phone, Standard Paper
24. Mobile RFID Forum of Korea (2005) WIPI API for Mobile RFID Reader Device, Standard Paper
25. Nokia. RFID Phones – Nokia Mobile RFID Kit, http://europe.nokia.com/nokia
26. Ohkubo M., Suzuki K., Kinoshita S. (2003) Cryptographic Approach to 'Privacy-Friendly' Tags. RFID Privacy Workshop 2003
27. Park W., Lee B. (2004) Proposal for Participating in the Correspondence Group on RFID in ITU-T. Information Paper. ASTAP Forum
28. Park B., Lee S., Youm H. (2006) A Proposal for Personal Identifier Management Framework on the Internet. ITU-T, COM17-D165
29. Park N., Kwak J., Kim S., Won D., Kim H. (2006) WIPI Mobile Platform with Secure Service for Mobile RFID Network Environment. Lecture Notes in Computer Science, vol. 3842, Springer, Berlin, pp. 741–748
30. Park N., Kim S., Won D., Kim H. (2006) Security Analysis and Implementation leveraging Globally Networked Mobile RFIDs. Lecture Notes in Computer Science, vol. 4217, Springer, Berlin, pp. 494–505
31. Sakurai Y., Kim H. (2006) Report for Business Models and Service Scenarios for Network Aspects of Identification (Including RFID). ITU-T, TSAG TD 314
32. Sarma S.E., Weis S.A., Engels D.W. (2002) RFID Systems, Security and Privacy Implications. Technical Report MIT-AUTOID-WH-014, AutoID Center, MIT, Cambridge, MA
33. Shepard S. (2005) RFID: Radio Frequency Identification. McGraw-Hill, New York, NY
34. Son M., Lee Y., Pyo C. (2006) Design and Implementation of Mobile RFID Technology in the CDMA Networks, Proceedings of ICACT 2006
35. Strandburg K.J., Raicu D.S. (2005) Privacy and Technologies of Identity: A Cross-Disciplinary Conversation, Springer, Berlin
36. Sullivan L. (2004) Middleware Enables RFID Tests. Information Week, No. 991
37. Thornton F. et al. (2006) RFID Security, Syngress, Rockland, MA
38. Tsuji T., Kouno S., Noguchi J., Iguchi M., Misu N., Kawamura M. (2004) Asset management solution based on RFID. NEC Journal of Advanced Technology 1(3): 188–193

39. Tsukada M., Narita A. (2006) Development Models of Network Aspects of Identification Systems (Including RFID) (NID) and Proposal on Approach for the Standardization. ITU-T, JCA-NID Document 2006-I-014

40. Weis S. et al. (2003) Security and Privacy Aspects of Low-Cost Radio Frequency identification Systems. First International Conference on Security in Pervasive Computing (SPC) 2003

41. Weis S.A., Sarma S.E., Rivest R.L., Engels D.W. (2003) Security and Privacy Aspects of Low-Cost Radio Frequency Identification Systems. Proceedings of First International Conference on Security in Pervasive Computing (SPC 2003)

42. Yoo S. (2005) Mobile RFID Activities in Korea. Contribution Paper of the APT Standardization Program

43. Yutaka Y., Nakao K. (2002) A Study of Privacy Information Handling on Sensor Information Network. Technical Report of IEICE

Appendix 1

The following is an example of owner privacy reference profile.

```xml
<?xml version="1.0" encoding="UTF-8" ?>
- <RPSMessage issueInstant="2006-01-20" issuer="ETRI" sendFrom="ETRI_RPS" sendTo="msdg"
    xmlns="urn:etri:names:RPS:1.0:rpsprotocol" xmlns:ds="http://www.w3.org/2000/09/xmldsig#"
    xmlns:rps="urn:etri:names:RPS:1.0:rps" xmlns:xsi="http://www.w3.org/2001/XMLSchema-instance"
    xsi:schemaLocation="urn:etri:names:RPS:1.0:rpsprotocol
    http://129.254.124.89:8082/RPS/rpsSchema/etri-rps-schema-rpsprotocol-1.0.xsd">
  - <ResponseMessage requestID="r_0601_1" responseID="s_0601_1">
    - <Status>
        <StatusCode>Success</StatusCode>
      </Status>
    - <RPSProfile IssueInstant="2006-01-20T12:42:54Z" Issuer="ETRI" MajorVersion="1" MinorVersion="1"
        ProfileID="p_0601.1" xmlns="urn:etri:names:RPS:1.0:rps"
        xmlns:xsi="http://www.w3.org/2001/XMLSchema-instance"
        xsi:schemaLocation="urn:etri:names:RPS:1.0:rps etri-rps-schema-rps-1.0.xsd">
      - <OPRP ProfileType="OPRP">
          <OwnerID>01025870285</OwnerID>
          <CompanyID>msdg</CompanyID>
          <ServiceCategoryID>12</ServiceCategoryID>
        - <ProfileStatement>
            <Item>code</Item>
            <PrivacyLevel>1</PrivacyLevel>
            <Item>owner</Item>
            <PrivacyLevel>9</PrivacyLevel>
            <Item>distribution</Item>
            <PrivacyLevel>5</PrivacyLevel>
          </ProfileStatement>
        - <Obligation due="2016-01-19T00:00:00Z" obligation_id="_c12345">
            <ObligationContent>delete information at 2016-01-19T00:00:00Z</ObligationContent>
          </Obligation>
        </OPRP>
      </RPSProfile>
    </ResponseMessage>
</RPSMessage>.
```

Appendix 2

The following is an example of access group profile by service information access
groups.

```
<?xml version="1.0" encoding="UTF-8" ?>
- <RPSMessage issueInstant="2006-01-20" issuer="ETRI" sendFrom="ETRI_RPS" sendTo="msdg"
    xmlns="urn:etri:names:RPS:1.0:rpsprotocol" xmlns:ds="http://www.w3.org/2000/09/xmldsig#"
    xmlns:rps="urn:etri:names:RPS:1.0:rps" xmlns:xsi="http://www.w3.org/2001/XMLSchema-instance"
    xsi:schemaLocation="urn:etri:names:RPS:1.0:rpsprotocol
    http://129.254.124.89:8082/RPS/rpsSchema/etri-rps-schema-rpsprotocol-1.0.xsd">
  - <ResponseMessage requestID="r_0601_1" responseID="s_0601_1">
    - <Status>
        <StatusCode>Success</StatusCode>
      </Status>
    - <RPSProfile IssueInstant="2006-01-20T12:42:58Z" Issuer="ETRI" MajorVersion="1" MinorVersion="1"
        ProfileID="p_0601.1" xmlns="urn:etri:names:RPS:1.0:rps"
        xmlns:xsi="http://www.w3.org/2001/XMLSchema-instance"
        xsi:schemaLocation="urn:etri:names:RPS:1.0:rps etri-rps-schema-rps-1.0.xsd">
      - <PAG ProfileType="PAG">
          <OwnerID>01025870285</OwnerID>
          <ServiceCategoryID>12</ServiceCategoryID>
          <AccessGroup>1</AccessGroup>
          <DPAG>NfK5vh93jHBCci2T7cgdrpz3IL4=</DPAG>
          <OPAG>eyGEismvNb4N2y1rn8OFGTTbhCA=</OPAG>
          <OPAG>/qf2V/VqKkSNp9S1Ne5eJ5yvPZo=</OPAG>
          <OPAG>9W1jUapxz/DevqAU0TUl5CA2GHo=</OPAG>
          <AccessGroup>2</AccessGroup>
          <DPAG>JWe+sKQ+FOSIrLzOuNiinXf955k=</DPAG>
          <OPAG>kvL9mYebDCRmq4ZIr7Y8SQMjecE=</OPAG>
          <OPAG>q4dEZ6fR/1/HGkreh9wOCYtFiq4=</OPAG>
          <OPAG>TBtSQJz2vjiWzxY/oXsy5NopPy4=</OPAG>
          <AccessGroup>3</AccessGroup>
          <DPAG>OOxGdINkl7nhcY79TLll8ljEPKc=</DPAG>
          <OPAG>UBq1RE6umtMrViVws2/2KOw3kM4=</OPAG>
          <OPAG>DdtYd8iW9D6HNOELAB5/HrkoicO=</OPAG>
          <OPAG>QXCsKieCoVFv6eE9cyKuSCwb1ZQ=</OPAG>
          <AccessGroup>4</AccessGroup>
          <DPAG>jXnHgzRVbIuW9eCLdRCDgXcZ8TI=</DPAG>
          <OPAG>Od+lUoMxjTGv5aP/Sg4yU+IEXkM=</OPAG>
          <OPAG>AP1LRUmhCUqukm72Lp29PNzC5FY=</OPAG>
          <OPAG>z7auELEXP7QWeJ+8wbPVraASIrg=</OPAG>
          <AccessGroup>5</AccessGroup>
          <DPAG>Kj/A5dh+7THb3GgI0AvOIdPeSy4=</DPAG>
          <OPAG>SvE09Bux0M4Fuk61O8Zq0tWL/mY=</OPAG>
          <OPAG>jfcPhZZOTUB4PQ+830OnlFUtj1I=</OPAG>
          <OPAG>5lb3Wwnh30yxrL+IcGL20gOZmGU=</OPAG>
          <OPL>9</OPL>
        </PAG>
      </RPSProfile>
    </ResponseMessage>
  </RPSMessage>
```

Using RFID-based "Touch" for Intuitive User Interaction with Smart Space Security

Zoe Antoniou*, Dimitris Kalofonos, and Franklin Reynolds

Abstract Home networks and networked consumer electronic devices are increasingly becoming a part of our everyday lives. One of the challenges in designing smart home technology is making these systems secure and, at the same time, easy-to-use for non-expert consumers. We believe that mobile devices equipped with a "touch" network interface and corresponding middleware are ideal for enabling users to intuitively setup and manage the security of their smart homes. In this chapter, we describe such a middleware for mobile phones based on Near Field Communication (NFC) technology. We propose a mobile middleware architecture called iTouchSec based on a higher level User-Interaction with Security (UI-SEC) middleware, called IntuiSec, and a lower level NFC middleware, called iTouch. We present the overall architecture, as well as the detailed design of the necessary NFC records that are exchanged over RF. Finally, we present our experience with an initial implementation of parts of the proposed middleware using actual NFC hardware and Symbian-based mobile phones.

1 Introduction

The availability of inexpensive wireless and wireline networking equipment has led to the rapid proliferation of home networks. Even though there is an ever increasing number of "smart" consumer electronics products, home networks are primarily used for traditional computer applications, e.g., web browsing, games, printing, data backups, home finances, etc. There are a number of standards and industry efforts such as UPnP [1] and DLNA [2] that may help enable networked home-entertainment applications, a step toward making "smart homes" a reality. UPnP

Z. Antoniou
Pervasive Computing Group, Nokia Research Center Cambridge, Cambridge, MA 02142, USA
e-mail: zoe.antoniou@nokia.com

and DLNA attempt to standardize the description of devices and services, as well as the protocols used to communicate with devices and services.

It is technically possible for astute users to build smart homes that integrate home computers, smart phones, mobile media players, home entertainment systems, remote-controlled security systems, and heating and lighting systems. However, even if products based on current UPnP standards are eventually available, the task of setting up and managing a home network; selecting and configuring products to maximize interoperability; setting up and managing the security of remote access services; administering user accounts; and securely managing access to devices, services, and content will remain beyond the ability of most users.

A user interested in a smart home network faces many of the same problems faced by a modern enterprise, albeit, on a much smaller scale. In both cases, products from many vendors have to be integrated, the internal network devices and services have to be shared, and connectivity to the public internet has to be provided without compromising the security of the internal network. However, the IT department of the Enterprise has the advantage of being able to hire experts to solve their problems and requiring their employees to spend the time to learn how to use the corporate computing systems. The typical consumer will not be interested in a "smart home" if it requires taking classes to learn how to use the TV or hiring a consultant to deploy and administer the networking and security policies of their smart home.

Mark Weisner is widely regarded as the father of "ubiquitous computing." In his well-known 1991 Scientific American article [3], he envisioned a world where computing and networking were seamlessly integrated into the minutia of everyday life. The IEEE began publishing its Pervasive Computing magazine in January of 2002. The editors of the magazine, who were very strongly influenced by Weisner's work, had a similar vision of "pervasive computing":

> The essence of this vision is the creation of environments saturated with computing and wireless communication, yet gracefully integrated with human users. Many key building blocks needed for this vision are now viable commercial technologies: wearable and handheld computers, high bandwidth wireless communication, location sensing mechanisms, and so on. The challenge is to combine these technologies into a seamless whole [3].

Seamless integration and ease of use are recurring themes in pervasive computing and smart home research. To paraphrase the editors of Pervasive Computing, many of the enabling technologies needed for smart homes are already available but we do not yet have simple, intuitive, and secure methods for integrating, deploying, or using products based on these technologies.

Pointing and touching are natural methods for humans to indicate the focus of their attention or to select a target. This observation contributed to the creation of modern point-and-click GUIs. The idea of using some method to point at a real-world target is not a new idea. Remote controls for TVs based on infrared (IR) technology have been available for decades and have convincingly demonstrated the utility of a "point-and-click" UI for specialized, real-world applications. Magnetic card readers, and more recently, RFID tags have been widely used to build interfaces based on touch (or "almost" touch). Over the years, these and other technologies have been used to create user interfaces for pervasive computing applications.

Our work has emphasized the use of Near Field Communication (NFC) technology [4] and mobile phones to build a touch-based interface for smart homes or other smart spaces. NFC emerged from Radio Frequency Identification (RFID) technology. NFC devices allow bidirection communication over a short operating range of a few centimeters. This allows devices to exchange information over the RF medium. We trust the short range of NFC communication to protect the exchange of secrets. We do not attempt to make the RFID channel more secure against snooping (eavesdropping) or spoofing (impersonation). Other researchers are developing techniques such as shaking [5] to authenticate local, ad hoc communication channels. NFC-compatible tags have a modifiable state and can support capacities of several kilobytes.

NFC-enhanced phones can have a central role in the user interaction with smart homes and other smart spaces. Mobile phones are widely deployed, and personal devices with GUIs, significant communication, storage and computational capabilities. By adding NFC interfaces to mobile phones we have created an intuitive "touch"-based user interaction modality for smart spaces.

In this chapter, we will describe iTouchSec, a "user-oriented" security framework for smart spaces. A smart space consists of infrastructure devices and visitor devices. Infrastructure devices are set up and managed by the smart space owners and host the smart space services and content. Visitors can use their devices to access and interact with the smart space. Although the proposed approach can find application in any smart space, this chapter focuses on the use cases of a smart home.

The primary goals of iTouchSec are to make it possible for end-users with little or no special training to deploy and manage their own networks with their own network security policies. Network security management for end-users can be particularly challenging due to lack of integration between the security mechanisms of different "network layers," security abstractions with subtle semantics, and complex tools with poorly designed user interfaces. Today, consumers who are interested in deploying secure home networks have to use completely different tools to deal with media access (WEP, WPA, etc.), perimeter security (NAT/firewall and DMZ), host firewalls (Windows or MacOS personal firewalls), remote connectivity (VPN, IPSec, etc.) and application specific security protocols and tools (such as https or the various apache web server security toolkits). iTouchSec is an integrated framework and network security architecture which attempts to deal with these challenges.

iTouchSec emphasizes the use of a new user interaction modality based on touching the physical object the user intends to interact with. This approach has made it possible to build a consistent user interface to smart objects derived from user's experience with physical objects. For example, iTouchSec users touch devices together to set up or connect to a network; a user can launch an application associated with a device by touching it; a user touches one device to another to discover it and establish a secure communication channel with it; a user can give permission to use a device to another user by a simple touch gesture.

Instead of exposing end-users to a wide variety of low-level security mechanisms, such as cryptographic protocols, and traditional security concepts, such as authentication, access control, and nonrepudiation – iTouchSec users deal with simple but powerful abstractions. One example is the "Passlet." Passlets, which will be described later in the chapter, can be thought of as security tokens that provide access rights. iTouchSec users do not explicitly set access control lists, program firewalls, or configure application-specific membership lists. The creation and distribution of Passlets is all that is needed. As alluded to in the previous paragraph, Passlets can be created and exchanged via "touch."

In the following sections, we describe the rationale, design, and implementation of the iTouchSec framework. We begin by discussing factors that motivated the design of the framework. This includes a discussion of security issues related to smart spaces, motivation for a touch-based user interface, and some smart home scenarios. We show through selected use cases how mobile devices implementing the proposed architecture can simplify the management of access control at the network and application level. This is followed by detailed descriptions of the iTouchSec architecture and its implementation. We conclude with a brief summary of related work and some closing thoughts.

2 Security and Threats in Smart Spaces

iTouchSec is intended to operate in a Smart Home environment. An example of a smart home is a house with both wired and wireless connectivity that contains numerous devices providing various services to users in the home. Such devices could be media servers, audio/visual (A/V) renderers, refrigerators – or any household appliance or consumer electronics device, as depicted in Fig. 1. Currently many of these devices are not yet network-capable, but the technology to make them so is available and there are various initiatives strongly advocating for such adoption in

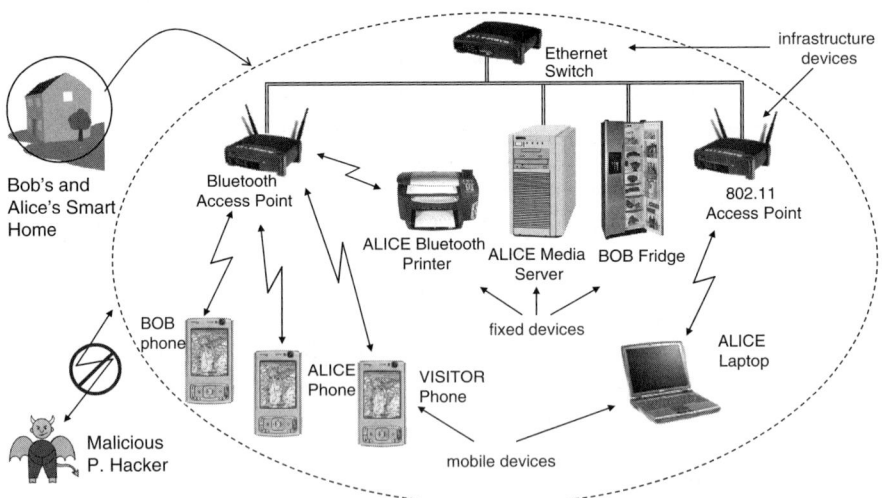

Fig. 1 An example of a smart home

the near future. For an example of such an initiative, see the Digital Living Network Alliance (DLNA) [2].

Smart homes are likely to have several occupants. iTouchSec assumes that each occupant would like to protect the privacy of his/her own devices whilst being able to grant selective access to other individuals. Collectively as a home, iTouchSec assumes that the occupants would like to prevent unauthorized nonhome occupants from gaining any access to the home network; this includes even connectivity access, such as the ability to acquire an IP address and passively read network traffic. iTouchSec, also, works under the assumption that home occupants would like to be able to grant visitors access to devices within their homes for temporary periods of time. Thus, the threats that iTouchSec aims to protect against are:

- Unauthorized connectivity access to the home network
- Unauthorized access to services on a home device

In terms of protecting against unauthorized connectivity access, iTouchSec focuses on wireless connectivity. Gaining illicit wired access is seen as a much less significant threat to home network security than gaining access over a wireless connection since the intruder must have physical access to the home in order to connect over a wire – i.e., they must break into a person's home before they can attach a malicious device with a wire. A device that is attached by wire is also more noticeable than a person snooping over a wireless connection and so could more easily be detected and removed. Furthermore, if an intruder does happen to acquire unauthorized wired connectivity to the home, he or she will still not be able to access the services offered by any of the home devices due to the second layer of protection at the service level. However, an extension to include wireline connectivity access control is possible.

iTouchSec does not intend to protect a household against sophisticated adversaries. Its aim is to provide a strong level of protection for non-expert users in a smart home environment and to do so in an intuitive manner. The framework presents a shift in the design of secure systems and with this it hopes to both decrease the number of unprotected wireless home networks and provide a foundation for smart home security frameworks. It does not aim to redesign existing cryptographic algorithms; it leverages well-established ones as part of the framework.

The trust framework of iTouchSec is built on the assumption that we can use NFC technology to reliably and safely introduce two devices to each other, i.e., exchange secrets between two devices. These secrets are used to guarantee the security of subsequent communication via WLAN or other communication channels. This assumption of safe "introductions" may not be true if:

- The NFC interface is malicious even though the device associated with the tag is genuine. This scenario, though possible, is unlikely when the tag is co-located with the network device. Otherwise, it is possible that the NFC interface is rewired to another network device.
- Both the NFC interface and its associated device are malicious. This could happen if an intruder sets up a malicious device in a smart space.

- The RF signal is intercepted or "spoofed": Special equipment can potentially be used to (a) eavesdrop on any information sent in the clear via RFID/NFC and (b) impersonate a genuine NFC/RFID interface and device.

The above threats illustrate the possibilities for a malicious device to steal personal information from users. The problems of establishing safe and secure communication between two previously unknown parties are fundamental to secure networking. Some of these attacks can be detected only with physical inspection of the participating devices and their surroundings.

Messages sent as part of the service access process can be secured using a combination of symmetric and asymmetric cryptography. The keys published in tags are public keys in the form of X509 certificates. Since it is expensive to use asymmetric cryptography, the mobile device generates a one-time use symmetric key for securing the sensitive data in the communication. This symmetric key is itself encrypted by the private (asymmetric) key and included in the message sent to set up communication between the mobile device and the network device. The network device first decrypts this symmetric key and then uses this symmetric key to decrypt the message sent by the mobile node. Optionally, parts of the message may also be digitally signed using keys derived during the process. Finally, to protect against replay attacks, messages can include a sequence number (or another monotonically increasing value such as time of service request).

3 RFID-Based "Touch": An Intuitive User Interaction Paradigm

Technology miniaturization has made it feasible to integrate NFC functionality into consumer mobile devices and has given birth to the *touch* paradigm. User input is received through straightforward gestures such as touching or pointing-and-clicking as part of everyday activities. An efficient user interface (UI) design can potentially replace sequences of multiple button clicks and menu selections by a simple touch or point-and-click action in order to complete a task. By exploiting the benefits of this paradigm, it is possible to enhance existing service discovery protocols in order to create smart space architectures.

Users do not want to employ technology but rather to interact with their environment. Even though mobile phones have become a commodity, a major part of mobile applications and services is hardly used by today's consumers. For example, basic functions such as calling or text messaging are easy to use and have been widely adopted. In contrast, browsing or file sharing require complex configurations and setup procedures and are less popular at present. NFC technology offers the tools for the realization of the touch paradigm which enables fast, convenient, and intuitive user interaction with smart objects, devices, services, and other users. This interaction modality coupled with service discovery provides a unique opportunity to boost the adoption of mobile services, in particular wireless proximity services. Rather than requiring new models of behavior, social interaction in proximity can build on familiar human activities such as giving, sharing, greeting, self-expression and acknowledgment, and so on.

4 Context Awareness

User activities in proximity can exploit the benefits of context information. Context information enables the right services to be delivered to the right user at the right time. Objects pertaining to a certain context can be active or passive at given moments in time depending on the situation of the user. Proximity-based services can be categorized in four context-aware categories:

(a) *People in places.* Social interaction, communication and collaboration in close physical range. Example applications are face-to-face content sharing, ad hoc collaboration, and group formation.
(b) *Me and my stuff.* Creation, management and storage of personal and third party content with particular focus on the home environment. For example, personal mobile devices (e.g., phones, PDAs) can access and control services and content on networked home devices (e.g., laptop, audio system, home appliances).
(c) *Smart spaces.* Accessing local content and services relevant to a particular location. Location-based services can find application both in the workplace and in the consumer market. For the former, an example is personal mobile devices discovering and interacting with other devices in a specific room or building (e.g., printers, projectors). Examples for the latter are service activation at point-of-sale locations and content download (e.g., download a movie preview from a poster).
(d) *Safe consumption.* Research and purchase of goods, content and services with perceived security and trust. All transactions have security and privacy requirements but special attention is needed when purchases and monetary transactions are involved such as ticketing and electronic wallet applications.

Location is an important element of context information that can be exploited. NFC technology can provide a convenient way to access location-aware, mobile services, and content through hot spots, e.g., NFC-equipped devices could easily read tags at point-of-sale locations. This can serve to compliment cellular coverage and provide the illusion of full mobility, thus, making it less necessary to assure real-time full mobility for all applications. Yet, commercial success of this business model is dependent on whether users are willing to wait to connect, pricing and sufficient coverage with clearly marked hot spots.

5 UI Interaction

Enabled by NFC technology, service discovery has a direct impact on the design of smart and intuitive user interfaces for pervasive computing. Currently, similar point-and-click interfaces are not flexible. Most RFID readers, bar code scanners, and IR remote controls are single purpose devices. In some cases, IR can be used for multiple purposes, but the interface can hardly be characterized as intuitive.

Traditional graphical user interface (GUI) displays receive input through mouse clicks and menu selections. With the NFC-touch paradigm the physical space becomes an extension to the GUI of a mobile device, where NFC-enabled physical objects and devices can be touched in order to activate associated services and applications in the same manner as clicking on an icon on a conventional display. An object may be selected, clicked upon, or dropped on different applications, which invokes different actions:

- *Selection* via the traditional GUI methods (e.g., mouse click, menu list) or NFC touch. This selection does not trigger a default action; the user needs to take a second manual step such as context-click in order to interrogate the object or trigger an associated application, e.g., user selects a printer that he/she visually discovers in a room to check its properties.
- *Select-and-launch*. A selection is associated with a default action which is the launch of the (most common) application. For instance, the user touches the NFC tag on a movie DVD and the browser is launched with the movie URL as an input.
- *Drag-and-drop*. Select object A, drag it, and drop it onto resource B, which launches the resource with the object as an input. This event pattern is very useful in order to associate an object with an application that it is not normally associated with. For example, use a mobile phone to transfer a file from a laptop to a local printer. The phone–laptop touch gesture represents the file select-and-drag. The phone–printer touch gesture represents the drop.

The term *TAP* is used to describe the user action of using a mobile device to touch another device. The mobile device being used is known as the *TAPing device* and the device that is subsequently touched is known as the *TAPed device*. During a TAP, the system leverages some form of Location Limited Channel (LLC) to perform a read or exchange transaction, such as the exchange of public keys, mutual verification of the Home Secret, transferal of link keys, and so on. A mobile device can, also, be used as an intermediary to perform a TAP between two nonmobile devices. The proposed architecture in this chapter uses NFC [4] as a means for realizing the TAPing interaction modality and implement out-of-band service discovery. NFC is a lower power radio, which can be exploited to discover nearby devices while keeping higher power radio turned off during quiescent operation. It also assumes that communications over NFC is inherently authentic and difficult to eavesdrop due to the very short range of operation (in the order of a few centimeters).

6 Example Use Cases

This section presents several basic use cases for easy user interaction with security in the smart home [6]. The scenario consists of a household with two home occupants – Bob and Alice.

The first use case involves bootstrapping new devices into the smart home. When Bob and Alice purchase a new device, they would like it to have permanent

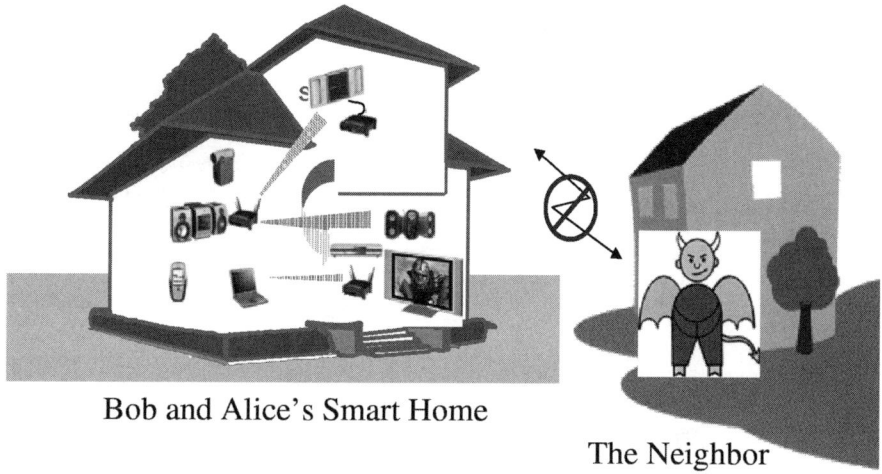

Bob and Alice's Smart Home

The Neighbor

Fig. 2 Bootstrapping and securing the Smart Home network

connectivity to the smart home network over a secure channel. At the same time they would like to protect their home from being accessed by nonauthorized users. The process for achieving this would ideally be straightforward, consistent, and intuitive, which is clearly not the case with today's methods of dealing WEP keys, MAC address filters, and so on. Figure 2 illustrates this.

Once the devices are connected and can all communicate with each other securely, Bob would like to prevent Alice from accessing his devices until he explicitly grants her access. He could also opt to have his devices have some default level of access to everybody. By way of example, Bob purchases a new media server whose content he wishes to keep private. Both Bob and Alice jointly purchased the A/V renderer in the living room so they both have access to it by default. Thus Bob can now stream content from his media server and display it on the A/V renderer, however, Alice cannot. Figure 3 illustrates this. The process of introducing a new device to the home network can be as simple as a touch gesture. For instance, Bob touches the new media server with his mobile phone which is already a home device and the media server joins the home network.

Alice later decides that she wants to retrieve content from the server and asks Bob to give her access. Bob agrees, but he only wants to give her permission to download music and movies from the server. He does not want her to access any other file types or to upload or delete anything. Figure 4 illustrates the situation after Bob grants Alice limited access to his media server. Granting permissions can also be realized with a simple touch gesture. For instance, using his mobile phone Bob selects the permission he wants to grant to Alice and touches her mobile phone. This gesture is enough to update Alice's permissions in the home network.

A further use case involves granting temporary access to visitors for use of specific services in the smart home. Again, by way of example, the fridge in Bob and

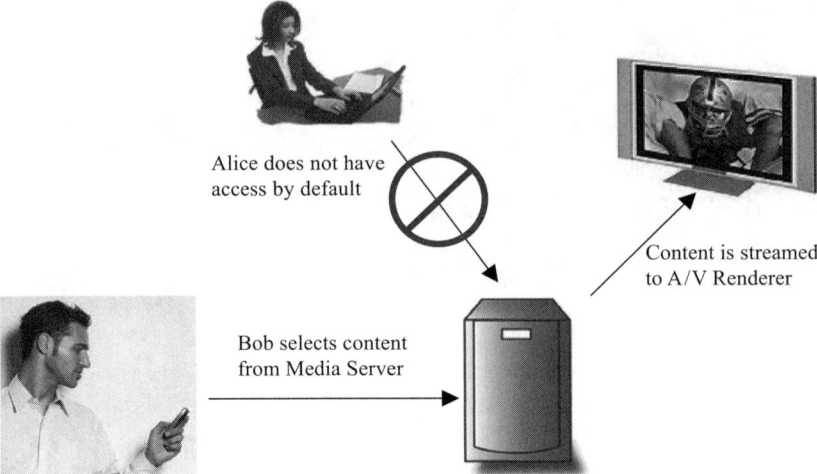

Fig. 3 Before Bob grants Alice access

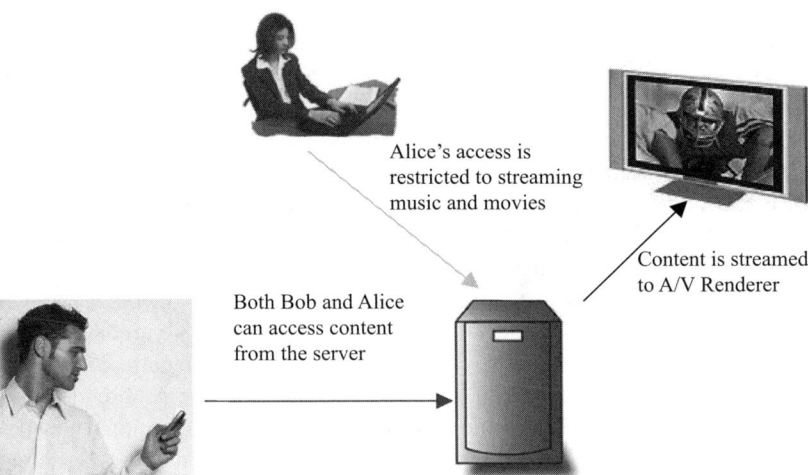

Fig. 4 After Bob grants Alice access

Alice's home breaks down. The repairman comes over and by default is not even able to connect to and browse their home network. However, Alice grants the repairman access for the day to selected functionality provided by the fridge so that he can perform the necessary repairs. At the end of the day, once the work is complete, the repairman automatically loses all access to Bob and Alice's smart home network. Figure 5 shows the repairman's access during the day.

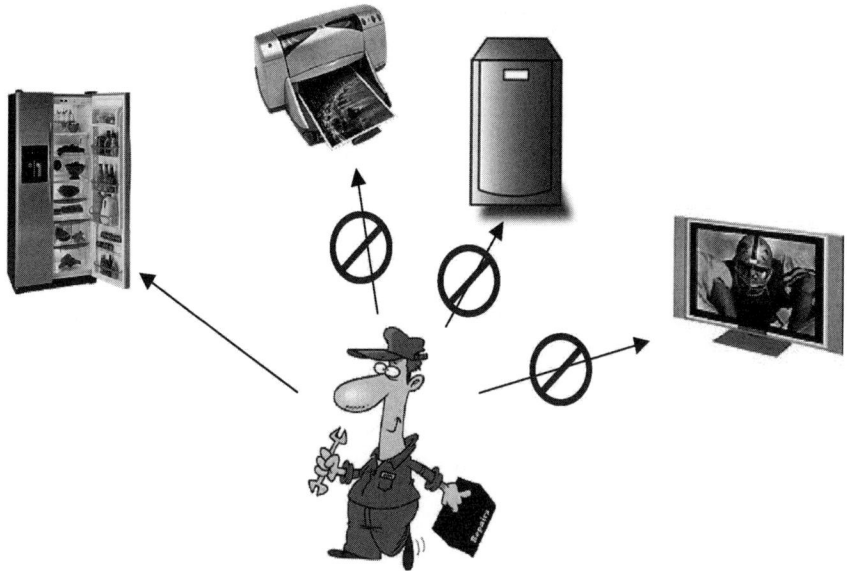

Fig. 5 The repairman has access for one day only to repair the fridge

7 iTouchSec Functional Specification and Architecture

The iTouchSec architecture (Fig. 6) is comprised of two main subsystems, iTouch and IntuiSec. iTouch [7, 8] is an NFC-enhanced middleware architecture for mobile devices that can provide a layer of indirection between a variety of wireless proximity connectivity technologies (in this case NFC or potentially Infrared, Camera, Laser, etc.) and user-domain applications and service discovery engines. Intuitive gestures, such as point-and-click or touch, can be used to bootstrap network connectivity, service discovery, give access rights, share, set up personal and social networks, launch an application, and so on. Data is formatted and exchanged as NFC packets, also referred to as records. iTouch consists of the core module, called Link Management, and multiple specialized modules for value-added dedicated tasks such as "out-of-band connectivity," "iTouch service discovery," and "iTouch security." The Link Management layer provides the basic read/write, send/receive functionality, as well as, composition and parsing of the NFC records.

The user interaction with the security layer of the proposed architecture is based on IntuiSec [9–11]. IntuiSec is a framework that enables non-expert users to easily create secure networks in smart spaces and interact with smart home security. IntuiSec provides a level of indirection between the concepts that users understand intuitively and the underlying security system settings through a middleware layer and user-level tools. IntuiSec functionality can be broken into:

- *Easy setup.* A process to take ownership of new devices and add them securely to the home network.

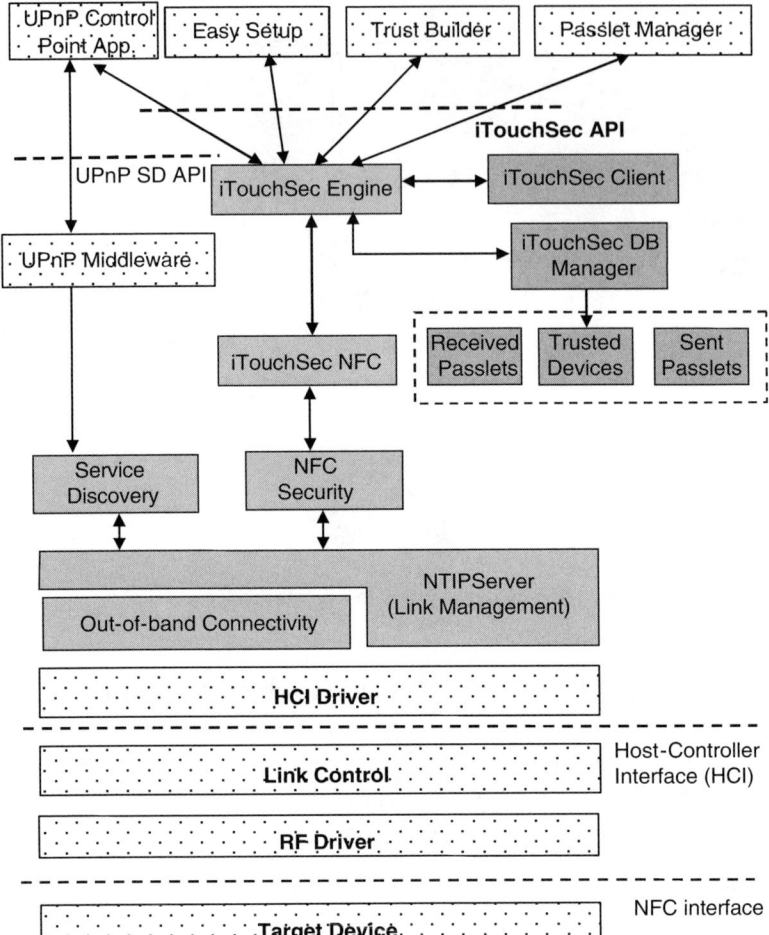

Fig. 6 iTouchSec architecture

- *Build trust.* A process by which users can intuitively establish trusted relationships between devices. This enables the user to securely authenticate and reliably verify the identity of a remote device.
- *Grant access.* A process by which users can easily grant access to both home occupants and visitors to use devices that they own.
- *Security visualization.* A set of APIs exposed by the framework to enable other applications to display security-related parameters using visualizations meaningful to non-expert users.

The proposed middleware architecture is depicted in Fig. 6, which resides in one mobile device. Note that the "Target Device" box represents a peer device with the same architecture.

8 Basic Concepts

iTouchSec makes certain assumptions that have an impact on the system design. It assumes that every device has NFC as a LLC. We believe this is realistic based on industry momentum, decreasing cost, and proliferation of this technology. It, also, assumes that communications over LLCs are inherently authentic and difficult to eavesdrop due to the very short range of LLCs (in the order of a few centimeters). iTouchSec also assumes that there is strong platform security that protects data in the device and that the hardware and software shipping with the devices is trustworthy. Finally, iTouchSec assumes that some distributed middleware with a request/response service discovery and invocation framework exists, e.g., UPnP [1] as proposed in DLNA [2].

iTouchSec defines *Home Devices* as devices which have common knowledge of a *Home Secret*. The term Home is used in its more general sense to signify any Smart Space, which can be the user's home. The Home Secret is given to a Home Device when the user "unlocks" the home network for it, a procedure we call *Easy Setup* and is described in Sect. 21. This may be performed optionally using a physical object reminiscent of a key, e.g., an RFID tag, called *PhyKey* (a PHYsical KEY). Home Devices have permanent connectivity access to the home network and they alone can be used to allow *Visitor Devices* to access the home network and any of its services. Home Devices are further distinguished into three different types (a) *Infrastructure Devices* (e.g., access points, DNS, and DHCP configuration servers) that provide network connectivity to other devices within the home; (b) *Mobile Devices* (e.g., mobile phones) that can be easily moved around and used to touch all types of Home Devices; and (c) *Fixed Devices* (e.g., printers) that cannot be easily moved around to touch other devices.

iTouchSec further distinguishes different *user roles*. The first user of a brand new device becomes its first *Owner*. This happens through the Easy Setup process described in Sect. 21. Owners, by default, have full access to any of the services on a device. Only an Owner can grant permission to other users to access services offered by the device and add other users to the device's Owner list. *Users* interact with service provider devices in the home through devices that act on their behalf. Therefore, *it is users and not devices that are the subjects of all actions in the iTouchSec smart home*. This enables access privileges to persist regardless of which device a user is using. Users can be either *Home Occupants* (i.e., owners of at least one Home Device) or *Visitors* (i.e., owners of only Visitor Devices). These devices do not have knowledge of the Home Secret and so do not have connectivity access to the home network by default. Visitors can only be granted access to the home network and its services by a Home Occupant.

Each Home Device stores the *Home Secret*. The *HomeID* is defined as the hash of the Home Secret and it is used to uniquely identify the smart home (HomeID = hash(Home Secret)). Alternatively, it is also possible instead of having a shared Home Secret to assume a keypair H_{PRI}, H_{PUB}, and define HomeID = hash(H_{PUB}). The Home Secret may be distributed using the PhyKey or by other means such as a manual input method. Each user is assigned a system-wide keypair

(UK_{PRI}, UK_{PUB}). The keypair (UK_{PRI}, UK_{PUB}) may be generated using different mechanisms, such as a username/password pair, biometric information, or a user RFID tag. The *UserID* is defined as the hash of UK_{PUB} (UserID = hash(UK_{PUB})) and it is used to uniquely identify each user in the framework. The UserID of the Owner of a device denotes that device's *OwnerID*. Each service provider device has a unique private/public keypair (PK_{PRI}, PK_{PUB}) that is preshipped with the device. The *ProviderID* is defined as the hash of PK_{PUB} (ProviderID = hash(PK_{PUB})) and it is used to uniquely identify service provider devices.

9 "Touch" to Setup and Connect

iTouchSec aims to provide a "buy-plug-and-play" user experience. When a user purchases a new device, the Easy Setup module (Fig. 6) guides him/her to perform the three simple steps described below and depicted in Fig. 7 for the case of a new fixed device:

1. The user performs the imprinting process, by which the new device comes to learn about its first Owner. Figure 7 shows a username/password entry system, however, this is not a requirement.
2. The user "unlocks" the home network by transferring the Home Secret to the device. This can be done with the insertion of the PhyKey, as shown in Fig. 7, however, the use of a PhyKey is again not a requirement. A PhyKey provides an analogy to existing keys that people use for their doors in the sense that the PhyKey "unlocks" the home network, just like the house key unlocks the front door.
3. The final step of the Easy Setup process requires the user to TAP an Infrastructure Device in order to acquire the necessary link level security parameters to connect to the Home Network. If the TAPing device is a Mobile Device, transfer of the

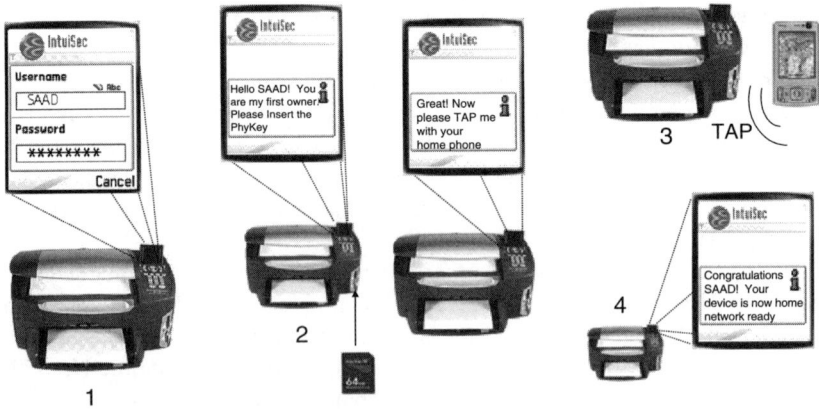

Fig. 7 Example easy setup process of a fixed device (e.g., an 802.11 printer)

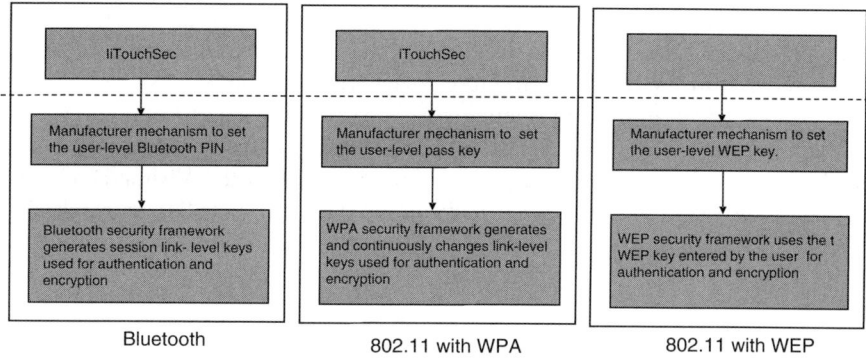

Fig. 8 An illustration of the ordered-secrets mechanism with wireless link-level security implementations

connectivity security parameters is then done over the NFC channel. If the device being set up is a Fixed Device, then an already bootstrapped mobile intermediary may be used to perform this third step by TAPing the Fixed Device. If the device being set up is an Infrastructure Device then this step is not required and the device may simply be plugged into the home network. As part of the TAP in this step, the device that is transmitting the connectivity security information verifies that the device being set up is a Home Device by prompting it to prove that it knows the Home Secret.

iTouchSec introduces a middleware layer that replaces the user interaction needed for creating and changing common secrets in smart home environments. This approach is applicable to a wide variety of underlying security systems, all of which support the concept of a user-level common secret as illustrated in Fig. 8. It also allows for access control privileges that expire without the need for manual revocation.

iTouchSec introduces an algorithm to set and update user-level common secrets based on an ordered list mechanism. An ordered list of user-level common secrets is generated randomly by the iTouchSec software running on an Infrastructure Device, which selects the first secret in the list and sets the user-level common secret with it. The Infrastructure Device periodically (e.g., daily) selects the next secret in the ordered list and again sets the user-level common secret with it. When the Infrastructure Device reaches the end of the ordered list, it generates a new list. This is expected to happen infrequently, e.g., with 1,000 secrets and daily rotation approximately every 3 years. The process of renewing the entire ordered list could take place at any time if, for example, the members of the smart home feel that the list has been compromised.

During Step 3 of the Easy Setup process, the list of ordered secrets is transferred to a Home Device that is being set up. The newly setup device then stores this list and moves down on it until it finds the currently active secret. Every time the Infrastructure Device changes the common secret, the client device momentarily loses access and will have to sequentially move down the list until it finds again the common secret used by the AP. When it exhausts the list of secrets, the device notifies the user to TAP the Infrastructure device to obtain the new list. This scheme also allows users to easily grant temporary access to the smart home network to Visitor Devices. For example, if a user wants to give a visitor access for 3 days and the secrets change daily, then iTouchSec transfers to the Visitor Device the current and the two following secrets in the ordered list using the Passlet mechanism described in Sect. 11.

10 "Touch" to Build Trust and Discover

Once the Home Devices are connected to the home network, they can proceed to search for and discover each other. But how can the user trust devices that were discovered over the network using a Control Point? The concept of building trust between devices is used to allow the user to credibly verify the identity of a device that he/she is communicating with over the network. In order to eliminate the need for Certificate Authorities (CA) in the smart home, iTouchSec offers a user-level tool called "Trust Builder" (Fig. 6) that leverages inherently secure TAPing based on NFC to establish trust. The premise is that a user can trust what he/she has seen and touched. This means a user's device obtains trustworthy information from a service provider device through a TAP, e.g., as the ProviderID, its OwnerID, and its HomeID, which can later be used to authenticate it over the network.

TAPing using the Trust Builder tool also provides limited device discovery functionality. During TAPing, along with the security information, some information about the owner and type of the TAPed device is also obtained. As a result, the Trust Builder tool creates a list of trusted devices and displays this information to the user.

11 "Touch" to Grant Access

Today, the process of granting access is in general cumbersome for non-expert users. For example, granting access to an 802.11 visitor device usually involves finding out its MAC address, editing the MAC address ACL of the AP, giving the common secret of the wireless network to the visitor device, editing the ACL of the resource to share, and sometimes creating a new user account for the visitor. At the end of the visit, the home occupant has to take actions to revoke access. As a major step in improving this scenario, iTouchSec introduces Passlets [9] which are *user-perceived* entities that carry *user-level* intent. Passlets are inspired by the concept of "capabilities." They behave like "tickets" or "passes" that give permission to the bearer to first connect to the home network and then use a subset of services provided by a networked device, without requiring that device to be configured in advance. For example, a user would use the iTouchSec "Passlet Manager" tool (Fig. 6) on his mobile phone to browse through his list of owned devices, select what to share, select the Passlet parameters (or use the default), and then just TAP the visitor device to grant access for as long as the Passlet prescribes.

The internals of Passlets will vary depending on the underlying security mechanisms that are being used and iTouchSec does not dictate a single implementation of Passlets. However, all Passlets designs contain link-level connectivity information necessary to connect to the home network and they are digitally signed using the private key of the owner of the device being shared. For example, a Passlet can contain the UserID of the user for whom the Passlet is being generated, the ProviderID of the Provider that the Passlet is providing access to, and connectivity information such as the MAC address or ESSID of the access point and the access point secrets, which are one or more user-level secrets required to connect to the access point for the specified period.

iTouchSec proposes Passlet sessions, which are established when the bearer of the Passlet first attempts to use it to gain access and disappear after the Passlet expires. Passlet can be manually revoked before their expiry time by the creator of the Passlet. The user simply browses the list of issued Passlets and selects the one to be revoked. A message that is digitally signed by the user is then sent to the iTouchSec middleware at the Provider device indicating that the Passlet should be revoked. The iTouchSec software at the Provider then adds the Passlet to a list of revoked Passlets. The revoked Passlets list is a nondecreasing list; however, its space requirements can be minimized through the use of a unique identifier for each Passlet and through the use of a global maximum for the Passlet duration.

12 iTouchSec NFC Module

The bulk of the security functionality of iTouchSec is implemented by the iTouchSec Engine (Fig. 6). The function calls that require sending and receiving information over the NFC interface are asynchronous and constitute the TAPing protocol. It is a set of commands used by the iTouchSec tools to perform Easy Setup, build Trust, and manage Passlets. All devices have a TAP server running on them that listens for incoming TAPs. The TAPing protocol is part of the iTouchSec NFC module (Fig. 6).

The iTouchSec NFC module is responsible for issuing outgoing commands and validating/consuming incoming commands and data. iTouchSec middleware and tools are agnostic to the specifics of the NFC communication and the NFC record format. The iTouchSec NFC module uses a set of methods provided by the NFC Security module (Sect. 13) to send and receive iTouchSec specific data (e.g., Home-IDs) and commands to the iTouchSec Security module. iTouchSec middleware performs all the NFC-related formatting and processing.

The iTouchSec NFC module implements the TAPing protocol. The first step in any and all TAPing transactions is for the devices to exchange and authenticate their respective HomeIDs. If the HomeIDs do not match, the transaction enters "unprivileged mode," where certain requests are not honored (e.g., GET_CONNECTIVITY_INFO). If they match then the devices proceed with the requested transaction(s). The main TAPing commands are:

- HOME_ID <homeID>: Sends the home ID to the TAPed device.
- GET_CONNECTIVITY_INFO: If the device making the request is authorized after exchanging the HomeID, the receiver of the request transfers a file containing the ordered list of common secrets described in Sect. 9.
- CONNECTIVITY_INFO: Sends the connectivity information as a response to the GET_CONNECTIVITY_INFO command.
- GET_DEVICE_INFO: Gets information of all Provider devices (e.g., UPnP devices) residing in the physical device: DeviceID, OwnerID, HomeID, Device-Name, DeviceType.
- DEVICE_INFO: Sends the device information as a response to the GET_DEVICE_INFO command.
- PASSLET <passlet>: Sends a passlet to the TAPed device. Passlets include fields such as UserID, DeviceID, DeviceName, ExpiryTime, PermissionLevel, and DeviceType.

The remaining commands are OK, ERROR, UNAUTHORISED, DONE, and CLOSE. Based on an incoming command, this module invokes the appropriate actions in the iTouchSec Engine.

13 NFC Security Module

In the proposed architecture there is a specialized NFC security module, named NFC Security (Fig. 6). This module provides public API functions that allow the security layer to use the NFC communication channel. For the proposed design, the API calls are:

- SetCommand(aCommand): This method passes a command from the iTouch-Sec NFC module to the wireless proximity communication channel in order to be sent.
- SetPasslet(aPasslet): This method passes a passlet from the iTouchSec NFC module to the lower layers in order to be sent.
- SetHomeID(aHomeID): This method passes the HomeID from the iTouchSec NFC to the lower layers in order to be sent.

14 Link Management

The Link Management module is the core of the NFC part of the middleware. In this module, a variable length packet format is defined for the frames communicated over RF. This format is based on NDEF guidelines (NFC Data Exchange Format) as defined by the NFC Forum [4]. In the heart of the Link Management is the NtipServer. The NtipServer operates in the asynchronous mode and can manage multiple clients on the host device. To support shared access to the NFC medium between multiple applications, each instance of an NFC session is bound to an endpoint selector (a unique string) identifying the NFC data type to be delivered to an application. Each application is expected to register for one or more endpoint selectors before it can use the NFC medium. Similarly, data read from a NFC tag or received from another NFC interface carries an associated selector. Record types are mapped to clients in an internal registration database. Simultaneous registration for the same selector by multiple applications is disallowed.

Incoming NFC packets are received by the NtipServer. These packets can contain one or more records. The NtipServer parses them and extracts the NFC record(s). Next, it determines the destination of these records based on the registration database. This allows clients to maintain a level of privacy and have complete control of the interpretation of the data. In the reverse direction, client applications can use the iTouchSec API to write to tags or send messages through the NFC interface. In this case, the NtipServer receives a set of data parameters through the API and composes an NFC record in the appropriate format before sending it over the air interface.

The Link Management module operates in the following states Fig. 9:

- INIT: In this state, the Link Management layer initializes itself and powers up the transceiver. Once completed it moves into the IDLE state.
- IDLE: This is the default state.
- BIND: An interim state to process requests from applications to bind an endpoint selector to a session. The application may provide an asynchronous callback routine.

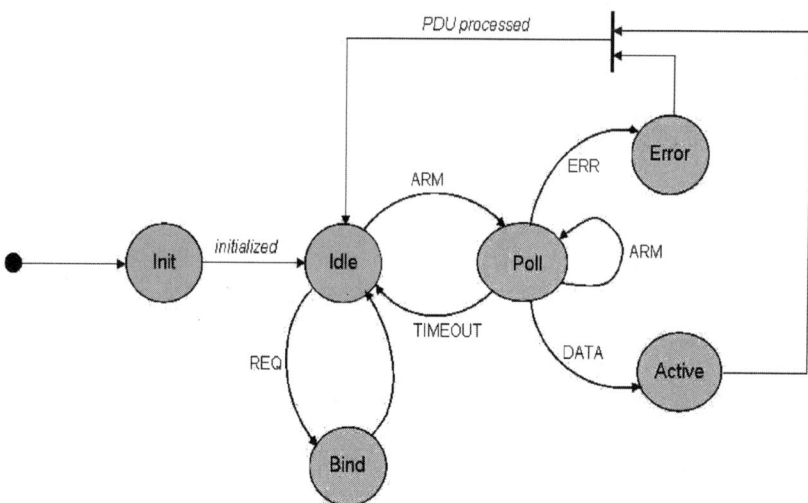

Fig. 9 Link management states of operation

- POLL: Upon explicit user input, the Link Management layer sends a command over the Host Controller Interface (Sect. 15) to arm the transceiver with a timeout interval. If a response arrives, an integrity check is performed on the data and the system transitions to either the ACTIVE or ERROR state.
- ACTIVE: The Link Management layer parses the records within a payload. This involves retrieving the endpoint selector from the data, identifying the corresponding session/application, and extracting the records. In the case of connectivity parameters, the out-of-band connection setup is initiated to wakeup the corresponding radio. If service discovery parameters are present, the system bootstraps the discovery as explained in Sect. 17.
- ERROR: The Link Control layer performs the first level of error checking, by verifying the payload length against the number stored in the header. A checksum may, also, be included with the payload to verify its integrity. If the check fails, the system enters the ERROR state. It is also possible to reach the ERROR state when a collision is detected. Error messages are handled by the application. The system handles collisions though we do not foresee its use in environments dense enough to involve frequent collisions.

15 Lower Layers

The Link Management module interfaces with the Host Controller Interface (HCI) driver (Fig. 6). The HCI driver resides on the mobile device and communicates with an NFC transceiver unit. The HCI interface can be USB interface, for instance, as they are becoming increasingly available on a variety of devices, and can enable a simple plug-in addition of the NFC functionality to the handheld. The Link Control layer controls the RF layer, instructing it to perform queries, and poll for a response from a peer NFC device in proximity. The NFC interface is controlled by the host via commands sent through the HCI driver. The design allows for the NFC transceiver to be "armed" at the press of a button on the host, activating the RF layer. At all other times, the RF unit is in idle mode, thus conserving power.

16 Out-Of-Band Connectivity

NFC is the out-of-band, secondary radio channel which can be used to set up a wireless connection by passing the primary channel connection parameters as part of the payload. The out-of-band-connectivity module awakens the corresponding radio based on the Protocol Descriptor that specifies the wireless interface (e.g., WiFi, Bluetooth) and the connection parameters. The configuration can be completed without the need for manual input. The result of the connection setup is a descriptor or communication endpoint, similar to a socket descriptor on UNIX systems that is passed as an opaque handle to the application.

Records containing connection setup parameters can declare their own record types. When such a record type is extracted from an NFC packet in the Link Management module, it is passed to the Out-of-Band module where it is parsed and processed to establish the connection.

17 Trusted Service Discovery

The term out-of-band discovery is defined as the process of obtaining discovery parameters needed to invoke a service outside the usual channels of processing service discovery queries. TAPing can be exploited as a means for out-of-band service discovery. Similar to the Out-of-band Connectivity module, the Service Discovery module can declare its own record type(s) that contain service discovery descriptions.

NFC can be exploited to discover nearby devices while keeping higher power radio turned off during quiescent operation. The user is prompted to bring his/her NFC-equipped mobile device in close proximity to the NFC-equipped target device in the environment and arm the NFC transceiver by a key press. This action triggers the reading of the service description. The discovery process then establishes a communication link with the discovered device, and bootstraps the mechanism for service invocation. Once the service interfaces are known to the mobile device, it may make remote procedure calls and subscribe for event notifications. Due to the short operating range of NFC communication, the client need to only wait for one response to be collected from the environment; thus reducing the discovery latency. This also avoids the response implosion problem associated with the traditional forms of discovery.

If the data read over the NFC interface is a service discovery record, a discovery bootstrap process is initiated. The description itself is partitioned into two logical parts: (1) a device description and (2) one or more service descriptions. A device description contains the physical and logical properties of the device such as device type, vendor information, and optionally a unique ID for identifying the device. Every device could be a provider of one or more services; hence the device description includes information (such as a URL) to locate the corresponding service descriptions. The service descriptions are retrieved and parsed to identify their interfaces. The client is then able to subscribe to event notifications, make remote procedure calls, and observe changes in the state of the device.

18 NFC Record

The proposed NFC tag record design (Fig. 10) is a flexible and extensible structure that can be used to exchange the iTouchSec commands and data in a variety of use cases. The information exchanged is the Payload. The Payload contains a Header

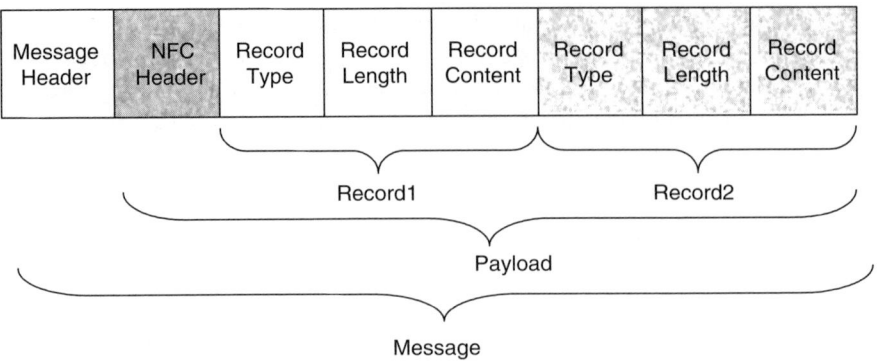

Fig. 10 The basic tag structure

and a record list with one or more records. The Header contains the length of the Payload and it is used to determine how much data must be read from a tag. Each record is a sequence of three elements, a triplet of (Type, Content-Length, Content). The record Type identifies the structure and semantics of the record by providing the Type name. The Content-Length identifies the length of the record Content. The record Content contains the actual data.

iTouchSec defines its own record type abbreviated as "IS." Incoming "IS" records are received from the Link Management module and are parsed to extract the iTouchSec commands and data. These are then passed to the iTouchSec NFC module. Likewise, outgoing commands and data are packaged into an "IS" record and are transmitted over the RF interface. For each "IS" record, there is a subrecord containing one of the TAPing commands. Further data subrecords are defined to convey parameters relevant to each commend (e.g., "psl" for passlet data or "hid" for the homeID). These data subrecords can follow the same triplet structure: Type, Length, Content. The command subrecord can either follow the triplet structure or it can be defined as the first byte to follow the "IS" type declaration (this is the format assumed in this section). The tag record format for all the current iTouchSec transactions is listed in Table 1.

19 Proof-Of-Concept Implementation

The implementation results presented in this section highlight the key functionality aspects of the proposed architecture. The implementation of the system builds on prototypes we have developed in projects iTouch [7, 8], IntuiSec [9–11], and Smart Sleeve [12]. Software components were developed for Symbian-based Nokia mobile phones, such as the Nokia 9500 Communicator and the Nokia 6620 Smartphone. The hardware support for the NFC interface is provided by a smart sleeve with an NFC add-on module. A demonstration of the experimental setup of the system is

Table 1 NFC record formats for iTouchSec commands

Record type	Subrecord type	Data subrecords [type, length, content]
IS	HomeID	[hid, hidLength, <homeID>]
IS	GetConnectivityInfo	
IS	GetDeviceInfo	
IS	ConnectivityInfo	[dn, dnLength, <DeviceName>]
		[mac, macLength, macAddress>]
		[ct, ctLength, ConnectivityType>]
		[lk, lkLength, <LinkKeys>]
IS	DeviceInfo	[did, didLength, <DeviceID>]
		[dn, dnLength, <DeviceName>]
		[hid, hidLength, <homeID>]
		[oid, oidLength, <ownerID>]
		[dt, dtLength, <DeviceType>]
IS	Passlet	[uid, uidLength, <UserID>]
		[un, unLength, <UserName>]
		[did, didLength, <DeviceID>]
		[dn, dnLength, <DeviceName>]
		[et, etLength, <ExpiryTime>]
		[per, perLength, <Permissions>]
		[dt, dtLength, <DeviceType>]
		[pid, pidLength, <PassletID>]
IS	Ok	
IS	Done	
IS	Error	
IS	Unauthorized	
IS	Close	

depicted in Fig. 11. The front-end tool is an application called iTouchConfig, which provides basic functions such as starting and stopping the NtipServer, reading from and writing to an RFID tag, and sending to and receiving from another NFC-enabled device. In addition, an NFC hot-button has been implemented to arm the NFC reader before each reading action.

In the rest of this section, we present some implementation examples that demonstrate iTouchSec functionality using NFC-based touch.

20 Service Discovery

The iTouchSec service discovery functionality can be realized through a number of technologies (e.g., UPnP, Web Services). This section presents a technical realization in the context of the UPnP framework. For details on UPnP the reader is referred to [1].

Connecting (both callbacks)...OK
Subscribed to ALL records.
Sending read NTIPHandleNewNtipRecordL; Size = 10
Record type = typeR
iTouchConfig Record Content = Michalakis
Function Success...OK
Sending read NTIPHandleNewNtipRecordL; Size = 18
Record type = myCard
7:46 PM Record Content = Dimitris Kalofonos

Exit

Fig. 11 A demonstration of the experimental setup

20.1 Brief UPnP Overview

Universal Plug and Play (UPnP) is a set of protocols including the Simple Service Discovery Protocol (SSDP), the Simple Object Access Protocol (SOAP), and the General Event Notification Architecture (GENA) originally developed by Microsoft Corporation and currently under development by the Universal Plug and Play Forum [1]. UPnP standardizes the protocols spoken between clients (called control points) and services. It leverages existing standards such as TCP/IP, HTTP, and XML.

Devices, services, and control points are the basic abstractions of the UPnP device architecture. The device model is hierarchical. In a compound device, the root device and any embedded devices are discoverable. Clients can address a root or an embedded device independently. SOAP servers in the device act as entry points for interacting and controlling it. Each service has a set of methods or actions with a set of optional input/output parameters and return values. The control point is the client and the device is the server. Control points can invoke actions on services. All UPnP devices that conform to UPnP Forum specifications follow the same basic pattern of operation: addressing, description, discovery, control, eventing, and presentation. SSDP is used for the service discovery process to (a) announce a device's presence to others and (b) search for devices and services. A device sends a multicast message either to advertise its presence to control points or to search for services in an UPnP network. Devices that hear this message respond with a unicast response message.

UPnP uses XML to describe device features and capabilities. For instance, the aforementioned advertisement message contains a URL that points to an XML file in the network describing the UPnP device's capability. By retrieving this XML file, other devices can learn about the advertised device's features, control it and interact with it.

20.2 SSDP Presence Announcement for the Printer

The printer SSDP presence announcement is shown in Table 2. A brief explanation of the various fields is given next. For a detailed description the reader is referred to [1]. The value of the NTS field identifies this SSDP message as a presence announcement of a new device or service. The Cache-Control field specifies the duration that the presence announcement is valid. The control point (user) device caches the complete service discovery record for the specified time frame. The USN field provides the device Universally Unique ID (UUID). This is one of the most important fields as it is used to uniquely identify a device. It may contain other information about the device type (e.g., root device, device type, service type). The Server field provides information on the operating system of the device, the product name, and version. The NT field has a potential search target description, i.e., how the control point can search for the discovered device or service. Finally, the Location field contains the URL from which the UPnP device description document can be retrieved. This is another essential field.

An example SSDP presence announcement for a printer is shown in Table 2. This would be exchanged between the printer and the discovering device (e.g., phone) over the NFC interface (Fig. 6). Once read, the announcement is extracted from the NFC packet by the NTIP server and passed to the service discovery module in order to bootstrap the service discovery process (see Sect. 17). The service discovery process is completed by the UPnP middleware and the printer control point is launched.

Table 2 SSDP presence announcement for the printer

Field	Description	Example data
NT	Search target	URN:schemas-upnp-org:device:Printer:1
USN	Device ID::NT	UUID:0e2fc7b3-4c09-4665-b4ae-f6f90448ba99::urn:schemas-upnp-org:device:Printer:1
Server	OS name/ver., UPnP/1.0, product name & version	Microsoft-Windows-NT/5.1 UPnP/1.0 UPnP-Device-Host/1.0
Location	Root device description URL	http://192.168.64.11:53891/upnp/device/Printer.xml
Cache-Control	Announcement expiration	1,800 s
NTS	Must be "ssdp:alive"	ssdp:alive

21 Easy Setup

One of the scenarios possible with the current implementation is setting up a new mobile device to become part of the home network. This is part of the Easy Setup process described in Sect. 9.

In Step 3 of this process, the user "touches" with the NFC interface of his mobile phone that of the Network Access Point (NAP) and establishes permanent connectivity for the phone to the home network. This one-touch action from the user triggers a series of steps in the middleware. Once the user TAPs the NAP, the iTouchSec NFC module issues an outgoing request command "GET_CONNECTIVITY_INFO." This is passed to the NFC Security module through the SetCommand(GET_CONNECTIVITY_INFO) method of its API. The NFC Security module packages it in the following record: [IS][GET_CONNECTIVITY_INFO] and passes it to the Link Management layer where it is further formatted with the necessary headers. Then it is sent through the HCI to the RF interface. The NFC packet is received by the NAP and forwarded through the middleware in the reverse order where it is finally received by the NAP iTouchSec NFC module. Provided that HomeID authentication succeeds, it issues a "CONNECTIVITY_INFO" reply command with the appropriate data parameters. The command and data are formatted into an NFC reply packet in the same fashion as the request packet and sent back to the mobile device. An example demonstration of this step for a WLAN NAP with SSID = "LitePad" is shown in Fig. 12. Once

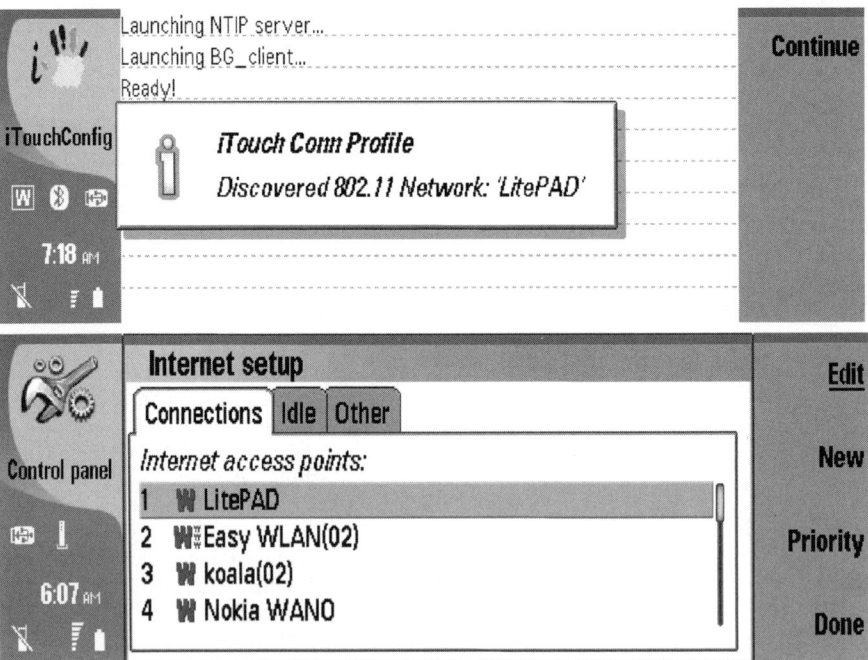

Fig. 12 Using NFC to establish connectivity to the home network

the NAP connectivity information is received by the mobile device, an information box pops up announcing the new WLAN network and "LitePad" is added to the list of available wireless networks on the device.

22 Building Trust

Another implementation example of the proposed architecture is the use of NFC-based touch to build trust with a device. The user interaction in this case is depicted in Fig. 13, where a snapshot of the GUI of the "Trust Builder" user-level tool is also shown.

The user launches the "Trust Builder" tool in her mobile phone, which prompts her to TAP the device she wishes to establish trust with. At this point, the following takes place. The iTouchSec NFC module issues an outgoing command "GET_DEVICE_INFO." It uses the SetCommand(GET_DEVICE_INFO) method of the NFC Security API to pass the command to the NFC Security module. The command is then packaged in the following record [IS][GET_DEVICE_INFO], which is then passed to the Link Management layer where it is further formatted with the necessary headers before it is sent through the HCI to the RF interface. The NFC packet is received by the TAPed device and forwarded through the middleware in the reverse order. The iTouchSec NFC module in response issues a "DEVICE_INFO" reply command with the appropriate data parameters as listed in Table 1. The command and data are formatted into an NFC reply packet in the same fashion as the request packet and sent back to the mobile device. Once the TAP is complete, information about the new trusted device appears in the user's trusted devices list as shown in Fig. 13.

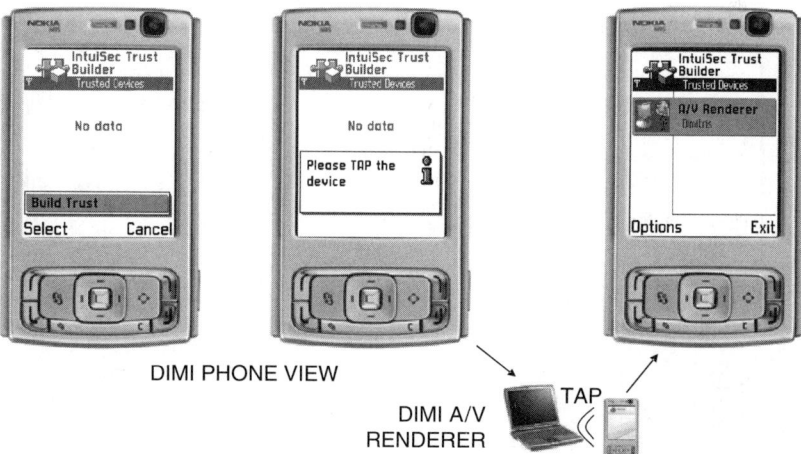

Fig. 13 User interaction for establishing trust with a device

23 Granting Access with Passlets

Another implementation example of the proposed architecture is the use of NFC-based touch to grant access to a visitor to services provided by one of the home devices, e.g., a printer. The user interaction in this case is depicted in Fig. 14, where a snapshot of the GUI of the "Passlet Manager" user-level tool is also shown.

The user launches the "Passlet Manager" tool and creates the Passlet he/she wants to grant to the guest. The tool then prompts the user to TAP the guest device. At this point, the following takes place. The iTouchSec NFC module issues an outgoing command "PASSLET." It uses the SetCommand(PASSLET) and SetPasslet (<passlet_data>) methods of the NFC Security API to pass the command and passlet_data to the NFC Security module. Command and passlet_data are then packaged in the following record [IS][PASSLET][<passlet_data>] with the Passlet parameters as listed in Table 1. The record is passed to the Link Management layer where it is further formatted with the necessary headers before it is sent through the HCI to the RF interface. The NFC packet is received by the guest device and forwarded through the middleware in the reverse order. The command and data are received by the iTouchSec NFC module and added to the "Received Passlets" database. An information box pops up announcing the new Passlet to the guest.

24 Related Work

There is growing recognition that there is a need to make security easier to use [13, 14]. In particular, although there are many proposals for smart home security [14–16], the issue of usability can be very challenging for nonexpert consumers.

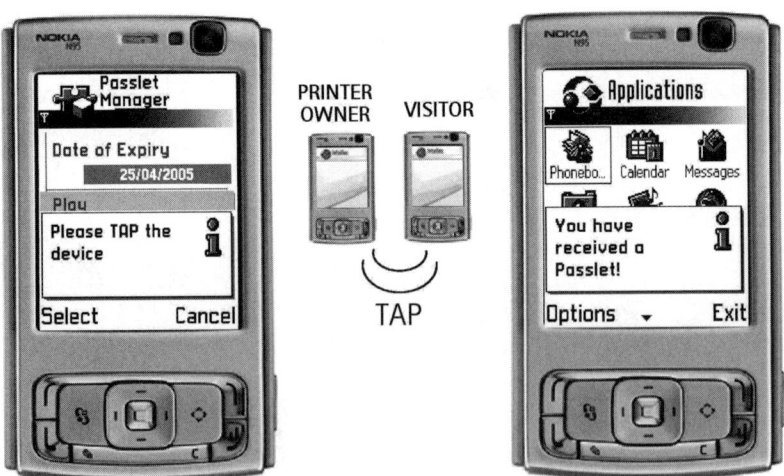

Fig. 14 User interaction for granting access to a visitor to print

This has led to research to make existing home security mechanisms easier to use for nonexpert [6–10], and standardization efforts, such as the work of Wi-Fi Alliance Easy Setup WG [17].

One of the most intuitive modalities to interact with smart devices and objects is through the use of "touch" interfaces (also known as LLCs, e.g., infrared, RFID, and NFC), e.g., as proposed in [7, 8, 14, 18–22]. LLCs are both intuitive and inherently secure channels and usage of infrared-based LLCs have been proposed to facilitate the user interaction with security [9–11, 23, 24]. Use cases for NFC-based security initializations using mobile devices are considered in the NFC forum [4].

There has been significant previous work in the area of technology for point-and-click user interfaces and the usability of point-and-click user interfaces in ubiquitous computing environments. Though the success of traditional IR remote controls is clear evidence of the usefulness of real-world point-and-click UIs, it should be mentioned that there has been research into the limitations of point-and-click interfaces [25]. HPLab's Cooltown [26] was an early research project that investigated real world, point-and-click user interfaces. Cooltown used IrDA to achieve limited, multiapplication, point-and-click functionality. Its appeal was limited by the short range of IrDA, the need to add powered IrDA transceivers to each target, and the need to carefully aim at the remote transceiver of the target. The work in [27, 28] discusses the use of laser pointers in ubiquitous computing environments. Because lasers generate a very precise beam, they do not have the range limitations of IrDA, but they require even more careful aim. They also require more expensive equipment for the targets.

Short range (sometimes called near-field) RFID tags can be used to create point-and-click interfaces. References [7, 8, 19, 20, 29] describe different ways to build RFID-tags-based pointing frameworks. Tags are very inexpensive and they do not need their own source of power as they can be powered by RFID readers.

Bar code stripes require no power and are even less expensive than RFID tags. Traditional bar code scanners are somewhat bulky but there are commercially available software packages that enable cell phones with cameras to scan 1D and 2D bar codes. Cell phone cameras were not designed for this sort of application and users must have patience and steady hands for these systems to work. The Target Recognition based on Image Processing (TRIP) [30] system also uses camera phones – but instead of bar code stripes they have new graphic patterns designed to be more easily captured and recognized using cell phone cameras.

A system for using RFID tags to invoke a variety of services was described by Riekki et al. [21]. Their system included pictographs associated with the RFID tags as a hint to the users of what service was associated with the tag. Their work did not focus on security concerns.

With the wide spread availability of Bluetooth-enabled devices, users are increasingly familiar with the idea of mutually authenticating devices – i.e., device pairing. Mayrhofer et al. [5] describe a clever use of accelerometers to mutually authenticate devices. The user picks up both devices, holds them together, and shakes. A similar motion pattern is detected by the accelerometers of both devices. Small variations in the pattern of motion leads to unique "signatures" that can be used to boot strap secure communication. Another, previously paired device could act as a proxy for devices that are too large or too fragile to shake. The "Loud and Clear" system [31] uses an audio channel to securely pair (authenticate) devices. iTouchSec currently uses RIFD to make security associations but other, bi-direction channels, such as IR could be easily added.

Hoepman [32] discusses "ephemeral pairing" protocols. These protocols are used for light-weight, short-term associations. An example of ephemeral pairing could be smart credit cards or public photoprinters. iTouchSec Passlets are easy to create and share and they have a "life-time." These features provide some of the user visible benefits of "ephemeral pairing," but iTouchSec is not optimized for these types of associations.

Feeney et al. [33] describe the use of point-to-point infrared communication channels to secure "spontaneous networks." Spontaneous networks are small to medium-sized ad hoc networks. Their goal to derive the "configuration infrastructure – naming, addressing, authentication, and key distribution" from face to face interactions between users using local, point-to-point communication channels (in their case, infrared) is similar to iTouchSec.

As wireless networking and sensing technology has improved, people have begun to study the use of tangible user interfaces (smart objects, surfaces, etc.) for interacting with digital information and smart spaces. The work of Ishii et al. [34] and Fishkin et al. [35] deal with the general challenge of creating new user interface design ideas. Unlike iTouchSec, their work does not focus specifically on problems of configuring networks and network security.

25 Conclusions

Today, end-users who want to set up a secure smart home have to learn how to use a bewildering array of tools and technologies. Wireless LAN access point security, various personal firewalls, VPNs, or other remote access technology must be mastered merely to prevent malicious access to the smart home. It is even more difficult to provide secure, limited access to specific applications or content within a smart home to visitors. The iTouchSec framework attempts to provide powerful tools that are simple to use and understand. Simple, high-level abstractions such as Passlets, and a new intuitive user interface for smart spaces based on touch, are the cornerstones of the iTouchSec framework.

Our experience using the framework and technology derived from the framework has been very favorable. We think the ideas and the framework can serve as a foundation for future work in user – centric security and user interfaces for pervasive computing environments.

The use of NFC technology has been invaluable in creating the iTouchSec touch-based user interface. It would be valuable to continue exploring the design and use of touch-based user interfaces. For example, it may be possible to distinguish between a brief touch, such as a tap, and a persistent touch. A tap and a persistent touch may trigger different actions. For example, if a user creates a persistent touch by placing a mobile phone next to another phone (or a camera, or a printer, etc.) it may be possible to use the NFC channel as the primary communication channel to transfer files – especially if there is no other communication channel available.

NFC technology is rapidly improving. For example, in 2006 HP announced "memory spots," tags with on-chip storage capacity of up to 4 Mbits and with data transfer rates of 10 Mbits per second. Such powerful tags enable practical new applications such as "smart documents" that can be easily read by humans and computers. These objects may have relatively short lifetimes or the "ownership" of the objects may be hard to establish. New and interesting user interface challenges arise from trying design systems that can deal with the number and variety of smart objects enabled by evolving NFC technology.

The design of successful user-friendly security tools requires a careful balance between simplicity, flexibility, and power. The design of the user interface of iTouch-Sec is biased toward simplicity and ease of use. To achieve this simplicity, it is lacking some features commonly found in commercial security tools. One such feature is the ability to associate security privileges with a role (e.g., administrator, controller) rather than a specific user. Another example is the ability to associate privileges with a group (e.g., accounting, engineering) rather than a list of specific users. Exploring strategies for providing these or other new security management features while preserving the ease of use of iTouchSec is another interesting direction for future research.

Acknowledgments The authors would like to thank Saad Shakhshir, Srikant Varadan, and Marios Michalakis for their contributions to the IntuiSec, iTouch, and Smart Sleeve projects, respectively, during their internships at Nokia Research Center. Special thanks also to our colleagues Kathy Chan, Venus Liong, Yinghua Ye, and Paul Wisner.

References

1. Universal Plug-and-Play (UPnP) Forum, UPnP Device Architecture 1.0.1, December 2003
2. Digital Living Network Alliance (DLNA), Digital Living Network Alliance Home Networked Device Interoperability Guidelines Expanded, March 2006
3. M. Weisner. The computer for the 21st century. Scientific American. 265(3): 94, September 1991
4. NFC Forum, www.nfc-forum.com
5. R. Mayrhofer and H. Gellersen. Shake Well Before Use: Authentication Based on Accelerometer Data, Proceedings of Pervasive, 2007
6. S. Shakhshir and D. Kalofonos. Usable Security in Smart Homes, Proceedings of the Eighth International Wireless Personal Multimedia Communications Symposium (WPMC'05), Aalborg, Denmark, 2005
7. Z. Antoniou and S. Varadan. iTouch: RFID middleware for boosting connectivity and intuitive user interaction in smart spaces, Nokia Research Center Technical Report (NRC-TR-2006–002), May 2006. http://research.nokia.com/tr/NRC-TR-2006–002.pdf
8. Z. Antoniou and S. Varadan. Intuitive Mobile User Interaction in Smart Spaces via NFC-Enhanced devices, Proceedings of IEEE ICCGI'07, Guadeloupe, March 2007

9. D. Kalofonos. IntuiWare Security Functional Specification, Nokia Research Center Technical Report (Internal), April 2004.
10. S. Shakhshir and D. Kalofonos. IntuiSec: A Framework for Intuitive User Interaction with Smart Home Security, Nokia Research Center Technical Report (NRC-TR-2006–003), April 2006. http://research.nokia.com/tr/NRC-TR-2006–003.pdf
11. D. Kalofonos and S. Shakhshir, IntuiSec: A Framework for Intuitive User Interaction with Smart Home Security Using Mobile Devices, Proceedings of IEEE PIMRC, September 2007
12. M. Michalakis, D. Kalofonos, and B. Shafai. An Experimental Hardware Extension Platform for Mobile Devices in Smart Spaces, Proceedings of International Conference on Pervasive Systems and Computing (PSC'06), Las Vegas, NV, 2006
13. C. Hurley, M. Puchol (Editor), R. Rogers, and F. Thornton. WarDriving: Drive, Detect, Defend, A Guide to Wireless Security, 1st edition, Syngress, Rockland, MA, 2004
14. P. Dourish, R. Grinter, J. Delgado de la Flor, and M. Joseph. Security in the wild: User strategies for managing security as an everyday, practical problem. Personal and Ubiquitous Computing. 8(6): 391–401, 2004
15. H. Nakakita, K. Yamaguchi, M. Hashimoto, T. Saito, and M. Sakurai. A study on secure wireless networks consisting of home appliances. IEEE Transactions on Consumer Electronics. 49(2): 375–381, 2003
16. C. Ellison. Home network security. Intel Technology Journal. 6(4): 37–48, 2002
17. Wi-Fi Alliance (WFA), www.wi-fi.org
18. Z. Antoniou and D. Kalofonos. NFC-based Mobile Middleware for Intuitive User Interaction with Security in Smart Homes, Proceedings of the Fifth IASTED International Conference on Communication Systems and Networks (CSN'06), Palma de Mallorca, Spain, August 2006
19. Z. Antoniou. A Touch is Worth A Thousand Clicks, Proceedings of IEEE International Conference on Wireless Information Networks and Systems, WINSYS part of ICETE, Portugal, August 2006
20. Z. Antoniou, G. Krishnamurthi, and F. Reynolds. Intuitive Service Discovery in RFID-Enhanced Networks, Proceedings of IEEE COMSWARE Conference, India, 2006
21. J. Riekki, T. Salminen, and I. Alakarppa. Requesting pervasive services by touching RFID tags. Pervasive Computing Magazine. 5(1): 40–46, 2006
22. T. Pering, R. Ballagas, and R. Want. Spontaneous marriages of mobile devices and interactive spaces. Communications of the ACM. 48(9): 53–59, 2005
23. D. Balfanz, G. Durfee, R. Grinter, D. Smetters, and P. Stewart. Network-in-a-Box: How to Set Up a Secure Wireless Network in Under a Minute, Proceedings of 13th USENIX Security Symposium, San Diego, CA, pp. 207–222, 2004
24. D. Balfanz, D. Smetters, P. Stewart, and H. Chi Wong. Talking to Strangers: Authentication in Ad-Hoc Wireless Networks, Proceedings of the Network and Distributed Systems Security Symposium (NDSS'02), San Diego, CA, 2002
25. A. Schmidt, H.-W. Gellersen, M. Beigl, and O. Thade. Developing User Interfaces for Wearable Computers – Don' t Stop to Point and Click, Intelligent Interactive Assistance and Mobile Multimedia Computing (IMC'2000), Rostock-Warnemünde, Germany, 9–10 November 2000
26. T. Kindberg, J. Barton, J. Morgan, G. Becker, D. Caswell, P. Debaty, G. Gopal, M. Frid, V. Krishnan, H. Morris, J. Schettino, and B. Serra. People, Places, Things: Web Presence for the Real World, Available at http://cooltown.hp.com/dev/wpapers/index.asp
27. M. Beigl. Point & Click - Interaction in Smart Environments. In: H.-W. Gellersen, ed., Handheld and Ubiqutious Computing, Lecture Notes in Computer Science, vol. 1707, ISBN 3-540-66550-1, pp. 311–314, Springer, Berlin, 1999
28. S. S. Intille, V. Lee, and C. Pinhanez. Ubiquitous Computing in the Living Room: Concept Sketches and an Implementation of a Persistent User Interface, UbiComp, 2003
29. T. Salminen and J. Riekki. Lightweight Middleware Architecture for Mobile Phones, http://www.mediateam.oulu.fi/publications/pdf/647.pdf
30. D. L. de Ipina, I. Vazquez, D. Garcia, J. Fernandez, I. Garcia. A Reflective Middleware for Controlling Smart Objects from Mobile Devices, SoC-EUSAI, 2005

31. M. T. Goodrich, M. Sirivianos, J. Solis, G. Tsudik, E. Uzun. Loud and Clear: Human Verifiable Authentication Based on Audio, Proceedings of ICDCS 2006: 26th Conference on Distributed Computing Systems, IEEE Computer Society, Los Alamitos, CA, p. 10, 2006
32. J. H. Hoepman. The Emphemeral Pairing Problem, Proceedings of Eighth International Conference on Financial Cryptography, pp. 212–226, Springer, Berlin, 2004
33. L. M. Feeney, B. Ahlgren, A. Westerlund, and A. Dunkels. Spontnet: Experiences in Configuring and Securing Small Ad Hoc Networks, Fifth International Workshop on Networked Appliances (IWNA5), October 2002
34. H. Ishii and B. Ullmer. Tangible Bits: Towards Seamless Interfaces Between People, Bits, and Atoms, Proceedings of CHI'97, pp. 234–241
35. K. Fishkin, A. Gujar, B. Harrison, T. Moran, and R. Want. Embodied User Interfaces for Really Direct Manipulation. Communications of the ACM. 43(9): 75–80, 2000

RFID Readers Deployment for Scalable Identification of Private Tags

Agusti Solanas* and Jesús Manjón

Scalability is frequently used as a magic incantation to indicate that something is badly designed or broken. Often you hear in a discussion "but that doesn't scale" as the magical word to end an argument.

Werner Vogels
CTO - Amazon.com

Abstract The deployment of the RFID technology can put the privacy of its users in jeopardy. With the aim of averting the fears of the RFID potential users, a plethora of security and privacy methods have been designed. However, due to the important growth of this technology, scalability problems have arisen and the proper deployment of the technology has become a challenge.

In this chapter, we provide a brief overview of the most relevant methods for providing security and privacy to the users of the RFID technology. We pay a special attention to the hash locks proposal and we recall a method based on the distribution of RFID readers, with the aim to provide security and privacy in a scalable fashion.

In order to test our method, we have developed a simulator that is presented in this chapter. By using this simulator we study the distribution of several kinds of readers on a variety of scenarios, and we report some of the obtained results.

1 Introduction

Scalability is the property of a system, a network, or a process, which indicates its ability to deal with growing amounts of work in a proper manner. Monolithic systems (i.e. systems that rely on a single entity) generally suffer from scalability

A. Solanas
CRISES Research Group, UNESCO Chair in Data Privacy,
Department of Computer Engineering and Maths, Rovira i Virgili University,
Tarragona, Catalonia (Spain)
e-mail: agusti.solanas@urv.cat

P. Kitsos, Y. Zhang (eds.), *RFID Security: Techniques, Protocols and System-on-Chip Design,* © Springer Science+Business Media, LLC 2008

problems, and most of the times distributed systems (i.e. systems that consist of a number of connected entities) have been used to solve the limitations of the former ones. These limitations are related not only to scalability but also to a higher number of issues:

- *Fault-tolerance*. A monolithic system relies on a single entity (e.g. a mainframe). If this entity fails, the whole system collapses. On the contrary, the failure of an entity in a distributed system can be handled by the other entities. Although the service that the system provides (e.g. a web service) could be affected by the failure of an entity (e.g. a server), the system does not crash immediately and gives space for a reaction.
- *Price*. The Grosch's law says that computer performance increases as the square of the cost. If you want to do it twice as cheaply, you have to do it four times faster.[1] This law was originally applied to mainframes in 1965 but it has been seen that it does not hold with the new microprocessors technology. In fact, Grosch's law contradicts Moore's law, and it is quite clear that in terms of cost, it is better to increase the number of units instead of using powerful ones.
- *Incremental growing*. Although monolithic systems can increase their capabilities, they do not grow naturally. On the contrary, distributed systems have been designed to simplify the addition of new entities into the system. The ability of a distributed system of being easily enlarged is a clear advantage.
- *Bottlenecks*. Computer systems can be connected to a network in order to serve queries from users. Monolithic systems have to manage all the queries that arrive at a given time instant. When the number of queries grows, they are stored in a queue prior to being served. The use of monolithic systems may lead to an increase in the delays due to their lack of parallelism. On the contrary, distributed systems are able to balance the load amongst the elements of the system and delays are reduced.

Radio Frequency IDentification (RFID) technology is specially sensitive in terms of scalability because the number of RFID elements is growing faster with each passing day. RFID protocols must be secure and private, but at the same time, they have to be scalable if they want to be a reality. As we will see in Sect. 2, there are lots of proposals to manage the RFID systems securely and guaranteeing users privacy, but there is a lack of scalability. RFID devices have an important presence in our daily life, even when we do not see them, and they will become ubiquitous in the near future. The spectacular market push of RFID technology is due to the interest by large retailers (e.g. Wal-Mart[2]), important manufacturers (e.g. Gillette, Procter & Gamble, etc.) and governments. As a result, almost every object is liable to carry an RFID tag. RFID devices can be seen as a proper substitute of bar codes since they

---

[1] "There is a fundamental rule, which I modestly call Grosch's law, giving added economy only as the square root of the increase in speed – that is, to do a calculation ten times as cheaply you must do it 100 times as fast."

[2] Wal-Mart started to explore the RFID technology in 2003 and devoted at least three billion dollars to implement it [7].

are mainly used to identify objects. Unlike bar codes, RFID devices allow objects to be identified without visual contact and help in improving and automating many processes e.g. supermarket checkouts, product inventories, etc. This is possible due to the ability of RFID tags for being read fast and in parallel. An RFID system consists of two main components:

- *RFID tags*. They are small passive devices with a variety of possible appearances from stickers to small grains embedded in official documents. A tag basically consists of a microchip and a metal coil, which acts as an antenna. In some cases, it can also contain a battery and some other microchips intended for increasing its computational power.
- *RFID readers*. They are active devices used to read the information stored in the tags. In a nutshell, readers emit a radio wave so that all tags in their range *answer* by broadcasting their embedded information (i.e. a set of bits). This information, generally known as *Electronic Product Code* (EPC), is usually the identifier of the object into which the tags are stuck.

In order to see the relation between RFID systems and scalability and to illustrate our view, let us consider the next example:

Example 1. We want to design an application to control the vehicles crossing the borders of some western European countries. The idea is that each vehicle should be monitored in order to determine in which country it is. To do so, we assume that each vehicle has an RFID tag that identifies it. We also assume that a set of RFID readers have been distributed along the border in order to detect the tags in the vehicles. To tackle this problem, we can adopt a centralised (*monolithic*) or a distributed architecture. Figure 1 shows a graphical representation of both approaches.

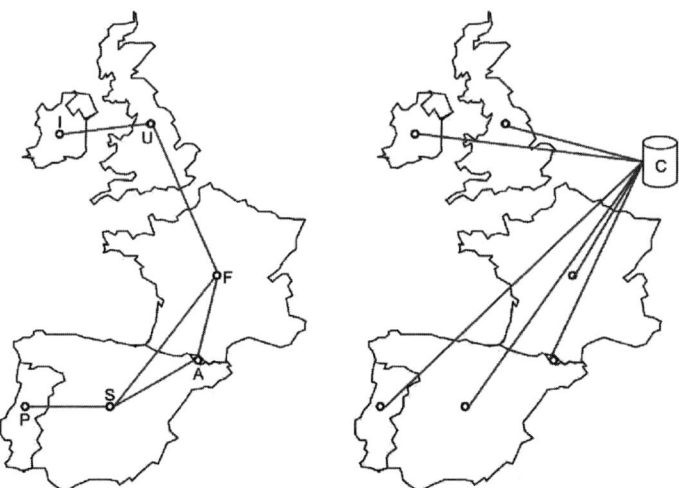

Fig. 1 Our border control example. *Left:* Adjacency graph for the distributed nodes of the example. *Right:* Centralised version of the example

In the centralised approach, all the border readers are connected to a central database C that contains all the possible IDs, let us say n IDs. Thus, when a vehicle crosses a border, a query to the centralised database is performed in order to determine whether the vehicle has permission to access the country. It is clear that this architecture does not scale because the delays will linearly grow as n grows and the number of queries that C has to manage will linearly grow as n grows also.

On the contrary, the distributed approach considers an adjacency graph $\mathcal{G} = (\mathcal{V}, \mathcal{E})$ where the vertexes $\mathcal{V} = \{P, S, A, F, U, I\}$ represent the distributed databases located in each country, and the edges $\mathcal{E} = \{(P \leftrightarrow S), (S \leftrightarrow A), (S \leftrightarrow F), (A \leftrightarrow F), (F \leftrightarrow U), (U \leftrightarrow I)\}$ represent the possible ways that a vehicle can take. Note that the distributed databases do not store n IDs. Each database stores the IDs of the vehicles which are in its influence area (i.e. it stores the IDs of the vehicles in its country and its adjacent countries). It is not necessary to store the IDs of the vehicles in I in the database in P, because it is impossible that a vehicle from I reaches P without crossing S. By applying this distributed strategy, the number of IDs that must be stored in each database is clearly lower that n and the system scales.

1.1 Contribution and Plan of the Chapter

In this chapter, we study one of the most important problems that the RFID technology has to face i.e. scalability. We consider the scalability problem of the RFID technology from a security and privacy point of view. We want to obtain secure and private RFID systems able to scale properly with the growing number of tags.

With the aim being to define a reference framework, in Sect. 2 we provide a brief overview of the private and secure RFID protocols proposed in the literature. Next in Sect. 3, we deeply analyse the hash locks approach and, we show how its scalability problems can be overcome by means of the distribution of RFID readers and a communication protocol amongst these readers. Section 4 presents our simulator. We show the main capabilities of the simulator and we explain how to use it. In Sect. 5 we report a number of experimental results that we have obtained using our simulator. We study a variety of scenarios and we show that the protocol that we propose scales properly. Finally, in Sect. 6 the chapter concludes by pointing out some future challenges.

2 Brief Comments on RFID Security and Privacy

There is a variety of methods for providing RFID tags with privacy. The proposed methods mainly depend on the computational power of the tags in which they will be run. In general, we can divide the methods considering the computational constraints of the tags:

1. *Elemental or basic tags*, which are not capable of performing cryptographic operations such as generating random values or computing hashes. This kind of tags can be disabled by using killing or sleeping commands. A different approach is the use of proxies [5, 18], but they require the users to have some technical background. The blocking approach [15] is another method that relies on the concept of *blocker tag* that simulates the full spectrum of possible tags. By doing this, it becomes very difficult for a reader to know which tags are really being carried by a given user. Sarma, Weis and Engels [19] proposed the re-labelling of tags in order to avert their tracking. Innoue and Yasuura [10] and Good et al. [6] proposed some variants of this approach. The use of pseudonyms has been also proposed [11] and re-encryption has been also considered [13].

2. *Symmetric-key tags*, which are capable of dealing with symmetric-key cryptography protocols. These tags can deal with more complex protocols like the OSK by Ohkubo, Suzuki and Kinoshita (OSK) [16], and some of its variants [1, 8]. The YA-TRAP protocol by Tsudik [22] is another candidate for being used with this kind of tags. Weis, Sarma, Rivest and Engels proposed the use of hash locks and presented the *deterministic hash lock* approach in [24], later Juels proposed an improvement called *improved randomised hash locks* in [12]. Henrici and Müller in 2004 [9] also proposed a privacy protocol for this kind of tags and Engberg, Harning and Jensen [4] provided an alternative based on zero-knowledge protocols.

3. *Public-key tags*, which are capable of managing public-key cryptography protocols. Some protocols have been proposed for this kind of tags based on Elliptic Curve Cryptography (ECC) [2], the Elliptic Curves Discrete Logarithmic Problem (ECDLP) [17], or Physical Unclonable Functions (PUF) [3, 23].

Symmetric-key tags are very well suited for most real-world applications, specifically the hash locks approach is a very interesting method for providing privacy to the RFID technology. We will briefly introduce the main idea of deterministic hash locks in Sect. 2.1. Next, we will focus on the improved randomised hash locks in Sect. 2.2.

2.1 Deterministic Hash Locks

Weis, Sarma, Rivest and Engels proposed in 2003 the use of hash locks in RFID devices. A first approach, called *deterministic hash locks*, was presented in [24]. A tag is usually in a "locked" state until it is queried by a reader with a specific temporary meta-identifier *Id*. This is the result of hashing a random value (*nonce*) selected by the reader and stored into the tag. The reader stores the *Id* and the *nonce* in order to be able to interact with the tag. The reader can unlock a tag by sending the *nonce* value. When a tag receives it, the value is checked. Another way of running this scheme is by using some *meta-keys*. Each tag is initialised with a (*Id*, meta-key) pair, then, in order to unlock the tag, the *meta-key* is used. The problem of this

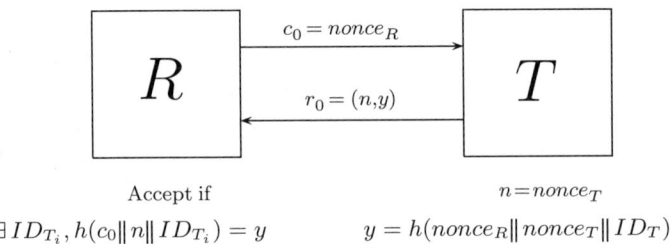

Accept if

$\exists ID_{T_i}, h(c_0 \| n \| ID_{T_i}) = y$

$n = nonce_T$

$y = h(nonce_R \| nonce_T \| ID_T)$

Fig. 2 Diagram of Juels–Weis improved randomised hash locks

solution is the cost of storing these pairs. Note that the hash approach does not suffer from this shortcoming. Another security problem that must be faced is how to securely send the meta-identifiers from readers to tags and vice versa.

2.2 Improved Randomised Hash Locks

A recent approach based on hash locks can be found in [12], where the *improved randomised hash locks* are presented. The basic operation of the improved randomised hash locks is depicted in Fig. 2 and is next briefly described:

1. A reader R sends a challenge c_0 to a tag T, where $c_0 = nonce_R$ is generated uniformly at random.
2. T generates its own nonce $nonce_T$ and hides its unique identifier ID_T by sending a response $r_0 = (nonce_T, h(nonce_R \| nonce_T \| ID_T))$.
3. To determine ID_T, R must perform an exhaustive search of the IDs in its database to compute $r_i = (nonce_T, h(nonce_R \| nonce_T \| ID_{T_i}))$ and compare the result with r_0. Once R finds an ID_{T_i} that satisfies $r_i = r_0$, the tag is identified.

In [14] it is proved that improved randomised hash locks offer strong tag privacy in front of eavesdroppers. The main limitation of this technique is scalability: indeed, the authors of [14] express their belief that, for RFID tags capable of only symmetric-key cryptography, their definition of strong privacy may require the reader to perform brute-force search to identify tags, which scales poorly. They also point out the need of definitions and protocols for RFID privacy that are weaker, but more *practical* and *useful*. We absolutely agree with their remarks and we concentrate on the definition of a protocol permitting the practical use of privacy schemes like the improved randomised hash locks in a scalable manner.

3 Our Approach to Scalability

With the aim being to provide scalability to hash-locks-like protocols, we recall a solution that can be divided in two main parts [20]:

- A *readers deployment*. Readers must be deployed properly in order to cover the tags movement area.
- A *protocol suite*. The protocol is used by the readers to exchange information on the location of the tags. By exchanging this information, the readers do not need to store huge amounts of tag IDs, but only the ones that are in their cover range or close to it.

In Sect. 3.1, we explain how to distribute the readers in order to let them collaborate to control the tags flow. Next, in Sect. 3.2 we detail our protocol.

3.1 Readers Deployment

We now describe the spatial distribution of readers and the distribution of IDs amongst them. Consider an area Ψ that can be covered by a number of readers. Assume that tags enter and leave Ψ through designated points called system access points (SAP) and system exit points (SEP), respectively.

Readers are placed according to a grid pattern depicted in Fig. 3. Let A_i be the square cell covered by the ith reader R_i. For the sake of clarity, we consider that all readers have the same cover range, so that all cells have the same size. However, our protocol has been designed to deal with variable cover range readers as we will show in Sects. 4 and 5. Further, we consider cells to be disjoint and to span the entire area Ψ. Formally,

$$\bigcup A_i = \Psi, \forall i. \tag{1}$$

$$A_i \cap A_j = \emptyset, \forall i, j | i \neq j. \tag{2}$$

It is assumed that the readers are able to locate a tag by collaborating. Although the technologies for locating a tag are beyond the scope of this work, we next discuss some relevant issues about tag location.

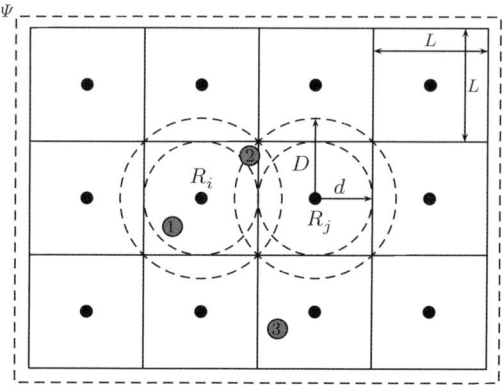

Fig. 3 Scheme of the coverage of a set of readers. The *numbers* are used to indicate different tag location situations [20]

Let D be the radius of the smallest circle containing a square cell and d the radius of the greatest circle inscribed in a square cell. Depending on the location of the tag, readers can face three different situations (see numbers in Fig. 3):

1. The distance between the tag and the reader R_i is less than d: In this case, the tag is located in the square area covered by R_i.
2. The distance between the tag and R_i is less than D and greater than d: In this case R_i needs the help of an adjacent reader R_j to determine the location of the tag. Note that it is not necessary to exactly locate the tag: determining its current cell is enough. Thus, it is possible to fulfil this task with only two readers.[3]
3. The distance between the tag and R_i is greater than D: In this case, the tag is off range of R_i.

3.2 Information Sharing Protocol Suite

Our method is based on sharing information between readers. The shared information is the tag ID and the ID of the reader which covers the cell in which a certain tag is located. The readers use three kinds of messages to share information: tag arrival, tag roaming and tag departure (*cf.* Table 1). This information is stored in the local cache of each reader involved in the message exchange.

In order for information to be shared without unnecessary replication, each reader removes from its cache the information related to tags which are no longer in the reader's cover range or in the cover range of an adjacent reader.

The suite consists of three protocols corresponding to the life cycle of a tag with respect to the system (i.e. arrival, roaming and departure):

1. *Arrival protocol.* It starts when a new tag enters the system. Upon arrival of a tag, a number of messages are sent in order to propagate the tag ID and the ID of the reader that acted as a SAP.
2. *Roaming protocol.* It is used when a tag moves from the cover range of a reader to the cover range of another reader. As a result, a number of ID propagation messages are generated and a request of ID deletion is also sent to the readers which are no longer accessible by the tag from its new location.

Table 1 Message types used by the information sharing protocol suite

Message	Meaning
$(T, \oplus)$	Tag T enters into the system
(T, R_i)	Tag T enters into the cover range of reader R_i
$(T, \ominus)$	Tag T leaves the system

---

[3] If the exact location was needed, at least three readers ought to be used.

3. *Departure protocol.* It is responsible for managing the departure of tags from the system. It generates a number of ID deletion notifications to rid the readers of the IDs of the departed tags.

3.2.1 Arrival Protocol

Let system access points (SAP)s be the entrances to the system. The number of SAPs and their locations is variable and user-definable, i.e. they depend on the nature and lay-out of the facility (airport, factory, store, etc.) served by the RFID system.

Each SAP is supposed to know all the possible tags which can enter the system through it. For example in a wholesale distribution centre, fresh fish enters through gate 1 served by SAP1, cleaning products enter through gate 2 served by SAP2, etc. Thus, SAP1 only needs to know the information about fresh fish and SAP2 only needs to know the information about cleaning products. Thanks to this division, the amount of information stored in the SAPs scales better and goods can enter the system in an orderly fashion.

We assume that a SAP consists of a reader connected to a computer that can efficiently access a database of tag IDs. Regarding the remaining readers (those which are not SAPs), they can be very simple devices with little storage and computational capabilities.

Note 1. This approach substantially differs from a centralised scheme in which all readers are connected to a back-end computing system. In our approach, only SAPs need a connection to the back-end and only the incoming tags are considered, which increases scalability.

For the sake of simplicity we describe the protocol with a single SAP and a single reader R_{in}. The generalisation to multiple SAPs is straightforward.

The arrival protocol is as follows:

Protocol 1 (Tag Arrival)

1. The protocol starts when a tag T is detected by a SAP (cf. Step 1 of Fig. 4). If T is not found in the SAP database then the SAP raises an alarm to inform that there is an unidentified tag trying to enter the system without permission.

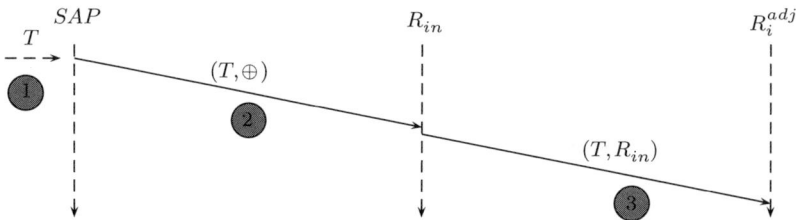

Fig. 4 Messages generated upon the arrival of an authenticated tag into the system

2. *If the tag is correctly identified, the SAP sends a message to R_{in} in order to inform the reader that a new tag T is going to enter the reader's cover range (cf. Step 2 of Fig. 4).*
3. *R_{in} adds the tag T to its cache; thus, when T reaches the cover range of R_{in}, the reader is able to authenticate T. After authentication, R_{in} sends a message to all its adjacent readers R_{in}^{adj} to inform them that T has entered R_{in}'s cover range and can roam to any of them (cf. Step 3 of Fig. 4).*
4. *In response to that message, the adjacent readers R_{in}^{adj} add T to their caches and record the name of the message originator (i.e. in this case R_{in}) who is the current owner of the tag.*

3.2.2 Roaming Protocol

The roaming protocol task is to share the tag ID information needed for updating the caches of the readers. By properly updating their caches, the readers can authenticate the tags in their cover ranges, and they can also leverage their resources (memory allocation, computational power). To that end, the ID information of the tags controlled by a given reader must be shared with adjacent readers, and non-adjacent readers must remove the ID information from their caches. Such a removal averts the uncontrolled growth of the readers' caches and makes the system scalable in terms of computational cost and memory space.

The roaming protocol is launched when any tag T moves from the cover range of a reader to another (adjacent) reader. The protocol is as follows:

Protocol 2 (Roaming)

1. *A tag T is detected by a reader R_i other than the owner of the tag (cf. Step 1 of Fig. 5), where we denote by owner of T the last reader that informed the rest of readers that T was in its cover range. Due to the spatial distribution of the readers, T must come from one of the adjacent readers to R_i, so R_i has in its cache the ID information of T and is able to identify it.*
2. *After identification, R_i sends a message to its adjacent readers R_i^{adj} in order to inform them that the new owner of T is R_i (cf. Step 2 of Fig. 5).*
3. *Upon message reception, the adjacent readers behave differently depending on their current cache information:*

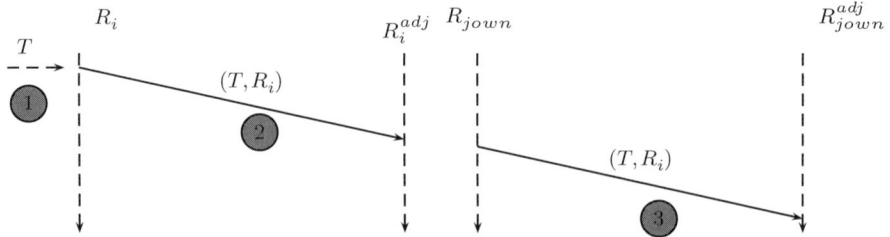

Fig. 5 Messages generated during the roaming protocol

(a) *If an adjacent reader $R_j \in R_i^{adj}$ has no information about T in its cache then it simply appends T and its owner information.*

(b) *If an adjacent reader $R_j \in R_i^{adj}$ has information about T in its cache but was not the previous owner of T, then it only needs to update the ID of the owner of T (i.e. in this case R_i).*

(c) *The adjacent reader R_{own} which was the previous owner of T must communicate to its adjacent readers R_{own}^{adj} that the new owner of T is R_i. To do so, R_{own} propagates the message from R_i to their adjacent readers (cf. Step 3 of Fig. 5).*

4. *When the adjacent readers of R_{own} receive the message, they behave differently depending on their adjacency relations:*

(a) *If a reader $R_k \in R_{own}^{adj}$ is adjacent to R_i then it does nothing.*

(b) *If a reader $R_k \in R_{own}^{adj}$ is not adjacent to R_i then it removes the information on T from its cache.*

5. *At the end of the protocol, only readers adjacent to the current owner keep information on T in their cache.*

3.2.3 Departure Protocol

In almost any RFID application, tags which have entered a controlled system must leave it. For example, in a supermarket, grocery, warehouse, etc., tags travel from the shelves to the checkout.

In order to control the departure of the tags from the system, the SEPs are used. A SEP is an area covered by a reader from which no tag can go back into the system (e.g. a checkout in a supermarket).

The departure protocol works as follows:

Protocol 3 (Departure)

1. *The protocol starts when a tag T is detected by a SEP (cf. Step 1 of Fig. 6).*
2. *The SEP informs its adjacent readers R_{SEP}^{adj} that T must be removed from their caches because there is no chance for T to go back (cf. Step 2 of Fig. 6). The adjacent readers of the SEP, including the previous owner R_{own} of T, erase the information on T from their caches.*

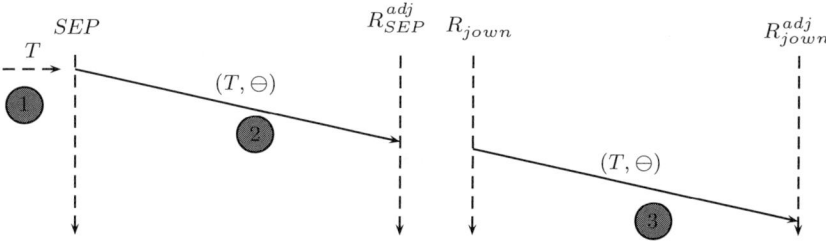

Fig. 6 Messages generated during the departure protocol

3. *The previous owner R_{own} of T propagates the removal message to its adjacent readers $R_{\mathrm{own}}^{\mathrm{adj}}$ (cf. Step 3 of Fig. 6).*

4. *The readers $R_{\mathrm{own}}^{\mathrm{adj}}$ remove any information on T from their caches, and no data remain in the system about the departed tag.*

4 Our Simulator

After the massive irruption of the RFID technology in the retail sector, many RFID manufacturers have developed a number of simulators to analyse the efficiency and reliability of their products. However, most of these simulators are basically centred in the study of a specific RFID device. Although these simulators are useful to determine the capabilities of a given device, they are useless for analysing the scalability of a whole RFID system. Moreover, it is difficult to find free open-source simulators and, more complex suites such as *iAnywhere* [21] are expensive commercial applications. Thus, one of the greatest problems when measuring the efficiency of a new protocol is the lack of tools.

With the aim being to overcome this limitation, we have implemented an RFID simulator that is able to analyse the scalability of a given protocol. Our simulator has been implemented in ANSI C and the graphical user interface (GUI) uses the GTK 2.0 libraries. Thus, the simulator is portable to different platforms such as Windows, Linux and Mac.[4]

Figure 7 shows the main parts/modules of the simulator. In the next sections, we provide the reader with a general idea on how the main modules of the simulator

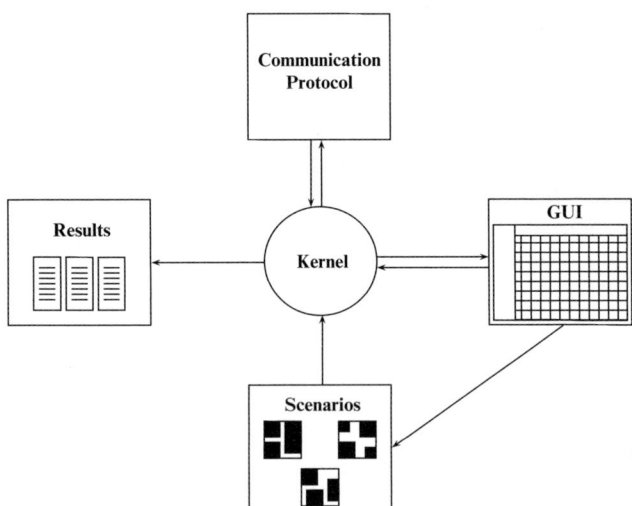

Fig. 7 Main modules of our RFID simulator

[4] Currently the simulator has versions in Windows and Linux.

work. Note that we will not go into details on how the modules have been implemented but we only give a conceptual approach.

4.1 The Kernel of the Simulator

The kernel is a central module of the simulator, it controls the flow of tags entering into and leaving from the system and interacts with the communication protocol module to simulate the readers network.

The entrance of tags into the system is controlled by two main parameters: the maximum number of tags and the frequency of arrival. The movement of the tags is also controlled by this module (in this version of the simulator the movement of the tags is random).

The kernel collects all the data and stores them in the results files. It also controls the number of tags in the system and stops the simulation once all the tags have left the scenario.

4.2 The Communication Protocol Module

This module defines the behaviour of the simulated readers and their interaction. It is an event-oriented module in the sense that it reacts to the events that the kernel raises. When a tag changes its location, the kernel raises a tag movement event and the communication protocol module must catch the event and update the state of the readers.

In this chapter we have considered the protocol described in Sect. 3. However, the simulator is able to deal with virtually any protocol of this kind.

4.3 Scenarios Management

An interesting feature of our simulator is the possibility to design and manage different scenarios. In order to study the scalability of a protocol completely, it is necessary to test it in different situations and observe its behaviour in bottlenecks, big areas without obstacles, narrow corridors, etc. By using our simulator it is simple to build new scenarios with a variety of features. (see Table 2)

Initially, a scenario is an empty space divided into a number of squares (e.g. 20×30 squares) that form a grid. A square is the smallest unit that we consider in the simulation (i.e. each tag is located in a square at a given time). The number of squares in each scenario can be defined by the user. Once the total number of squares is defined, the user can place SAPs and SEPs, simulating the doors that would exist in a real environment. Then, the readers are distributed amongst the squares and their cover ranges are configured.

All the scenarios can be saved and reused as times as needed, and they are easily exportable to applications such as Excel or Matlab.

4.4 The Graphical User Interface

The graphical user interface (GUI) allows the users to interact with the simulator. The GUI has three main objectives:

1. *Help the user in building and managing scenarios.* During the definition of scenarios, a crucial point is the placement of the readers and their configuration. The simulator allows to modify the cover range of each reader in a simple way and gives the possibility of studying the behaviour of a protocol on the same scenario but with different configurations of the readers. In Fig. 8 we can observe that each reader (marked with a black circle) is configured with a given cover range (e.g. at the corridors located on the left of the scenario, readers have a cover range of a single square, while in the room located on the right the cover range of the readers is 25 squares).

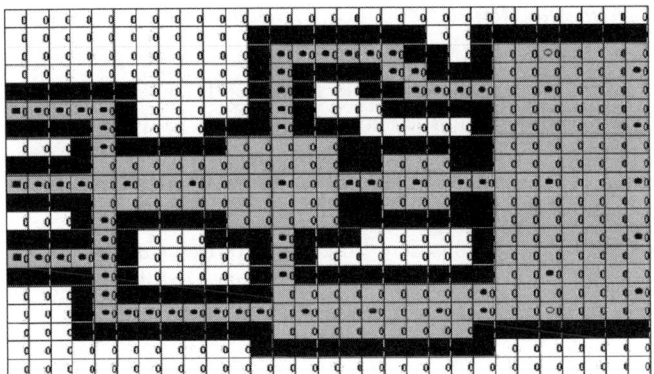

Fig. 8 A view of an scenario with different readers

Table 2 Symbols used to create scenarios

Symbol	Meaning
•	Reader covering a cell
■	Wall
⊡	System Access Point (SAP)
⊙	System Exit Point (SEP)

2. *Allow the user to monitor the evolution of a simulation.* If desired, the user can monitor the evolution of the simulation. The movement of the tags and the reader's cache size are shown at each simulation step.

3. *Permit the user to define the desired outputs.* As we show in the outputs module section, there is a number of results that can be obtained from the simulation and the user is able to select which one he/she wants to get.

4.5 The Outputs Module

During a simulation, information about the number of tags in each cell and the state of the caches of the readers is collected. Files are created at different steps of the simulation and they later allow an accurate study of the evolution of the caches of the readers over time.

In this version of the simulator, we have mainly focused on the size of the caches of the readers. However, it is easy to extend the simulator to let it gather a variety of information related to the simulated protocol.

4.6 Some Additional Features

In addition to the described features, the simulator has other complementary ones to make easy the study of the results of the simulation.

As shown in Fig. 9, it is possible to plot the path that a particular tag has followed. This path is seen graphically while the simulation takes place, and it can be checked after the simulation. Before the simulation starts, it can be checked whether a scenario is correct. A scenario is considered to be correct when it has access and exit points, it is surrounded by walls (i.e. it is finite), and it is totally covered by readers. If a scenario is not correct the simulation will not start and the program will

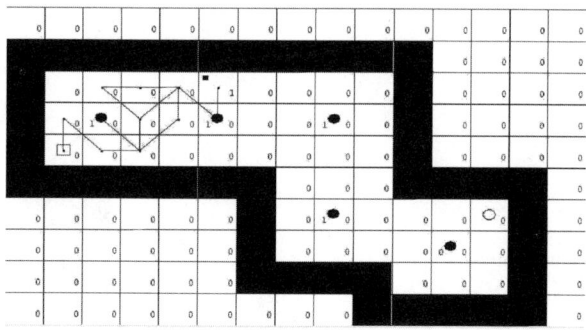

Fig. 9 Example of the path of a tag

warn the user about the error. Finally, the simulator allows to observe the connections amongst the readers. In our scalability protocol, these connections are essential since the readers exchange information with all their neighbours.

5 Experimental Results

In this section, we report a number of experiments that we have carried out by using our simulator. We have created five scenarios which have interesting features such as corridors, bottlenecks, small and big rooms, etc. We have configured the readers to have different cover ranges, thus, we can study the load that the most powerful readers have to manage. Considering a random movement of the tags, we have mainly concentrated on the analysis of the size of the caches of the readers.

In Sect. 5.1, we recall our border control example explained in the first section, and we show how to simulate it by using our simulator. Next, in Sect. 5.2 we analyse the proposed scenarios in detail.

5.1 Our Example

The simulation of our example is very easy and it will help us to clarify how a simulation is done. First, the scenario is built by using the GUI and the readers, the SAPs and the SEPs are placed in it. Figure 10 shows the scenario that represents our border example depicted in Fig. 1. We have configured all the readers to have the same cover range. By doing this, we guarantee that the load of each simulated local database is similar (i.e. we assume that the number of vehicles does not depend on the surface of the country[5]). We have placed two SAPs, one in P and the other in I, and a SEP in A. These points are placed arbitrarily and they are intended for an illustration purpose only.

Fig. 10 Example represented in the simulator

---

[5] This assumption is unrealistic, but it is good enough for our example.

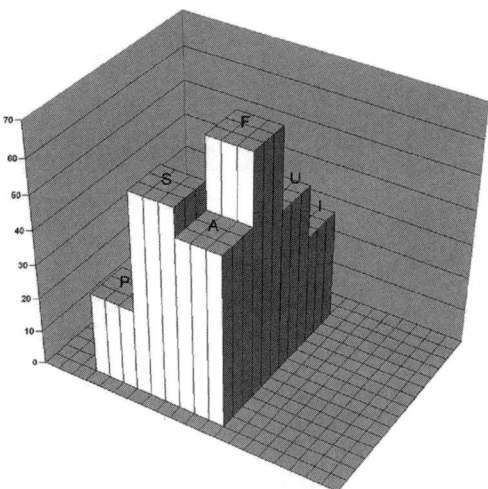

Fig. 11 Results of the simulation of the example

Once the scenario is built, we start a simulation with a few tags (e.g. 500 tags) and we obtain some results on the size of the caches of the readers. Figure 11 shows an example of the obtained results when simulating our example. From the study of these results we can emphasise some conclusions:

1. Although A is connected to very loaded nodes (i.e. S and F), it has a lower load because the SEP is located in its cover range.
2. P and I have a low load because they are only connected to another node.
3. S and F are the most loaded readers because they are connected to three other readers, thus, they have to store in their caches the information of all these readers.

5.2 Some Other Scenarios

In [20], we studied the scalability of the protocol described in Sect. 3 considering that all the readers had the same cover range (i.e. one square). Although this study was useful to show that the scalability of the RFIDs system is clearly improved, it is necessary to consider more complex scenarios in which the cover range of the readers is diverse. Moreover, it is also interesting to determine the influence of the cover range on the scalability of the system. To do so, we have designed five scenarios (*cf.* Fig. 12) and we have covered them with a variety of readers. The scenarios that we have built can be summarised as follows:

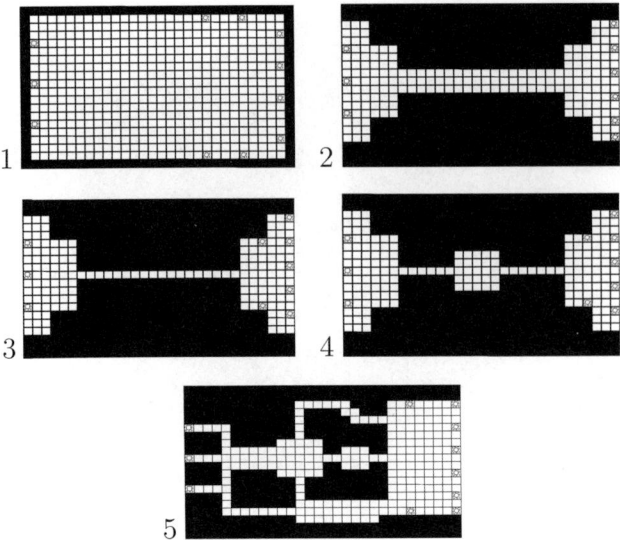

Fig. 12 Scenarios used in our simulation study

1. *Empty scenario.* This scenario has three SAPs and eight SEPs, there are no obstacles or walls in it, thus, it is useful to show the influence of the readers' cover range on the scalability of the system.
2. *Scenario with a wide corridor.* This scenario has three SAPs and six SEPs. By using this scenario, we analyse how the corridors and their associated bottlenecks affect the scalability of the system. Moreover, we consider different options for covering the area with heterogeneous readers.
3. *Scenario with a narrow corridor.* This scenario has three SAPs and seven SEPs. It is a scenario similar to the previous one, but in this case the corridor is narrow. Thus, the corridor can only be covered by one-square range readers.
4. *Scenario with a narrow corridor and a small room in the middle.* This scenario has three SAPs and seven SEPs. In this case we can study the bottlenecks that appear at the entrance and the exit of the small room in the middle of the corridor.
5. *Complex scenario with corridors and small and big rooms.* This scenario has three SAPs and eight SEPs. It is the most realistic of all the analysed scenarios. It has been designed to study how the protocol behaves in a real and complex scenario.

5.2.1 Empty Scenario Analysis

This scenario has 600 squares which have been covered by readers having a cover range between 1 and 25 squares. Figure 13 shows the size of the caches for different cover ranges when the maximum number of tags into the system is 1,000. It can be

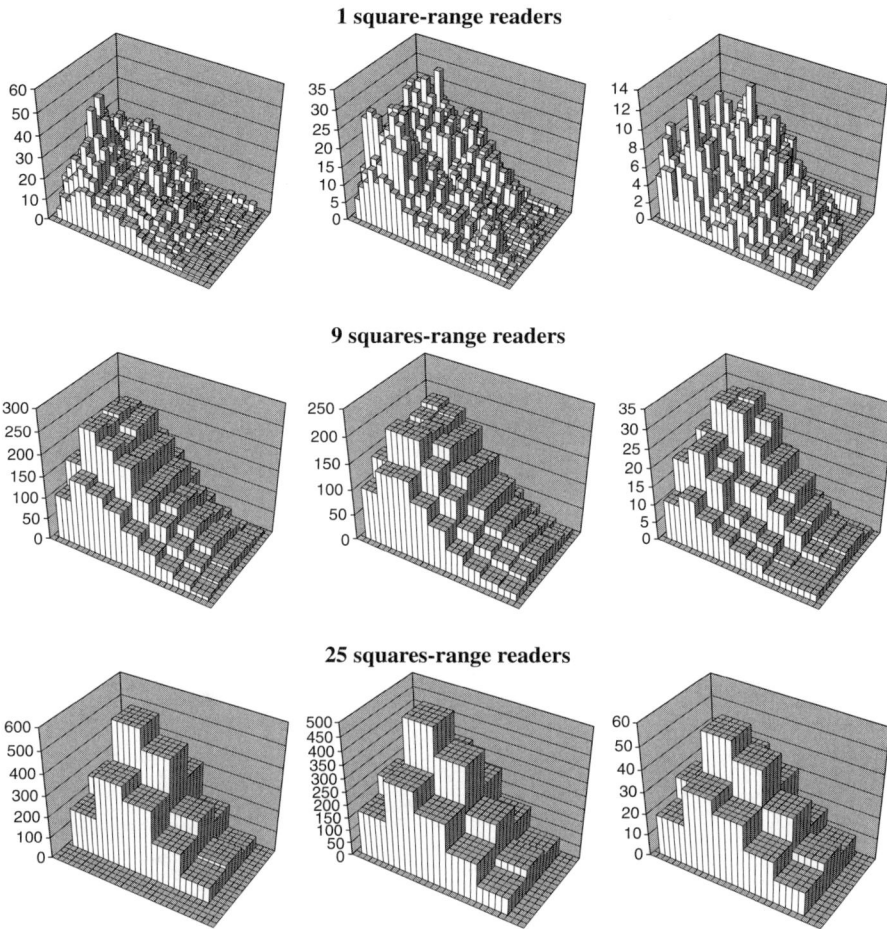

Fig. 13 Evolution over time (from *left* to *right*) of the caches of the readers in the first scenario covered by 1, 9 and 25 squares-range readers with a maximum of 1,000 tags in the system

seen that most of the tags are located close to the SAPs. This happens because the tags move randomly and there are no obstacles in the scenario. Moreover, once a tag reaches a SEP it leaves the system. Hence, the density of tags which are close to a SEP is very low. The average cache size of the readers grows with the number of covered squares. Specifically, when the cover range is 1 square, the average is 13.9, when the cover range is 9 squares, the average is 116.8 and, when the cover range is 25 squares, the average is 261.7. In all cases the cache size is much lower than the total number of tags (i.e. 1,000 tags). The same behaviour can be observed when the number of tags grows to 10,000 and 50,000 (*cf.* Figs. 14 and 15).

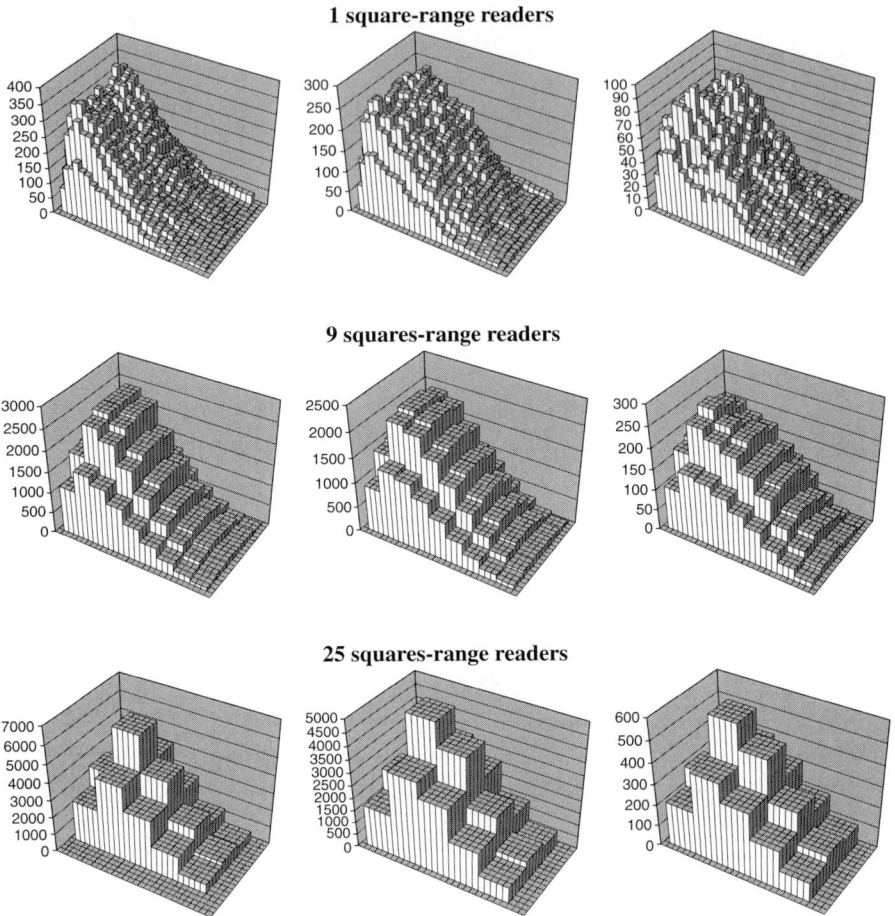

Fig. 14 Evolution over time (from **left** to **right**) of the caches of the readers in the first scenario covered by 1, 9 and 25 squares-range readers with a maximum of 10,000 tags in the system

5.2.2 Analysis of Scenarios with Corridors

In this scenarios (i.e. 2–4) we are specially interested in analysing the bottlenecks which appear at the entrance and at the exit points of the corridor, and how the cover range of the readers affects their cache size. To do so we have introduced 10,000 tags into the system and, as Fig. 16 shows, the behaviour of the caches follows a similar pattern than in the previous scenario. Most of the tags are located at the entrance of the system near to the SAPs. On the other hand, the density in the corridor is very uniform and the only *dangerous* point is the entrance to the corridor. If we analyse the average size of the caches we observe that in the second scenario with 1 square cover range readers the average is 323, if we use 9 squares cover range readers the average is 1,842.9 and 2,265 when 25 squares cover range readers are used. In the

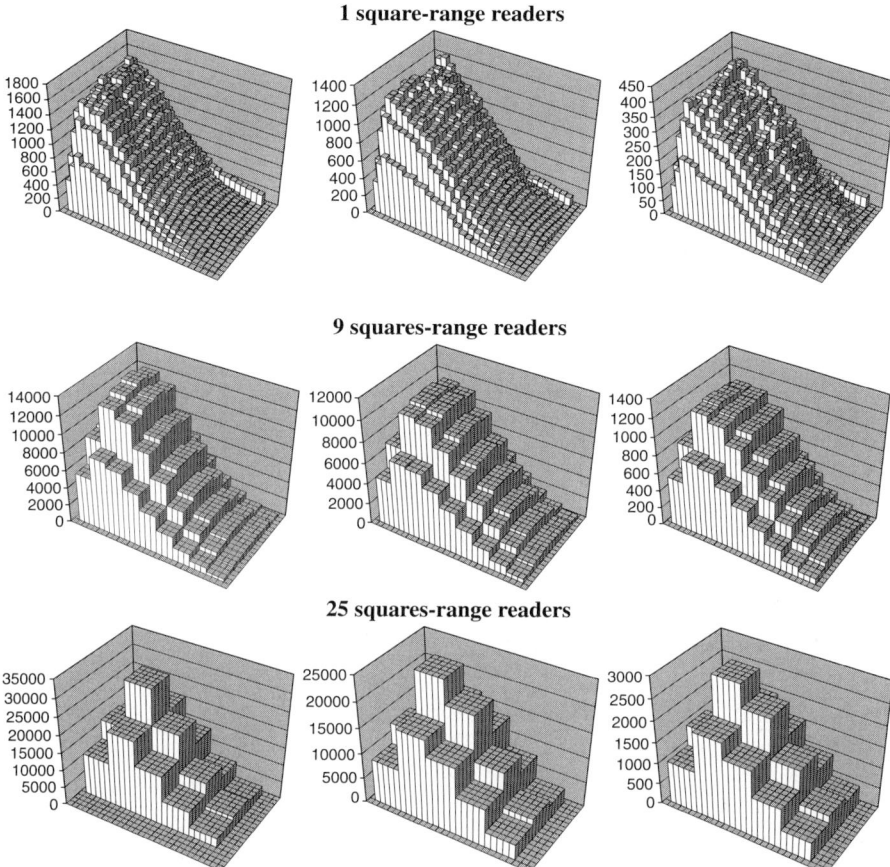

Fig. 15 Evolution over time (from *left* to *right*) of the caches of the readers in the first scenario covered by 1, 9 and 25 squares-range readers with a maximum of 50,000 tags in the system

third and fourth scenario, where a mixture of readers has been used, the average size of the caches is 2,282.7 and 1,064.2, respectively (Figs. 17 and 18).

The most relevant result that must be emphasised in these scenarios is that the use of 25 squares cover range readers does not increase substantially the average size of the caches, thus, it seems that the use of higher range readers in bottlenecks should be a good choice.

5.2.3 Complex Scenario Analysis

This scenario has been designed to show that the mixture of different cover range readers scales properly in a complex environment full of corridors, bottlenecks and big and small rooms. Figure 19 shows the cache sizes of the readers used in this

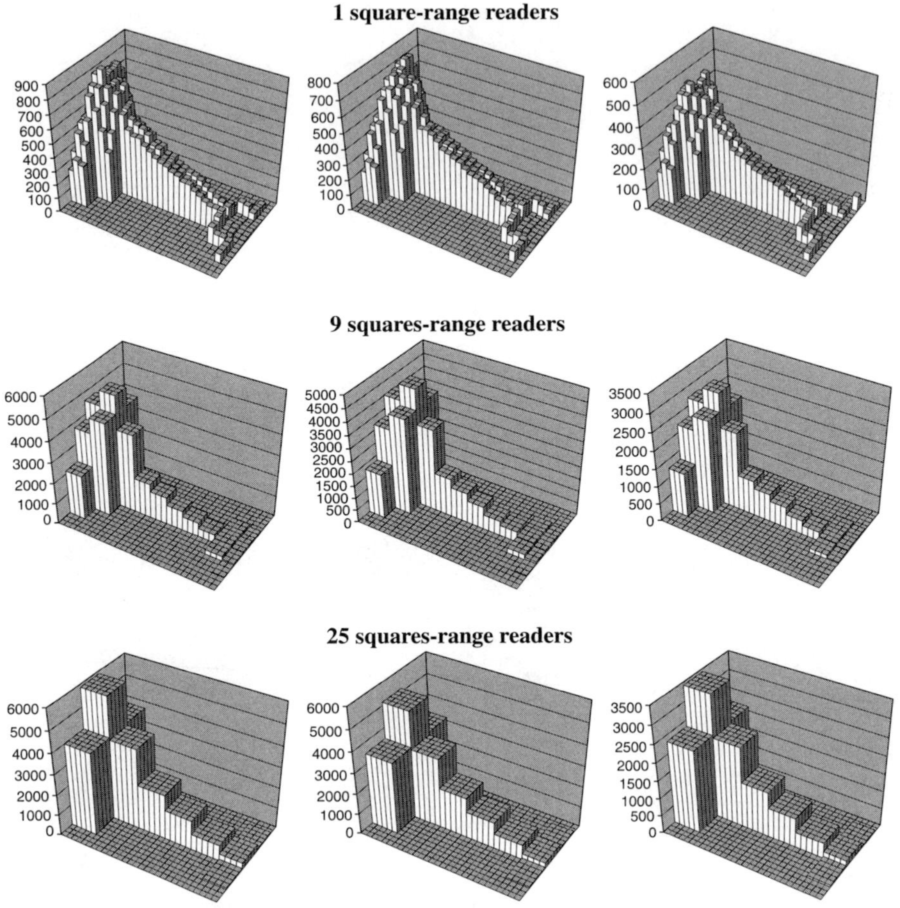

1 square-range readers

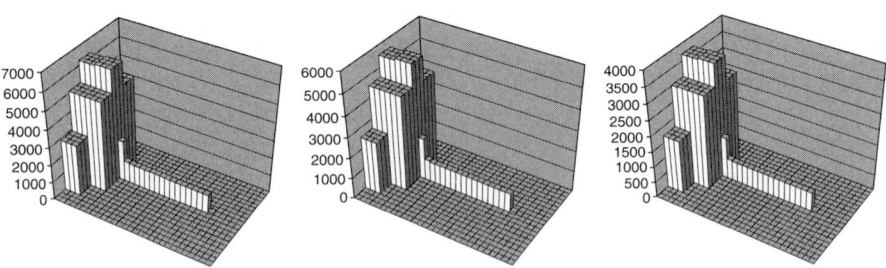

9 squares-range readers

25 squares-range readers

Fig. 16 Evolution over time (from *left* to *right*) of the caches of the readers in the second scenario covered by 1, 9 and 25 squares-range readers with a maximum of 10,000 tags in the system

Fig. 17 Evolution over time (from *left* to *right*) of the caches of the readers in the third scenario with a maximum of 10,000 tags in the system

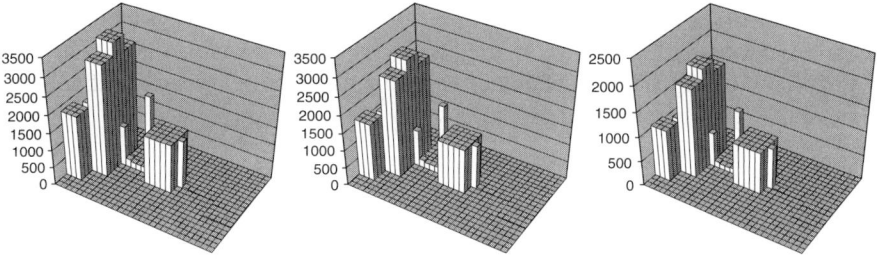

Fig. 18 Evolution over time (from *left* to *right*) of the caches of the readers in the fourth scenario with a maximum of 10,000 tags in the system

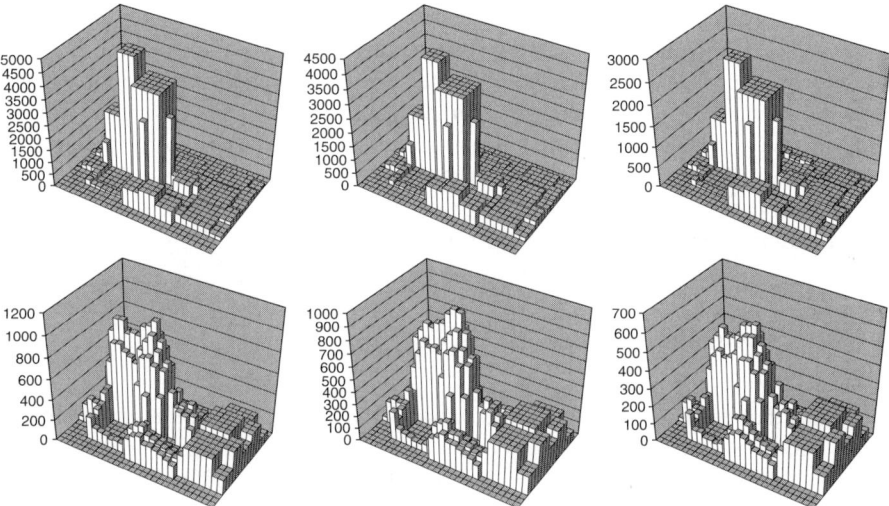

Fig. 19 Evolution over time (from *left* to *right*) of the caches of the readers in the fifth scenario with a maximum of 10,000 tags in the system. In the first row, most of the readers have cover ranges of 9 or 25 squares while in the second row most of the readers have a 1 square cover range

scenario. We have analysed two different mixtures of readers. First, we have studied the scenario covered by readers having a cover range of 9 and 25 squares (*cf.* first row of Fig. 19). Second, we have modified the readers configuration and we have mainly used 1 square cover range readers (*cf.* second row of Fig. 19). We have introduced 10,000 tags into the system to study the relation between the cover range of the readers, their cache size and the scalability.

We observe that in the first configuration the average cache size is 930.16 while in the second it is 331.80. In both cases the system scales properly (i.e. in the worst case the cache size is ten times smaller than the cache that a centralised system would need). It is clear that the use of readers having a higher cover range makes the average cache size to grow. However, we have shown that it is possible to use a mixture of reader configurations while maintaining the scalability of the system. Thus, it is possible to find the best configuration in terms of scalability (i.e. cache size) and cost (i.e. number of readers used). In our study, we have used 65 readers in the first configuration and 137 in the second one. Depending on the scenario and on the prices of the readers, it would be easy to find the most efficient configuration in terms of scalability and cost.

6 Conclusions

In this chapter, we have tackled the problem of providing RFID systems with scalability. We have shown that scalability is a major problem that must be addressed if we want the RFID systems massive deployment to be a reality.

We have briefly analysed the most relevant techniques for providing RFID systems with security and privacy, and we have recalled a proposal for making some of these techniques scalable. We have presented our RFID simulator in detail and we have shown several experimental results that proof the usefulness of our simulator.

From the experimental results we can conclude that the proposed protocol is scalable even when the cover range of the readers increases. Moreover, we have seen that it is useful to place high cover range readers in the bottlenecks because they do not increase the average cache size significantly.

7 Future Work

The work that we are planning for the future is mainly centred in the improvement of the simulator. We want to design a new module for controlling the movement of the tags (currently it is random). We plan to allow the user to define predetermined paths which the tags will be forced to follow.

We also plan to modify the kernel in order to let it work with physical constraints associated with the used readers. Moreover, we will modify the simulator to allow the existence of overlaps amongst the readers.

Some improvements on the communication protocol are also envisaged. We believe that the number of messages exchanged by the readers could be reduced by changing the adjacency graphs used.

Disclaimer and Acknowledgements

The authors are solely responsible for the views expressed in this paper, which neither necessarily reflect the position of UNESCO nor commit that organisation. This work was partly supported by the Spanish Ministry of Education through projects TSI2007-65406-C03-01 "E-AEGIS" and CONSOLIDER CSD2007-00004"ARES", and by the Government of Catalonia under grant 2005 SGR 00446.

The authors thank Dr. Antoni Martínez-Ballesté for his comments and suggestions.

References

1. G. Avoine and P. Oechslin. A scalable and provably secure hash based RFID protocol. In F. Stajano and R. Thomas, editors, *The Second IEEE International Workshop on Pervasive Computing and Communication Security – PerSec 2005*, pp. 110–114, IEEE, IEEE Computer Society Press, Washington, DC 2005

2. L. Batina, J. Guajardo, T. Kerins, N. Mentens, P. Tuyls, and I. Verbauwhede. An elliptic curve processor suitable for rfid-tags. *Cryptology ePrint Archive*, 2006/227, 2006. http://eprint.iacr.org/

3. L. Batina, J. Guajardo, T. Kerins, N. Mentens, P. Tuyls, and I. Verbauwhede. Public-key cryptography for RFID-tags. In *Printed handout of Workshop on RFID Security. RFIDSec 06*, pp. 61–76, 2006. http://homes.esat.kuleuven.be/lbatina/Batina_RFID06_2.pdf

4. S.J. Engberg, M.B. Harning, and C.D. Jensen. Zero-knowledge device authentication: Privacy and security enhanced RFID preserving business value and consumer convenience. In *Second Annual Conference on Privacy, Security, and Trust*, 2004

5. C. Floerkemeier, R. Schneider, and M. Langheinrich. Scanning with a purpose – supporting the fair information principles in RFID protocols. In *Proceedings of Second International Symposium on Ubiquitous Computing Systems UCS*, 2004

6. N. Good, J. Han, E. Miles, D. Molnar, D. Mulligan, L. Quilter, J. Urban, and D. Wagner. Radio frequency identifcation and privacy with informarion goods. In *Workshop on Privacy in the Electronic Society (WPES)*, 2004

7. C.C. Haley. Are you ready for RFID? *Internetnews.com*, November 2003. http://www.internetnews.com/wireless/article.php/3109501

8. M. Hellman. A cryptanalitic time-memory tradeoff. *IEEE Transactions on Information Theory*, IT-26: 401–406, 1980

9. D. Henrici and P. Müller. Hash-based enhancement of location privacy for radio-frequency identification devices using varying identifiers. In Ravi Sandhu and Roshan Thomas, editors, *IEEE International Workshop on Pervasive Computing and Communication Security? PerSec 2004*, pp. 149–153, Orlando, Florida, USA, IEEE, IEEE Computer Society Press, Washington, DC, March 2004

10. S. Inoue and H. Yasuura. RFID privacy using user-controllable uniqueness. In *RFID Privacy Workshop,* MIT, Cambridge, MA, November 2003

11. A. Juels. Minimalist cryptography for low-cost RFID tags. In *In Fourth International Conference on Security in Communication Networks-SCN 2004*, volume 3352, pp. 149–164, Springer, Berlin, 2004. Lectures Notes in Computer Science

12. A. Juels. RFID Security and Privacy: A Research Survey. Manuscript, September 2005

13. A. Juels and R. Pappu. Squealing euros: Privacy protection in RFID-enabled banknotes. In R. Wright, editor, *Financial Cryptography'03*, volume 2742, pp. 103–121, Springer, Berlin, 2003. Lectures Notes in Computer Science

14. A. Juels and S.A. Weis. Defining strong privacy. *IACR eprint*, April 2006

15. A. Juels, R.L. Rivest, and M. Syzdlo. The blocker tag: Selective blocking of RFID tags for consumer privacy. In V. Atluri, editor, *Eighth ACM Conference on Computer and Communications Security*, pp. 103–111, 2003

16. M. Ohkubo, K. Suzuki, and S. Kinoshita. Efficient hash-chain based RFID privacy protection scheme. In *International Conference on Ubiquitous Computing – Ubicomp, Workshop Privacy: Current Status and Future Directions*, 2004

17. T. Okamoto. Probably secure and practical identification schemes and corresponding signature schemes. In E.F. Brickell, editor, *Advances in Cryptology – CRYPTO'92*, volume 740 of *LNCS*, pp. 31–53, Springer, Berlin, 1992

18. M. Rieback, B. Crispo, and A.S. Tanenbaum. RFID guardian: A battery-powered mobile device for RFID privacy management. In Colin Boyd and Juan Manuel Gonzlez Nieto, editors, *Australasian Conference on Information Security and Privacy – ACISP 2005*, volume 3574, pp. 184–194, Springer, Berlin, 2005. Lecture Notes in Computer Science

19. S.E. Sarma, S.A. Weiss, and D.W. Engels. RFID systems, security and privacy implications. Technical Report, MIT-AUTOID-WH-014, AutoID Center, MIT, 2002
20. A. Solanas, J. Domingo-Ferrer, A. Martínez-Ballesté, and V. Daza. A distributed architecture for scalable private RFID tag identification. *Computer Networks*, 51(9): 2268–2279, June 2007. http://dx.doi.org/10.1016/j.comnet.2007.01.012
21. Sybase. http://www.ianywhere.com/products/rfid_anywhere.html
22. G. Tsudik. YA-TRAP: Yet another trivial RFID authentication protocol. In *Fourth Annual IEEE International Conference on Pervasive Computing and Communications Workshops (PERCOMW'06)*, pp. 640–643, 2006
23. P. Tuyls and L. Batina. Rfid-tags for anti-counterfeiting. In D. Pointcheval, editor, *Topics in Cryptology-CT-RSA 2006*. Springer, Berlin, 2006. Lecture Notes in Computer Science
24. S. Weis, S. Sarma, R. Rivest, and D. Engels. Security and privacy aspects of low-cost radio frequency identification systems. In W. Stephan D. Hutter, G. Müller and M. Ullmann, editors, *International Conference on Security in Pervasive Computing – SPC 2003*, volume 2802, pp. 454–469. Springer, Berlin, 2003

Part III
Encryption and Hardware Implementations

Public-Key Cryptography for RFID Tags and Applications

Lejla Batina*, Jorge Guajardo, Bart Preneel, Pim Tuyls, and Ingrid Verbauwhede

Abstract RFID tags are small wireless devices expected to be pervasive in the future. In addition to their rigorous constraints featuring an extremely low-power budget and small die size, they also give rise to serious security and privacy issues. Typical security services include authentication, key management, and encryption. Although some experts have given up on the feasibility of public-key solutions for RFID, assuming it is too expensive and too power hungry, there exists a firm line of research exploring the limits of compact public-key implementations for RFID devices.

An emerging application is the use of RFID tags for anticounterfeiting by embedding them into a product. Public-Key Cryptography (PKC) offers an attractive solution to the counterfeiting problem and thus, exploring possible implementation options for this application is attractive.

In this chapter, we discuss PKC-based solutions well suited for RFIDs. Our focus is on cryptographic solutions based on elliptic curves. We describe low-cost Elliptic Curve Cryptography (ECC) processors supporting security algorithms and protocols for RFID. We also investigate which PKC-based identification protocols are useful for these anticounterfeiting applications. We argue that identification of RFID tags can reach high security levels. In particular, we elaborate how secure identification protocols based on the DL problem on elliptic curves can be implemented on an RFID tag in a bit more than 10,000 gates. We describe various cases of elliptic curves over $\mathbb{F}_{2^p}$ with p prime and over composite fields $\mathbb{F}_{2^{2 \cdot p}}$. Some of the implementations described in this chapter make RFID tags suitable for anticounterfeiting purposes even in the off-line setting. Finally, we compare different implementation options and explore the cost that side-channel attack countermeasures would have on such implementations.

L. Batina
Katholieke Universiteit Leuven, ESAT/COSIC, Belgium
e-mail: Lejla.Batina@esat.kulewen.be

P. Kitsos, Y. Zhang (eds.), *RFID Security: Techniques, Protocols and System-on-Chip Design,* © Springer Science+Business Media, LLC 2008

1 Introduction

RFID technology has a broad spectrum of applications. First, RFID tags were meant to be used mainly for inventory, identification, supply chain management, access control, payment systems, and similar applications but nowadays they are envisioned in many other areas as well such as medical care, vehicle tracking, counterfeit detection, etc. In short, RFID tags are meant to be a ubiquitous replacement for bar codes with some added functionality.

An emerging application is the use of RFID tags for anticounterfeiting by embedding them into a product. However, there is a risk related to naively using those tags for several applications. In particular, if no appropriate cryptographic measures are taken, the privacy of a user carrying tagged items can be severely compromised. In order to enable these applications and at the same time minimize the risks, PKC offers attractive solutions.

The suitability of public-key algorithms for RFID is an open research problem. There exists a common belief that PKC is too slow and too expensive for low-cost RFID applications. Thus, most protocol proposals deal only with symmetric-key cryptography [1]. In particular, the difficulty for PKC lies in extreme resource limitations such as area resources that should occupy less than $1\,mm^2$, low-power ($<500\,\mu W$ or $I < 10\,\mu A$ at $1.5V$ [90]), very little memory, short bandwidth, and low energy. From the implementation point of view the preferred platforms should rely on custom hardware as pure hardware solutions are energy and cost effective. In addition, hardware implementations usually offer more side-channel attack resistance, which is an important consideration as shown in [42, 74, 75, 78].

The fact that tags have very constrained resources (memory, power, speed, area) but need security measures, poses very interesting challenges to the security community. First, it is natural to investigate whether existing cryptographic algorithms can be implemented on a tag. Second, it encourages research for new protocols and algorithms targeted at resource-constrained devices. Moreover, the research community lacks consensus as to the feasibility of implementing public-key cryptoalgorithms on (high-end) RFID tags. For example, Tuyls and Batina [87] claim that PKC on a tag is possible and Avoine et al. [2] states: "Unfortunately asymmetric cryptography is too heavy to be implemented on a tag."

An interesting example is RFID-based identification which requires authentication as a cryptographic service. This property can be achieved by symmetric as well as asymmetric primitives. Most of the previous work dealt with implementations of symmetric ciphers. The most notable example is the work of Feldhofer et al. [24], which considered the implementation of AES on an RFID tag. On the other hand, several authors discussed the suitability of PK algorithms for RFID [62, 87, 91]. Recently, the work of Wolkerstorfer [90] showed that ECC-based PKC might be feasible on RFID tags by implementing the Elliptic Curve Digital Signature Algorithm (ECDSA) on a small IC. Afterward, many authors presented PK solutions for RFID applications that mostly rely on ECC.

In this chapter, we describe possible solutions for PKC on RFID tags from the hardware point of view. We also discuss PKC-based identification protocols in this context. In particular, we elaborate on the feasibility of identification protocols based on ECC.

The remainder of this chapter is organized as follows. Section 2 provides an overview of related work. In Sect. 3, we list some RFID-relevant ECC-based protocols and we provide background information on ECC, arithmetic required to implement curve operations, and what is known about the security of the discrete logarithm (DL) problem in elliptic curves. In Sect. 5, we discuss design criteria for low-footprint and low-power implementations on all the levels of ECC-based protocols. We also elaborate on a suitable selection of parameters and algorithms. In Sect. 6, we outline corresponding architectures. Possible improvements and future directions are listed in Sect. 7. The application of PKC for RFID to the counterfeiting problem is discussed in Sect. 8. We end with conclusions and directions for future work in Sect. 9.

2 Related Work

In recent years, low-power and compact implementations became an important research area with the constant increase in the number of hand-held devices such as mobile phones, smart cards, PDAs, etc. Schroeppel et al. [83] presented a design for ECC over binary fields that was optimized for power, space, and time in order to provide digital signatures. The processor in [83] had an area complexity of 191 kgates. The work of Goodman and Chandrakasan [33] also dealt with energy-efficient solutions. They proposed a domain-specific reconfigurable cryptographic processor (DSRCP) for ECC over both types of finite fields. At 50 MHz, the processor operates at a supply voltage of 2 V and consumes at most 75 mW of power. In ultra-low-power mode (3 MHz at VDD = 0.7 V), the processor consumes at most 525 µW. Özturk et al. [76] introduced modulus scaling techniques that are applicable for ECC over a prime field to develop a low-power elliptic curve processor (ECP) architecture. They obtained an ECC processor using a 166-bit long prime of size 30 kgates with a performance of 31.9 ms for point multiplication.

The work of Gaubatz et al. [29] discusses the necessity and the feasibility of PKC protocols in sensor networks. For those applications the situation is similar as for RFID. Namely, there are many cryptographic services that could benefit from PKC such as key-exchange, broadcast authentication, encryption etc., but constraints are very hard to meet. In [29], the authors investigated implementations of two algorithms for this purpose, i.e., Rabin's scheme and NTRUEncrypt. The conclusion is that NTRUEncrypt features a suitable low-power and small footprint solution with a total complexity of 3 kgates and power consumption of less than 20 µW at 500 kHz. On the other hand, they showed that Rabin is not a feasible solution. In [30], the authors compared the previous two algorithm implementations with an ECC solution for wireless sensor networks. The architecture of the ECC processor occupied an area of 18,720 gates and consumes less than $400\,\mu W$ of power at 500 kHz. The field used was a prime field of order $\approx 2^{100}$.

Some more efforts for PKC processors for RFID tags include the results of Wolkerstorfer [91] and Kumar and Paar [55]. Wolkerstorfer [91] investigated RFID authentication by means of the ECDSA protocol [46]. The chip has an area complexity of about 23 kgates and it features a latency of 6.67 ms for one point multiplication at 68.5 MHz. However, it can be used for both types of fields, e.g., $\mathbb{F}_{2^{191}}$ and $\mathbb{F}_{p_{192}}$. The results of Kumar and Paar [55] include an area complexity of almost 12 kgates and a latency of 18 ms for one point multiplication over $\mathbb{F}_{2^{131}}$ at 13.56 MHz. The operating frequency might be too high for RFID applications and without power consumption values the results cannot be properly evaluated. Note that with such a high frequency the power consumption is increased. Furthermore, power and area have the most impact on the feasibility of RFID implementations.

The work of Sakiyama et al. [81] presents a low-cost Modular Arithmetic Logic Unit (MALU) for Elliptic/Hyperelliptic Curve Cryptography (ECC/HECC) suitable for these applications. The best solution for the MALU among various trade-offs supporting ECC field arithmetic features 2,171 gates with an average power consumption of less than $40\,\mu W$. The result is obtained by hardware resource sharing of the datapath and the usage of composite fields for ECC. The follow-up work presented in [5] describes a low-power and low-footprint processor using the MALU for ECC suitable for sensor networks. The best solution features 6,718 gates for the arithmetic unit and control unit (data memory not included) in 0.13 μm CMOS technology over the field $\mathbb{F}_{2^{131}}$, which provides a reasonable level of security for the time being. In this case, the consumed power is less than $30\,\mu W$ when the operating frequency is 500 kHz. In [35] Guajardo et al. present an improvement to the architecture of the MALU. This work presents a constant factor decrease in the area of the multiplier combined with decrease in the critical path of the multiplier from linear in the number of gates to logarithmic.

Recently, McLoone and Robshaw [62] reconsider the hardware cost of PKC for ultra-constrained area and power applications. Their implementation allows for a tag that can participate in an authentication protocol a *limited* number of times. As a result, the tag can store in memory the responses to a limited number of authentication and thus, the hardware costs are minimal.

An ECC solution to provide identification for an RFID tag by means of Schnorr's identification scheme was discussed in [87] and the complete implementation is described in [6]. The authors show how secure identification protocols based on the DL problem on elliptic curves can be implemented on a constrained device such as an RFID tag requiring between 8,500 and 14,000 gates, depending on the implementation characteristics. They investigate the case of elliptic curves over $\mathbb{F}_{2^p}$ with p prime and over composite fields $\mathbb{F}_{2^{2 \cdot p}}$. The implementations in the paper make RFID tags suitable for anticounterfeiting purposes even in the off-line setting.

If only resistance against passive attacks is needed, the Schnorr Identification scheme can be used as it is known to be secure against passive attacks under the discrete logarithm assumption. An alternative for providing more security is to use Okamoto's identification scheme [73], which is secure against passive, active and concurrent attacks under the DL assumption. In contrast to the work of McLoone and Robshaw, Tuyls and Batina [87] consider tags that are able to participate in an authentication protocol an *unlimited* number of times.

The work in [7] describes the smallest known hardware implementation for ECC/HECC. The two solutions for PKC are based on arithmetic on elliptic/hyperelliptic curves. One solution relies on ECC over binary fields $\mathbb{F}_{2^n}$ where n is a composite number of the form $2p$ (p is a prime) and the other on HECC on curves of genus 2 over $\mathbb{F}_{2^p}$. This implies the same arithmetic unit for both cases which supports arithmetic in a field $\mathbb{F}_{2^p}$. This is the most compact solution which results in acceptable performance, featuring an area of less than 5 kgates (without data registers, so just ALU and control included) with an average power consumption of less than $10\,\mu W$. However, without knowing the number of gates that are required for registers, it is hard to properly evaluate this solution. We compare known implementations with respect to the performance, gate complexity, and power consumption in more detail in Sect. 6.1.

Among various RFID applications in this chapter we discuss anticounterfeiting. The use of RFID tags for anticounterfeiting purposes was proposed by Juels [48]. By associating an RFID tag (which contains specific product and reference information) with the product, one aims to verify the authenticity of the product. Loosely speaking the verification is performed as follows. When a product passes a reader, the reader checks whether the necessary and authentic product and reference information is present on the tag. For this purpose it runs a protocol with the tag. If the necessary information is there and verified to be authentic, the product is declared to be genuine and otherwise not. However, by capturing the necessary authentication information (obtained, e.g., by eavesdropping the protocol between the tag and the reader), and by storing it in a new chip, the attacker has effectively made a clone of the original tag that cannot be distinguished from the original tag by a reader. In order to make the cloning of a tag unfeasible, it should not be possible to derive the tag secrets by active or passive attacks. In Sect. 8, we discuss these issues in detail.

3 Elliptic Curve Cryptography

Elliptic curve cryptosystems, introduced independently by Miller [68] and Koblitz [51], are widely used and recommended for encryption and digital signatures, among other applications. Both, curves defined over binary fields $\mathbb{F}_{2^m}$ and fields $\mathbb{F}_q$ with $q = p^m$ where p is an odd prime number, are recommended for use. In standards such as the IEEE P1363 [43] fields for $q = p$ and $q = 2^n$ where $p \approx 2^n$ and $n \geq 160$ are recommended. Although, it appears that elliptic curve systems over both prime ($\mathbb{F}_p$) and binary ($\mathbb{F}_{2^n}$) fields provide the same level of security, fields $\mathbb{F}_{2^n}$ are better suited for hardware implementations. In particular, arithmetic in $\mathbb{F}_{2^n}$ can be implemented more efficiently in dedicated or reconfigurable hardware than arithmetic over $\mathbb{F}_p$, because of the natural representation of binary field elements as vectors of zeros and ones. Note that the opposite is true for microprocessors and microcontrollers that come already equipped with optimized integer multipliers. However, we will concentrate on the hardware point of view as the application at hand (e.g., RFID tags) requires specialized hardware and, in general, does not admit a large integer multiplier due to cost considerations.

This section is meant as an introduction to elliptic curves in general. In particular, we discuss the group law on the group of points of an elliptic curve defined over a binary field, we define the Discrete Logarithm Problem on an Elliptic Curve (ECDLP), we describe the different choices of finite fields and the element representation, and we discuss algorithms well suited for point multiplication in constrained environments. For extensive treatments of ECC we refer the reader to [12, 22, 38].

3.1 The Group Law and the ECDLP

In the remainder of this chapter, we will only consider nonsupersigular elliptic curves defined over binary fields $\mathbb{F}_{2^m}$. Note that virtually all elliptic curves used in practice are nonsupersingular curves due to the MOV attack [65] (see also [26]) on supersingular elliptic curves. Supersingular elliptic curves have recently found applications in identity-based cryptography (see for example [14]) but this is not the subject of this chapter.

A nonsupersingular elliptic curve E defined over $\mathbb{F}_{2^n}$ is defined as the set of solutions $(x, y) \in \mathbb{F}_{2^n} \times \mathbb{F}_{2^n}$ to the equation:

$$y^2 + xy = x^3 + ax^2 + b, \tag{1}$$

where $a, b \in \mathbb{F}_{2^n}, b \neq 0$, together with the point at infinity $\mathcal{O}$.

ECC relies on the group structure induced on an elliptic curve. It turns out that the set of points of an elliptic curve together with a binary operation, named point addition, has the structure of an Abelian group. In particular, the point addition (resp. point doubling) operation in affine coordinates is defined as follows. Let $P_1 = (x_1, y_1)$ and $P_2 = (x_2, y_2)$ be two points on an elliptic curve E. Assume $P_1, P_2 \neq \mathcal{O}$ and $P_1 \neq -P_2$. The sum $P_3 = (x_3, y_3) = P_1 + P_2$ is computed as follows ([12], p. 57): If $P_1 \neq P_2$,

$$\begin{aligned} \lambda &= (y_2 + y_1) \cdot (x_2 + x_1)^{-1} \\ x_3 &= \lambda^2 + \lambda + x_1 + x_2 + a \\ y_3 &= \lambda(x_1 + x_3) + x_3 + y_1. \end{aligned} \tag{2}$$

If $P_1 = P_2$,

$$\begin{aligned} \lambda &= y_1/x_1 + x_1 \\ x_3 &= \lambda^2 + \lambda + a \\ y_3 &= (x_1 + x_3)\lambda + x_3 + y_1. \end{aligned} \tag{3}$$

Finally, we note that the special point $\mathcal{O}$ plays the role of the identity element in the group of points of E. Thus, for any $P \in E$, it is true that $P - P = \mathcal{O}$ and $P + \mathcal{O} = P$.

Note that (2) and (3) imply that a point doubling or addition operation requires one field inversion (I), two field multiplications (M), and one squaring (S).

The Elliptic Curve Discrete Logarithm Problem (ECDLP) is defined as follows. Let $E/\mathbb{F}_q$ be an elliptic curve defined over $\mathbb{F}_q$ and let $P \in E(\mathbb{F}_q)$ be a point of order m, where the order of P is defined as the smallest integer m such that $mP = \mathcal{O}$. Note that the operation mP means add P to itself $m - 1$ times using (2) and (3). Let $Q \in E/\mathbb{F}_q$, so that $Q = aP$ for some integer a, $2 \leq a < m$. The ECDLP is then given P and Q find a such that $Q = aP$. This problem is believed to be hard, i.e., it is still unknown if there exists an algorithm to solve it in less than fully exponential time. We will discuss in more detail what is known about the security of the ECDLP for different parameters in Sect. 4.

3.2 Projective Coordinates and Point Multiplication

Equations (2) and (3) define the point addition and doubling operations in affine coordinates. In the search to decrease the complexity of the point operation, researchers have come up with several other point coordinate representations [20, 59, 87], all of which belong to the projective coordinate type. The common idea in projective coordinates is to avoid the computation of a finite field inverse in every addition or doubling operation. This is achieved at the cost of additional finite field multiplications, squarings, and additions. Thus, whether is better to use affine or projective coordinates will heavily depend on the complexity of the inverse operation. In the particular case of RFID applications, where area must be minimized (this implies for example that a dedicated inverter is not an option), projective coordinates are almost always the coordinate of choice. Table 4 provides a comparison of different projective coordinates and the number of operations that each require. In general, the López-Dahab [59] coordinates (or a variation [87]) are the best suited to small implementations such as those typically required in RFID applications.

While the basic point addition and doubling operations are key to the performance of an elliptic curve implementation, the way in which one computes aP, for an integer a and a point $P \in E$ is just as important and can help in the overall area reduction of an ECP. In addition, the particular implementation of the point multiplication operation can help in improving its resistance to side-channel attacks such as SPA, timing analysis, and DPA [53, 54, 78]. Although, from a conceptual point of view the simple double-and-add algorithm [66, p. 614] could be thought to be the best choice, it turns out that a variation, which only makes use of the x-coordinate of the point P during point multiplication is better to reduce the area requirements. The algorithm, known as the Montgomery Ladder, was introduced originally by Montgomery [69] and it has been further optimized in recent years [47]. Algorithm 1 maintains the relationship $P_2 - P_1$ as invariant. It uses a representation where computations are performed on the x-coordinate only in affine coordinates (or on the X- and Z-coordinates in projective representation). That fact allows one to save registers which is one of the main criteria for obtaining a compact solution.

Algorithm 1 Montgomery ladder algorithm for point multiplication incorporating López–Dahab projective coordinates [59]

Require: an integer $k = (k_{l-1}, ..., k_1, k_0)_2$ and a point $P(x, y)$
Ensure: $Q = kP$

  $X_1 \leftarrow x, \ Z_1 \leftarrow 1, X_2 \leftarrow x^4 + b, Z_2 \leftarrow x^2.$
  for i from $l - 2$ downto 0 **do**
    if $k_i = 1$ **then**
      $x(P_1) \leftarrow x(P_1 + P_2), x(P_2) \leftarrow x(2P_2)$
    else
      $x(P_2) \leftarrow x(P_2 + P_1), x(P_1) \leftarrow x(2P_1)$
    end if
  end for
  Return $x(P_1)$

3.3 Finite Field Element Representation in $\mathbb{F}_{2^n}$

The choice of a "suitable" parameter n in $\mathbb{F}_{2^n}$ for ECC can have a high impact on the performance of the overall implementation. In principle, one can choose between n prime or composite. Choosing for composite n and, thus, using so-called composite fields has advantages for implementations. For example, a part of the arithmetic in the subfield can be precalculated and stored in a memory to boost performance. However, with respect to cryptographic security it is typically recommended to use fields $\mathbb{F}_{2^p}$ where p is a prime or fields of composite degree $n = 2p$, where p is prime. A detailed discussion of the security of elliptic curves over of such "nonstandard" fields is deferred to Sect. 4.

Most literature considers the case of elliptic curves over $\mathbb{F}_{2^p}$ with p prime (as recommended by standards [43]). However, recent work also considers composite fields $\mathbb{F}_{2^{2p}}$ for RFID applications [4, 6]. In this case, $\mathbb{F}_{2^{2p}}$ is a field of quadratic extension over $\mathbb{F}_{2^p}$, so we can write $\mathbb{F}_{2^{2p}} = \mathbb{F}_{2^p}[x]/(f(x))$, where f is an irreducible polynomial of degree two. In this case, each element from the field $\mathbb{F}_{2^{2p}}$ can be represented as $c = c_1 t + c_0$ where $c_0, c_1 \in \mathbb{F}_{2^p}$. Then, a multiplication in $\mathbb{F}_{2^{2p}}$ requires three multiplications and four additions in $\mathbb{F}_{2^p}$ using a method due to Karatsuba [50]. The clear advantage in using composite fields for ECC or using HECC [52] is the fact that the field arithmetic can be performed in a smaller field. This implies reduction of the arithmetic unit by a factor of two.

In terms of finite field element representation, we use a polynomial basis representation as they are believed to facilitate flexible and generic implementations of ECC. Also, there is no significant improvement with respect to area compactness when using normal bases. In a polynomial basis, the basis elements have the form $1, \omega, \omega^2, \ldots, \omega^{n-1}$ where ω is a root of an irreducible polynomial $f(x)$ of degree n over $\mathbb{F}_2$. In this basis representation, the elements of $\mathbb{F}_{2^n}$ are polynomials of degree at most $n - 1$ over $\mathbb{F}_2$, and arithmetic is carried out modulo the irreducible polynomial $f(x)$ of degree n over $\mathbb{F}_2$. According to this representation an element of $\mathbb{F}_{2^n}$ is a polynomial of length n and can be written as: $A(x) = \sum_{i=0}^{n-1} a_i x^i = a_{n-1}x^{n-1} + a_{n-2}x^{n-2} + \cdots + a_1 x + a_0$ where $a_i \in \mathbb{F}_2$.

To summarize, implementing an elliptic curve based system requires the designer to choose among, finite fields, finite field representations, elliptic curve coordinate representation, and the point multiplication algorithm. All such algorithms affect decisively the performance of the overall system and thus, it is important to make informed choices. In particular, for RFID and other low-cost applications, one needs ECC algorithms that minimize the memory requirements and that require the fewest field operations. For example, squaring can be considered as a special case of multiplication and subtraction is actually an addition in the case of binary fields. Hence, the main operations to deal with are additions and multiplications. We will discuss our final choice of algorithms in Sect. 5. We end this chapter with a discussion about the security of curves over different fields and a survey of known attacks and their complexities.

4 Security of ECC over Nonstandard Finite Fields

We talk about nonstandard fields because recent work has focused on reducing the complexity of EC hardware implementations by reducing the field sizes and considering curves defined over fields of composite degree. For example, the implementations in [4, 6] target curves over 131-bit and 139-bit binary fields (both 131 and 139 are prime numbers) both of which are smaller bit sizes than the recommended 160-bit standard security [43]. In addition, the authors also considered composite fields of the form $\mathbb{F}_{2^{2p}}$ where p is prime and the resulting fields are of order $\approx 2^{134}$ and 2^{142}, respectively. We refer to these fields as "nonstandard" as their bit-sizes are smaller than those *commonly* considered to be secure (i.e., 160-bits and up) and they include composite degrees, some of which are considered unsuitable for crypto applications due to the Weil Descent attack [25,27,32]. Note that the choice of fields in [4, 6] is naturally related to the fact that smaller operand bit-lengths increase the computational efficiency of cryptographic operations. This practice is not favored in the crypto community because of the *reduced security* offered by the resulting system. Note that we do not argue that the DL problem on an EC over $\mathbb{F}_{2^{131}}$ is easier to solve than over a $\mathbb{F}_{2^{163}}$ field. However, in this section, we analyze the security of EC over nonstandard fields based on current state of the art attacks and conclude that *such fields offer acceptable security* for many RFID applications including anticounterfeiting. We based our definition of *acceptable* on the dollardays cost measure used in [57] and quantify it explicitly. We emphasize a point that has been often made in the past couple of years as we have seen a migration toward larger operand sizes: security is a risk assessment exercise. Thus, we conclude based on our cost-based analysis that the risk of using EC-131 is acceptable for many RFID anticounterfeiting applications.

4.1 Case Study: Security of a 131-Bit EC Implementation

For random elliptic curves $E(\mathbb{F}_q)$, the best known attack is Pollard's Rho algorithm
[77], with complexity $(\sqrt{\pi 2^{n-1}})/2$, where $q \approx 2^n$ and there is always a cofactor at
least 2 in the group order of $E(\mathbb{F}_q)$. Assuming this to be the best attack, Lenstra
and Verheul [58] estimated that an EC defined over a 132-bit prime field provided
security equivalent to 952-bit RSA system (and a 70-bit block cipher) in the year
2000. In this context equivalent means that an attack against either RSA or EC
requires approximately the same computational effort (measured in MIPS-years) to
be successful. The year 2000 refers to the fact that in the model of [58], both EC
over 132-bit field and 952-bit RSA provided the same security level in the year
2000 as DES did in 1982. We will write "DES(yyyy)" to mean "the security of DES
in in the year yyyy." Thus, the notion of "equivalent key size" in [58] is based on
computational effort[1] (measured in MIPS-years) required for a successful attack.
On the other hand, the work in [57] takes a cost-based approach to equivalence
(measured in *dollardays*). The result is that according to [57], 131-bit EC should
be considered *cost* equivalent to a 66-bit block cipher and 694-bit RSA offering
security equivalent to DES(1982) in the year 1996. In what follows, we will base
our discussion on the cost-based approach of [57].

It might be tempting at this point to conclude that based on the previous paragraph a 131-bit EC does not provide adequate security for RFID applications. However, a closer look will indicate otherwise. First, note that the notion of security in
both [57, 58] is based on the assumption that DES provided adequate commercial
security in the year 1982 and, in particular, that a DES key could be recovered in
one day with an investment of about US$40 million (40M dollardays). Note that the
US$40 million figure refers to the initial investment and not to the cost per key. For
example, with an initial investment of US$300,000 [13] estimates that at a rate of
1 key per 3 h and with a life expectancy of 3 years, each DES key could be recovered at a cost of US$38 per key. Nevertheless, whoever intends to undertake such an
attack must make an initial investment of US$300,000.

Clearly, not every application will be worth investing US$40 million dollar-
days to recover a key. In fact, only very large organizations or intelligence agencies would be able to invest such large sums of money to recover a key according
to [3, 13]. To put the security of 131-bit EC into perspective, assume that your application requires security equivalent to DES(1993) or equivalently that using the
model of [57] you would like security against an organization willing to invest
about $\frac{100 \times 10^6}{2^{2(1993-1980)/3}} \approx US\$300,000$ (a medium organization or illegitimate business
according to [13]) Then, EC-131 would provide security equivalent to 858-bit RSA
and equivalent to DES(1993) until 2007. In the previous discussion, we have not
taken into account the fact that if certain hardware-based attacks are taken into account [88], it is recommended [3] to add 8–10 more bits to an EC defined over

---

[1] Note that Lenstra and Verneul [58] also discuss an alternative cost-based approach on which
"equivalence" can be based. However, the authors in [58] advocate the computational approach as
allowing more rigorous analysis and as being less subjective.

binary fields to obtain security equivalent to an EC defined over a prime field of similar size. However, as we will show in Sect. 4.2, recent estimates show that EC over 131-bit binary fields provide security against medium and large organizations in the near to medium term, even assuming dedicated hardware attacks.

4.2 What is Known Today?

Table 1 summarizes the results of the most recent EC challenge solutions [19]. In 2004, the ECC2-109 challenge was solved [19]. It was estimated that it required about 1.6×10^{16} iterations and if running on a single dedicated Athlon XP 3200+PC, it would have taken 1,200 years. Based on this attack we estimate the cost of breaking an EC over $\mathbb{F}_{2^{131}}$. First, note that the curve we used over $\mathbb{F}_{2^{131}}$ has order $\approx 2^{130}$. Hence, breaking the DL problem in $E(\mathbb{F}_{2^{131}})$ is by a factor of $\sqrt{2^{21}} \approx 1,400$ times harder to break than the ECC2-109 challenge. Using the *dollardays* cost measure, the previous figures imply that the ECC2-109 challenge cost was $1,200\,\text{years} \times 365\,\text{days} \times 100\,\text{dollars} \approx 43$ million dollardays[2] in 2004 and the 131-bit EC challenge would be in the billion of dollardays even today. Note also that the 43 million dollardays figure implies that 131-bit EC offers comparable security to DES(1982). Thus, we feel that our selection of field order provides medium-term security which is sufficient for many applications intended for RFIDs including anticounterfeiting.

Remark 1. Since we are *not* using a signature scheme and thus, no hash function, attacks on the signature scheme exploiting collisions on the hash function are not applicable to our setting.

Table 1 Solved elliptic curve DL challenges

Year	Source	Challenge name	Field type	Field size (bits)	Estimated # iterations	Actual # iterations	Iterations per second	Machine days
2000	[19,39]	ECC2K-108	$\mathbb{F}_{2^m}$ (Koblitz)	109	1.5×10^{15}	2.3×10^{15}	$160,364$	166,000 days on 1 Alpha 500 MHz workstation
2002	[19]	ECCp-109	$\mathbb{F}_p, p$ prime	109	2.1×10^{16}	–	–	–
2004	[19]	ECC2-109	$\mathbb{F}_{2^m}$	109	1.6×10^{16}	–	–	1,200 years on 1 Athlon XP 3200+

---

[2] We have assumed that an Athlon XP 3200+ cost in 2003–2004 US$100 which is a conservative estimate based on [79].

Remark 2. Very recently there have been two proposals [16, 36, 37] for hardware machines intended for solving the DL problem in EC. In [37], the authors estimate that breaking the Certicom ECC-109 challenge (over prime fields) would take 30 days at a cost of US\$3,000,000. Even if we were to assume that the same cost was required for breaking an EC implementation over $\mathbb{F}_{2^{131}}$, such an EC would require resources in the order of 90 million dollardays, thus comparable to DES(1982). The work in [16] proposes a design specifically tailored to binary fields. The authors estimate the time required to break the DL problem in an EC over $\mathbb{F}_{2^n}$ as $(\frac{\sqrt{\pi \times 2^n}}{2} + 2^{n/3})/100 \times 10^6$ where 100×10^6 is the number of point additions per second that their FPGA-based processor can perform for an EC defined over $\mathbb{F}_{2^{79}}$. Then, assuming conservatively that the processor has the same throughput over $\mathbb{F}_{2^{131}}$, and an FPGA unit price[3] of US\$50, the DL problem over $E(\mathbb{F}_{2^{131}})$ could be solved at a cost of $(\sqrt{\frac{\pi \times 2^{130}}{2}} + 2^{130/3})/100 \times 10^6 \times 3600 \times 24\,\text{days} \times 50\,\text{dollars} \approx 267$ million dollardays, thus also comparable to DES(1982).

4.3 Security of EC Over Composite Fields

In [25], the Weil descent attack is introduced against EC defined over binary fields of composite degree $n = k \cdot m$. At the time, it appeared that this work effectively rendered all composite field implementations of EC insecure. However, closer examination has demonstrated that composite fields with degree $n = 2p$ (i.e., an extension of degree two), where p is prime, remain secure against Weil Descent attacks and its variants. Table 2 summarizes what is known at the present moment. Note that EC over composite fields $\mathbb{F}_{2^{2p}}$, p prime have been previously proposed in the literature for efficient implementations [21, 83] and their security proved against known attacks [21].

5 Design Criteria for a Low-Cost and Low-Power ECC Processor

In this section, we discuss possible criteria used to design low-cost implementations. Looking at the hierarchy of operations as given in Fig. 1 the conclusion is that one needs to address low-power design principles at all levels. In particular, all levels of the ECC hierarchy should be simplified and accordingly minimized while still attaining an acceptable performance level for the target applications. Here we do not consider resistance to side-channel attacks.

---

[3] Reference [37] estimate the cost of a Xilinx XC3S1000 FPGA at US\$50 for low quantities. The Virtex 4 used in [16] is a higher complexity FPGA, thus this price assumption should provide us with a conservative estimate.

Table 2 Composite field degrees considered insecure and history of related attacks

Year	Source	History of attacks
1998	[25]	Weil Descent Attack introduced. Fields $\mathbb{F}_{2^{k \cdot m}}$ are potentially weak.
1999	[27]	Attack applied to EC over $\mathbb{F}_{2^{4n}}$.
2000	[31]	Algorithm for attacking hyperelliptic curves of genus >4 with complexity better than Pollard's Rho.
2001	[63]	Weil Descent attack against EC over fields $\mathbb{F}_{2^p}$, p prime is infeasible. Only a small fraction of EC over $\mathbb{F}_{2^{155}}$ are susceptible to the Weil Descent attack.
	[85]	It is shown that curves over $\mathbb{F}_{2^{4n}}$ should not be used and that curves over $\mathbb{F}_{2^{5n}}$ are still secure but attack using Weil Descent method offers better complexity than Rho method.
	[45]	Weil Descent attack is shown to work on curves defined over $\mathbb{F}_{2^{62}}$, $\mathbb{F}_{2^{93}}$, $\mathbb{F}_{2^{124}}$, $\mathbb{F}_{2^{155}}$. Attack on curves over $\mathbb{F}_{2^{155}}$ is only applicable to insignificant fraction of curves.
	[21]	Based on the analysis method of [63], it is shown that fields $\mathbb{F}_{2^{2p}}$, p prime are not susceptible to the Weil Descent attack. Specific instances are $\mathbb{F}_{2^{178}}$, $\mathbb{F}_{2^{226}}$, $\mathbb{F}_{2^{1018}}$, and $\mathbb{F}_{2^{1186}}$.
2002	[32]	Very efficient algorithm to reduce the ECDLP to the DL problem in a Jacobian of a hyperelliptic curve over $\mathbb{F}_q$. Index calculus method to solve the DL problem on hyperelliptic curves of genus ≥ 4.
	[61]	Shown that the Weil Descent attack is not applicable to ANSI X9.62 Standard curves $\mathbb{F}_{2^{176}}$, $\mathbb{F}_{2^{208}}$ and $\mathbb{F}_{2^{272}}$, $\mathbb{F}_{2^{304}}$, and $\mathbb{F}_{2^{368}}$. However, if efficient algorithm is found to compute isogenous curves from among most vulnerable ones, the Weil Descent attack yields better complexity than the Rho method.
	[28]	The attack from [32] to a much larger number of elliptic curves over certain composite fields of even characteristic. Larger proportion than previously thought of EC over $\mathbb{F}_{2^{155}}$ should be considered weak.
2003	[40]	Further generalization of [32]. Larger number of EC curves defined over $\mathbb{F}_{2^{155}}$ to attack.
2004	[67]	It is shown that EC defined over fields $\mathbb{F}_{2^{5k}}$ are weak, in the sense that Weil Descent attacks are faster than Pollard's Rho ones. In particular, curves over $\mathbb{F}_{2^{210}}$ can be solved a factor of 2^{13} faster than with Pollard's Rho (and for one quarter of these curves 2^{20} times faster). EC over fields $\mathbb{F}_{2^{4k}}$ are weak but not as weak as those defined over $\mathbb{F}_{2^{5k}}$.
2006	[64]	Analysis strongly suggests that finite fields $\mathbb{F}_{2^n}$ where n is divisible by 3, 5, 6, 7, or 8, should not be used to implement EC cryptographic protocols.

Note that choosing to implement elliptic curves over a binary field instead of using a prime field has other advantages besides area reduction. In particular, a field $\mathbb{F}_{2^n}$ offers far more options to be explored as there are many choices for bases, irreducible polynomials, fields (prime extension vs. composite degree), etc. These implementation options can be used to our advantage in reducing area, power, and attaining acceptable performance for the application at hand. Based on Fig. 1, we would like to optimize at every level of the hierarchy following a bottom-up approach. This implies: speeding-up the finite-field arithmetic which includes addition, multiplication, squaring, and inversion if using affine coordinates, choosing a "good" representation (i.e., coordinates that are more efficient) and accelerating the

Fig. 1 The hierarchy of ECC operations

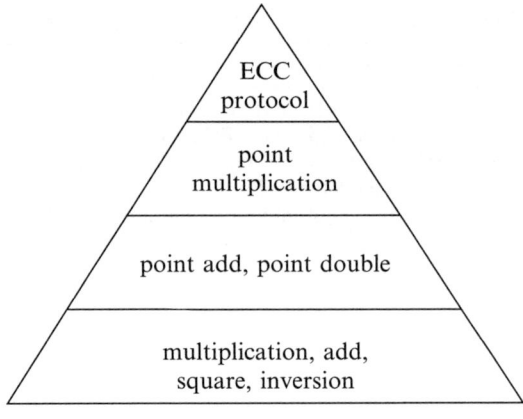

Fig. 2 An architecture for a low-cost ECC processor

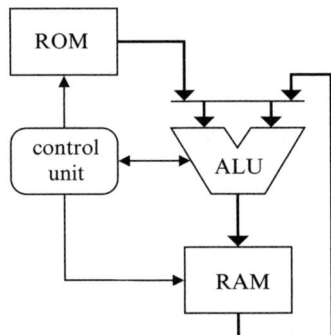

scalar point multiplication operation. Observe that by choosing projective instead of affine coordinates, one can completely avoid the finite field inversion operation. Actually, in this case just one inversion is necessary for conversion of a projective result to affine representation, but this can be done via multiplications. Therefore, modular multiplication is the most critical operation for an efficient implementation. This is a typical design choice in many works targeting low-cost, low-power ECC implementations for RFID and sensor networks [4–7, 56]. Thus, much of the remainder of this chapter focuses on the design of a processor with a finite field multiplier as its core.

In Fig. 2, an ECC processor is roughly depicted. The point multiplication (PM) is often realized as a sequence of instructions stored in the control unit. The basic Galois field operations are executed on an arithmetic unit (ALU). In ROM the ECC parameters and the constants can be stored. On the other hand, RAM contains all input and output variables and it therefore communicates with both, the ROM and the ALU. We deal with each of those levels separately and on all levels we discuss some algorithmic and architectural choices leading to a feasible RFID solution.

5.1 Minimizing Memory Requirements

As previously, discussed, point multiplication can be performed via the Montgomery ladder [47, 69] as shown in Algorithm 1. Looking at Algorithm 1 the basic operations are point addition and doubling. These operations can be performed according to the formulas introduced in [59]. The original formulas in [59] require two intermediate variables but it is possible to eliminate one more intermediate register at the cost of performing more operations. The resulting sequence of point operations is shown in Algorithm 2. In this way one can make a trade-off between performance and area as point operations require now 6 and 8 multiplications for point addition and doubling (instead of 5 and 6 multiplications), respectively. Here, we count squarings as multiplications.

Algorithm 2 requires only one intermediate variable T, which results in five registers in total. Namely, the required registers are for the storage of the following variables: X_1, X_2, Z_1, Z_2, and T. Also, the algorithm shows the operations and registers required if the key-bit $k_i = 0$. The other case is completely symmetric and it can be performed accordingly. More precisely, if the addition operation is viewed as a function $f(X_2, Z_2, X_1, Z_1) = (X_2, Z_2)$ for $k_i = 0$, due to symmetry for the case $k_i = 1$ we get $f(X_1, Z_1, X_2, Z_2) = (X_1, Z_1)$ and the correct result is always stored in the first two input variables. This is possible due to the property of scalar multiplication based on Algorithm 1. The complexity of operations and memory requirements is as follows. When Algorithm 1 deploys Algorithm 2 for point addition and doubling, the following number of operations in $\mathbb{F}_{2^n}$ are performed:

$$\begin{aligned} \#\text{registers} &= 5 \\ \#\text{multiplications} &= 14\lfloor \log_2 k \rfloor + 2 \\ \#\text{additions} &= 3\lfloor \log_2 k \rfloor + 1 \end{aligned}$$

Algorithm 2 EC point addition and doubling: operations that minimize the number of registers [4]

Require: $X_1, Z_1, X_2, Z_2, x_4 = x(P_2 - P_1)$	**Require:** $b \in \mathbb{F}_{2^n}, X_1, Z_1$
Ensure: $X(P_1 + P_2) = X(P_3) = X_3, Z_3$	**Ensure:** $X(2P_1) = X(P_5) = X_5, Z_5$
1: $Z_3 \leftarrow X_2 \cdot Z_1$	$Z_5 \leftarrow Z_1{}^2$
2: $X_3 \leftarrow X_1 \cdot Z_2$	$Z_5 \leftarrow Z_5{}^2$
3: $Z_3 \leftarrow X_3 + Z_3$	$Z_5 \leftarrow b \cdot Z_5$
4: $Z_3 \leftarrow Z_3{}^2$	$X_5 \leftarrow X_1{}^2$
5: $X_3 \leftarrow X_3 \cdot X_2$	$X_5 \leftarrow X_5{}^2$
6: $X_3 \leftarrow X_3 \cdot Z_1$	$X_5 \leftarrow X_5 + Z_5$
7: $T \leftarrow x_4 \cdot Z_3$	$Z_5 \leftarrow X_1{}^2$
8: $X_3 \leftarrow X_3 + T$	$Z_5 \leftarrow Z_5 \cdot Z_1$
9:	$Z_5 \leftarrow Z_5 \cdot Z_1$

Here we give details for only one option, e.g., Montgomery ladder and López-Dahab coordinates because both algorithms have nice properties that allow for a compact solution. Furthermore, the Montgomery ladder provides simple side-channel resistance due to the balanced sequence of operations for both possible choices of the key-bit. In Sect. 6, a detailed comparison is provided for other choices of coordinates and point multiplication algorithms.

5.2 Reduction of Control Logic

The point multiplication and point addition/doubling are implemented as finite state machines. When reading and writing from/to specific registers the best option is to reuse the same registers for both point operations. In this case, it is necessary to always check the value of the key-bit k_i because the sequences of instructions for point operations are different. This strategy increases the size of the control logic block but memory requirements are minimized as discussed above.

In traditional solutions, after computing $Q = k \cdot P$, one is required to transform back to affine coordinates and compute the y-coordinate of Q. However, a different solution is advocated in [6]. A simple solution is to send both the end values of registers containing P_1 and P_2 in Algorithm 1 to the verifier so that the verifier himself can recover the y-coordinate of Q. This would incur in the sending of four finite field elements, corresponding to the projective coordinate representation of P_1 and P_2. Alternatively, the protocol can be run by only using the x-coordinates of all points involved. Note that this was first observed by Miller in his seminal paper [68]. In either case, the projective coordinates sent to the verifier should be masked with a random value to avoid the attack described in [70]. This requires two extra multiplications at the end of the point multiplication which adds negligible overhead in comparison to the rest of the computation.

5.3 Simplifying The Arithmetic Unit

From Algorithm 2 it is evident that one has to implement at least finite field multiplication and addition operations. Squaring can be considered as a special case of multiplication in order to minimize the area and inversion is usually avoided by use of projective coordinates as discussed in Sect. 3. The single inversion that is necessary for conversion of projective to affine coordinates can be computed at the base reader's side. Note also that, if necessary, this inversion can be calculated via multiplications using Fermat's little theorem or the Itoh–Tsujii algorithm [34, 44]. In this way, the area remains almost intact as only some small control logic has to be added.

Addition of two elements $C = A + B \in \mathbb{F}_{2^n}$ is performed via an n–bitwise logical XOR operation. The standard way to compute the product $C = A \cdot B \in \mathbb{F}_{2^n} \cong$

$\mathbb{F}_2[x]/f(x)$, and $A = \sum_{i=0}^{n-1} a_i x^i$, $B = \sum_{j=0}^{n-1} b_j x^j$, $f = x^n + \sum_{i=0}^{s} f_i x^i$, $s < n$, is the one that uses convolution [11],

$$
C = \left(\sum_{i=0}^{n-1} \sum_{j=0}^{n-1} a_i b_j x^{i+j} \right) \bmod f = \left(\sum_{j=0}^{n-1} b_j \left(\sum_{i=0}^{n-1} a_i x^i \right) x^j \right) \bmod f
$$

$$
= \sum_{j=0}^{m-1} \left(b_j A x^j \right) \bmod f . \tag{4}
$$

This represents the most compact solution, where the $b_j A x^j$ partial products from (4) are computed iteratively and reduction modulo f of the degree n partial product polynomial is performed on each of the n iterations. The digit serial multiplication algorithm [86] may be considered as a generalization of this. Rather than processing the coefficients b_j of $B \in \mathbb{F}_{2^n}$ serially, a number of them are processed in parallel. In this case, there is some room for a trade-off between gate count and performance. This is an important consideration in low-frequency implementations over relatively small (composite) fields as discussed here.

In particular, $B = \sum_{j=0}^{n-1} b_j x^j$, rather than being considered as n coefficients of $\mathbb{F}_2$ is considered as being composed from $d = \lceil \frac{n}{D} \rceil$ *words*, each word containing D elements of $\mathbb{F}_2$ (see [9] for the corresponding generalization to any field characteristic). Now $B = \sum_{k=0}^{d-1} \tilde{b}_k x^{Dk}$ and each $\tilde{b}_k = \sum_{l=0}^{D-1} b_{Dk+l} x^l$, $0 \le k \le d-1$. Thus,

$$
C = \sum_{k=0}^{d-1} \left((\tilde{b}_k A) \bmod f \right) x^{kD} \bmod f \tag{5}
$$

can be calculated in d iterations and the $\tilde{b}_k A$ partial products are calculated recursively. A variant of the Song–Parhi method [86], is illustrated as Algorithm 3. When $D = 1$ then $d = n$ and $\tilde{b}_k = b_j \in \mathbb{F}_2$, then this method reverts to Horner multiplication. Squaring $C = A^2 \in \mathbb{F}_{2^n}$ is a special case of multiplication [60]. It is well known that $A^2 = \sum_{i=0}^{n-1} a_i x^{2i}$ which can then be reduced modulo f to a field element in $\mathbb{F}_{2^n}$.

Algorithm 3 Digit serial multiplication in $\mathbb{F}_{2^n}$ [86]

Require: polynomials $A = \sum_{i=0}^{n-1} a_i x^i$, $B = \sum_{k=0}^{d-1} \tilde{b}_k x^{kD}$ where $\tilde{b}_k = \sum_{l=0}^{D-1} b_{Dk+l} x^l$ and $f \in \mathbb{F}_2[x]$, irreducible of degree n
Ensure: $C = A \cdot B \bmod f(x)$
  $C \leftarrow 0$
  for k from 1 to $d-1$ **do**
    $C^{(k)} \leftarrow x^D (C^{(k-1)} + \tilde{b}_{d-1}^{(k-1)} A) \bmod f$
    $B^{(k)} \leftarrow x^D B^{(k-1)}$
  end for
  $C^{(d)} = (C^{(d-1)} + \tilde{b}^{(d-1)} A) \bmod f$
  Return C

---

6 Example of ECC Processor for RFID Applications

A more detailed example of the ECP for RFID as in [6] is shown in Fig. 3. The operational blocks are as follows: a Control Unit(CU), an Arithmetic Unit (ALU), and Memory (RAM and ROM). As mentioned previously, the ECC parameters and the constants x_4 and b in Algorithm 2 are stored in ROM. On the other hand, RAM contains all input and output variables and it therefore communicates with both, the ROM and the ALU.

The Control Unit controls scalar multiplication and point operations. In composite field implementations it also controls the operations in extension fields. In addition, the controller commands the ALU which performs field multiplication, addition, and squaring. When the START signal is set, the bits of $k = \sum_{i=0}^{n_k-1} k_i 2^i$, $k_i = \{0,1\}$, $n_k = \lceil \log_2 k \rceil$, are evaluated from MSB to LSB resulting in the assignment of new values for P_1 and P_2, dependent on the key-bit k_i. This is processed in an n-bit shift register. When all bits have been evaluated, an internal counter gives an END signal. The result of the last P1 calculation is written to the output register and the VALID output is set. In addition to controlling scalar multiplication and point operations, the CU also controls the operations in extension fields in the case of composite fields. The CU consists of a number of simple state machines and a counter and its area cost is small. The processor memory consists of the equivalent to seven n-bit ($n = p$) registers for ordinary fields and nine n-bit ($n = 2p$) registers for composite fields.

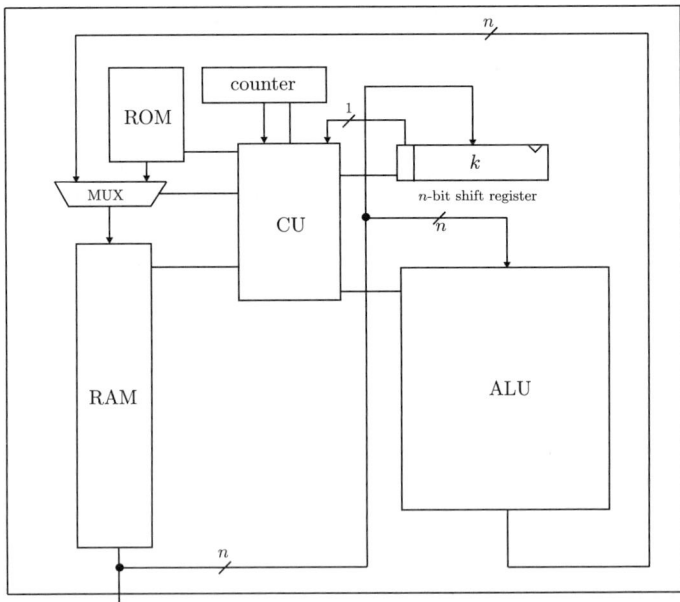

Fig. 3 ECP architecture [4]

Table 3 Basic operations in $\mathbb{F}_{(2^p)^2}$ expressed in operations in $\mathbb{F}_{2^p}$

Operation	Addition	Multiplication	Squaring
$\mathbb{F}_{2^p}$	2ADD	3MUL+4ADD	2SQR+ADD

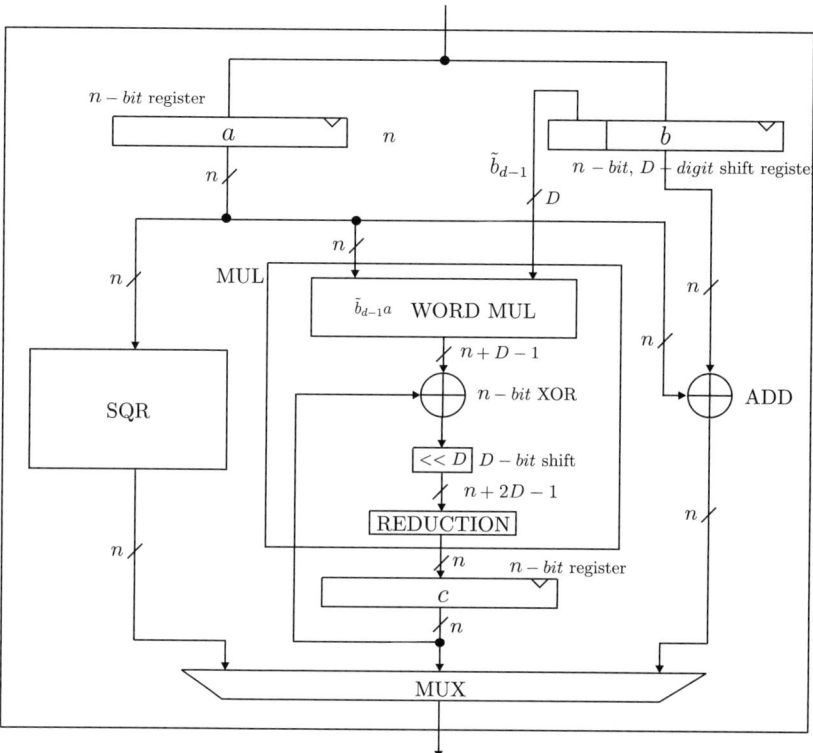

Fig. 4 ALU architecture [4]

The number of cycles required for one point multiplication for this architecture is $(n_k - 1)(15\lceil\frac{(n_k-1)}{D}\rceil + 3)$ and $9(n_k - 1)(\lceil\frac{(n_k-1)}{D}\rceil + 1)$ with and without squarer, respectively. The formulas for composite fields can be obtained directly from here using Table 3.

The largest contribution in area to the overall design comes from the ALU illustrated in Fig. 4. It consists of three n-bit registers a, b, and c, where b is also a D-bit shift register, and circuitry for implementing addition, squaring, and multiplication in $\mathbb{F}_{2^n}$. Load and store operations between the ALU and memory cost a single clock cycle. The ADD block consists of n XOR gates, and the SQR block consists of at most $(3n)/2$ XOR gates (for particular irreducible polynomials this is known to be even cheaper [92]) and computes $\mathbb{F}_{2^n}$ additions and squarings in a single clock cycle once data has been loaded into the ALU. The MUL block implements an iteration of Step 3 of Algorithm 3 in a single clock cycle. Multiplication is calculated then in $d = \lceil \frac{n}{D} \rceil$ clock cycles. For composite fields, the field arithmetic translates to the arithmetic in the subfield as shown in Table 3. We refer to [35] for a detailed treatment of the area and delay complexity of different finite field multipliers.

6.1 Comparison and Discussion

In [87], the feasibility of the ECC version of Schnorr's identification protocol in an RFID system was investigated and area and latency estimates were provided. The detailed numbers for several types of coordinates and point multiplication algorithm were given in [6]. Table 4 summarizes the number of cycles required for basic operations and for a whole point multiplication in an EC over $\mathbb{F}_{2^p}$. In Table 4 is assumed that for the Jacobian coordinates one has to perform n doublings and $n/2$ addition operations on average and that for the López–Dahab coordinates the Montgomery ladder is used which requires n iterations. Note that the modified formulas presented in [87] provide simple side-channel attack resistance if implemented without the use of a dedicated squarer. In this case, the formulas of [87] are almost three times as slow as the standard formulas from [59] which do *not* provide side-channel resistance. Previous works are summarized in Table 5. We caution against comparing as different authors use different technologies, field sizes, and tools. However, it is clear that systems based on NTRU, for example, require resources comparable to private-key cryptography algorithms. In addition, somewhat unexpectedly, hash algorithms such as SHA-1 require large amounts of resources almost comparable to the smallest ECC implementations. In particular, the implementation in [49] does

Table 4 Cycle count for EC operations over $\mathbb{F}_{2^p}$. L: Load, C: Computation, S: Store, $d = \lceil \frac{p}{D} \rceil$, $n = \lceil \log_2 k \rceil$

Operation	L	C	S	Total cycles
$\mathbb{F}_{2^p}$ addition	2	1	1	4
$\mathbb{F}_{2^p}$ squaring	1	1	1	3
$\mathbb{F}_{2^p}$ multiplication	2	d	1	$d+3$
EC operations assuming a squarer				
Projective coordinate type		Addition	Doubling	Total $k \cdot P$
Jacobian projective $(X/Z^2, Y/Z^3)$ [20] $a \neq 0$		$15d + 88$	$5d + 46$	$12.5nd + 90n$
Jacobian projective $(X/Z^2, Y/Z^3)$ [20] $a = 0$		$14d + 82$	$5d + 46$	$12nd + 87n$
López–Dahab $(X/Z, Y/Z)$ [59] $b \neq 120$		$4d + 23$	$2d + 22$	$6nd + 45n$
López–Dahab $(X/Z, Y/Z)$ [59] $b = 1$		$4d + 23$	$d + 19$	$5nd + 42.5n$
Modified López–Dahab $(X/Z, Y/Z)$ [87] $b \neq 1$		$6d + 29$	$3d + 28$	$9nd + 57n$
EC operations assuming no squarer				
Modified López–Dahab $(X/Z, Y/Z)$ [87] $b \neq 1$		$7d + 29$	$8d + 28$	$15nd + 57n$

Table 5 Performance and area of different algorithms and implementations

Ref.	Algorithm	Finite field/ parameter size	Area [gates]	Techn. [μ m]	Op. frequency [kHz]	Perf.
[30]	NTRUEnc.	$N = 167, p = 3, q = 128$	3,000	0.13	500	58.45 ms
[24]	AES	Block size = 128 bits	3,595	0.35	100	10.2 ms
[49]	SHA-1	Data size = 512 bits	4,276	0.13	500	0.81 ms
[6]	EC	$\mathbb{F}_{(2^{67})^2}$	12,944	0.25	175	2.39 s
[6]	EC	$\mathbb{F}_{2^{131}}$	14,735	0.25	175	430 ms
[30]	EC	$\mathbb{F}_{p_{100}}$	18,720	0.13	500	410.5 ms
[91]	EC	$\mathbb{F}_{2^{191}}$ and $\mathbb{F}_{p_{192}}$	23,000	0.35	68,500	6.7ms
[76]	EC	$\mathbb{F}_{p_{166}}$	30,333	0.13	20,000	31.9 ms
[5]	EC	$\mathbb{F}_{2^{131}}$	8,104	0.13	500	106 ms

not include the area required for storage, which brings the *overall* area of the implementation in the range of 10,000 gates (see for example [23]). Finally, we can observe that there is a wide range of options when it comes to implementing ECC.

7 Possible Improvements

Here we discuss on all levels some options that could result in an even more compact solution while maintaining similar performance. Memory requirements can be further reduced by using the so-called "common Z projective coordinate" as proposed in [56]. In this way one more register can be saved as only one Z-coordinate has to be stored. Furthermore, by using a unidirectional circular shift register file and reusing some registers, the authors have suggested some additional savings in storage.

For the savings in control there is probably not much to do as this part contributes the least to the area of the processor. Therefore, any further suggestions to optimize control would not help much unless combined with some improvements in memory and/or arithmetic unit.

Considering an arithmetic unit, a substantial shrinking is done by Sakiyama et al. in [81]. They showed that modular addition can also be supported by the same hardware logic as multiplication[4] as also explained in [5]. This operation requires additional multiplexers and XORs. However, the cost of this solution is much cheaper compared to the case of having a separate modular adder. This type of hardware sharing is very important for such low-cost applications. The proposed datapath is scalable in the digit size d which can be determined by exploring the best combination of performance and cost. The strong part of this architecture is that it uses the same cell(s) for finite field multiplication and addition without a large overhead in multiplexors. More details about the architecture and in particular about the cells inside are given in [5].

---

[4] A similar idea was independently introduced in [9].

8 Applications: Counterfeiting

Counterfeiting refers to making fraudulent replicas of goods. In recent years, such an activity has become widespread in countries with weak judicial systems and virtually nonexistent IP protection laws. Note that the counterfeiting problem affects the whole world not just poor countries. In fact, global economic damage across all industries due to the counterfeiting of goods is estimated at over $600 billion annually [80]. In the US, seizure of counterfeit goods has tripled in the last 5 years and in Europe over 100 million pirated and counterfeit goods were seized in 2004. Fake products cost businesses in the United Kingdom approximately $17 billion [80]. In India 15% of fast-moving consumer goods and 38% of auto-parts are counterfeit. Other industries where many goods are being counterfeited are the toy-industry, content and software, cosmetics, publishing, food and beverages, tobacco, apparel, sports goods, cards, etc.

The above mentioned examples show that counterfeiting may lead to a highly reduced income for companies and to a damaged brand. Maybe more unexpectedly, counterfeiting has an impact on our safety as well. In some cases this is even tragic. Counterfeited spare parts of planes have caused planes to crash [41]. Counterfeiting of medicines poses a real growing threat to our health. Thousands of people die because of taking medicines containing ingredients that are very dangerous or taking medicines that do not contain any active ingredient at all. This kind of problem is very large in South-East Asia where many fake antimalaria drugs containing no active ingredient are being distributed. The World Health Organisation (WHO) estimates that counterfeit drugs account for 10% of the world pharmaceutical market representing currently a value of $32 Billion. In developing countries these numbers are often much higher: in China and Colombia 40% of drugs are counterfeit, in Vietnam 33% of the antimalaria drugs are fake, and in Nigeria 50% of the drugs are counterfeit. According to the Food and Drug Administration, in the US there has been a rise of 600% in counterfeit drugs since 1997. In Europe, counterfeit drugs have increased by 45% since 2003.

In order to solve these problems, industries and governments have to take measures that reduce the size of the counterfeiting problem drastically. On top of several legislation, procedural measures and control mechanisms (which are necessary), good technological counter-measures are needed in order to have a serious impact. More precisely, technological components are required that allow control mechanisms to verify the authenticity of a product undoubtedly in an economic way. Many technologies have been developed to thwart the counterfeiting problem: holograms, watermarks, security threats, security films, bar codes, taggants, security inks, etc. Radio Frequency Identification is a new technology that is being considered for those purposes [48]. A major advantage of RFID tags is that no line of sight communication is required for read-out in contrast to other systems such as bar-code based systems. This might speed up the authentication procedure drastically. By locating an RFID tag containing specific product and reference information on a product, one aims to verify the authenticity of the product. Loosely speaking the verification is performed as follows. When a product passes a reader, the reader checks whether the necessary reference information is present on the tag. For this purpose the tag and the reader run a protocol. If the required information is there and verified to be authentic, the product is declared to be genuine and otherwise not. However, by eavesdropping on the channel between the tag and the reader, recording their communications, and storing this information in a new chip, the attacker can effectively make a clone of the original tag that cannot be distinguished from the original tag by the reader. In order to make tag cloning infeasible, it should not be possible to derive the tag secrets by:

- Active attacks or attacks in which the adversary actively participates in the protocol, e.g., by installing a fake reader and inserting, changing, or dropping messages in the protocol.
- Passive attacks or attacks in which the adversary only eavesdrops the communication between the reader and the tag.
- Physical attacks or attacks in which the target of the adversary is the tag itself using sophisticated physical tools (photo-flash lamps, focused ion beams, etc.).

By using good cryptographic protocols, active and passive attacks can be thwarted. Recently, a lightweight version of a protocol protecting against nonphysical attacks was developed in [48]. We stress, however, that it is rather easy to *physically* clone a tag. This means that an attacker can capture the RFID tag, investigate it, use physical means to read out its memory and its security sensitive data (identification number, reference information, keys, etc.), and produce a new tag with exactly the same data in memory. When this tag is embedded into a product, it is impossible for a reader to distinguish an authentic product from a fake one. Note that this is not just an academic thought experiment. For example, Bono et al. [15] have shown how an RFID transponder device manufactured by Texas Instruments and used in many car keys can be successfully cloned with off-the-shelf equipment and minimal RF expertise and Carluccio et al. [17, 18] show how to build cheap RFID readers which could be used for tracing individuals via RFID chips embedded in passports (see [89] for yet another example).

In order to protect an RFID tag against this type of cloning attack, one can attempt to prevent read out of its memory by using several protective measures [71, 84]. These measures often increase the price of the tag so much that it becomes unacceptably high for its main application. In order to thwart physical cloning attacks, Tuyls and Batina [87] proposed to integrate physical unclonable functions (PUFs) with RFID tags. The secret key material in the tag is extracted from these structures. The protocols presented in [87] are only secure against passive attacks. In [6], the efficiency of protocols (Okamoto-identification protocol) that are also secure against active and concurrent attacks is investigated. In what follows, we summarize the protocols in [6, 87] and discuss some implementation issues.

8.1 Model

We consider RFID tags embedded in a product or its package for detection and prevention of product counterfeiting. The tag is manufactured and embedded into the product by a legitimate authority which is assumed to be trusted. We consider an active attacker that knows the position of the tag in the product or its package, so she can remove the tag from the package to investigate it. We also assume that the attacker can (passively) eavesdrop on the channel between a reader and the tag, or can install a fake reader that communicates with the tag (active attack). Finally, we assume that the attacker can physically attack the tag; i.e., she can try to read out its memory. The goal of the attacker is to produce a fake RFID tag containing reference information such that it can only be distinguished from a real tag with small probability. Clearly, by embedding such a fake tag into a fake product, the fake product is identified as an authentic one.

8.2 Off-line Authentication

We distinguish between on-line and off-line authentication. Off-line authentication is the most attractive one from a practical point of view but also the most challenging one, as costs grow much more in this case. The particular case of on-line authentication was considered in [87] and it does not make use of PKC, thus we do not discuss it any further.

In [87] a PUF-Certificate-Identity-based Identification scheme was proposed. For the sake of completeness we describe it briefly here but refer to [87] for the details. Given the following algorithms and definitions:

- A tag with identity I and a PUF
- A standard identification scheme $\mathcal{SI} = (K_g, P, V)$, where K_g denotes the key generation algorithm, and P, V denote the interactive protocols run by the prover and verifier, respectively
- A secure signature scheme $\mathcal{SS} = (\mathrm{SK_g}, \mathrm{Sign}, V_f)$, with $\mathrm{SK_g}$ denoting the key generation algorithm, Sign denoting the signing algorithm and V_f the verification algorithm run by a verifier

a PUF-Certificate-Identity-based Identification scheme $(\mathrm{MK_g}, \mathrm{UK_g}, \hat{P}, \hat{V})$ can be constructed as follows.

During *enrollment* the issuer uses $\mathrm{SK_g}$ as the master-key generation algorithm $\mathrm{MK_g}$ for the secure signature scheme. The result is the issuer's master-key pair (mpk, msk). The algorithm $\mathrm{UK_g}$ creates for each tag a public-secret key pair (pk, sk) using the algorithm K_g for the SI-scheme. The issuer runs a protocol with the tag to determine the PUF's challenge c and helper data w such that the PUF response $x(c)$ maps onto the secret key sk. The helper data w are written into the ROM (EEPROM) memory of the tag. Finally, the issuer, using his master secret-key *msk* to sign, creates the following certificate that is also stored in the ROM of the tag Cert ← $(pk, \text{Sign}(msk, pk||ID))$.

During *authentication* the algorithms $\hat{P}$ and $\hat{V}$ are run as follows. The tag (in the role of the prover) sends the certificate Cert to the reader. If Cert is valid, the tag and the reader run the SI-protocol. If the tag passes this protocol too, the reader decides that the tag is authentic and otherwise not. Note that in order to run this last step, the tag has to challenge its PUF and use the helper data to obtain the secret key *sk* from the measured response $y(c)$.

The security of the scheme depends on three factors (1) the security of the PUF as a secure storage of the secret key, (2) the security of the identification scheme used, and (3) the security of the signature scheme used. It was shown in [87] that if the PUF is unclonable and a good Fuzzy Extractor is used for key extraction, the PUF provides a secure way of storing secret keys. The security of the scheme against impersonation attacks depends on the security of the identification scheme used against those attacks. Therefore, it is of crucial importance to understand which trade-off is being made between efficiency and security.

8.3 Schnorr's ID Protocol Based on ECDLP

The protocol of Schnorr [82] is shown in Fig. 5. In this case, a tag (prover) proves its identity to a reader (verifier) in a three-pass protocol. As it can be observed from the protocol, the critical operation is the point multiplication. This can be achieved via the processor presented in Sect. 6.

1. **Common Input:** The set of system parameters in this case consists of: (q, a, b, P, n, h). Here, q specifies the finite field, a, b, define an elliptic curve, P is a point on the curve of order n and h is the cofactor. In the case of tag authentication, most of these parameters are assumed to be fixed.
2. **Prover-Tag Input:** The prover's secret a such that $Z = -a \cdot P$.
3. **Protocol:** The protocol involves exchange of the following messages:

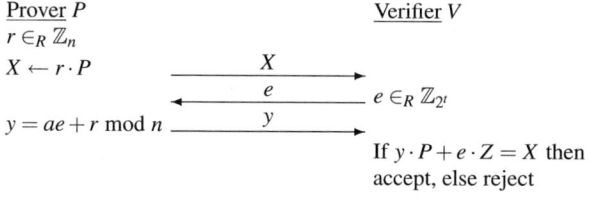

Fig. 5 Schnorr's identification protocol

1. **Common Input:** The set of system parameters in this case consists of: $(q, FR, a, b, P_1, P_2, n, h)$. Here, q specifies the finite field, FR is a field representation, a, b, define an elliptic curve, P_i is a point on the curve of order n and h is the cofactor. In the case of tag authentication, these parameters are assumed to be fixed.
2. **Prover-Tag Input:** The prover's secret (s_1, s_2) such that $Z = -s_1 P_1 - s_2 P_2$.
3. **Protocol:** The protocol involves the exchange of the following messages:

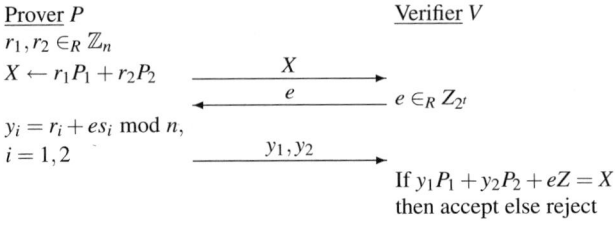

Fig. 6 Okamoto's identification protocol

8.4 Okamoto's ID Protocol Based on ECDLP

Note that Schnorr's protocol is only resistant against passive attacks under the discrete logarithm assumption. Another protocol that is also resistant against active and concurrent attack under the discrete logarithm assumption is Okamoto's identification protocol [73]. Reference [6] investigates therefore the efficiency of the implementation of this protocol, which we summarize here. Note that we are considering Okamoto's identification protocol (Fig. 6) as it provides security against active adversaries and it is based on the hardness of the DL problem. Other protocols found in the literature include Beth's identification protocol [10] and the XDL-IBI scheme in [8]. Beth's protocol only requires one point multiplication but it remains an open problem to prove its security against active adversaries. The XDL-IBI scheme also requires only one point multiplication but is only secure against passive adversaries and concurrent attacks (under a modified assumption). Thus, it seems that by analyzing both Schnorr's and Okamoto's we cover the efficiency of all *available* ID protocols based on the hardness of the DL problem. Protocols based on the hardness of integer factorization also exist (see [8, 72] for a thorough classification of ID-based protocols) but their performance and scalability properties are worse, in general, than those based on ECDLP as the comparisons in [30] show.

9 Conclusions

This chapter discusses the feasibility of implementing PKC for RFID applications. It also provides an overview of previous architectures. The work done in the past shows ECC on RFID to be a viable solution. This is important as it allows much more sophisticated protocols based on PKC than currently being considered for use in RFID.

Usual options deal with ECC over $\mathbb{F}_{2^p}$ where p is a prime number of bit-length between 130 and 140 and ECC over composite fields. The authors described different ALU architectures to obtain more compact and still feasible solutions. The main criteria used to meet the extreme low-power requirements typical of RFID applications were based on area minimization and reduction of the operating frequency. The best architecture with respect to area and performance is already close to 10 kgates. We have also discussed the feasibility of public-key-based secure identification protocols for RFID tags. As an example the implementation of Schnorr's and Okamoto's identification protocols were compared. As expected, Okamoto's protocol is more expensive than Schnorr's identification protocol. The trade-off is that Okamoto's protocol also provides security against active attacks.

We expect hyperelliptic curves to be feasible as well on an RFID tag. In order to compare its efficiency with ECC implementations more research is needed. During our research, it became apparent that there seems to be no better way to foresee the real impact of design decisions than to actually implement the whole design including memory. Furthermore, we expect the memory part of the processor to be crucial in meeting the power requirements.

Acknowledgments Lejla Batina is funded by research grants of Katholieke Universiteit Leuven and FWO projects (BBC and SESOC). This work was supported in part by the IAP Programme P6/26 BCRYPT of the Belgian State (Belgian Science Policy) and by the IBBT-QoE project of the IBBT.

References

1. M. Aigner and M. Feldhofer. Secure symmetric authentication for RFID tags, March 8–9, 2005, Graz, Austria. *Telecommunication and Mobile Computing – TCMC*, 2005
2. G. Avoine, E. Dysli, and P. Oechslin. Reducing time complexity in RFID systems. In B. Preneel and S. E. Tavares, editors, *Selected Areas in Cryptography – SAC 2005*, volume LNCS 3897, pp. 291–306, Springer, Berlin, 2005
3. S. Babbage, D. Catalano, L. Granboulan, A. Lenstra, C. Paar, J. Pelzl, T. Pornin, B. Preneel, M. Robshaw, A. Rupp, N. Smart, and M. Ward. ECRYPT Yearly Report on Algorithms and Keysizes (2004). Technical Report D.SPA.10, ECRYPT – European Network of Excellence in Crpytology, March 1, 2005. Revision 1.0. Available at http://www.ecrypt.eu.org/documents.html
4. L. Batina, J. Guajardo, T. Kerins, N. Mentens, P. Tuyls, and I. Verbauwhede. Public key cryptography for RFID-tags. *Printed Handout of Workshop on RFID Security – RFIDSec 06*, July 2006. Available at http://events.iaik.tugraz.at/RFIDSec06/Program/index.htm
5. L. Batina, N. Mentens, K. Sakiyama, B. Preneel, and I. Verbauwhede. Low-cost elliptic curve cryptography for wireless sensor networks. In L. Buttyán, V. D. Gligor, and D. Westhoff, editors, *Security and Privacy in Ad-Hoc and Sensor Networks – ESAS 2006*, volume LNCS 4357, pp. 6–17, Springer, Berlin, September 20–21, 2006
6. L. Batina, J. Guajardo, T. Kerins, N. Mentens, P. Tuyls, and I. Verbauwhede. Public-key cryptography for RFID-tags. In *IEEE Conference on Pervasive Computing and Communications Workshops, PerCom 2007 Workshops, IEEE International Workshop on Pervasive Computing and Communication Security – PerSec 2007*, pp. 217–222, New York, March 19–23, 2007, IEEE Computer Society Press, Washington, DC, 2007

7. L. Batina, N. Mentens, K. Sakiyama, B. Preneel, and I. Verbauwhede. Public-key cryptography on the top of a needle. In *International Symposium on Circuits and Systems – ISCAS 2007*, pp. 1831–1834, IEEE, New York, NY, May 27–20, 2007

8. M. Bellare, C. Namprempre, and G. Neven. Security proofs for identity-based identification and signature schemes. In C. Cachin and J. Camenisch, editors, *Advances in Cryptology – Eurocrypt 2004*, volume LNCS 3027, pp. 268–286, Springer, Berlin, 2004

9. G. Bertoni, J. Guajardo, S.S. Kumar, G. Orlando, C. Paar, and T.J. Wollinger. Efficient GF(p^m) arithmetic architectures for cryptographic applications. In M. Joye, editor, *Topics in Cryptology – CT-RSA 2003*, volume LNCS 2612, pp. 158–175, Springer, Berlin, 2003

10. T. Beth. Efficient zero-knowledge identification scheme for smart cards. In C.G. Günther, editor, *Advances in Cryptology – EUROCRYPT'88*, pp. 77–84, 1988

11. T. Beth and D. Gollmann. Algorithm engineering for public key algorithm. *IEEE Journal on Selected Areas in Communications*, 7(4): 458–465, May 1989

12. I. Blake, G. Seroussi, and N.P. Smart. *Elliptic Curves in Cryptography*. London Mathematical Society Lecture Notes Series, Cambridge University Press, Cambridge, 1999

13. M. Blaze, W. Diffie, R.L. Rivest, B. Schneier, T. Shimomura, E. Thompson, and M. Wiener. Minimal Key Lengths for Symmetric Ciphers to Provide Adequate Commercial Security – A Report by an Ad Hoc Group of Cryptographers and Computer Scientists, January 1996. Available at http://theory.lcs.mit.edu/~rivest/publications.html

14. D. Boneh and M.K. Franklin. Identity-based encryption from the weil pairing. *SIAM Journal of Computing*, 32(3): 586–615, 2003

15. S. Bono, M. Green, A. Stubblefield, A. Juels, A. Rubin, and M. Szydlo. Security analysis of a cryptographically enabled RFID device. In P. McDaniel, editor, *USENIX Security Symposium – Security'05*, pp. 1–16, Usenix, Berkeley, CA 2005

16. P. Bulens, G.M. de Dormale, and J.-J. Quisquater. Hardware for collision search on elliptic curve over $GF(2^m)$. In *Special-Purpose Hardware for Attacking Cryptographic Systems – SHARCS'06*, Cologne, Germany, April 03–04, 2006

17. D. Carluccio, T. Kasper, and C. Paar. Implementation details of a multi purpose ISO 14443 RFID-tool. *Printed Handout of Workshop on RFID Security – RFIDSec 06*, July 2006. Available at http://events.iaik.tugraz.at/RFIDSec06/Program/index.htm

18. D. Carluccio, K. Lemke, and C. Paar. Electromagnetic side channel analysis of a contactless smart card: First results. *Printed Handout of Workshop on RFID Security – RFIDSec 06*, July 2006. Available at http://events.iaik.tugraz.at/RFIDSec06/Program/index.htm

19. Certicom Corp. Certicom ECC Challenge. Available at http://www.certicom.com/index.php?action=res\,ecc_challenge

20. D.V. Chudnovsky and G.V. Chudnovsky. Sequences of numbers generated by addition in formal groups and new primality and factorization tests. *Advances in Applied Mathematics*, 7(4): 385–434, 1986

21. M. Ciet, J.-J. Quisquater, and F. Sica. A secure family of composite finite fields suitable for fast implementation of elliptic curve cryptography. In C. Pandu Rangan and C. Ding, editors, *Progress in Cryptology – INDOCRYPT 2001*, volume LNCS 2247, pp. 108–116. Springer, Berlin, 2001

22. H. Cohen and G. Frey. *Handbook of Elliptic and Hyperelliptic Curve Cryptography*. Discrete Mathematics and Its Applications 34. Chapman & Hall/CRC, Boca Raton, FL, 2005

23. M. Feldhofer and C. Rechberger. A case against currently used hash functions in RFID protocols. *Printed Handout of Workshop on RFID Security – RFIDSec 06*, July 2006. Available at http://events.iaik.tugraz.at/RFIDSec06/Program/index.htm

24. M. Feldhofer, S. Dominikus, and J. Wolkerstorfer. Strong authentication for RFID systems using the AES algorithm. In M. Joye and J.-J. Quisquater, editors, *Cryptographic Hardware and Embedded Systems – CHES 2004*, volume LNCS 3156, pp. 357–370, Springer, Berlin, 2004

25. G. Frey. How to disguise an elliptic curve (Weil descent). Presentation given at the second Elliptic Curve Cryptography Workshop (ECC'98). Slides available at http://www.cacr.math.uwaterloo.ca/, September 14–16, 1998

26. G. Frey, M.Müller, and H.-G. Rück. The tate pairing and the discrete logarithm applied to elliptic curve cryptosystems. *IEEE Transactions on Information Theory*, 45(5): 1717–1719, 1999

27. S.D. Galbraith and N.P. Smart. A cryptographic application of Weil descent. In M. Walker, editor, *Cryptography and Coding – IMA Int. Conf.*, volume LNCS 1746, pp. 191–200. Springer, Berlin, 1999. The full version of the paper is HP Labs Technical Report, HPL-1999-70

28. S.D. Galbraith, F. Hess, and N.P. Smart. Extending the GHS Weil descent attack. In Lars R. Knudsen, editor, *Advances in Cryptology – EUROCRYPT 2002*, volume LNCS 2332, pp. 29–44, Springer, Berlin, 2002

29. G. Gaubatz, J.P. Kaps, and B. Sunar. Public key cryptography in sensor networks – revisited. In C. Castelluccia, H. Hartenstein, C. Paar, and D. Westhoff, editors, *European Workshop on Security in Ad-Hoc and Sensor Networks – ESAS 2004*, volume LNCS 3313, pp. 2–18, Springer, Berlin, August 6, 2004

30. G. Gaubatz, J.-P. Kaps, E. Öztürk, and B. Sunar. State of the art in ultra-low power public key cryptography for wireless sensor networks. In *IEEE International Workshop on Pervasive Computing and Communication Security – PerSec 2005*, Kauai Island, Hawaii, March 2005

31. P. Gaudry. An algorithm for solving the discrete log problem on hyperelliptic curves. In B. Preneel, editor, *Advances in Cryptology – EUROCRYPT 2000*, volume LNCS 1807, pp. 19–34, Springer, Berlin, 2000

32. P. Gaudry, F. Hess, and N.P. Smart. Constructive and destructive facets of weil descent on elliptic curves. *Journal of Cryptology*, 15(1): 19–46, 2002

33. J. Goodman and A.P. Chandrakasan. An energy-efficient reconfigurable public-key cryptography processor. *IEEE Journal of Solid-State Circuits*, 36(11): 1808–1820 November 2001

34. J. Guajardo and C. Paar. Itoh–Tsujii inversion in standard basis and its application in cryptography and codes. *Designs, Codes and Cryptography*, 25(2): 207–216, 2002

35. J. Guajardo, Sandeep S. Kumar, T. Kerins, and P. Tuyls. Finite field multipliers for area constrained environments. Presented at the second Benelux Workshop on Information and System Security – WISSEC 2007, September 20–21, 2007

36. T.E. Güneysu. Efficient Hardware Architectures for Solving the Discrete Logarithm Problem on Elliptic Curves. Diplomarbeit, Chair for Communication Security – Ruhr-Universität Bochum, January 31, 2006. Available at http://www.crypto.rub.de/theses.html

37. T. Güneysu, C. Paar, and J. Pelzl. On the security of elliptic curve cryptosystems against attacks with special-purpose hardware. In *Special-Purpose Hardware for Attacking Cryptographic Systems – SHARCS'06*, Cologne, Germany, April 03–04, 2006

38. D. Hankerson, A. Menezes, and S. Vanstone. *Guide to Elliptic Curve Cryptography*, Springer, Berlin, 2004

39. R. Harley. *Elliptic Curve Discrete Logarithms Project*. Available at http://pauillac. inria.fr/~harley/ecdl/, December 2000. Website

40. F. Hess. The GHS attack revisited. In Eli Biham, editor, *Advances in Cryptology – EUROCRYPT 2003*, volume LNCS 2656, pp. 374–387, Springer, Berlin, 2003

41. D.M. Hopkins, L.T. Kontnik, and M.T. Turnage. *Counterfeiting Exposed: Protecting Your Brand and Customers*. Business Strategy, Wiley, New York, NY, 2003

42. M. Hutter, M. Feldhofer, and S. Mangard. Power and EM attacks on passive 13.56 MHz RFID devices. In P. Paillier and I. Verbauwhede, editors, *Cryptographic Hardware and Embedded Systems – CHES 2007*, volume LNCS 4727, pp. 320–333, Springer, Berlin, September 10–13, 2007

43. *IEEE P1363-2000: IEEE Standard Specifications for Public Key Cryptography*, 2000. Available at http://standards.ieee.org/catalog/olis/busarch.html

44. T. Itoh and S. Tsujii. A fast algorithm for computing multiplicative inverses in $GF(2^m)$ using normal bases. *Information and Computation*, 78: 171–177, 1988

45. M. Jacobson, A. Menezes, and A. Stein. Solving elliptic curve discrete logarithm problems using Weil descent. *Journal of the Ramanujan Mathematical Society*, 16: 231–260, 2001

46. D. Johnson and A. Menezes. The elliptic curve digital signature algorithm (ECDSA). Technical Report CORR 99-34, Department of Combinatorics & Optimization, University of Waterloo, Canada, February 24, 2000. http://www.cacr.math.uwaterloo.ca

47. M. Joye and S.-M. Yen. The montgomery powering ladder. In B.S. Kaliski Jr., Ç.K. Koç, and C. Paar, editors, *Cryptographic Hardware and Embedded Systems – CHES 2002*, volume LNCS 2523, pp. 291–302, Springer, Berlin, 2002

48. A. Juels and S.A. Weis. Authenticating pervasive devices with human protocols. In V. Shoup, editor, *Advances in Cryptology – CRYPTO 2005*, volume LNCS 3621, pp. 293–308, Springer, Berlin, 2005

49. J.-P. Kaps and B. Sunar. Energy comparison of aes and sha-1 for ubiquitous computing. In X. Zhou, O. Sokolsky, L. Yan, E.-S. Jung, Z. Shao, Y. Mu, D.C. Lee, D. Kim, Y.-S. Jeong, and C.-Z. Xu, editors, *Emerging Directions in Embedded and Ubiquitous Computing – EUC 2006 Workshops: NCUS, SecUbiq, USN, TRUST, ESO, and MSA*, volume LNCS 4097, pp. 372–381, Springer, Berlin, August 1–4, 2006

50. A. Karatsuba and Y. Ofman. Multiplication of multidigit numbers on automata. *Soviet Physics – Doklady*, 7: 595–596, 1963

51. N. Koblitz. Elliptic curve cryptosystems. *Mathematics of Computation*, 48: 203–209, 1988

52. N. Koblitz. A family of Jacobians suitable for Discrete Log Cryptosystems. In S. Goldwasser, editor, *Advances in Cryptology: Proceedings of CRYPTO'88*, volume LNCS 403, pp. 94–99, Springer, Berlin, 1988

53. P. Kocher. Timing attacks on implementations of Diffie-Hellman, RSA, DSS and other systems. In N. Koblitz, editor, *Advances in Cryptology: Proceedings of CRYPTO'96*, volume LNCS 1109, pp. 104–113, Springer, Berlin, 1996

54. P. Kocher, J. Jaffe, and B. Jun. Differential power analysis. In M. Wiener, editor, *Advances in Cryptology – CRYPTO'99*, volume LNCS 1666, pp. 388–397, Springer, Berlin, 1999

55. S. Kumar and C. Paar. Are standards compliant elliptic curve cryptosystems feasible on RFID? *Printed Handout of Workshop on RFID Security – RFIDSec 06*, July 2006. Available at http://events.iaik.tugraz.at/RFIDSec06/Program/index.htm

56. Y.K. Lee and I. Verbauwhede. A compact architecture for montgomery elliptic curve scalar multiplication processor. In *Workshop on Information Security Applications – WISA 2007*, LNCS, Springer, Berlin, August 27–29, 2007

57. A.K. Lenstra. Key lengths. In H. Bidgoli, editor, *Handbook of Information Security*. Wiley Publishing, To appear. Electronically published on June 30, 2004. Available at http://cm.bell-labs.com/who/akl/index.html

58. A.K. Lenstra and E. Verheul. Selecting cryptographic key sizes. *Journal of Cryptology*, 14(4) : 255–293, December 2001

59. J. López and R. Dahab. Fast multiplication on elliptic curves over $GF(2^m)$. In Ç.K. Koç and C. Paar, editors, *Cryptographic Hardware and Embedded Systems – CHES*, volume LNCS 1717, pp. 316–327, Springer, Berlin, 1999

60. E.D. Mastrovito. *VLSI Architectures for Computation in Galois Fields*. PhD Thesis, Dept. Electrical Engineering, Linköping University, Linköping, Sweeden, 1991

61. M. Maurer, A. Menezes, and E. Teske. Analysis of the GHS Weil descent attack on the ECDLP over characteristic two finite fields of composite degree. *LMS Journal of Computation and Mathematics*, 5: 127–174, 2002

62. M. McLoone and M.J.B. Robshaw. Public key cryptography and RFID tags. In *Topics in Cryptology – CT-RSA 2007*, Volume LNCS 4377, springer, Berlin 2007

63. A. Menezes and M. Qu. Analysis of the Weil descent attack of Gaudry, Hess and Smart. In D. Naccache, editor, *Topics in Cryptology – CT-RSA 2001*, volume LNCS 2020, pp. 308–318, Springer, Berlin, 2001

64. A. Menezes and E. Teske. Cryptographic implications of Hess' generalized GHS attack. *Applicable Algebra in Engineering, Communication and Computing*, 16(6): 439–460, 2006

65. A. Menezes, T. Okamoto, and S.A. Vanstone. Reducing elliptic curve logarithms to logarithms in a finite field. *IEEE Transactions on Information Theory*, 39(5): 1639–1646, 1993

66. A. Menezes, P. van Oorschot, and S. Vanstone. *Handbook of Applied Cryptography*, CRC Press, Boca Raton, FL 1997

67. A. Menezes, E. Teske, and A. Weng. Weak fields for ECC. In T. Okamoto, editor, *Topics in Cryptology — CT-RSA 2004*, volume LNCS 2964, pp. 366–386, Springer, Berlin, 2004

68. V.S. Miller. Use of elliptic curves in cryptography. In H.C. Williams, editor, *Advances in Cryptology – CRYPTO'85*, volume LNCS 218, pp. 417–426, Springer, Berlin, 1985

69. P. Montgomery. Speeding the Pollard and elliptic curve methods of factorization. *Mathematics of Computation*, 48: 243–264, 1987

70. D. Naccache, N.P. Smart, and J. Stern. Projective coordinates leak. In C. Cachin and J. Camenisch, editors, *Advances in Cryptology – EUROCRYPT 2004*, volume LNCS 3027, pp. 257–267, Springer, Berlin, 2004

71. M. Neve, E. Peeters, D. Samyde, and J.-J. Quisquater. Memories: A survey of their secure uses in smart cards. In *second International IEEE Security In Storage Workshop (IEEE SISW 2003)*, pp. 62–72, Washington DC, USA, 2003

72. G. Neven. *Provably Secure Identity-Based Identification Schemes and Transitive Signatures.* PhD Thesis, Faculteit Toegepaste Wetenschappen – Departement Computerwetenschappen, Afdeling Informatica. Katholieke Universiteit Leuven, Leuven, Belgium, 2004

73. T. Okamoto. Provably secure and practical identification schemes and corresponding signature schemes. In E.F. Brickell, editor, *Advances in Cryptology – CRYPTO'92*, volume LNCS 740, pp. 31–53, Springer, Berlin, 1992

74. Y. Oren and A. Shamir. Power analysis of RFID tags. Original Announcement at RSA Conference 2006, February 14th, 2006. Webpage available at http://www.wisdom.weizmann.ac.il/~yossio/rfid/

75. Y. Oren and A. Shamir. Remote password extraction from RFID tags. *IEEE Transactions on Computers*, 56(9): 1292–1296, 2007

76. E. Özturk, B. Sunar, and E. Savaş. Low-power elliptic curve cryptography using scaled modular arithmetic. In M. Joye and J.J. Quisquater, editors, *Cryptographic Hardware in Embedded Systems – CHES 2004*, volume LNCS 3156, pp. 92–106, Springer, Berlin, 2004

77. J.M. Pollard. Monte Carlo methods for index computation (mod p). *Mathematics of Computation*, 32: 918–924, 1978

78. T. Popp, S. Mangard, and E. Oswald. Power analysis attacks and countermeasures. *IEEE Design and Test of Computers – Design and Test of ICs for Secure Embedded Computing*, 24(6): 535–543, November–December 2007

79. PriceWatch.info. Price History of Athlon 3200+. Available at http://www.pricewatch.info/item/16909

80. RFID and UHF: A Prescription for RFID Success in the Pharmaceutical Industry. White paper, ADT/Tyco Fire and Security, Alien Technology, Impinj Inc., Intel Corporation, Symbol Technologies Inc., and Xterprice, June 2006

81. K. Sakiyama, L. Batina, N. Mentens, B. Preneel, and I. Verbauwhede. Small-footprint ALU for public-key processors for pervasive security. *Printed Handout of Workshop on RFID Security – RFIDSec 06*, July 2006. Available at http://events.iaik.tugraz.at/RFIDSec06/Program/index.htm

82. C.-P. Schnorr. Efficient identification and signatures for smart cards. In Gilles Brassard, editor, *Advances in Cryptology – CRYPTO '89*, volume LNCS 435, pp. 239–252, Springer, Berlin, 1989

83. R. Schroeppel, C.L. Beaver, R. Gonzales, R. Miller, and T. Draelos. A low-power design for an elliptic curve digital signature chip. In Burton S. Kaliski Jr., Çetin Kaya Koç, and Christof Paar, editors, *Cryptographic Hardware and Embedded Systems – CHES 2002*, volume LNCS 2523, pp. 366–380, Springer, Berlin, 2002

84. S.P. Skorobogatov and R.J. Anderson. Optical fault induction attacks. In B.S. Kaliski Jr., Ç.K. Koç, and C. Paar, editors, *Cryptographic Hardware and Embedded Systems – CHES 2002*, volume LNCS 2523, pp. 2–12, Springer, Berlin, 2002

85. N.P. Smart. How secure are elliptic curves over composite extension fields? In B. Pfitzmann, editor, *Advances in Cryptology – EUROCRYPT 2001*, volume LNCS 2045, pp. 30–39, Springer, Berlin, 2001

86. L. Song and K.K. Parhi. Low energy digit-serial/parallell finite field multipliers. *Kluwer Journal of VLSI Signal Processing Systems*, 19(2): 149–166, 1998

87. P. Tuyls and L. Batina. RFID-tags for anti-counterfeiting. In D. Pointcheval, editor, *Topics in Cryptology – CT-RSA 2006*, volume LNCS 3860, pp. 115–131, Springer, Berlin, February 13–17 2006

88. P.C. van Oorschot and M.J. Wiener. Parallel collision search with cryptanalytic applications. *Journal of Cryptology*, 12(1): 1–28, 1999

89. J. Westhues. *Demo: Cloning a Verichip.* http://cq.cx/verichip.pl, Last updated: July 2006

90. J. Wolkerstorfer. Is elliptic-curve cryptography suitable to secure RFID tags?, 2005. *Workshop on RFID and Lightweight Crypto*, Graz, Austria

91. J. Wolkerstorfer. Scaling ECC hardware to a minimum. In ECRYPT workshop – Cryptographic Advances in Secure Hardware – CRASH 2005, September 6–7 2005. Invited talk

92. H. Wu. Bit-parallel finite field multiplier and squarer using polynomial basis. *IEEE Transactions on Computers*, 51(7): 750–758, 2002

New Designs in Lightweight Symmetric Encryption

C. Paar, A. Poschmann*, and M.J.B. Robshaw

Abstract In this article, we consider new trends in the design of ultra-lightweight symmetric encryption algorithms. New lightweight designs for both block and stream ciphers as well as the underlying hardware design rationale are discussed. It is shown that secure block ciphers can be built with about 1,500 gate equivalences and, interestingly, it seems that modern lightweight block ciphers can have similar hardware requirements to lightweight stream ciphers.

1 Introduction

The bulk of cryptographic work is done using symmetric primitives. While we might appeal to asymmetric cryptography to establish a shared key [39], cryptographic data processing is almost always done using a symmetric encryption, authentication, or hashing algorithm.

In this article, we will consider some new trends in the design of lightweight symmetric encryption algorithms. There are two distinct types of symmetric encryption; *stream ciphers* and *block ciphers* and the essential difference between them can be described as follows:

- A block cipher transforms blocks of *plaintext* into *ciphertext* under the action of a *key*. This is typically a relatively complicated transformation but, apart from the reused key, the encryption of one block is independent of another.
- A stream cipher generates a *keystream* by sampling a constantly evolving *cipher state*. The state is typically initialized under the action of a *key* and an *initialization vector*. The sampling operation and the operation used to update the state are

A. Poschmann
Horst Görtz Institute for IT Security, Embedded Security Group (COSY),
Ruhr-Universität Bochum, Germany
e-mail: poschmann@crypto.rub.de

P. Kitsos, Y. Zhang (eds.), *RFID Security: Techniques, Protocols and System-on-Chip Design,* © Springer Science+Business Media, LLC 2008

usually computationally lightweight. The *plaintext stream* is then encrypted by combining it directly – typically using bitwise exclusive-or – with the keystream to give the *ciphertext stream*

Stream ciphers themselves can be divided into *synchronous* and *self-synchronizing* stream ciphers. For the first, the cipher state is updated independently of the generated ciphertext. For the second, the self-synchronizing stream cipher state update includes the generated ciphertext. The two types of stream cipher have very different error-propagation and synchronization properties [39] but, for the purposes of this article, we need no more detail. The vast majority of contemporary proposals are synchronous, but (secure) self-synchronizing stream ciphers appear to be rather difficult to design [17].

It is well established that a block cipher can be used to give a stream cipher. The NIST modes of operation provide three ways of doing this and are known as the *cipher feedback*, *output feedback*, and *counter* modes [42]. Interestingly, the cryptographic folklore suggests that stream ciphers of a dedicated design should be more efficient than block ciphers and, therefore, more efficient than stream ciphers based on block ciphers. Such an advantage might manifest itself in increased encryption speeds, or more compact and power-efficient implementations. However, as stream cipher cryptanalysis and block cipher design have advanced, this advantage has been somewhat eroded. This is something we will return to in our conclusions.

Since there is an algorithmic distinction between block and stream ciphers, we will address the two primitives separately. First, we will consider the state of the art in low-cost block cipher design. Then we will consider low-cost stream cipher designs. Since block ciphers can be used to give stream ciphers, the most efficient block cipher proposals will, in some sense, set the bar against which the most efficient stream cipher proposals should be compared. To set the stage, we will consider some of the basic building blocks for cryptographic primitives and compare their efficiency in hardware.

Before starting out we mention some particular considerations that apply to deployments in constrained environments such as low-cost tags for RFID applications. Very often, such applications require only a moderate level of security and reflect the very limited financial gains available to an attacker. The security demanded for typical industry applications such as electronic commerce or internet communication may not be suitable for some constrained devices and, since increased security levels translate directly into more physical space in silicon and increased deployment costs, security in excess of what is required is both costly and unwelcome. An appropriate security level will only be revealed by risk assessment and cost–benefit analysis, but 80-bit security may well be adequate in many such applications.

Generally speaking, applications for constrained devices are unlikely to require the encryption of large amounts of data. Implementations can therefore be optimized for the space they occupy or the power they consume without too much practical impact. So while security will often be the main consideration for some cryptographic primitive, the physical space required for an implementation will typically be the primary physical consideration, closely followed by peak and average power consumption, and timing requirements being a less-important third metric. Interestingly, the lack of large amounts of encrypted data helps reduce exposure to a range of attacks that manipulate time–memory–data trade-offs [1, 9, 11]. Further savings can be made in some applications when the cryptographic key is fixed at the time of device manufacture. In such cases there would be no need to rekey a device which rules out both a range of key manipulation attacks [7] as well as the consumption of additional resources.

2 Hardware Efficiency of Cryptographic Building Blocks

Hardware efficiency can be measured in many different ways; the length of the critical path (or maximum frequency), latency, clock cycles, power/energy consumption, throughput, and area requirements all have a significant influence on the viability of an implementation.

One particular problem in passive RFID applications is that the tags face strict power constraints. A rule-of-thumb is that the power consumption should be less than 15μW and the power consumption is given by the voltage times current consumption. For chips built in CMOS[1] technology the power consumption is the sum of two parts: static and dynamic power consumption. The static power consumption is roughly proportional to the area, i.e., the larger the area the higher the power consumption. The dynamic part is proportional to the switching activity, which is proportional to the operating frequency.

To lower power consumption, RFID applications are typically clocked at a low frequency, e.g., 100 kHz or 500 kHz. In this frequency range the static power consumption is dominant. RFID applications usually have harsh cost constraints and the silicon area of the chip is directly proportional to the cost. Therefore, a good way to minimize both the cost and the power consumption is to minimize the area requirements. It has become common to use the term *hardware efficient* as a synonym for small area requirements.

Area requirements are usually measured in μm², but this value depends on the fabrication technology and the standard cell library. In order to compare the area requirements independently it is common to state the area as *gate equivalents* (GE). One GE is equivalent to the area which is required by the two-input NAND gate with the lowest driving strength of the appropriate technology. The area in GE is derived by dividing the area in μm² by the area of a two-input NAND gate.

2.1 Architecture Strategies

Generally speaking, there are three major hardware architecture options for block ciphers: parallel (loop unrolled), round-wise, and serial. A *parallel*, or loop unrolled,

---

[1] *Complementary Metal Oxide Semiconductor*, the most widely used technology.

block cipher implementation performs several round operations of the encryption/decryption process within one clock cycle. Usually parallel implementations are *pipelined*, i.e., registers are inserted in the critical path so as to increase the maximum clock frequency. While parallel implementations have high throughput rates, this is rarely the focus for RFID applications. Rather, the high area and power demands mean that parallel implementations of block ciphers and stream ciphers are rarely suited for passive RFID applications.

In a *round-wise* implementation, one round function of a block or a stream cipher is processed within one clock cycle. The decreased throughput comes at the benefit of decreased area and power consumption. From a low power and low area perspective, round-wise implementations are best suited for stream ciphers and make a reasonable option for block ciphers. For example PRESENT [13] has been implemented in a round-wise manner.

To lower power consumption and area requirements, implementations can be *serialized*; here only a fraction of one round is processed in a clock cycle. Up to a certain point this strategy can significantly decrease the area and the power consumption and the impressive results by Feldhofer et al. on the AES [41] are achieved by serialization [23]. However, it might not always be a suitable implementation strategy since the savings can sometimes be canceled by the overheads in additional control logic. Nevertheless, from a low-power and low-area perspective, serial implementations appear to be best suited for RFID-like implementations in the case of block ciphers. The natural way of implementing stream ciphers is in a bit serial fashion.

2.2 Internal State Storage

Ciphers have an internal state which we might refer to as *cipher state* and *key state*. When a block cipher is used, the cipher state is initialized by the plaintext (or ciphertext) and modified under the action of the *key* (and therefore the key state). When a stream cipher is used, the cipher state is initialized by the *initialization value* and the *key*. Stream ciphers then use the initialized cipher state to output the keystream. Block ciphers have a fixed number of rounds and the final internal state serves as the ciphertext. Note that independent of the implementation strategy, see above, the internal cipher state has to be saved at each round.

In software environments kilobytes of RAM and ROM are available. In low-cost tag applications this is not the case. Although most RFID tags have a memory module, for cryptographic algorithms there is only the barest minimum of storage capacity available. Furthermore, read and write access to the memory module (usually EEPROM) is very power consuming. As a consequence it is preferable to store all intermediate values and variables in registers rather than in external memory.

Registers typically consist of *flipflops*. Compared to other standard cells, flipflops have a rather high area and power demand. For example, when using the *Virtual Silicon* (VST) standard cell library based on the *UMC L180* 0.18μ *1P6M Logic process*

(UMCL18G212T3), flipflops require between 6 and 12 GE to store a single bit. As a consequence, to store an internal state of say 144 bits (64-bits block state and 80-bits key state), at least 864 GE are required. Storage of the internal state typically accounts for at least 50% of the total area and power consumption. Therefore stream and block cipher implementations for low-cost tag applications should aim to minimize the storage required.

2.3 Combinatorial Elements

The term *combinatorial elements* includes all the basic Boolean operations such as NOT, NAND, NOR, AND, OR, and XOR. It also includes some basic logic functions such as multiplexers (MUX). The gate count for these basic operations is typically independent of the library used. For the *Virtual Silicon* (VST) standard cell library based on the *UMC L180 0.18μ 1P6M Logic process* (UMCL18G212T3) the figures for two-input gates with the lowest driving strength is given below. Note that in hardware XOR and MUX are rather expensive when compared to the other basic Boolean operations.

Gate	NOT	NAND	NOR	AND	OR	XOR	MUX
GE	0.5	1	1	1.33	1.33	2.67	2.67

2.4 Feedback Shift Registers

A common building block for stream ciphers is the *Feedback Shift Register* (FSR). An FSR inputs and outputs one bit per cycle and the input bit is a function of the previous state. Depending on the feedback function FSRs are either *Linear Feedback Shift Registers* (LFSR) or *Non-linear Feedback Shift Registers* (NFSR).

The hardware implementation of a bit-wise LFSR will consist of flipflops, to hold the register state, and XOR gates to compute the feedback. An LFSR is a reasonably hardware efficient building block. The feedback path, which often consists of a moderate number of binary XOR gates, will typically account for a few dozens GE, while the shift register consisting of flipflops cause a larger gate count. It is important to mention that while hardware efficiency will strive to minimize both the size of the register state and the number of XORs in the feedback path (sometimes called *feedback taps*), a cipher design using short registers or very few feedback taps may become susceptible to cryptanalysis.

NFSRs might use more complex Boolean functions or even *substitution boxes* (see below) in the feedback function. These tend to have a higher gate count then a series of XORs but it is the size of the register, and therefore the number of flipflops, that accounts for the bulk of the area. Therefore the hardware complexity of NFSRs is typically only slightly higher than that of LFSRs.

2.5 Confusion and Diffusion

Shannon [45] was the first to formalize the ideas of *confusion* and *diffusion* as two attractive properties in the design of a secure cipher. In practice, almost all block ciphers are product ciphers, i.e., they are based on subsequent operations of confusion and diffusion. In a block cipher, confusion is often identified with a substitution layer (see below) while diffusion is usually identified with a permutation or "mixing" layer. In reality is not always easy to separate and identify the components that contribute to confusion or diffusion.

Some ciphers use arithmetic operations as a diffusion and confusion technique, but this can significantly increase the area and power consumption. Arguably the most common confusion method is based on S-boxes (see Sect.2.6). A small change in the input to an S-box leads to a complex change in the output. In order to spread these output changes over the entire state quickly, a dedicated diffusion layer has to be applied. The classical way of doing this is to use bit *permutation*. In hardware, bit permutations can be realized with wires and no transistors are involved. They are therefore a very efficient component. Note that more complex diffusion techniques, such as the mix-column layer used in the AES, are also possible. Even though they have cryptographic advantages, they come at a higher hardware cost.

2.6 S-box Design

Many block ciphers, and some stream ciphers, use S-boxes to introduce nonlinearity. In software S-boxes are often implemented as *look-up tables* (LUT). In hardware these LUT can have a large area footprint[2] or they pose technological problems since a mix of combinatorial logic and ROM cannot always be easily achieved with a standard hardware design flow. Hence a purely combinatorial realization is often more efficient.

If combinatorial implementations do not exploit any internal structure in the S-box, then the area requirements will grow rapidly with the number of input and output bits. The more output bits an S-box has, the more Boolean equations will be required. And the more input bits an S-box has, the more complex these equations are likely to be. An interesting interaction between cryptography and hardware implementation can be observed here: In order to withstand differential and linear cryptanalysis [8, 38], high nonlinearity of S-boxes is required, which directly translates into a high gate count. A close look on the hardware efficiency of the S-boxes in AES [41], DES [40], and PRESENT [13] illustrates this.

AES uses a bijective 8-bit S-box, i.e., eight input bits are mapped to eight output bits. In [47], the hardware properties of several implementations of AES S-boxes, each illustrating different design goals, are compared. It turns out that the AES S-box

---

[2] Note that LUTs with a large memory footprint in software can be vulnerable to side-channel attacks based on *cache misses*.

realized as Boolean logic requires about 1,000 GE while there is no implementation that requires less than 300 GE. These figures also include the inverse S-box.

DES uses eight different S-boxes that map six input bits to four output bits. In [35], the authors state that in their DES ASIC design the S-boxes require in total 742 GE. However, taking into account that Boolean terms can be shared between the eight different S-boxes, it is not surprising that the area requirements for a single 6-bit to 4-bit S-box typically is around 120 GE. This can also be observed in implementations of DESXL and DESL, which will be introduced below. Both algorithms use 6-bit to 4-bit S-boxes but, in contrast to DES, a single S-box is repeated eight times. Therefore only one instance of the S-box has to be implemented in a serialized design, which requires 128 GE.

In [34], the area requirements of so-called *SERPENT-type* S-boxes are described. These are a special subset of 4-bit to 4-bit S-boxes fulfilling certain criteria and the authors found that the area requirements for this type of S-box varied between 21 and 39 GE. As an example, PRESENT uses a single, bijective 4-bit to 4-bit S-box which can be implemented with 21 GE. However, in [13], the authors state that a single S-box requires 28 GE. This deviation is caused by the fact that synthesis results depend heavily on the technology of the standard cells that are used.

3 Lightweight Block Ciphers

A block cipher can be viewed as a family of permutations indexed by a key k. For a block cipher that operates on b-bit blocks, the permutations are of the set of all b-bit inputs. There is a wide variety of design philosophies for block ciphers and the state of the art is well advanced. All the block ciphers of interest to us in this chapter are *iterated* and consist of the repeated application of a *round function*. At each round some key-related information is used to influence the computation, and this key information is derived from the user-supplied key k using a key schedule.

The computation in a single round usually follows one of two topologies. These have been termed a *Feistel* cipher or a *substitution–permutation network* which we will denote by SPN. This article is not an appropriate place to discuss block cipher design in detail, but since the choice of topology has some influence on the efficiency of an implementation, we briefly distinguish between them.

Suppose that we denote the cipher state at the start of round i by $L_i||R_i$ where $||$ denotes bitwise concatenation, and the round key to be used as k_{i+1}. If we further denote the round function in a Feistel cipher by f and the round function in an SPN cipher by g, then loosely speaking we have the following equations for a single round of encryption where the two arguments to f and g are the cipher state and the round key:

$$\text{FEISTEL CIPHER: } L_{i+1} = R_i \quad R_{i+1} = L_i \oplus f(R_i, k_{i+1})$$

$$\text{SPN CIPHER: } L_{i+1}||R_{i+1} = g(L_i||R_i, k_{i+1}).$$

At first sight, if our goal is compact hardware implementation then there appear
to be two major advantages of Feistel ciphers when compared to SPN ciphers:

1. The round function f would be identical for encryption and decryption
2. Only part of the cipher state – one half in a classical Feistel cipher – is processed
   each round

The first property suggests that hardware implementations of Feistel ciphers would
reuse the same datapath for encryption and decryption, with only the control logic
being adapted. The second property suggests that fewer gates will be required to
realize one round of encryption since only part of the cipher state is processed.
However, it is notable that the many important block cipher today, e.g., the *Advanced
Encryption Standard* (AES) [41] and the most compact block cipher, PRESENT [13],
are both SPN ciphers. So do Feistel networks really hold an intrinsic advantage?

It appears not. The first potential advantage of a Feistel cipher, given above, is
rarely relevant since for many tag-based applications decryption is not required. For
example, when used for authentication in a challenge–response protocol the block
cipher needs only to be used in the encryption direction. Also, if a block cipher were
to be used as a stream cipher, e.g., when operating in counter or output feedback
mode [42] then again, it is only used in the encryption direction. As for the second
advantage given above, while it is true that the function f in a Feistel cipher might
be more compact than the function g in a substitution–permutation network, this is
a little misleading. Feistel ciphers will probably require more encryption rounds to
achieve the same level of mixing as an SPN cipher and this can lead to a significant increase in execution time and energy requirements. But, in addition, Feistel
ciphers also require additional gates *after* the application of f to mix the untrans-
formed state with the transformed state. Usually the bitwise-XOR is used for this
which costs approximately 2.5–3 GE per bit. This is an implementation overhead
not required by SPN ciphers and further suggests that the minimum datapath for
a serialized implementation is likely to be better with an SPN cipher than with a
Feistel cipher. However, it should be noted that the state is the overwhelming proportion of the space requirements for both types of cipher. Thus any difference in
the implementation of the round functions is likely to have a limited overall effect.

3.1 State of the Art

There has been an increased interest in the design of lightweight block ciphers.
The benchmark implementation against which all others should be measured is the
implementation of the AES by Feldhofer et al. [23]. This works show that it is
possible to implement the AES in about 3,400 GE, a significant achievement. Yet
this is well above the amount of space that we would like to devote to an encryption
primitive in many RFID applications. At the same time, the AES arguably offers
more security in an RFID-based application than we really need. A new dedicated
design might provide a more suitable cost/security trade-off.

Table 1 Comparison of some particularly compact block cipher designs

	Key size	Block size	Cycles per block	Throughput at 100 kHz (kbps)	Logic process	Area GE	Area Rel.
DES [35]	56	64	144	44.4	0.18 μm	2,309	1.47
AES-128 [22]	128	128	1,032	12.4	0.35 μm	3,400	2.17
DESL [35]	56	64	144	44.4	0.18 μm	1,848	1.18
DESX [35]	184	64	144	44.4	0.18 μm	2,629	1.67
DESXL [35]	184	64	144	44.4	0.18 μm	2,168	1.38
PRESENT-80 [13]	80	64	32	200	0.18 μm	1,570	1
PRESENT-128 [13]	128	64	32	200	0.18 μm	1,886	1.20

A look at some older ciphers is quite illuminating. It is well known that DES was designed with hardware efficiency in mind, and DES still has very competitive hardware implementation properties. Implementations of around 3,000 GE [48] exist while a serialized implementation can be realized with around 2,300 GE [35]. The key length of DES limits its usefulness in many applications and makes proposals such as DESXL (2,168 GE) of some considerable interest [35]. This will be discussed further below. Implementation requirements for the *Tiny Encryption Algorithm* TEA [49, 50] are not known, but a crude estimate is that TEA needs at least 2,100 GE while XTEA needs at least 2,000 GE. These are "back-of-an-envelope" figures where we assume that a 32-bit bitwise exclusive-or requires 80 GE, a 32-bit integer addition requires 148 GE, and a 192-bit flipflop requires 1,344 GE. All these estimated figures do not take into account control logic which might significantly increase the required area. Four dedicated proposals for low-cost implementation are MCRYPTON [37], HIGHT [29], SEA [46], and CGEN [44], though the latter is not primarily intended as a block cipher. MCRYPTON has a precise hardware assessment and requires 2,949 GE, HIGHT requires around 3,000 GE while SEA with parameters comparable to PRESENT requires around 2,280 GE. All these figures are given in Table 1.

3.2 Two Dedicated Proposals: DESXL and PRESENT

3.2.1 DESXL

DESXL [35] is, as the name suggests, based on the *Data Encryption Standard* (DES) [40]. DES is a 64-bit block cipher with a 56-bit key. Both its history and structure are well known and details of the algorithm can be found in [40]. Unlike many modern ciphers DES was designed with good hardware properties in mind. However, even when adopted in the mid-1970s, DES was criticized for its short key length of 56 bits and this has only become a more pressing problem over the years [33]. Therefore, DESX was proposed by Rivest as a DES variant with higher resistance to brute-force attacks. This involves a process of prewhitening and post-whitening. DESX encryption using the 184-bit key $k||k_1||k_2$ can be described as

$$\text{DESX}_{k||k_1||k_2}(x) = k_2 \oplus \text{DES}_k(k_1 \oplus x).$$

The effectiveness of this simple technique was demonstrated by Kilian and Rogaway [32].

DESXL is derived from DESX but has two modifications. First the initial and final permutations (IP and IP^{-1}) in DES are omitted. Second, and more crucially, the eight original S-boxes of DES are replaced by a single S-box that is used eight times. The new S-box was chosen by randomly generating all S-boxes that fulfilled the original DES criteria with some additional conditions being added [35]. The goal was a single S-box that offers greater resistance to attacks such as differential and linear cryptanalysis than the original eight S-boxes of DES. The S-box is given below and more details are available in [35].

DESL (and DESXL) S-box: S															
14	5	7	2	11	8	1	15	0	10	9	4	6	13	12	3
5	0	8	15	14	3	2	12	11	7	6	9	13	4	1	10
4	9	2	14	8	7	13	0	10	12	15	1	5	11	3	6
9	6	15	5	3	8	4	11	7	1	12	2	0	14	10	13

3.2.2 Implementation of DESXL

Since DESXL is built around DES, an implementation of DES optimized for constrained environments is needed. An example is given in Fig. 1. This design, which is presented in [35], consists of five core modules: *mem_left*, *mem_right*, *keyschedule*, *controller*, and *sbox*.

Controller. The controller manages all control signals in the ASIC based on a finite state machine.

Keyschedule. This module generates the round keys of DES and consists of a 56-bit register, an input multiplexer, and an output multiplexer.

mem_left. This module consists of eight 4-bit registers, each composed of D-flipflops. Here the memory modules were designed as a shift register so that the output of a 4-bit block can be used as the new input to the following block. At the end of the chain, the current 4-bit block is provided and can be processed without an additional output multiplexer. This results in a saving of 48 GE.

mem_right. This module is similar to the *mem_left*. It consists of eight 4-bit wide registers, but it has different input and output signals. Instead of a 4-bit wide output it has a 6-bit wide output which accounts for the *expansion* function in DES. This design in a shift register manner saves even more area (72 GE) than in the *mem_left* module because, in this case, a 6-bit wide output multiplexer can be saved. Altogether 120 GE can be saved using this memory design when compared to regular approaches.

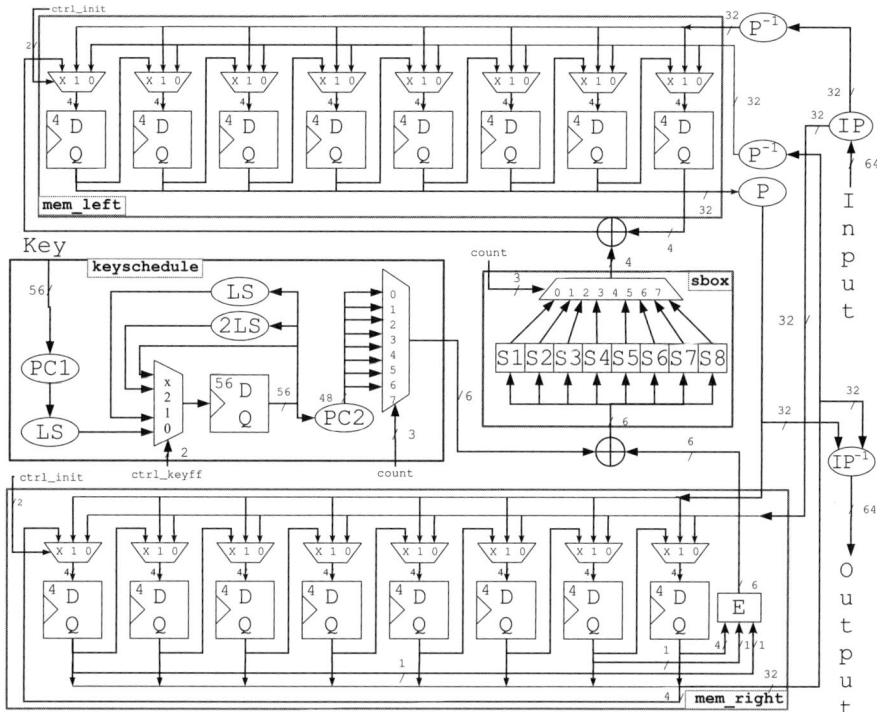

Fig. 1 Datapath of a serialized DES ASIC

sbox. For DES this module consists of the eight DES S-boxes and an output multiplexer. The S-boxes are realized in combinatorial logic, i.e., as a sum of products [20].

In the description that follows, it is assumed that the reader is familiar with the specifications of DES [40]. The 56-bit *key* is stored in the key flipflop register, after the PC1 and LS1 permutations have been applied, while the plaintext is mixed using the DES *initial permutation* and split into two 32-bit inputs to the modules *mem_left* and *mem_right*, respectively. The input of *mem_left* is modified by the inverse of the *P* permutation and stored in the registers of the modules *mem_left* and *mem_right* in one cycle. The output of the last register in *mem_right* is both stored in the first register of *mem_right* and expanded to six bits. After a bitwise exclusive-or operation with the appropriate block of the current round key, the expanded value is processed by the *sbox* module, which is selected by the *count* signal provided by the *controller* module. The result is bitwise exclusive-ored with the output of the *mem_left* module and stored in the first flipflop of the *mem_left* module. This is repeated eight times until all 32 bits of the right half are processed. By reducing the datapath from a 32-bit bus to a 4-bit bus, only $(6 \times 10) + (4 \times 10) = 100$ transistors (25 GE) are needed for the bitwise exclusive-or operations, compared to $(48 \times 10) + (32 \times 10) = 800$ (200 GE) transistors in a nonserial design. This saving comes at the cost of two additional multiplexers, one for the round key (72 GE) and one for the S-box output (48 GE). However, the second multiplexer is avoided in the final specification of DESXL.

Once all eight 4-bit blocks of both halves have been processed, they are concatenated to two 32-bit wide outputs of the modules *mem_left* and *mem_right*. The output of the module *mem_left* is transformed by the P permutation and stored as the new content of the *mem_right* module, while the output of the *mem_right* module is stored as the new content of the *mem_left* module. This execution flow repeats for another 15 rounds. Finally, both outputs from *mem_left* and *mem_right* are concatenated to give a 64-bit wide output and after IP^{-1} the ciphertext is generated.

The results of implementing this DES architecture are given in [35, 43] and summarized in Table 1. The registers comprise the majority of the chip size (33.78%), followed by the S-boxes (32.11%) and multiplexers (31.19%). Since the chip size of registers and multiplexers cannot be reduced it is natural to consider the space occupied by the S-boxes. And since there are no better logic minimizations of the original DES S-boxes, the designers of DESXL decided to use a new, single S-box repeated eight times. This resulted in a proposal called DESL.

For implementation, the main difference between DESL and DES lies in the f-function. The eight original DES S-boxes are replaced by a single S-box (see Table in Sect. 3.2.1) which is repeated eight times. This has implications for the design of the *sbox* module. As can be seen in Fig. 2, in a serialized design the S-box module is dramatically simplified. Another minor difference is that DESL omits the initial permutation IP and its inverse IP^{-1} for the sake of simplicity.

Adding the prewhitening and postwhitening to these algorithms obviously has an impact on the space required. Since we assume that all keys have to be stored on the RFID-tag in a nonvolatile memory and both the prewhitening and the postwhitening key never change, no additional flipflops are required for this operation. Therefore only two additional XOR-gates of 64 bits are required to perform the prewhitening and postwhitening. The gate count difference between DESXL and DESL (and also between DESX and DES) is 320 additional GEs. Figure 2 depicts the datapath of a serialized DESXL ASIC.

For the implementation of all these variants [35, 43] *Synopsys Design Vision V-2004.06-SP2* was used to map the design to the *Artisan UMC* 0.18μm *L180 Process 1.8-Volt Sage-X Standard Cell Library* and *Cadence Silicon Ensemble 5.4* for the Placement & Routing-step. *Synopsys NanoSim* was used to simulate the power consumption of the back-annotated verilog netlist of the ASIC. The implementation of DES requires 2,309 GE and 144 clock cycles are required to encrypt one 64-bit block of plaintext. For one encryption at 100 kHz the average current consumption is 1.19 μA and the throughput reaches 5.55 kB s^{-1}. For the serialized DESL ASIC implementation, the area requirement was 1,848 GE and, again, 144 clock cycles were needed to encrypt one 64-bit block of plaintext. For one encryption at 100 kHz the average current consumption was 0.89 μA with throughput reaching 5.55 kB s^{-1}.

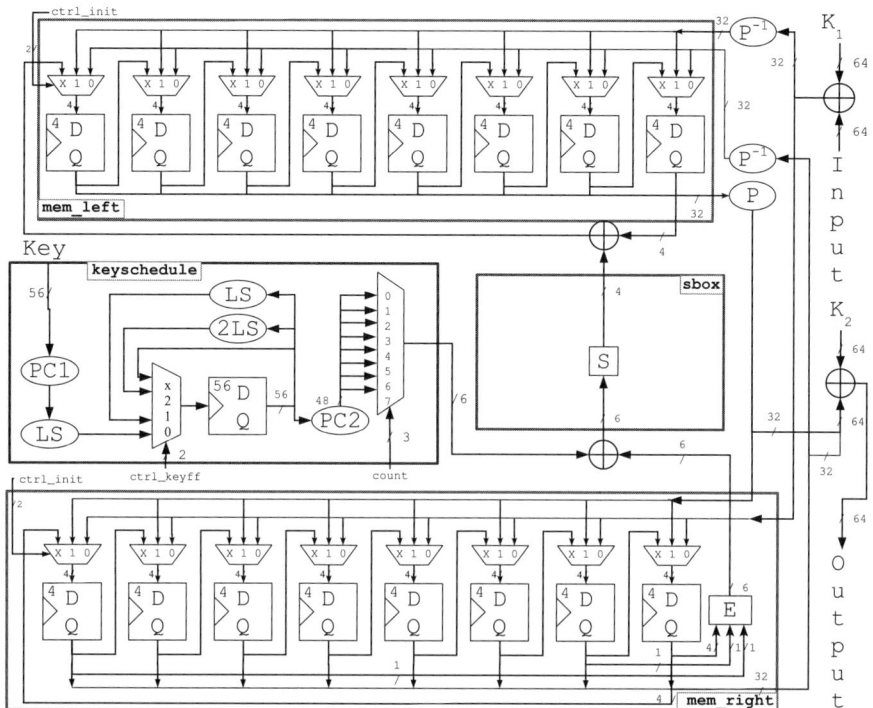

Fig. 2 Datapath of a serialized DESXL ASIC

3.2.3 Description of PRESENT

The block cipher PRESENT was designed with security, efficient implementation, and simplicity in mind. PRESENT is a 64-bit SPN block cipher with an 80-bit key. This is sometimes referred to as PRESENT-80 to differentiate it from PRESENT-128 which uses 128-bit keys. Encryption and decryption with PRESENT have roughly the same physical requirements and the encryption subkeys can be computed *on-the-fly*. PRESENT is described in pseudocode in Fig. 3 while details and design rationale can be found in [13]. The topology over two rounds is illustrated in Fig. 4.

PRESENT uses a single 4-bit to 4-bit S-box which is applied 16 times in parallel in each round. This was a direct consequence of the pursuit of hardware efficiency. Since a bit permutation is used as a linear diffusion layer, AES-like diffusion techniques [15] were not an option for PRESENT. Therefore some additional conditions were placed on the S-boxes to improve the so-called *avalanche of change*. Despite this, the S-box is particular well suited to efficient hardware implementation and is given below in hexadecimal notation.

x	0	1	2	3	4	5	6	7	8	9	A	B	C	D	E	F
$S[x]$	C	5	6	B	9	0	A	D	3	E	F	8	4	7	1	2

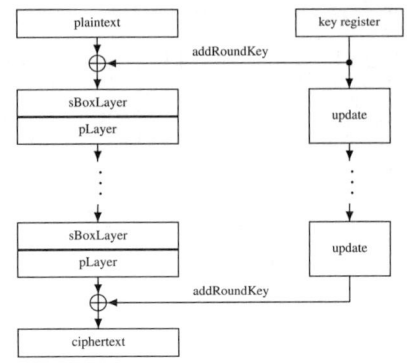

generateRoundKeys()
for $i = 1$ to 31 **do**
   addRoundKey(STATE,K_i)
   sBoxLayer(STATE)
   pLayer(STATE)
end for
addRoundKey(STATE,K_{32})

Fig. 3 A top-level algorithmic description of PRESENT

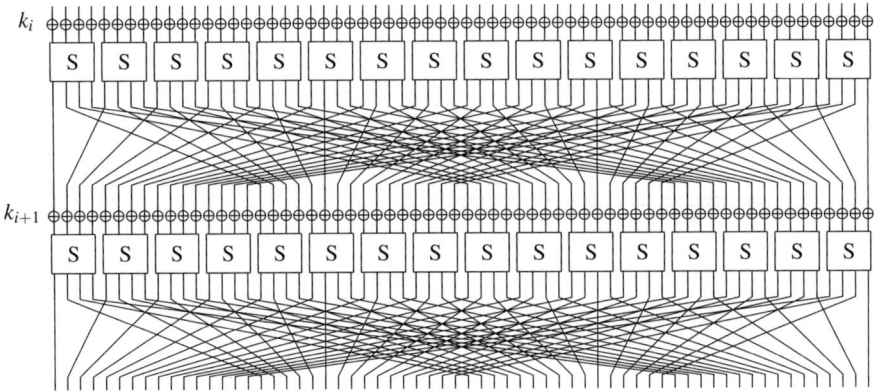

Fig. 4 Two round topology of PRESENT

3.2.4 Implementation of PRESENT

PRESENT-80 was implemented in VHDL and synthesized for the *Virtual Silicon* (VST) standard cell library based on the *UMC L180 0.18μ 1P6M Logic process*. The authors used *Mentor Graphics Modelsim SE PLUS 5.8c* for simulation and *Synopsys Design Compiler* version *Y-2006.06* for synthesis and power simulation [13]. Figure 5 shows the datapath of an area-optimized encryption-only PRESENT-80, which performs one round in one clock cycle, i.e., a 64-bit width datapath. The implementation requires 32 clock cycles to encrypt a 64-bit plaintext with an 80-bit key, occupies 1,570 GE and has a simulated power consumption of 5μW.

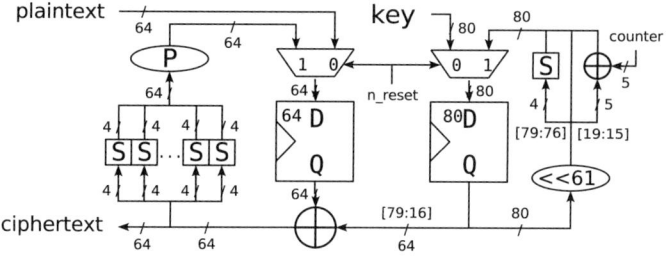

Fig. 5 The datapath of an area-optimized version of PRESENT-80

Module	GE	(%)	module	GE	(%)
Data state	384.39	24.48	KS: key state	480.49	30.61
s-Layer	448.45	28.57	KS: S-box	28.03	1.79
p-Layer	0	0	KS: Rotation	0	0
Counter: state	28.36	1.81	KS: counter-XOR	13.35	0.85
Counter: combinatorial	12.35	0.79	Key-XOR	170.84	10.88
Other	3.67	0.23			
			sum	*1,569.93*	*100*

The bulk of the area is occupied by flipflops for storing the key and the data state, followed by the S-layer and the key-XOR. Bit permutations are simple wiring and will increase the area only when the implementation is taken to the place and route-step. The main goal of the implementation was a small footprint in hardware, however, it has also been synthesized in a power-optimized implementation. For an additional 53 GE the power consumption is only $3.3\,\mu\text{W}$. Estimates also suggest that PRESENT-128 would occupy an approximate area of 1,886 GE. Beside a very small footprint PRESENT has a rather high throughput giving good energy-per-bit.

4 Lightweight Stream Ciphers

As was mentioned in Sect. 1, it is often suggested that stream ciphers of a dedicated design might be well suited to exacting conditions. Perhaps they offer particularly aggressive performance on some platform or perhaps they offer a particularly compact footprint in hardware. While the advanced state of block cipher research is beginning to challenge this view – it is not clear that a well-designed block cipher might not outperform a stream cipher in both regards – there have been considerable recent advances in the design of stream ciphers.

A stream cipher generates a keystream by sampling an evolving cipher state. Typically, for a dedicated stream cipher, the state is initialized under the action of a secret key k and an *initialization vector* (IV) v. By using an initialization vector different keystreams can be generated without a change of key and there is a mechanism – sending a new public value v – by which sender and receiver can synchronize with one another. Depending on the design of the stream cipher, state initialization might be closely related to keystream generation or it might be an entirely different process. It is well known that a keystream generated by a stream cipher with a finite cipher state must (eventually) repeat and its *period* must be sufficiently large. In practice, many stream ciphers come with an explicit upper-bound on the amount of keystream that can be generated after which the initialization vector, the key, or both are changed.

4.1 State of the Art

In contrast to block ciphers, the field of stream ciphers is very fragmented and there are no stream ciphers with the international profile of DES and the AES. As a consequence, the state of the art of the design and analysis of stream ciphers is not as well-developed as that of block ciphers. At first sight this might be a little surprising. After all, there is a very rich theory [36] surrounding the use of *Linear Feedback Shift Registers* (LFSRs) and for many years these have been used in the construction of stream ciphers. Yet, in some sense this has helped lead to the fragmentation of the field. Since stream ciphers can be built component-wise using these building blocks, it can be tempting for application developers to design subtly different stream ciphers that can be deployed in a proprietary manner. Despite this fragmentation, however, there are two stream ciphers of particular note. Somewhat ironically, given the reputation of stream ciphers for compact implementation, these two ciphers both occupy more space than most contemporary block ciphers.

The first of these two stream ciphers has been, and continues to be, used in many applications. RC4 was designed by Rivest in the mid-1990s and was provided as a proprietary cipher by RSA Data Security. The adoption of RC4 in industry has been widespread, e.g., [16, 30], and even though its confidentiality was compromised nearly a decade ago it remains, at core, a sound cipher. That said, it is showing its age and a variety of recommendations on how best to use the cipher help to protect the user from some unfortunate irregularities in the keystream. But independently of that, and in the applications of interest to us here, RC4 is completely unsuitable for environments with very restricted space. It should be noted that RC4 was primarily designed as a software-friendly cipher. The second stream cipher of note, SNOW 2.0 [18], is ISO-standardized and sports a contemporary design that meets modern performance requirements. While SNOW 2.0 might not be found in many products or applications, its success has lead to the design of variants such as SNOW 3G. However, SNOW 2.0 is also unsuitable for applications when space is limited. Both RC4 and SNOW 2.0 have a very large cipher state and rough estimates suggest that they would occupy around 12,000 and 7,000 GE, respectively.

To look for more compact stream ciphers we might turn to a host of proposals based on LFSRs and these typify the usual industry designs of the 1980s and 1990s. With LFSRs, however, it is difficult to achieve good performance, compact design, and good security at the same time. One of the most prominent stream ciphers of the era was A5/1. Its ensemble of three LFSRs with an irregular clocking mechanism require a very satisfying 1,000 GE [3], though unfortunately its cryptanalysis was equally impressive [10]. The cipher E0 [12] used in Bluetooth is another prominent shift-register-based stream cipher and estimates for its implementation suggest that over 1,600 GE are required.

The difficulty of designing a secure stream cipher is, to a great extent, a function of the way it operates. As well as a key the cipher typically uses an initialization value. Thus the cipher can be repeatedly initialized with different IVs that are known to an attacker while the secret key remains unchanged. Also, the key and IV are typically used to initialize the state of the stream cipher, and after this time the state evolves without any influence from the secret key. These are very special attributes to a cipher and it is not surprising that they lead to very special design demands. These properties have also lead to a series of time–memory–data trade-offs giving us lower bounds on the sizes of the state of a stream cipher. They also help us understand better the relationship between the sizes of the key, the IV, and the process of state initialization [1, 11, 19].

4.2 The eSTREAM Project

The eSTREAM project [17] is part of the ECRYPT Network of Excellence, and the goal of the project is to deliver a small portfolio of promising stream ciphers. The project is expected to end by May 2008.

At the start, the eSTREAM project identified stream ciphers for use in two very different ways. The first, labeled Profile 1, was for stream ciphers that could provide fast throughput in software. The second, labeled Profile 2, is of more interest to us here and required that stream ciphers be suitable for use in highly constrained environments. For Profile 2 (compact hardware) submissions, the must-satisfy values for the key and the IV lengths were 80 bits and at least one of 32 or 64 bits, respectively. Note that these values reflect the reasonable belief that the security level can, to some extent, be compromised so as to gain an implementation advantage.

The last round of the eSTREAM project featured 16 ciphers, eight for software and eight for hardware. At the time of writing, all have resisted cryptanalysis though some have been modified from their original submission. For entry in the final round of eSTREAM the main criterion after security for the hardware ciphers was the amount of space required for an implementation. For all eight final Profile 2 submissions there was good evidence that their implementation would require less space than an implementation of the AES and, in particular, the AES implementation of Feldhofer et al. that requires around 3,400 GE [22]. It appears that the candidates fall into three rough bands as shown in Table 2 : those that are slightly better than the AES, those that are certainly better than the AES, and two that appear to offer an exceptional performance profile.

Cryptanalysis may well take its toll on some of these finalists, but the space requirements for Grain v1.0 and Trivium are striking. When looking beyond the space

Table 2 Approximate gate count of some Profile 2 stream ciphers of the eSTREAM project

Algorithm	≈GE	Algorithm	≈GE	Algorithm	≈GE
Mickey v2.0 [2]	3,400	Decim v2.0 [4]	3,000	Grain v1.0 [28]	1,300
Pomaranch v3.0 [31]	3,300	Edon80 [24]	2,900	Trivium [14]	2,300
F-FCSR-H v2.0 [6]	3,200				

requirements we also find that these two algorithms are particularly amenable to low-cost hardware implementation; they offer great flexibility in their implementation so as to get a wide-range of performance metrics [25] as well as low-power implementations [21].

4.3 Two Dedicated Proposals: Grain and Trivium

4.3.1 Grain

Despite its version number Grain v1.0 [28] is the second version of the cipher; the first was broken [5] during the first phase of the eSTREAM project. Grain v1.0 consists of two feedback shift registers, one of which uses linear feedback while the other uses nonlinear feedback. Grain v1.0 offers 80-bit security, a level of security that is widely viewed as appropriate for the lower value applications often associated with RFID tag-based deployments. There is a variant – referred to as Grain-128 [27] – that offers 128-bit security as the name implies. A schematic overview of Grain v1.0 is given below.

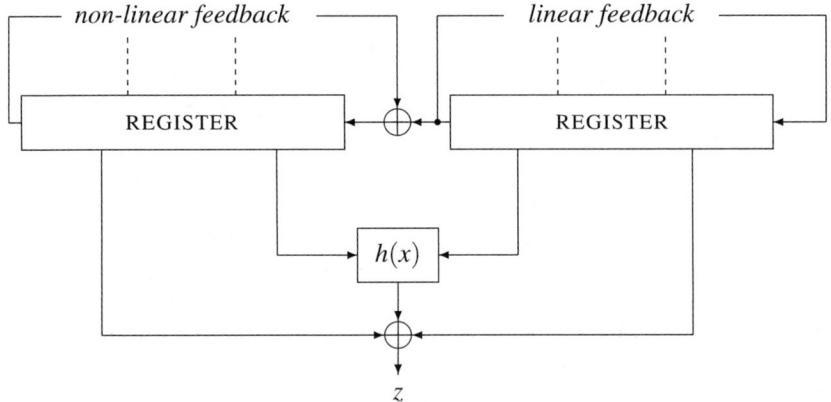

Each register is 80 bits in length and the function h is a boolean function that takes input bits from the two registers. It has been carefully chosen to provide particular cryptographic properties; more details can be found in [26, 28].

The hardware performance of Grain v1.0 has been studied closely in a variety of papers. One particularly nice feature of the algorithm, and one that is shared to some extent by Trivium, is the wide-ranging performance profile. In [25], for instance, it is shown that implementations oriented toward RFID tags, where reduced space is the driving metric, offer implementations requiring around 1,200 GE. Figures in [21] are derived from implementations that strive to minimize energy consumption; here the size might increase to 3,360 GE but with an electric current of merely 0.8 μA when clocked at 100 kHz.

4.3.2 Trivium

A second promising candidate from the eSTREAM project is Trivium [14]. Like Grain v1.0, it offers a wide rage of implementation options. A schematic overview of Trivium is given below.

Trivium uses three shift registers of lengths 93, 84, and 11, 1 respectively. From each register, internal bits are used as feedforward and feedback into the updating of the state, more details can be found in [14]. This bit extraction naturally divides each register into three parts and we have the following values which define the position of the bit extraction; $A_1 = 66$, $A_2 = 3$, $A_3 = 24$, $B_1 = 69$, $B_2 = 9$, $B_3 = 6$, $C_1 = 66$, $C_2 = 21$, and $C_3 = 24$. The function f takes the third- and second-to-last bits of each register, say x and y, and outputs the bitwise AND so $f(x,y) = xy$.

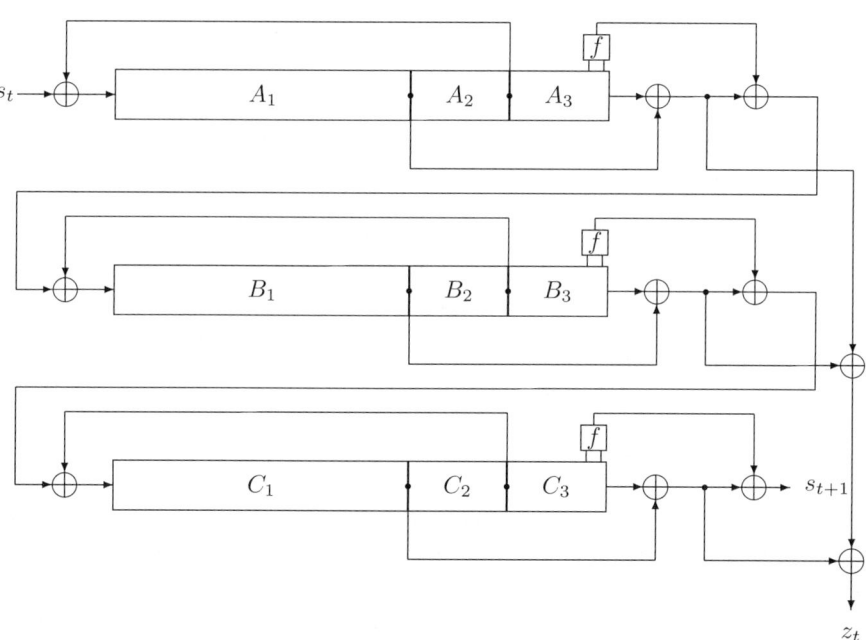

The total register size in Trivium is 288 bits and this translates into a larger implementation than Grain v1.0. Implementations oriented toward RFID tags in [25], for instance, offer implementations requiring around 2,300 GE, but the structure of Trivium allows for greater parallelization and hence a greater throughput. Figures in [21] are derived from implementations that strive to minimize energy consumption; here the size might increase to around 3,090 GE but with a current consumption of 0.68 µA when clocked at 100 kHz.

5 Conclusions

In this article, we have considered some innovative approaches in the design of low-cost symmetric encryption algorithms. This field of research is very active and much progress is being made in the design and implementation of both compact block and compact stream ciphers.

However, it is interesting to ask whether we really need both types of algorithms? When looking at the performance of Grain v1.0 and Trivium, one is tempted to say "yes." We get what appears to be very good encryption at roughly half the cost in space of using the AES. However, this is not necessarily a fair comparison since the AES was designed to be suitable for both software and hardware and offers three very high levels (128-, 192- and 256-bit) of security. By contrast Grain v1.0 and Trivium were exclusively designed with hardware implementation in mind and intended to offer only a single level of 80-bit security. It seems that the block cipher PRESENT offers a better point of comparison, but the space requirements for PRESENT are around 1,500 GE (see earlier). This is very close to those for Grain v1.0 and Trivium.

To be fair, a little more space is required to accommodate counter mode (say) if we were to use PRESENT as a stream cipher, but even so it is a reasonable indication that all three contemporary ciphers of a dedicated design require similar amounts of space. Perhaps this is not too surprising. Any given security level gives immediate requirements on the amount of state that we need. This means that we need to be using around 900 GE as the working space for any symmetric cipher that aims to offer, say, 80-bit security. We then need to add the space required for the different cipher operations and it seems that optimized ciphers (either stream or block) are able to make do with between 300 and 500 GE as an operational overhead.

Whether or not the specific proposals of Grain v1.0, Trivium, or PRESENT survive the efforts of cryptanalysts, there is no real reason to suppose that a compact block cipher necessarily requires more space than a compact stream cipher offering comparable security. To the authors of this article, the distinguishing feature between block and stream cipher implementation is unlikely to be the space required. Instead, any significant implementation difference may turn out to be the levels of parallelism and opportunities for low-energy optimization that are afforded. Depending on the design, these aspects appear to be more easily exploited with a stream cipher than with a block cipher. In the end, this may turn out to be the feature that distinguishes the two classes of symmetric encryption most when considering their design and implementation. Providing convincing evidence to the contrary is left as an open problem for the reader.

References

1. S. Babbage. A Space/Time Trade-off in Exhaustive Search Attacks on Stream Ciphers. *IEE European Convention on Security and Dectection*, 408, 1995
2. S. Babbage and M. Dodd. *MICKEY 2.0*, 2006 Available via www.ecrypt.eu.org/stream
3. L. Batina, J. Lano, N. Mentens, S. Berna Örs, B. Preneel, and I. Verbauwhede. Energy, Performance, Area Versus Security Trade-offs for Stream Ciphers. *State of the Art of Stream Ciphers 2004 (SASC 2004)*, Workshop Record, pp. 302–310, 2004. Available via www.ecrypt.eu. org/stream
4. C. Berbain, O. Billet, A. Canteaut, N. Courtois, B. Debraize, H. Gilbert, L. Goubin, A. Gouget, L. Granboulan, C. Lauradoux, M. Minier, T. Pornin, and H. Sibert. *DECIM v2.0*, 2006. Available via www.ecrypt.eu.org/stream
5. C. Berbain, H. Gilbert, and A. Maximov. Cryptanalysis of Grain. In M. Robshaw, editor, *Proceedings of FSE 2006*, volume 4047 of LNCS, pp. 15–29, Springer, Berlin, 2006
6. T. Berger, F. Arnault, and C. Lauradoux. F-FCSR-H v2.0. Available via www.ecrypt.eu.org/stream
7. E. Biham. New Types of Cryptanalytic Attacks Using Related Keys. In T. Helleseth, editor, *Proceedings of Eurocrypt'93*, volume 765 of LNCS, pp. 398–409, Springer, Berlin, 1994
8. E. Biham and A. Shamir. Differential Cryptanalysis of the Full 16-Round DES. In *Proceedings of CRYPTO*, pp. 487–496, 1992. Also available via citeseer.ist.psu.edu/biham93differential.html
9. A. Biryukov and A. Shamir. Cryptanalytic Time/Memory Trade-offs for Stream Ciphers. In T. Okamoto, editor, *Proceedings of Asiacrypt 2000*, volume 1976 of LNCS, pp. 1–13, Springer, Berlin, 2000
10. A. Biryukov, A. Shamir, and D. Wagner. Real-Time Cryptanalysis of A5/1 on a PC. In B. Schneier, editor, *Proceedings of FSE 2000*, volume 1978 of LNCS, pp. 37–44, Springer, Berlin, 2000
11. A. Biryukov, S. Mukhopadhyay, and P. Sarkar. Improved Time-memory Trade-offs with Multiple Data. In B. Preneel and S. Tavares, editors, *Proceedings of SAC 2005*, volume 3897 of LNCS, pp. 110–127, Springer, Berlin, 2005
12. S.I.G. Bluetooth *Specification of the Bluetooth System*, 2003. Available via www.bluetooth.org/spec, version 1.2
13. A. Bogdanov, G. Leander, L.R. Knudsen, C. Paar, A. Poschmann, M.J.B. Robshaw, Y. Seurin, and C. Vikkelsoe. PRESENT – An Ultra-Lightweight Block Cipher. In *Proceedings of CHES 2007*, volume 4727 of LNCS, pp. 450 – 466, Springer, Berlin, 2007
14. C. de Cannière and B. Preneel. *Trivium*. Available via www.ecrypt.eu.org/stream
15. J. Daemen and V. Rijmen. *The Design of Rijndael*, Springer, Berlin, 2002
16. T. Dierks and C. Allen. *The TLS Protocol*. Available via www.ietf.org/rfc/rfc2246.txt
17. ECRYPT Network of Excellence. *The Stream Cipher Project: eSTREAM*. Available via www.ecrypt.eu.org/stream
18. P. Ekdahl and T. Johansson. A New Version of the Stream Cipher SNOW. In K. Nyberg and H. Heys, editors, *Proceedings of SAC 2002*, volume 2595 of LNCS, pp. 47–61, Springer, Berlin, 2002
19. H. Englund, M. Hell, and T. Johansson. A Note on Distinguishing Attacks. In T. Helleseth, P. Kumar, and O. Ytrehus, editors, *Proceedings of 2007 IEEE Information Theory Workshop on Information Theory for Wirless Networks*, pp. 87–90, 2007

20. *Espresso*.   Available   via   `http://embedded.eecs.berkeley.edu/pubs/downloads/`
    `espresso/index.htm`
21. M. Feldhofer. Comparison of Low-Power Implementations of Trivium and Grain. *State of the Art of Stream Ciphers 2007 (SASC 2007)*, Workshop Record, February 2007. Available for download via `http://www.ecrypt.eu.org/stream/`
22. M. Feldhofer, S. Dominikus, and J. Wolkerstorfer. Strong Authentication for RFID Systems Using the *AES* algorithm. In M. Joye and J.-J. Quisquater, editor, *Proceedings of CHES 2004*, volume 3156 of LNCS, pp. 357–370, Springer, Berlin, 2004
23. M. Feldhofer, J. Wolkerstorfer, and V. Rijmen. AES Implementation on a Grain of Sand. *Information Security, IEE Proceedings*, 152(1): 13–20, 2005.
24. D. Gligoroski, S. Markovski, L. Kocarev, and M. Gusev. *Edon80*. Available via `www.ecrypt.eu.org/stream`
25. T. Good and M. Benaissa. Hardware Results for Selected Stream Cipher Candidates. *State of the Art of Stream Ciphers 2007 (SASC 2007)*, Workshop Record, February 2007. Available via `www.ecrypt.eu.org/stream`
26. M. Hell. *On the Design and Analysis of Stream Ciphers*. PhD Thesis, Lund University, 2007
27. M. Hell, T. Johansson, A. Maximov, and W. Meier. A Stream Cipher Proposal: Grain-128. In *IEEE International Symposium on Information Theory – ISIT 2006*, 2006. Also available via `www.ecrypt.eu.org/stream`
28. M. Hell, T. Johansson, and W. Meier. *Grain – A Stream Cipher for Constrained Environments*, International Journal of Wirelerss and Mobile Computing, 2(1): 86–93, 2007. Available via `www.ecrypt.eu.org/stream`
29. D. Hong, J. Sung, S. Hong, J. Lim, S. Lee, B. S. Koo, C. Lee, D. Chang, J. Lee, K. Jeong, H. Kim, J. Kim, and S. Chee. HIGHT: A New Block Cipher Suitable for Low-Resource Device. In L. Goubin and M. Matsui, editors, *Proceedings of CHES 2006*, volume 4249 of LNCS, pp. 46–59, Springer, Berlin, 2006
30. IEEE. *802.11 LAN/MAN Wireless LANS*, 2007. Available via `standards.ieee.org/getieee802/`
31. C. Jansen, T. Helleseth, and A. Kholosha. *Pomaranch v3.0*. Available via `www.ecrypt.eu.org/stream`
32. J. Kilian and P. Rogaway. How to Protect DES Against Exhaustive Key Search (an Analysis of DESX). *Journal of Cryptology: The Journal of the International Association for Cryptologic Research*, 14(1): 17–35, 1996. Available for download at `citeseer.ist.psu.edu/article/kilian96how.html`
33. S. Kumar, C. Paar, J. Pelzl, G. Pfeiffer, and M. Schimmler. Breaking Ciphers with COPA-COBANA – A Cost-Optimized Parallel Code Breaker. In *Workshop on Cryptographic Hardware and Embedded Systems – CHES 2006, Yokohama, Japan*, Springer, Berlin, 2006
34. G. Leander and A. Poschmann. On the Classification of 4-Bit S-boxes. In C. Carlet and B. Sunar, editors, *Proceedings of WAIFI 2007*, volume 4547 of LNCS, Springer, Berlin, 2007
35. G. Leander, C. Paar, A. Poschmann, and K. Schramm. New Lighweight DES Variants. In *Proceedings of Fast Software Encryption 2007 – FSE 2007*, volume 4593 of LNCS, pp. 196–210, Springer, Berlin, 2007
36. R. Lidl and H. Niederreiter. *Introduction to Finite Fields and their Applications*. Cambridge University Press, Cambridge, MA Revised edition, 1994
37. C. Lim and T. Korkishko. mCrypton – A Lightweight Block Cipher for Security of Low-cost RFID Tags and Sensors. In M. Yung, J. Song, and T. Kwon, editor, *Workshop on Information Security Applications – WISA'05*, volume 3786 of LNCS, pp. 243–258, Springer, Berlin, 2005
38. M. Matsui. Linear Cryptanalysis of DES Cipher. In T. Hellenseth, editor, *Advances in Cryptology – EUROCRYPT'93*, volume of 0765 LNCS, pp. 286 – 397, Springer, Berlin, 1994
39. A.J. Menezes, P.C. van Oorschot, and S.A. Vanstone. *Handbook of Applied Cryptography*. CRC Press, Boca Raton, FL, First edition, 1996
40. National Institute of Standards and Technology. *Data Encryption Standard (DES)*. Federal Information Processing Standards (FIPS) Publication 46-3, October 1999

41. National Institute of Standards and Technology. *Advanced Encryption Standard (AES)*. Federal Information Processing Standards (FIPS) Publication 197, November 2001. Available via csrc.nist.gov
42. National Institute of Standards and Technology. *SP800-38A: Recommendation for Block Cipher Modes of Operation*. Available via csrc.nist.gov, December 2001
43. A. Poschmann, G. Leander, K. Schramm, and C. Paar. New Lighweight Crypto Algorithms for RFID. In *Proceedings of The IEEE International Symposium on Circuits and Systems 2007 – ISCAS 2007*, pp. 1843–1846, 2007
44. M.J.B Robshaw. Searching for Compact Algorithms: CGEN. In P.Q. Nguyen, editor, *Proceedings of Vietcrypt 2006*, volume 4341 of LNCS, pp. 37–49, Springer, Berlin, 2006
45. C.E. Shannon. Communication Theory of Secrecy Systems. *Bell System Technical Journal*, 28(4): 656–715, 1949
46. F.X. Standaert, G. Piret, N. Gershenfeld, and J.-J. Quisquater. SEA: A Scalable Encryption Algorithm for Small Embedded Applications. In J. Domingo-Ferrer, J. Posegga, and D. Schreckling, editors, *Smart Card Research and Applications, Proceedings of CARDIS 2006*, volume 3928 of LNCS, pp. 222–236, Springer, Berlin, 2006
47. S. Tillich, M. Feldhofer, and J. Großschädl. Area, Delay, and Power Characteristics of Standard-Cell Implementations of the AES S-Box. In *Proceedings of Embedded Computer Systems: Architectures, Modeling, and Simulation – SAMOS 2006*, volume 4917 of LNCS, pp. 457 – 466, Springer, Berlin, 2006
48. I. Verbauwhede, F. Hoornaert, J. Vandewalle, and H. De Man. Security and Performance Optimization of a New DES Data Encryption Chip. *IEEE Journal of Solid-State Circuits*, 23(3): 647–656, 1988
49. D. Wheeler and R. Needham. TEA, a Tiny Encryption Algorithm. In B. Preneel, editor, *Proceedings of FSE 1994*, volume 1008 of LNCS, pp. 363–366, Springer, Berlin, 1994
50. D. Wheeler and R. Needham. TEA Extensions. October 1997. Available via www.ftp.cl.cam.ac.uk/ftp/users/djw3/ (Also Correction to XTEA. October, 1998)

Hardware Implementation of Symmetric Algorithms for RFID Security

Martin Feldhofer[*] and Johannes Wolkerstorfer

Abstract This book chapter provides an overview about hardware implementations of symmetric crypto algorithms for RFID security. Hardware design for RFID tags is challenging due to the fierce constraints concerning power consumption and chip area. After a general overview about RFID security, the requirements for passive RFID tags will be worked out. Different design measures for low-resource hardware implementations will be presented and their efficiency will be analyzed. The implementation part of this chapter presents a survey of implemented algorithms that are optimized for application in passive RFID tags. The evaluated algorithms include the block ciphers AES, TEA, and XTEA and the commonly used hash functions SHA-256, SHA-1, and MD5. These algorithms are compared with the new upcoming stream ciphers Grain and Trivium. The comparison of the achieved results favors the use of the AES algorithm for application of symmetric cryptography in RFID security.

1 Introduction

The usage of RFID technology in many different applications like logistics and supply-chain management is increasing fast. In addition to the manifold advantages there are some doubts about the missing security concepts in existing RFID standards. Unprotected RFID tags are prone violations of privacy and offer the possibility of unlimited copying of a tag. It is (was) widely believed that passively powered RFID tags are not capable to implement strong cryptographic algorithms because of

M. Feldhofer
Institute for Applied Information Processing and Communications,
Graz University of Technology, Austria
e-mail: Martin.Feldhofer@iaik.tugraz.at

P. Kitsos, Y. Zhang (eds.), *RFID Security: Techniques, Protocols and System-on-Chip Design,* © Springer Science+Business Media, LLC 2008

their limited power supply. In contrast, this chapter tries to give conclusive evidence that hardware implementations of strong symmetric primitives are possible on passive RFID tags.

In the recent years, many papers about RFID security protocols have been published that should help RFID system designers to make their RFID systems more secure. Most of the published articles make assumptions of the required resources on the passive RFID tag that do not hold in the real world. Some of them require high memory resources or very high computing power. Only a few publications are available that give details about the implementation of a security concept. It is difficult to implement strong cryptography on RFID tags and to meet the tight area and power requirements. The development of hardware modules for RFID security is challenging due to these stringent requirements. Hence, the hardware design requires sophisticated methodologies for saving resources on the tag. The combination of low-power design methods on all levels of implementation is necessary. This involves the application level, the protocol level, the algorithmic level, the architecture level, and the circuit level.

Providing security for RFID systems is far from trivial. Established approaches for protecting other networks cannot be applied without modification because of the special characteristics of RFID systems. The speciality of RFID systems lies in the properties of communication which makes the communication link (the air interface) very susceptible to eavesdropping and active attacks. Additionally, the different computing capabilities of RFID readers which have nearly unlimited computing power and the passive RFID tags which are very limited make the implementation of security protocols challenging. Computation of cryptographic primitives is typically very cost intensive compared to the simple tasks of an unprotected tag. Hence, security measures have to be designed very carefully to take the special requirements of RFID tags into account.

Nevertheless, the application of cryptographic protocols is the only useful approach for protecting RFID tags against threats like eavesdropping, unauthenticated access, illicit change of messages, and cloning of tags. Cryptography can help to provide the following security assurances which are necessary for protecting against attacks.

1.1 Security Assurances

Confidentiality means to keep data secret for parties that are not allowed to have access to the message while the transmission channel can be insecure. Encryption of data with a key is the best method to achieve confidentiality over the unprotected RF link of an RFID system. This implies that all communicating parties have access to authenticated keys. A common assumption is that only the communication from the RFID reader to the tag has to be protected because of its high transmission power. Experiments show that also the tag-to-reader communication via load modulation in the HF frequency range can be observed meters away from the passive tag with off-the-shelf equipment and appropriate filtering methods of the carrier signal. Also the backscatter mechanism in UHF systems where the energy is partly reflected by the tag has to be protected because it can be eavesdropped with little effort.

A further security assurance is *authentication* which means to proof a claimed identity during communication by means of cryptography. Typically, challenge–response protocols are used for this. Thereby, the party which wants to proof its identity is asked to encrypt a time-varying challenge which was generated and sent by the verifier. This encrypted challenge is sent back as response and the verifier can check this result with the appropriate key. In addition to single-side authentication either by the tag or by the reader it is also possible to have mutual authentication which means that both parties proof their identity to each other. Mutual authentication requires a three-pass protocol where reader and tag encrypt a challenge of the other party. In the context of RFID systems, authentication is the most important security feature because it can prevent cloning of tags and unauthorized access to the tag.

Data integrity prevents the threat of unauthorized manipulation of data. It is similar to authentication because a manipulation can also be seen as a change of origin of the data. In cryptographic protocols, integrity is normally achieved by means of keyed hash-functions which produce a fingerprint of the message that can only be generated if the secret authentication key is known. The last security assurance is *nonrepudiation* which prevents to deny commitments that were previously given by a party. Applications of nonrepudiation are not very important for RFID applications and hence will be left out here.

1.2 Overview of Security Primitives

All the above security assurances and protocols use different cryptographic primitives as underlying building blocks. There are three different categories of cyrptographic primitives: unkeyed primitives, symmetric primitives, and asymmetric primitives. The best known *unkeyed primitives* are hash functions and random-number generators. Hash functions produce a fixed-length message digest from a message of arbitrary length and are mostly used to provide data integrity. The first important property of hash functions is that it should not be possible to find two messages such that their hashing produces the same hash value. This property is known as collision resistance. Preimage resistance is another feature of hashes that assures to be hard to generate a message that produces a given message digest. Another type of unkeyed primitives is random number generators, which are used for generation of random values, e.g., in challenge–response protocols.

Symmetric primitives or alternatively symmetric-key ciphers have the concept that the sender and the receiver of a message use the same key (secret key) or a key that is easily derivable from a secret. The transformations are called encryption at the sender and decryption at the receiver. Block ciphers and stream ciphers are the best known symmetric primitives but keyed hash functions (MACs) are also part of this category. The major advantage of symmetric primitives is their limited computational complexity compared to public-key primitives. This allows efficient implementations on various platforms like on small microcontrollers or in dedicated hardware modules. Often it is possible to scale the implementation in terms of performance and resource complexity. The security of symmetric-key algorithms bases on the establishment and the management of secret keys which is also the major drawback of such systems. It is possible either to use preshared keys that are exchanged in a secure environment or to use special key-exchange protocols that mostly rely on asymmetric cryptography. Many RFID protocols are based on symmetric primitives and this chapter mainly covers their efficient implementation in hardware. *Asymmetric primitives* use different keys for encryption and decryption functionality. Although these pairs are generated at the same time, the private key is kept secret while the public key is published. In difference to symmetric keys, it is not possible to generate the private key from the public key due to underlying hard mathematical problems. For encryption of data the sender uses the public key of the receiver. Only the owner of the private key can decrypt that message. The major drawback of public-key cryptography is the high computational complexity. Most algorithms rely on finite-field arithmetic with word sizes up to several hundred bits and require large amounts of memory which is costly on small microcontrollers or in dedicated hardware. Until a few years ago it was believed that public-key cryptography is not possible on RFID tags. Nevertheless, newer publications also propose the use of public-key primitives for protecting RFID devices.

In this chapter about symmetric cryptography, we always try to rely on standardized algorithms and protocols. Standardized solutions should always be preferred because the security of the algorithm has been scrutinized by a large number of cryptanalytic experts. There exist several bad examples of using proprietary protocols and algorithms which were shown to be insecure finally. For example, Bono et al. [1] presented an attack against a digital transponder which was used in a payment system relying on a proprietary algorithm. The protection level of a total system depends on the weakest link in the security chain. It is not useful to protect a backend database with a high level of security while using low security in the communication between RFID reader and tags. Especially in RFID systems where passive tags may have long life cycles without the possibility of reconfiguration and maintenance, standardized algorithms with a appropriate key length should be preferred.

1.3 Security Protocols in RFID Systems

In this chapter, we focus on the use of symmetric-key primitives in RFID systems and we will describe their efficient implementation in hardware. Because unkeyed primitives have similar properties than symmetric-key primitives (many of them base on the use of a unkeyed primitive) we also include them in our observations. The applications of symmetric cryptography in RFID systems are manifold. Prevention of cloning allows to protect goods in the supply chain from forgery.

Additionally, it is possible to proof the identity of the communicating parties. Other possible applications involve confidentiality of stored data on the tag or the confidentiality of transmitted data via the air interface.

Protocol designers can choose from primitives like block ciphers and stream ciphers, hash functions, MACs, universal hash functions, or pseudorandom number generators (PRNGs). Some of these primitives can be efficiently turned into others. A block cipher can be turned into a hash function by using it in an appropriate mode of operation. For universal hash functions to be used in the setting for RFID security protocols, a cryptographically secure PRNG is needed to generate new keying material. Standardized and trusted PRNGs are in turn again based on block ciphers or hash functions. This implies that for RFID security protocols using PRNGs on top of primitives like ciphers or hash functions, no additional circuit of substantial size is required. MACs are mainly based on hash functions, ciphers, or universal hash functions. Summing up this short overview, we conclude that for RFID authentication protocols based on symmetric cryptography either hash functions or ciphers are the most suitable basic building blocks. Subsequently, we will thus focus on these. Hash functions are conceptually simpler than block ciphers, since they do not need a key. In the RFID security community, it is (or was) commonly assumed that hash functions are therefore also the better choice from the implementation point of view. As a consequence, most of the proposed protocols for protecting RFID tags base on hash function implementations. Weis et al. [34] proposed the hash-lock scheme and the improved randomized hash-lock scheme. Thereby, the tag is authenticated by the tuple $(r_t, h(\text{ID}, r_t))$ where r_t is a random value generated by the tag for tracking prevention and the hash value is generated over ID and r_t. In the protocol of Henrici et al. [19] the tag sends the hash value of its ID together with a transaction number to the reader and authenticates the tag to the reader. The reader responds with a random value which is used to refresh the tag identifier on every successful transaction. Lee and Verbauwhede [24] propose a protocol where a hash function $h(K||r_s)$ of a key K and a random value r_s is used for mutual authentication of reader and tag. Additionally, r_s is used for updating the key in the tag. Dimitriou [7] uses in his authentication protocol a MAC h_k which is based on hash functions. Random values generated by reader and tag are used for mutual authentication and refreshing the key in the tag. Rhee et al. [30] suggest a hash-based challenge–response protocol where the secret key in the protocol is the ID. The tag does not need to update the secret key which avoids attacks by interrupting sessions. In the protocol of Choi et al. [4], tags have a common secret key K and a tag-specific secret S which are used for mutual authentication based on hash functions. A counter c is incremented in the tag on each access for prevention of replay attacks.

There are only a few protocol designers which base their RFID authentication protocols on symmetric-key encryption primitives. Feldhofer et al. [14] use simple challenge–response authentication for unilateral and mutual authentication of RFID tags and readers. The random values r_t and r_r are the challenges from the tag and the reader which are encrypted using the function E_K which is in their case Advanced Encryption Standard (AES). They mention the problem of key distribution and key management when using symmetric authentication methods but do not provide any detailed solution for it. Because this chapter mainly deals with hardware design for passive RFID tag, we mainly restrict the applied protocols to challenge–response protocols with single-side authentication of the reader or the tag and mutual authentication where each party proves its identity to the other party.

1.4 Description of Important Symmetric Algorithms and Protocols

This section describes the most commonly used symmetric primitives in cryptography. This selection is a limited excerpt of available hash functions, block ciphers, and stream ciphers. We will present a short historical view of the algorithms and their use in standards and applications.

1.4.1 Hash Functions

Hash functions belong to the category of unkeyed cryptographic primitives and are mainly used for providing data integrity. This allows to detect data manipulation due to technical reasons or due to an attacker. A hash can be used for a MAC which allows to simultaneously verify data integrity and authenticity of a message. Due to the demanding design of hash functions regarding computational power and memory resources it might be favorable to minimize the use of hash functions in RFID applications and consider the use of block ciphers in an appropriate mode of operation instead.

For a long time, the design of cryptographic hash functions has received little research effort. Most of the MD4-family hash functions were believed to be secure. So far, the best known hash functions are the MD4-family hash functions SHA-1, SHA-256, MD4, MD5, and RIPEMD. All designs base more or less on the (today insecure) MD4 algorithm which was invented by Ron Rivest in 1990. The MD5 algorithm is used only in a few applications, SHA-1 is widely employed in many security protocols and applications like SSH, S/MIME, SSL, TLS, and IPSec or the keyed-hash message authentication code (HMAC-SHA1). Since the publication of an attack against a reduced version of the SHA-1 algorithm [33], the crypto community continues to improve the attacks [3] and endeavors to design new, more secure hash algorithms. So far NIST recommends to use the more secure SHA-2 variants (SHA-256, SHA-384, SHA-512) or to switch to alternative designs like Whirlpool, which is standardized in ISO/IEC 10118-3 [22], or the Tiger algorithm.

1.4.2 Block Ciphers

Block ciphers are symmetric-key primitives which operate on so-called blocks with a fixed number of bits. They transform the input block into a ciphertext of same size. Most block ciphers iterate the same transformation function several times using in each iteration a slightly varied round key derived from the secret key. Decryption is the inverse operation which allows to recover the plaintext from the ciphertext using the same secret key.

The Data Encryption Standard (DES) algorithm, which was standardized in the year 1976 by NIST has been the most important block cipher for several decades [27]. Due to its limited key size of 56 bits it was first extended by Triple-DES and is now replaced by AES which has larger key sizes (128, 192, and 256 bits). The Rijndael algorithm was selected as the winner of an open selection process initiated by NIST and has been standardized as the AES algorithm in the year 2001 [28]. During this selection process the four algorithms MARS, RC6, Serpent, and Twofish have also been selected as finalists. None of the algorithms was discovered to have severe weaknesses. Other interesting block ciphers for application in embedded systems are the Tiny Encryption Algorithm (TEA) and its extension (XTEA). Skipjack and IDEA are also suitable. TEA is based on very simple and fast operations that allow small software implementation.

1.4.3 Stream Ciphers

The name stream cipher comes from the fact that in such symmetric ciphers the plaintext bits are encrypted one after another by applying an exclusive-OR operation with a key stream that depends on the current internal state. The main difference to block ciphers is that a stream cipher does not operate on large blocks of fixed length but on individual bits. However, depending on the mode of operation, a block cipher can also be operated as a stream cipher. Typically, stream ciphers require less hardware resources. Although the RC4 algorithm is widely applied in security protocols like SSL and WEP, it is recommended not to be used anymore in new applications due to weaknesses. Other stream ciphers are currently used in GSM technology, e.g., A5/1 and A5/2 (both of which are broken).

The European Network of Excellence ECRYPT has started a contest called eSTREAM [9] in order to identify new stream ciphers for future standardization. In addition to a high security level, the ciphers should be suited for efficient implementation in software and/or hardware. So far there have been several interesting stream cipher designs published. The most promising approaches targeting hardware implementation are Trivium, Grain, Mickey-128, and Phelix. The designs that aim for efficient software implementation and which are in the final evaluation phase are Dragon, HC-256, LEX, Phelix, Py, Salsa20, and SOSEMANUK.

2 Requirements for Securing Passive RFID Tags

Passive RFID systems are very constrained systems. RFID tags have to have very low silicon area and must cope with a tightly limited power budget. These limitations have two major reasons: economical reasons and technical reasons. On one hand there are billions of RFID tags in the field. Thus, the production cost has to be low. Limiting the silicon area achieves this goal. On the other hand, the available power for operating passive RFID tags is limited because the power for operation has to be supplied over the air interface. The less power a passive RFID tag consumes, the longer is the operational range of the devices. In the following, we will scrutinize these observations and derive requirements that can be treated as upper bounds for the implementation of symmetric cryptography on RFID tags.

2.1 CMOS Technology and Area

The maximum silicon area of a passive RFID tag is not limited by technical considerations. From a purely technical point of view, tags can be as large as they want to be. However, economic considerations limit the size of RFID tags drastically. Production cost of integrated circuits rises more than linearly with the size of the circuit. Reasons for that is the yield achieved by the CMOS production process. The yield is the percentage of working parts in relation to all produced parts. Assuming a constant defect density when producing CMOS circuits, one can expect a yield that decreases exponentially with the circuit size because it is more likely that a defect affects a larger die. Thus, smaller RFID tags can be produced more efficiently. The market price of an RFID tag is not determined by the silicon area but by the functionality it provides. Thus, tags integrating the same functionality on smaller silicon area can be assumed to gain higher profit margins.

   Besides minimizing the circuit size on a given CMOS process technology, it is also of interest whether it is beneficial to go for more modern process technologies. Is it more efficient to produce an RFID tag on a modern 45-nm CMOS process than on good old 0.35-μm CMOS process? RFID tags do usually not make use of the most recent CMOS technologies like 45-nm CMOS. This has again technical and economical reasons. The technical reasons are determined by two facts. First, very fresh CMOS technologies do not feature nonvolatile memories. Nonvolatile memories like EEPROM or Flash are necessary to store information like the unique identifier or the electronic product-code on the tag. Second, newer CMOS technologies are susceptible to higher supply voltages and it is more difficult to realize mixed-signal circuits with good analog performance. The voltages induced by the (electro-) magnetic field emitted from the interrogator can exceed the maximum voltages allowed by these technologies. Producing RFID tags on newer process technologies requires more sophisticated strategies to limit these voltages by shunt regulators. Sophisticated shunt regulator again use a considerable amount of silicon area because they have to turn the received energy into heat. Besides technical limitations, also economical considerations play a role. The costs for the photolithographic masks that are needed for the CMOS production rise sharply with smaller process technologies. Thus, it is more costly to ramp-up production on smaller process technologies. It is also more costly to modify some of the masks during production to increase the yield or to do minor modifications of the design. Another economical argument against very new CMOS technologies is the vast amount of produced tags. CMOS circuits are produced in batches of wafers. Usually, a batch contains about ten wafers. Ten 300-mm wafers will result in more than a million RFID tags. This might be more than the whole annual demand. Another economical argument against recent CMOS technologies is astonishingly the small size of the RFID tags: When producing RFID tags on more recent CMOS technologies, their silicon area gets smaller. When the same machines are used to cut the dies from the wafer, the (scribe line) area where the wafer is sawn increases in proportion to the used silicon area. Moreover, tags below a certain size – let us say below $0.20\,\text{mm}^2$ – are more difficult to handle during production steps where they are bonded to straps or antennas.

Mapping all the previous considerations on available production processes gives the following picture: In the past, most produced RFID tags were operating in the HF range. Most of them were produced on CMOS technologies in the 0.35–$0.25\,\mu\text{m}$ range. These process technologies are comparatively cheap, have nonvolatile memory options, and have good analog performance. More recent products are fabricated on 180- and 130-nm CMOS processes. Up-to-date technologies like 90-, 60-, and 45-nm CMOS suffer from the drawbacks described above and are commonly not used for RFID tags (at the present).

The size of RFID tags is not limited by technical constraints – as already mentioned – but only by economical ones. Therefore, there is no upper limit for the silicon area of cryptographic hardware added to RFID tags. The added functionality defines the added value of the RFID tag. The larger this added value is, the higher is the market price of the tag and thus the higher the production cost can be. So size limitations of cryptographic circuits depends on their added value. A complexity of 1,000 GEs can be added to an RFID tag without perceptible additional production cost. One gate equivalent compares to the silicon area required by a NAND gate with two inputs. Gate equivalents are a measure that assesses the circuit complexity independently of the used CMOS technology. Moving circuits of certain complexity to smaller process technologies reduces therefore the area as shown in Table 1.

Table 1 outlines that cryptographic circuits of medium complexity (roughly 5,000 GEs) can be implemented on all relevant CMOS technologies without exceeding an area of $0.33\,\text{mm}^2$. Later analysis of various symmetric algorithms show that most of the symmetric cryptographic primitives can be realized with 5,000 GEs or even less. On advanced CMOS technologies having feature sizes of 180-nm and below,

Table 1 Circuit complexity for RFID security

Technology	1 gate equivalent (GE) [μm^2]	1,000 GE [mm^2]	5,000 [mm^2]	10,000 GE [mm^2]
$0.35\,\mu\text{m}$	55	0.055	0.28	0.55
$0.25\,\mu\text{m}$	24	0.024	0.12	0.24
180 nm	12	0.012	0.06	0.12
130 nm	5.2	0.0052	0.026	0.052
90 nm	2.4	0.0024	0.012	0.024

this complexity becomes comparatively small in relation to overall RFID tag sizes of 0.1–0.25 mm^2. Thus moving to more advanced CMOS technologies facilitates the implementation of cryptographic functionality.

2.2 Power and Energy

Implementing cryptographic functionality on RFID tags is limited not only by the available silicon area, but also by the available power for computing cryptographic operations. The energy for operating passive RFID tags is provided by the interrogator over the air interface. Most relevant for implementing cryptographic circuits are HF systems working on 13.56 MHz and UHF systems operating at 900 MHz. Both systems have in common that only a small fraction of the energy emitted at the interrogator's antenna is received by the RFID tag. Both systems induce a voltage in the antenna of the RFID tag. This voltage heavily depends on the distance of the RFID tag to the emitting antenna. In HF systems, the available power drops with the third power $P(d) = P_0/d^3$ of the distance d between tag and interrogator. Due to this sharp decline, the operation range of HF fields is well defined. In distances up to 1 m, a long-range reader can power HF tags. Exceeding this maximum power distance by some centimeters provides too less power for safe operation of the tag. The power regulator of the tag will recognize this situation and reset the tag. The powering distance of UHF systems is not as sharp as those of HF systems. UHF fields decline only quadratically (roughly speaking) with the distance d to the interrogator antenna: $P(d) = P_0/d^2$. This is due the fact that UHF systems operate in the far field, which begins in distances of a few centimeters distance from the interrogator's antenna. UHF systems have larger powering distances. They can provide enough power for operation of RFID tags up to 10 m. The powering distance heavily depends on the environment. In harsh industrial environments containing many metal objects, the superposition of electromagnetic waves can result in holes where no power is available – sometimes even close to the antenna of the interrogator. Superposition can also lead to over-distances where tags are powered in distances beyond 15 m that are not intended for communication.

The characteristics of HF and UHF fields define the constraints for energy and power consumption of RFID tags. When implementing cryptographic circuits on RFID tags it is desired that the cryptographic functionality does not limit the operating range. Thus, the tags have to cope with the limited power and energy budget. Before discussing these limits it is necessary to clarify whether power consumption or energy consumption is more important for passive RFID tags. The energy used for a (cryptographic) operation depends on the average power P_{avg} and the duration t of the computation: $E = P_{avg}t$. Energy efficiency is of utmost importance where devices are powered by battery. Once the energy storage capacity of the battery is drained, the device cannot work anymore. The situation for passively powered RFID tags is different. The average power transmitted from the interrogator to the tag is small but in theory the interrogator can supply the power arbitrarily long. Thus, the

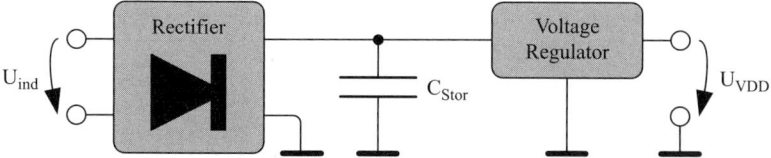

Fig. 1 Power supply of passive RFID tags

energy does not play an important role as long as there is enough time available to do the computation. Figure 1 shows the power supply of a passive RFID tag. The voltage induced by the antenna is rectified by a diode circuit. The charge is stored in the capacitor C_{Stor}, which buffers the energy. A voltage regulator lowers the voltage to supply a constant voltage U_{VDD} for the digital part of the RFID tag. U_{VDD} is usually 1.5 V for 0.35-μm CMOS technologies and around 1.0 V for newer technologies. The storage capacitor C_{Stor} is charged often in 13.56 MHz HF fields 27 million times a second. In UHF fields more than a billion times. The digital hardware draws comparatively seldom current from the storage. CMOS power consumption is determined mostly by dynamic power consumption which concentrates around clock edges. For keeping the average power consumption low, clock frequencies are kept low too. Most digital RFID circuits are not clocked above 100 kHz. Thus, the capacitor C_{Stor} has to provide power for a few nanoseconds of dynamic power consumption every 10 μs. In the meantime it is charged by the field. Every activity of the circuit draws current from C_{Stor} resulting in a drop of voltage. If the voltage in C_{Stor} drops below a threshold the RFID tag resets. The described properties of the circuit providing the supply voltage for the digital part lead to the following considerations for the power consumption:

- Keep the average power consumption P_{avg} low.
- No clock cycle should consume excessive power.

The first consideration stems from the observation that power is transferred continuously but slowly from the interrogator to the tag. This requires to keep the average power consumption low. But the total energy can be high if there is enough time because $E = P_{avg}t$. The second consideration regards the prevention of brown-out resets. RFID tags gather between one significant clock edge and the next the required energy. The storage capacitor cannot be made as large to store energy for very long periods due to its large silicon area demands. Thus, a single clock cycle may consume only as much energy as is available. When one clock cycle exceeds the available budget the voltage at the capacitor C_{Stor} drops below the threshold and the circuit will reset and render all computations useless. Therefore, every clock cycle should roughly consume the same amount of power to prevent such a situation.

The average power P_{avg} consumed by the digital circuit is proportional to the average current: $P_{avg} = I_{avg}U_{VDD}$. This is due to the fact that the supply voltage U_{VDD} is constant. Most often it is more convenient to measure the power consumption indirectly by stating the average current I_{avg}. This is also motivated by circuit simulators that output I_{avg} directly. Concrete power budgets can be given when analyzing

Table 2 Available power budget for crypto in HF and UHF systems

RFID system	Power distance	P_{avg}	I_{avg}
HF (13.56 MHz)	1 m	22.5 μW	15 μA (at $U_{VDD} = 1.5$ V)
UHF (900 MHz)	5 m	4 μW	4 μA (at $U_{VDD} = 1.0$ V)

the power situation in 1 m distance from HF long-range readers and in 5 m distance from UHF readers. These are distances, where secure RFID tags still should work properly. Table 2 gives a rough estimate. Exact values depend strongly on the particular situation. This comprehends the national regulations for the maximum radiated power from RFID antennas, the size and orientation of the interrogator's antennas, the antenna of the tag, the analog part of tag, and so on. Nevertheless, Table 2 gives a clear picture that cryptography on UHF systems is much harder to realize than on HF systems. It is thus more likely that secure RFID systems show up first in the HF frequency and then later in UHF. On both systems, only tailored cryptographic hardware can meet the requirements.

2.3 Throughput and Latency

The throughput and latency requirements of RFID systems are not as difficult to achieve as the power requirements. RFID systems provide only slow data rates compared to other digital communication techniques. HF systems communicate with data rates ranging from 6 to 106 kbps. UHF systems communicate faster and allow 5 to 320 kbps. In terms of cryptographic throughput this means that for instance 128-bit blocks of the AES algorithm can be transferred 40–2,500 times per second. Thus, AES hardware has to be able to compute one encryption within 25 ms in the slowest case or 400 μs in the fastest case. When the cryptographic hardware is clocked at 100 kHz this compares to 2,500 and 40 clock cycles to compute an encryption. These throughput considerations do not take communication overhead into account which will lower the throughput demands and relax the implementation of cryptographic hardware a little.

Latency is a performance measure like throughput, which measures the time to compute one encryption. Latency is of importance where only one cryptographic operation is computed – for instance in challenge–response protocols. These protocols could be mapped directly on the reader-talks-first communications scheme of passive RFID systems, where every communication is initiated by the interrogator. The reader issues the challenge and the tag answers after a defined time with the according response. The time for answering requests is standardized. In HF systems, this time is only 300 μs, in UHF protocols only 25 μs. These times compare to 30 and 2.5 clock cycles in a 100-kHz circuit. These are very demanding requirements. Even cryptographic hardware built for Gigabit networks can eventually not fulfill these requirements. Thus, the latency requirements have to be released somehow.

A very simple and effective approach was proposed by Feldhofer et al. [14]. They propose to separate the transmission of the challenge from picking up the response by sending two reader commands that are timely relocated. This gives a tag more time to do the cryptographic operation. This approach does not deteriorate the bulk authentication rate because multiple tags can be challenged in an interleaved scheme, where in a first phase many tags are challenged and afterward the responses are collected. This approach allows longer times to compute the cryptographic algorithm. It can be as long as it is acceptable to wait for a response if only one tag is in the field. Thus, tens of milliseconds seem appropriate, which relates to more than 1,000 clock cycles at 100 kHz.

2.4 Robustness Against Side-Channel Attacks

Cryptographic circuits should not be assessed only by their performance, their power consumption, and the required silicon area. Every optimization is useless when the circuit is not secure. Side-channel attacks turned out in the last years to be a severe threat against cryptographic hardware [26]. Side-channel analysis (SCA) attacks derive secret keys from side-channel information like the power consumption during computation. It makes use of statistical methods to correlate the power consumption of the circuit's model processing a distinct key with measured power traces. Unprotected circuits are prone to SCA. Most often, hundreds of power traces are enough to retrieve the secret key from cryptographic hardware without tampering it. The question is whether passive devices are also prone to side-channel attacks. In contrast to wired devices, passive RFID tags pull their power out of the reader fields. Thus, it could be assumed, this restrains SCA. But this is not true as has been shown recently. Michael Hutter et al. researched the electromagnetic side-channel of passive RFID devices in [20]. They found out that passive RFID tags are as prone to side-channel attacks as their wired counterparts. Thus, secure implementations of cryptographic algorithms have to include countermeasures against side-channel attacks.

3 Low-Resources Design Methodology

The stringent requirements imposed by passive RFID technology require also a stringent design methodology for implementation of a cryptographic algorithm [12]. The methodology should adhere to approved design principles. It has to evaluate different design options carefully. When designing hardware for passive RFID tags, low die size and low-power consumption are the major goals while high data throughput is of minor importance. It is inevitable to study the pros and cons of a measure to reach all design objectives. For example, a specific step can reduce the silicon area of the chip but can increase the power consumption significantly. In this section, we describe various aspects which require attention during the design.

The chosen hardware architecture for an algorithm mainly determines the properties of an implementation. Well-considered design decisions on architectural level help to meet requirements like low-power consumption, low die size, and acceptable throughput rates. Most of the published hardware implementations of symmetric algorithms have Gigabit throughput rates as optimization goal. This is in contrast to passive RFID systems where low-power consumption is the most important goal.

Along with optimizations on the architectural level it was very important to consider various implementation details to achieve the ambitious design objectives. Many high-volume products make use of full-custom optimizations to improve critical parts of a circuit by having full control over the logic styles, the dimension of transistors, and the layout. Here, we will concentrate on a semicustom design flow that does not make use of full-custom optimizations. The resulting circuits are implemented using standard cells only. Full-custom optimizations are the last resort to meet requirements. We apply them rarely to keep the development time within limits and to facilitate quick migrations to more recent CMOS technologies. Therefore, all optimizations are done on the register-transfer level – namely in the HDL model. This facilitates a fast semicustom design flow including synthesis, place, and route for different CMOS technologies to produce a first-time-right silicon.

3.1 Design for Low-Power Consumption

The most important design goal for passive RFID tags is to reduce the power consumption because this limits the operating range of the tag. There exist several techniques to achieve this goal. In principle, all these techniques try to minimize the dynamic power consumption by minimizing switching activity. The objective of design for low power is to have only that nodes showing activity which are necessary to compute the algorithm. Surprisingly many nodes in digital circuit show activity that do not contribute to the desired computation but cause undesired power consumption. As discussed before the energy of a computation and the duration of the computation are of minor interest. Thus, a simple but very effective measure for lowering the average power consumption P_{avg} is to stretch out the computation in time. Doing the same computation in longer time lowers P_{avg}. Thus, most digital circuits of RFID devices are not clocked faster than 100 kHz. Another very basic approach is to serialize the computation by stretching it out over more clock cycles. Both measures trade in principle lowered P_{avg} for increased t, where the time $t = \sharp_{cyc} t_{cyc}$ is defined by the number of cycles $\sharp_{cyc}$ and the cycle time $t_{cyc} = 1/f_{CLK}$, which is the reciprocal of the clock frequency f_{CLK}.

The total power consumption of a CMOS circuit is the sum of static and dynamic power consumption. The static power consumption caused by the leakage current. It mainly depends on the size of the circuit and is very small. It can be more or less ignored for our considerations because area minimization is anyway a design goal.

Equation (1) presents influences on the dynamic power consumption. The design measures for lowering the power consumption result from minimizing the factors in this equation.

$$P_{dyn} = C_L\,U_{VDD}{}^2\,f_{CLK}\,p_{sw}. \tag{1}$$

The load capacitance C_L of the chip increases as more gates are placed on the die. Thus, smaller circuits inherently save power. A reduction of the supply voltage (U_{VDD}) is also very effective because it has quadratic influence. These two factors tend to be predetermined by the low die-size constraint and the operating conditions of the chip. The supply voltage is most often chosen to be only slightly above $2V_{threshold}$. At low supply voltages, CMOS logic gets slow but that does not matter for digital circuits which are clocked slowly. The remaining two factors in (1) for minimization are p_{sw} and f_{CLK}. The reduction of the clock frequency f_{CLK} was already discussed. Lowering the switching activity p_{sw} of the circuit can be reduced by various means. Avoiding glitching activity and signal activity in unused circuit parts are effective to lower p_{sw}. Short critical paths, clock gating, and sleep logic are therefore effective measures that will be examined later. Before discussing these techniques on gate level, it is more relevant to investigate possibilities on architectural level.

3.1.1 Architectures for Low Power

Most cryptographic algorithms offer many architectural options for hardware implementation. One may ask whether there are general considerations on architectural level which are independent of the actual algorithm to reduce the power consumption. It turns out that the serialization of computations discussed above has important influence on the architectural level. Serialization of algorithms means to split the computation of operations on n bits into k smaller computations operating on w-bit chunks of the data (so $n = kw$). In other words this means that the width of the datapath is shortened from n bits to $w = n/k$ bits. Obviously, the smaller the width w of the datapath is, the smaller is the signal activity. The original width n of the datapath has here a slightly different definition than usual: It counts all the bits that need to be stored during a computation cycle. For instance, $n_{AES-128} = 256$ because AES-128 needs to store a 128-bit state and a 128-bit round key. The number of active nodes in the datapath is assumed to be proportional to the datapath width n. When reducing the datapath size to $w = n/k$, overhead has to be added that is proportional to $k + a_w w$. For instance, k clock-gating cells are necessary to activate only one out of k w-bit registers. Additional w-bit wide (multiplexer) circuitry is needed to select the input for the datapath from one of the k w-bit registers. But these multiplexers are surely not as big as the rest of the datapath. Thus the factor $a_w < 1$. A first estimate of the resulting power consumption will thus be $P(w) = P_0 \cdot w + P_0 \cdot (k + a_w \cdot w)$. Minimizing this functions yields $P'(w) = 0$: $w = \sqrt{\frac{1}{1+a_w}} \cdot \sqrt{n}$. Deriving the optimum word size from this formula yields w to be proportional to $\sqrt{n}$. For AES, where $n = 256$, this will result in $w = 16$. This word size is just an estimate. The optimal word width of the datapath of course depends heavily on the algorithm. Nevertheless, word sizes of 8–16 bits seem appropriate for hardware implementations of block ciphers. For hash functions, which have larger n, 16-bit wide to 32-bit wide datapaths are a good choice. The calculated optimal word width for a datapath has to be adapted for a specific algorithm when the used operations (e.g., size and number of S-boxes) do not directly fit the estimated value.

3.1.2 Clock Gating

Clock gating is a very effective measure for reducing the effective clock frequency. Figure 2 shows the basic concept of clock gating. Figure 2a shows how a register is built in a synchronous circuit with a multiplexer at the input which is used to feed back the old value into the register to store the register's content over more than a clock cycle. Loading new values is controlled by the enable signal. When using clock gating, see Fig. 2b, the enable signal activates the clock-gating cell which prevents glitching. As clock edges occur only when new data is stored in the registers, there is no need to have multiplexers at the input of each register to feedback the old value. The saved multiplexers can reduce the silicon area of the total circuit. Our implementations of symmetric algorithms base on synchronous circuit design making heavy use of clock gating. Datapath registers and nearly all control logic registers are clocked only when there is a potential signal change. It turned out to lower the power consumption significantly. Thereby, parts of the circuit are virtually switched off when they are not in use.

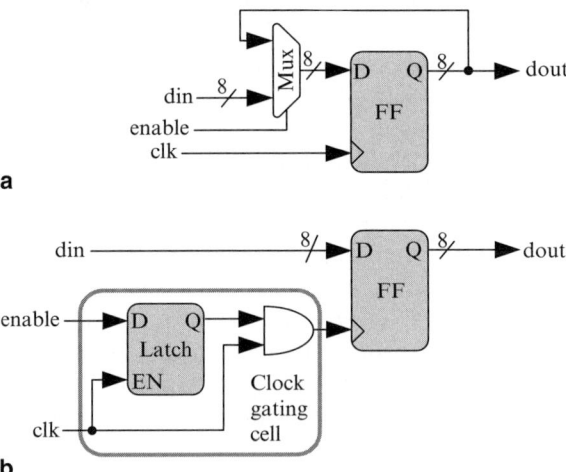

Fig. 2 (**a**) Traditional synchronous circuit. (**b**) Clock gating mechanism

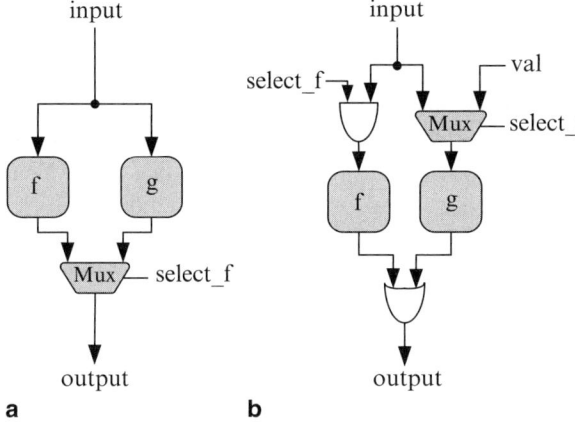

Fig. 3 (**a**) Traditional circuit. (**b**) Sleep logic

3.1.3 Sleep Logic

The switching activity p_{sw} of the circuit can be reduced by using a method called sleep logic. Whenever the output of a combinational circuit is not needed, changes of the input data will nevertheless cause switching activity and hence power consumption inside the module although the computed data is not needed. Figure 3a shows the traditional way of selecting the output of two different functions $f()$ and $g()$ using a multiplexer. Both functions are computed and one of them is selected at the output. In order to prevent this undesired switching activity, the inputs of the combinational circuit are masked using AND gates or the input of a function is set to a value which generates a zero output. Figure 3b shows details how to compute either function $f()$ or function $g()$. The property that $f(0) = 0$ and that $g(\text{val}) = 0$ is used to chose the desired result without the need of more complex multiplexers. A sleep signal that disables the AND gates prevents all switching activities of the combinational logic behind the gates because the input to the following combinational logic does not change.

3.1.4 Logic Depth

The logic depth of combinational circuits is also worth a short discussion. Digital circuits for passive RFID tags are usually clocked very slowly to keep the power consumption low. Hence, the clock period $t_{cyc} = 1/f_{max}$ can be very long, which allows a very deep multilevel logic. For instance, a 100-kHz clock allows a critical path of up to 10,000 ns. This compares to several hundred full adder cells. Although, long critical paths, do not deteriorate the performance of RFID tags, they can influence the power consumption negatively. In ordinary circuits with long critical paths, there are huge differences in the length of the combinational paths. As a consequence, the runtime of signals is very different and the probability of glitches increases sharply. Glitches have a major influence on the power consumption. When the timing differences are in the order of rise and fall times, glitching nodes can be fully charged and discharged. Signal activity caused by glitching does not contribute to the functionality because the signal transitions are reversed in the same cycle. But they do contribute to the power consumption. Nodes that are prone to glitching can have switching activities $p_{sw} > 1$ which is higher than those of clock signals. Therefore it is advisable to keep the combinational paths within limits. Besides saving power, this strategy also raises the reusability of RFID crypto modules: They can be used in other applications where higher clock frequencies and higher throughput requirements are demanded.

3.2 Design for Low Die Size

The design of a VLSI circuit for low die size using standard cells involves considerations from algorithmic level down to the circuit level. For implementing a cryptographic algorithm, the decision which word width should be used and which options of an algorithm should be implemented is important. For instance, it could make sense for a specific application that only encryption is necessary or that a special mode of operation is required. The smallest useful word size should be chosen because then it is possible to reuse the hardware resources as much as possible.

Implementing the required memory resources for a cryptographic hardware module is important because storage requires mostly the largest chip size. Using a dedicated RAM instead of a flip-flop-based approach is a possible solution to minimize the required hardware resources. Normally, RAM macros are more area efficient because of their regular layout. We think that dedicated RAM is only more efficient when memory sizes of several kilobits are needed. This is not the case for algorithms of symmetric cryptography. From the evaluated algorithms the SHA-256 implementation requires the highest amount of memory using 1024 bits. Additionally, a dedicated RAM circuit would consume much more power because RAMs use power-consuming precharging of bit lines for every access.

3.3 Different Circuits – A Metric for Comparison

A comparison between different implementations of crypto algorithms is only possible when a clear optimization goal is stated. Many publications use the area-delay product or the power-delay product to compare their results with each other. The multidimensional optimization problem to keep the cost parameters chip area, power consumption, energy consumption, number of clock cycles, clock speed, and others at a minimum depends on the field of application. Different applications (e.g., a HF application vs. an UHF application) might favor different solutions. Two parameters are often orthogonal while others are correlated. In battery-powered devices, for example, the energy consumption has to be minimized while other parameters like chip area and throughput are not as crucial. In RFID systems, we have two major design objectives which do not correlate. These are the reduction of power consumption and the reduction of the chip area. Important as well is the number of clock cycles for executing the algorithm. Of minor relevance are the energy consumption and the maximum clock frequency.

$$(\mathrm{I_{mean}\,[\mu A]},\ \mathrm{Area\,[GE]},\ \mathrm{Clock\,cycles\,[\#]}). \tag{2}$$

In (2) we define a triple of parameters for optimizing hardware modules in RFID systems. It allows comparing different implementations. This triple does not map all cost factors to a single one-dimensional value (a real metric). Instead, the individual cost factors remain unweighted. This allows calculating compound metrics like the area-delay product by others. The first parameter of (2) is the mean current consumption which is given in μA. The average current depends approximately linearly on the clock frequency. The supply voltage is usually not a free design parameter because it is chosen to be the minimum anyway. In our case, 100 kHz and 1.5 V are nominal operation conditions. The second parameter is the chip area of the implementation given in gate equivalents. The third parameter is the number of clock cycles to execute the algorithm. The cycle count and the circuit size measured in gate equivalents are nearly independent of the process technology. On the contrary, the power values given in this chapter are only valid for a 0.35-μm CMOS process technology. They cannot be mapped to a different technology easily.

4 Implementation of Symmetric-Key Algorithms

The implementation of symmetric-key algorithms for passive RFID tags is far from trivial. Due to the stringent requirements for low-power consumption, low chip area, and the limited number of clock cycles the design has to be highly optimized. The above-mentioned design strategies for hardware implementations of passively powered devices help to reduce the large design space of a specific cryptographic algorithm to a manageable subset. In this section, each algorithm will be shortly described and some related work to the specific algorithms will be presented. The given implementation details will show the design decisions during the analysis phase. This analysis of different algorithms in this section should allow a fair comparison of the presented algorithms. All algorithms are implemented on the same 0.35-μm standard cell CMOS process technology and have the same design goal to achieve best results for passively powered RFID tags. All presented modules can operate as stand-alone modules and have an AMBA bus interface. They do not rely on any external resources like memory macro cells. In the following, we will analyze the block ciphers AES-128, TEA, and XTEA where the results for AES come from [14, 16]. Afterward the hash function modules SHA-1, SHA-256, and MD5 will be investigated. The results for the hash functions mainly base on the work in [11]. We finish with a detailed analysis of the stream ciphers Grain and Trivium which bases on the work in [10]. For a comparison with asymmetric algorithms in general we make a short excursion to the realization of an ellipticcueve Cryptography (ECC) module. Although this is little bit off-topic in a symmetric algorithm chapter this helps to understand the problematic of asymmetric algorithms for passive RFID tags.

4.1 AES

AES is a symmetric block cipher which was originally named Rijndael. It was invented by Joan Daemen and Vincent Rijmen and became the new Federal Information Processing Standard FIPS-197 [28] standard in the year 2001 after a contest organized by the National Institute of Standards and Technology (NIST). Due to its flexible design, the AES algorithm can be implemented on various platforms very efficiently. Software implementations on small 8-bit microcontrollers and 32-bit processors are also suitable as dedicated hardware implementations. The free availability of the algorithm allowed the application of AES in numerous standards like IPSec, TLS, and wireless LAN according to the IEEE802.11i standard [21].

The main design principle of the authors of AES was the simplicity [6], which results in its efficient implementation on various platforms under different constraints. This is realized by using only a small set of basic operations which base on finite-field arithmetic and by means of symmetry properties at different levels. The first symmetry level is realized by applying the same round function to the fixed data block of 128 bits several times. All operations of the AES algorithm transform a block of data which is referred to as the *State*. It is organized as a 4×4 matrix of bytes. Depending on the key size which could be either 128, 192, or 256 bits the number of round transformation for encryption and decryptions are 10, 12, or 14. In Fig. 4, the AES algorithm is described in a graphical way. Within this chapter we focus on 128-bit keys because this is the mainly deployed key length.

T he round transformation modifies the State from the initial value which is the input plaintext to get the output ciphertext after the last round. The round function

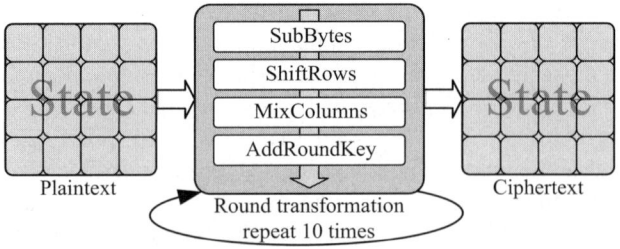

Fig. 4 Description of the AES algorithm

consists of the transformations SubBytes, ShiftRows, MixColumns, and AddRound-Key for encryption and InvSubBytes, InvShiftRows, InvMixColumns, and AddRoundKey for decryption. These transformations are key-dependent, linear, and nonlinear operations defined over different finite fields. The bytes of the State are either individually, row-wise, or column-wise involved in the operations. An exception of the round transformation is the initial round and the last round. The initial round consists only of the AddRoundKey operation while in the last round the Mix-Columns operation is omitted. While encryption calculates the ciphertext from the plaintext the decryption function of AES computes the plaintext from the ciphertext. There are only minor differences in encryption and decryption where the operations are executed in reverse order and the round keys are used backward. In the following, only the AES encryption operations are described in more detail.

SubBytes is the only nonlinear operation of AES. It consists of a byte-by-byte substitution of all 16 bytes of the State using the same look-up table often referred to as S-box. The values of this S-box can be calculated by an inversion in the binary extension field $GF(2^8)$ followed by an affine transformation. There are many possibilities to implement the S-box. For software implementations, the 256 entries are mostly stored in a look-up table. For hardware realizations, the design space is even larger. A thorough analysis for implementing the S-box for low-power applications and low chip area can be found in [32].

ShiftRows simply rotates each row of the State using a specific offset. This offset is given by the row index of the State which means that in the first row no rotation is performed, the second row is rotated by one byte, and so on. In comparison to software implementations where appropriate addressing of the bytes in the subsequent operation is sufficient, in hardware ShiftRows can also be implemented using appropriate wiring of the signals.

MixColumns operates on columns of the State where each column is interpreted as a polynomial of degree three with coefficients from the field $GF(2^8)$. Mix-Columns is defined as a multiplication of this polynomial with a constant polynomial $(03 \cdot x^3 + 01 \cdot x^2 + 01 \cdot x + 02)$ modulo the irreducible polynomial $x^4 + 1$. The operation can also be expressed as a matrix multiplication of a constant 4×4 matrix with the input column.

AddRoundKey adds a round key to the State using an exclusive-or operation over all 128 bits where in each round another key is used. The first 128-bit round key is equal to the cipher key. All other keys are computed from the previous round key using the key expansion operation. This operation uses the S-box functionality and some constants referred to as Rcon. In systems with sufficient memory resources it is common to calculate and store the whole key schedule in advance. The other possibility is the on-the-fly key expansion where the current round key is calculated on demand during the encryption or decryption operation which allows saving memory resources.

There are many implementations of AES published in literature. Most of them target on high clock speed and high data throughput using 128-bit architectures. As target technology either FPGAs or ASICs are used. Only a few energy-efficient implementations which use 32-bit architectures are available [25]. The description of an AES implementation suitable for passive RFID tags can be found in [16]. The presented architecture in this section is based on this paper. An exhaustive survey of related work for AES which would be beyond the scope of this section can be found in [15].

4.1.1 Architecture of AES Module for RFID Tags

Following the design guidelines described above in this chapter the selected word width should be the square root of the number of flip-flops used to store the AES State and the round keys. This would result in a word width of 16 bits. This would lead to the need for two S-boxes which require a larger chip area. Hence, the decision was made to use an 8-bit architecture instead. The drawback of an 8- bit architecture is the increased number of clock cycles per encryption. As this is not the major design goal for RFID tags this trade-off is acceptable.

The architecture of the proposed 8-bit AES module is depicted in Fig. 5. The three main parts of the module are the controller, the RAM, and the datapath. The controller is responsible for sequencing the ten rounds of an AES encryption and decryption, it addresses the RAM accordingly, and it generates the control signals for the datapath. It is realized as a hard-wired finite-state machine which optimizes the power consumption and reduces the chip area. The RAM stores the 128-bit State and the current 128-bit round key. These 256 flip-flops are organized in 32 8-bit registers to suit the 8-bit architecture of the module. Storing 32 bytes during AES calculation is the smallest possible configuration that is required by an in-place implementation of all operations without overwriting a State nor a key byte that is used afterwards. The RAM has only a single port which eases silicon implementation. Because of the small size of the RAM and the requirement for a standard-cell implementation the RAM is register-based instead of using a RAM hard-macro. It makes extensive use of clock gating to minimize the power consumption. This means that only one byte is clocked at a certain clock cycle.

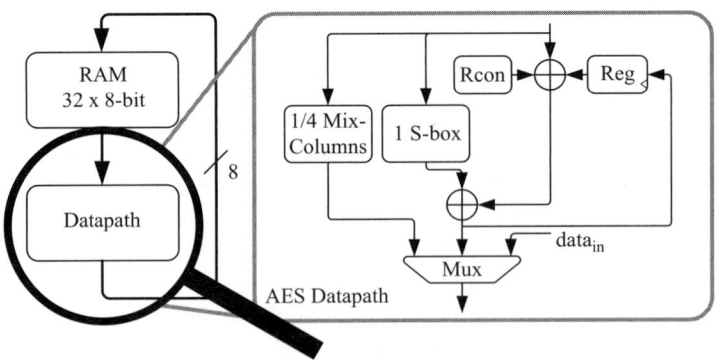

Fig. 5 Architecture of the AES module

The datapath of the AES module contains combinational logic to calculate the AES transformations SubBytes, MixColumns, and AddRoundKey (see Fig. 5). The ShiftRows transformation is degraded to a simple addressing operation. It is implemented by the controller during the execution of SubBytes by writing to the appropriate RAM address after the S-box substitution. The largest part of the AES datapath is the S-box which allows several options for implementation. The most obvious option is a 256×8-bit ROM to implement the 8-bit look-up. Due to the high-power consumption ROMs cannot be used for low-power designs. Hence, a more appropriate option is to calculate the substitution via combinational logic using finite-field arithmetic as presented in [36]. A feature of the "calculating" S-box is that it is possible to insert a pipeline register stage. This shortens the critical path of the S-box and lowers glitching probability and hence saves power. Moreover, in the presented design, the pipeline register in the S-box is used as intermediate storage. During the substitution of one byte, the next byte is read from the memory. The substituted byte is written to the current read address. By choosing the read addresses properly this procedure combines efficiently the SubBytes and the ShiftRows operation.

Another important solution to achieve the ambitious low-power and low chip area goals is the calculation of the MixColumns operation. The implemented submodule calculates only one-fourth of the MixColumns operation. We used this property to reduce the MixColumns multiplier circuitry to one-fourth of its original size. The resulting circuit calculates one output column in four clock cycle after a load phase of three cycles. The $\frac{1}{4}$-*MixColumns* circuit contains, besides the combinational circuit, three 8-bit registers to store three input column bytes. These registers have to be filled in the preloading phase before the first output can be calculated. The fourth input is taken directly from the RAM. Consequent output values are calculated by shifting the RAM output value to the registers and selecting the next value from RAM. A complete MixColumns operation to transform the entire State takes 28 clock cycles. The critical path of the MixColumns circuit is even shorter than the S-box.

Remaining components of the datapath are the submodule Rcon which is used during the key expansion, some XOR gates, and an 8-bit register. The XOR gates are required for round key generation and are used to combine the State with the round key during the AddRoundKey transformation. The 8-bit register is also necessary during the AddRoundKey operation and the key update. An encryption of a plaintext block works as follows. Before encryption or decryption is started, the plaintext block and the initial key has to be loaded into the RAM of the AES module. In the RFID tag application, the plaintext block is the 128-bit challenge which was received from the reader. The communication between the reader and tag is byte-oriented which fits nicely into the 8-bit architecture of the AES module. Hence, every received byte can be directly stored in the AES module and no intermediate memory is necessary. The cryptographic key is obtained in a similar way from the tag's nonvolatile memory resource. The memory-mapped interface has an address range for data and for keys and two additional registers for the control and the status word. Now the AES algorithm can be executed. It starts with a modification of the State by an AddRoundKey operation using the unaltered cipher key. Ten AES rounds follow by applying the transformations SubBytes, ShiftRows, MixColumns,

Table 3 Components and their complexity of the AES module

Module/component	I_{mean}		Chip area	
	(μA at 100 kHz)	(%)	(GE)	(%)
RAM	1.55	51.7	2,065	58.9
S-box	0.4	13.3	345	9.8
MixColumns	0.25	8.3	350	10.0
Register	0.1	3.3	58	1.7
Adder	0.05	1.7	80	2.3
Controller (FSM)	0.55	18.3	490	14.0
Others	0.1	3.3	115	3.3
Total	3.0	100	3,503	100

and AddRoundKey. Only the last round lacks the MixColumns operation. Round-keys are calculated just in time. This is usually called on-the-fly key schedule. The round key is derived from its predecessor by using the S-box, the Rcon, and the XOR functionality of the datapath.

4.1.2 Results of AES Implementation

The AES-128 encryption and decryption module was realized on a 0.35-μm CMOS process technology. The results were obtained by transistor level simulations and proved by a real silicon implementation and measurements. The chip has a complexity of 3,503 GEs where one gate equivalent compares to a NAND2 input gate having 55 μm^2 chip area. The mean current consumption is 3.0 μA when running the circuit at a frequency of 100 kHz and having a supply voltage of 1.5 V. Encryption and decryption of a 128-bit data block requires 1,044 clock cycles. The complexity of each component is listed in Table 3. It can be seen that the RAM circuit requires by far the largest amount of chip area, and power consumption. The S-box implementation and the MixColumns module are, in addition to the controller, the other major parts of the circuit. It can be seen that the presented solution achieves the high demands for integrating a symmetric cipher into passive RFID tags. These requirements are the low-power consumption and low die-size while conforming the requirements concerning encryption speed. The presented AES implementation serves as benchmark for the other modules presented in this chapter.

4.2 TEA/XTEA

The Tiny Encryption Algorithm (TEA) was invented by David Wheeler and Roger Needham who published the algorithm in the year 1994 [35]. TEA is a Feistel block cipher that is famous for its simple description and implementation (see Fig. 6). It works on 64-bit blocks of data (denoted as V) and uses a 128-bit key. The num-

Fig. 6 Description of XTEA algorithm

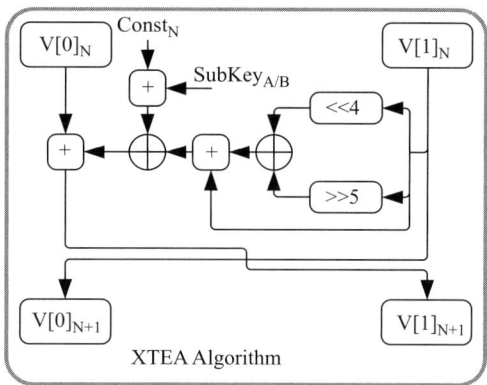

XTEA Algorithm

ber of rounds is suggested to be 32 and the used operations are additions modulo 32, XOR operations, and shift operations. No explicit permutation and substitution boxes are necessary. The rather simple key schedule uses the same mixing strategy in each round. A magic constant that is summed up is used to prevent attacks that are based on the symmetry of the rounds. Although some weaknesses are found in the TEA algorithm it seems that it is resistant against differential cryptanalysis. The most problematic weakness is that every key has three equivalent keys which reduces the effective key size to 126 bits. Because of this and other problems there are a number of revisions. The most popular one is called XTEA, which consists of some minor extensions which make the cipher more secure. The main differences lie in the more complex key schedule where the used part of the key depends on the accumulated magic constant, the usage of 64 rounds, and a rearrangement of the operations addition, XOR, and shift. The authors of the algorithm argue that TEA and XTEA are very suitable for implementation under fierce constraints like RFID systems. Hence, in the following the implementation of the XTEA which is optimized for low-power applications like RFID tags is presented. Only the architectural details of XTEA are presented because the TEA implementation looks very similar. For a comparison the results of both algorithms are shown in detail. A comparison with the literature is difficult because to the best of our knowledge there is only one implementation available [23].

4.2.1 Architecture of XTEA Module for RFID Tags

The architecture of the XTEA module follows the concept of a microcontroller approach where dedicated registers are used as memory elements and the combinational circuit is designed in particular for the required operations of XTEA. In Fig. 7 an overview of the XTEA datapath architecture is presented. The required memory resources are separated in 32-bit registers. In contrast to a RAM circuit where every memory element can be addressed individually, the internal structure of XTEA allows that each register output is used for a dedicated combinational

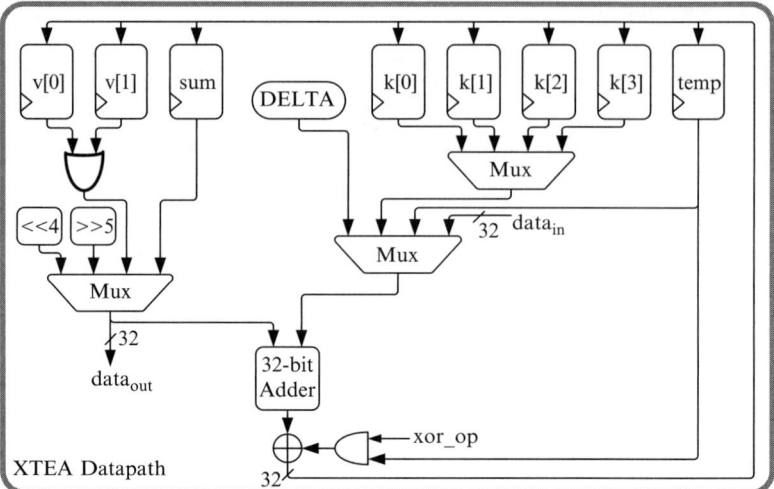

Fig. 7 Implemented architecture of XTEA

logic module. Each register is clock gated and allows that only one 32-bit register is clocked per cycle. This reduces the mean power consumption significantly. The output of each register can be disabled which allows that all multiplexers can be implemented as an OR tree. This eases the control logic and reduces glitching activity. In addition to two state registers and four registers for the key, one register is used to accumulate multiples of the constant Delta and one register is used to store intermediate results. This register builds some kind of overhead but it is necessary during execution of the algorithm. The shifter circuit allows a selection of the input signal unshifted, shift left by four positions, and shift right by five positions. The input from the sum register is also selectable but needs not to be shifted. The 32-bit constants Delta are implemented as an unstructured mass of standard cells. Their actual realization is decided by the synthesizer. The arithmetic-logic unit consists of a 32-bit adder/subtractor where for encryption only addition is necessary but decryption also requires subtraction. A straight-forward implementation using a ripple-carry adder is used. In series to the adder a further XOR gate is used which allows to combine the output of the temporary storage register with the result of the addition. The execution of the algorithm is controlled by a finite-state machine which is implemented as hard-wired logic. It generates the control signals for the datapath and sequences the 64 rounds of the same round function which requires 11 clock cycles per round. After loading data and keys via the AMBA bus interface, encryption or decryption is started. In every clock cycle a source register, destinations register, and the operation is selected which means that the execution works similar to a software realization.

Table 4 Implementation results of TEA and XTEA algorithm

Module/component	Chip area	
	(GE)	(%)
Eight 32-bit register	1,592	60.4
Arithmetic-logic unit (ALU)	347	13.2
Constant	5	0.2
Shifter	179	6.8
Multiplexer	180	6.8
Controller (FSM)	258	9.8
Others	75	2.8
XTEA total	*2,636*	*100*
TEA total	*2,633*	*99.9*

4.2.2 Results of TEA/XTEA Implementation

Table 4 shows the results of chip area for implementation of the algorithms XTEA and TEA. Both algorithms have nearly the same chip area with 2,636 and 2,633 GEs. The largest parts in the circuit are the eight 32-bit registers which are necessary for storing the state (2), the constant accumulator, the key (4), and a temporary register. The arithmetic-logic unit is mainly a 32-bit adder/subtractor and requires 347 GEs. The shifter, several multiplexers, and the control logic are only minor parts of the circuit. The power consumption figures where XTEA requires $3.86\,\mu A$ at $100\,kHz$ and TEA requires $3.79\,\mu A$ show that the 32-bit implementation requires a little more power as the 8-bit AES implementation. The number of clock cycles for encryption of one 64-bit block of data is 705 for XTEA and 289 for TEA which is in the same order of magnitude as the AES encryption of a 128-bit block of data.

4.3 SHA-1/SHA-256/MD5

In this section, the hash functions SHA-1, SHA-256, and MD5 will be analyzed. Because SHA-256 has the same security level as AES-128 and there exist some cryptanalytic attacks against the hash functions SHA-1 and MD5 we will only go into the details of SHA-256. Due to the very similar internal structure of the three algorithms most of the arguments for implementation can be directly applied to SHA-1 and MD5 too. Only the achieved results of all implementations will be presented at the end of this section. SHA-256 [29] is, like SHA-1, an iterated cryptographic hash function based on a compression function which was designed by the National Security Agency (NSA) and standardized by the NIST. The internal state of 256 bits is initialized using initialization vectors (IVs) as specified in [29]. SHA-256 updates the state of eight 32-bit variables A, ..., H according to the values of 16 32-bit words $M_0, \ldots, M_{15}$ of the message (SHA-1 has only five state variables). The compression function consists of 64 identical step transformations as presented

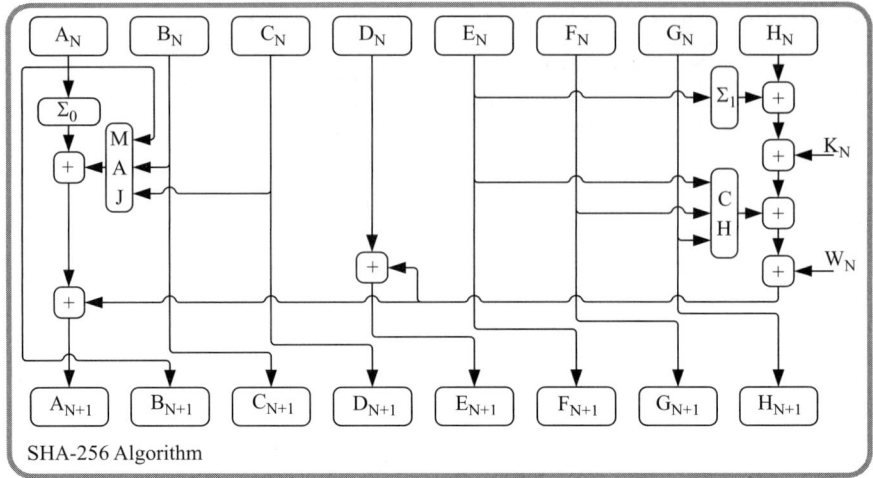

SHA-256 Algorithm

Fig. 8 One step of the state update transformation of SHA-256

in Fig. 8. The step transformations employ the bitwise Boolean functions Maj and Ch, and two GF(2)-linear functions $\Sigma_0(x) = ROTR^2(x) \oplus ROTR^{13}(x) \oplus ROTR^{22}(x)$ and $\Sigma_1(x) = ROTR^6(x) \oplus ROTR^{11}(x) \oplus ROTR^{25}(x)$. The ith step uses a fixed constant K_i which is a distinct 32-bit word for each step and the ith word W_i of the expanded message. The message expansion works as follows. An input message is split into 512-bit message blocks where in the last block a padding mechanism is used. The message expansion takes as input a vector M with 16 words and outputs a vector W with N words. The words of W_i, the expanded vector, are generated from the initial message M according to the following formula:

$$W_i = \begin{cases} M_i & \text{for } 0 \leq i \leq 15 \\ \sigma_1(W_{i-2}) + W_{i-7} + \sigma_0(W_{i-15}) + W_{i-16} & \text{for } 16 \leq i \leq 63 \end{cases}.$$

The functions $\sigma_0(x)$ and $\sigma_1(x)$ are defined as follows: $\sigma_0(x) = ROTR^7(x) \oplus ROTR^{18}(x) \oplus SHR^3(x)$ and $\sigma_1(x) = ROTR^{17}(x) \oplus ROTR^{19}(x) \oplus SHR^{10}(x)$. After 64 steps, the feed-forward operation is applied. It is done word-by-word by modular addition of the previous chaining values (the IVs in the case of the first block) to the current state variables. Although many hash hardware architectures have been proposed so far, none of the published work focus on low die-size and low-power-consumption requirements as needed for contactlessly powered devices like RFID tags. Nearly all of these architectures focus on GBit throughput rates and do not mind low-power consumption at all. Especially the implementations using FPGAs as target technology make extensive use of pipelining and unrolling techniques. Only a few publications of ASIC hash implementations are available so far. Papers of Dadda et al. [5], Dominikus [8], Ganesh and Sudarshan [17], and Satho and Tadanobu [31] are somehow comparable with the presented design but were not optimized for application in passive RFID tags.

4.3.1 Architecture of SHA-256 Module for RFID Tags

Implementing the SHA-256 (and also other MD4 family hash functions like SHA-1, MD5, and MD4) algorithm as a 32-bit architecture is the only useful word width because of the design of the algorithm. High-level simulations with a data word size of 8 bits showed that the performance is unacceptably bad which was not astonishing as the algorithms were designed for 32-bit platforms. The architecture of the SHA-256 module consists of a datapath and a controller circuit. The datapath of the proposed 32-bit SHA-256 module is depicted in Fig. 9. The main parts of the module are the RAM circuits, the dedicated logic functions for the SHA-256 transformations, two temporary storage registers, and one 32-bit adder. The controller module which is not shown in the figure is implemented as a finite state machine that generates the control and address signals for the datapath. All RAM parts have a register-based implementation which allows the use of clock gating to minimize power consumption. The major design principle is that only 32 flip-flops are clocked within the same clock cycle. This averages the power consumption of the circuit. This represents the most important design difference to existing SHA-256 implementations where in every clock cycle all registers of a RAM module need a clock pulse because of the pipelined structure. The RAM consists of the three different parts: message expansion RAM (W-RAM), state variables A–H (State-RAM) and the chaining variables (H-RAM). The W-RAM stores the 16 32-bit words W_i necessary for the message schedule. This RAM module is single ported to ease silicon implementation and reduce controlling complexity. The State-RAM contains the eight state variables A–H which are used during the step transformation. Because of the internal structure of

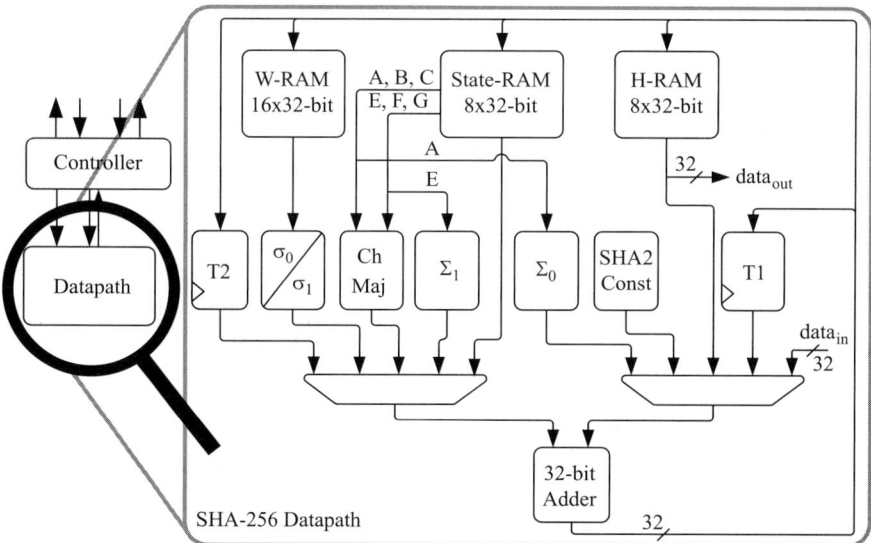

Fig. 9 Architecture of low-power SHA-256 datapath

the algorithm it was necessary to implement this 8×32-bit RAM having a separated read and write port. The synthesis results show that the additional hardware for this dual-port RAM is negligible but the throughput is augmented significantly. The third RAM part, the H-RAM, stores the 8×32-bit chaining variables. It is updated only at the beginning and at the end of each step transformation. The output of the datapath comes directly from the H-RAM.

The dedicated hardware modules in the datapath perform the SHA-256 functions σ_0, σ_1, Ch, Maj, Σ_0, and Σ_1. The inputs of these functions come either from the output of the RAM or they are directly routed from the state variables A–H to the corresponding module. The sequence of 64 constant 32-bit words are stored in a look-up table which was generated by the synthesizer. The two 32-bit registers $T1$ and $T2$ are used during step transformation to store intermediate results. Again clock gating is used to reduce the power consumption while the registers are not required. All RAM circuits, the dedicated hardware modules, the look-up table, and the registers have the mechanism to disable the output of the module which sets the 32-bit output of the module to zero. This sleep-mode logic reduces the switching activity of the combinational logic behind these gates to a minimum. Additionally, the two large multiplexers degrade to a mere an OR tree. A further big difference to existing hash architectures is the use of a single 32-bit adder. The critical path of the circuit is heavily reduced to produce less signal activity and hence has lower power consumption.

A calculation of a hash value works as follows. Before the step transformation starts the initial hash value has to be loaded into the H-RAM module. Then the data input including padding is stored in the W-RAM. After transferring data from the H-RAM to the state variables the step transformation starts. According to the SHA-256 algorithm the state variables are updated using the dedicated functions, the intermediate storage registers and the appropriate SHA-256 constant from the look-up table. The message schedule is also executed in each step which allows to generate the required message expansion in place without any additional memory. At the end of the 64 rounds the hash value is calculated using the state variables and the old chaining variables. Now the hash value can be read at the data output or the next message block has to be processed with the same procedure except the initialization of the H-RAM.

Allowing a comparison to the hash functions SHA-1 and MD5 the implementation of their datapaths are presented in Figs. 10 and 11. The main differences lie in fact that these algorithms have different state size and their compression function requires different combinational logic modules. For instance, SHA-1 requires only five state variables, five chaining variables, and only one additional register for intermediate results. Only four 32-bits constants are required and different rotation functions are used. The MD5 module even has only four state variables and four chaining variables. Hence, the required number of flip-flops is reduced to 24×32-bit. Also simpler combinational logic is used but 64 32-bit constants are required.

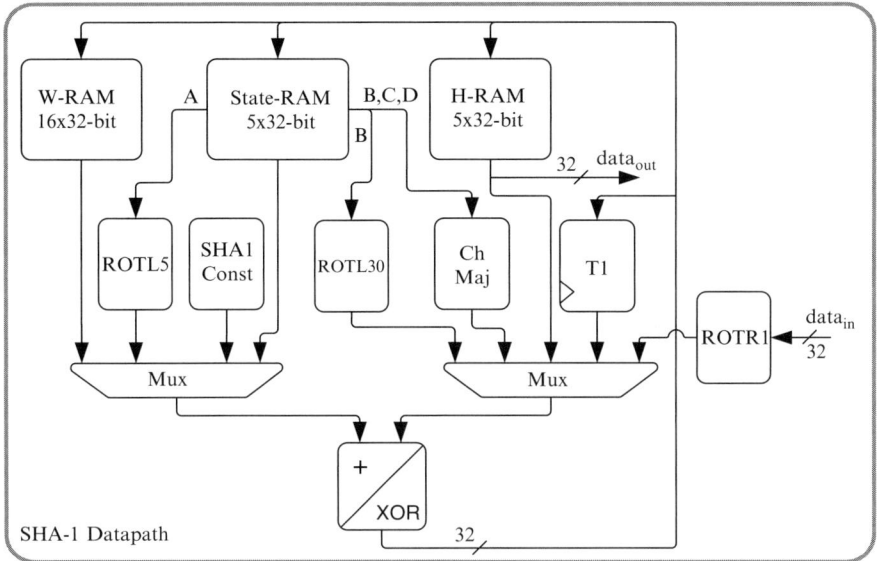

Fig. 10 Architecture of low-power SHA-1 datapath

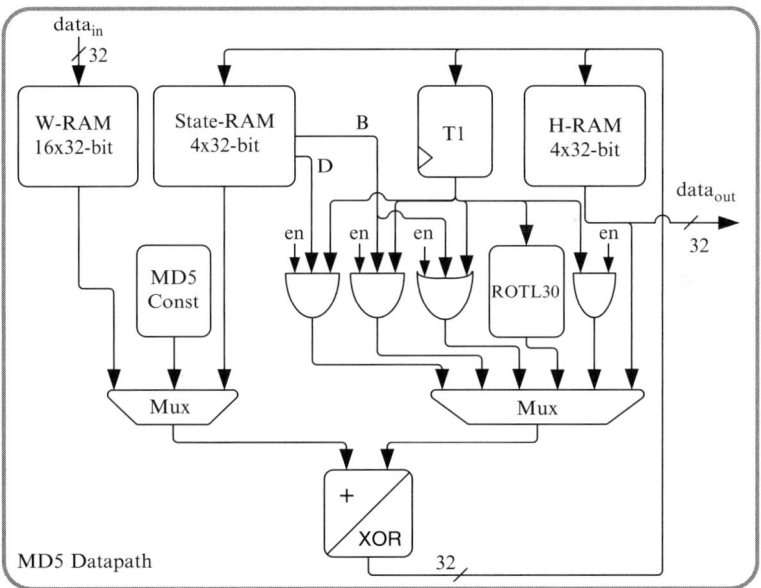

Fig. 11 Architecture of low-power MD5 datapath

Table 5 Synthesis results of all components of the low-power SHA-256 module

Module/component	I_{mean}		Chip area	
	(µA at 100 kHz)	(%)	(GE)	(%)
State-RAM	1.53	26.1	1,984	18.3
W-RAM	1.17	20.0	3,881	35.7
H-RAM	0.16	2.7	2,427	22.3
SHA-256 constants	0.07	1.2	612	5.6
32-bit Registers	0.59	10.1	2×197	3.6
σ_0 & σ_1	0.30	5.1	343	3.2
Σ_0 & Σ_1	0.07	1.2	300	2.8
Ch & Maj	0.07	1.2	174	1.6
32-bit Adder	1.01	17.2	156	1.4
Multiplexer & others	0.50	8.2	233	2.1
Controller	0.41	7.0	364	3.3
SHA-256 total	*5.86*	*100*	*10,868*	*100*
SHA-1 total	*3.93*	*66.8*	*8,120*	*74.7*
MD5 total	*3.16*	*53.7*	*8,001*	*73.6*

4.3.2 Results of SHA-256 Implementation

The implementation of the SHA-256 architecture requires a current consumption of 5.86 µA at a frequency of 100 kHz on a 0.35-µm CMOS process technology with a supply voltage of 1.5 V. Performing a hash calculation on a 512-bit block of data requires 1,128 clock cycles. Although not designed for high speed, the circuit has a maximum clock frequency of 50 MHz and achieves a data throughout of up to 22.5 Mbps. The required hardware complexity is 10,868 GEs. The complexity of each component is also listed in Table 5.

4.4 Grain

The stream cipher Grain was designed by Martin Hell, Thomas Johansson, and Willi Meier [18]. Their main goal was to design an algorithm which is very easy to implement in hardware and requires only small chip area. It is a bit-oriented synchronous stream cipher which means that the key stream is generated independently from the plaintext. In general, a stream cipher consists of two phases. The first phase is the initialization of the internal state using the secret key and the IV. After that, the state is repeatedly updated and hence used to generate key-stream bits. The main elements of the stream cipher Grain are two 80-bit shift registers where one has a linear feedback (LFSR) and the other a nonlinear feedback (NFSR). The key size is specified with 80 bits and additionally an initial value of 64 bits is required. Unfortunately, the initial version of Grain (Version 0) had a weakness in the output function which was discovered during the first evaluation phase. This paper uses Grain (Version 1) for implementation which solves the security issues of the initial version.

Fig. 12 Grain stream cipher

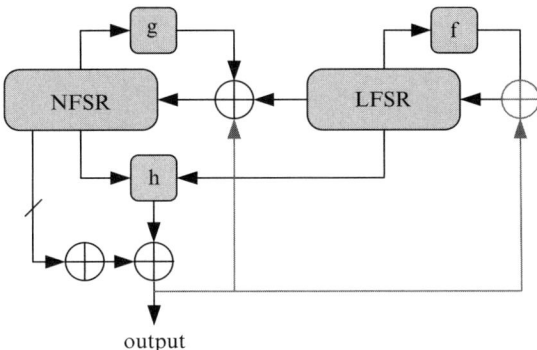

output

The basic structure of the algorithm can be seen in Fig. 12. Two polynomials of degree 80, $f(x)$ and $g(x)$, are used as feedback function for the feedback shift registers LFSR and NFSR. The output function $h(x)$ uses as input selected bits from both feedback shift registers. Additionally, seven bits of NFSR are XORed together and the result is added to the function $h(x)$. This output is used during the initialization phase as additional feedback to LFSR and NFSR (grey lines in the figure indicate this feedback). During normal operation this value is used as key stream output.

4.4.1 Architecture of Grain Module for RFID Tags

The hardware module of Grain was implemented with a 16-bit AMBA APB interface in a 0.35-μm CMOS process technology. This interface fits to the 16-bit datapath architecture. The reason for implementing a 16-bit word size was the low-power design approach as presented in Sect. 3. The details of the datapath are shown in Fig. 13. It can be seen that the feedback shift registers NFSR and LFSR shift 16 bits per clock cycle. Only a single register is clocked at the same point in time via clock gating which eases the input of the key and the initial value because the same 16 input wires are connected to all registers. Additionally, it reduces the mean power consumption significantly. This comes at the expense of having a temporary register which stores intermediate results. Additionally, all combinational circuits like the feedback functions $g\_function, f\_function$, and $h\_function$ have to be implemented in radix-16. The inputs of these functions are selected bits from the registers and are not shown in detail in this figure. The $h\_function$ in the datapath description also includes the XOR operations of the output function in the algorithm description. The output of the module is registered and instead of the key stream the encrypted result of the data input is stored in the register. Instead of a multiplexer that selects the correct feedback function for the temporary register, AND gates are used to enable and disable the appropriate inputs. Producing a 16-bit encryption result after initialization requires 13 clock cycles.

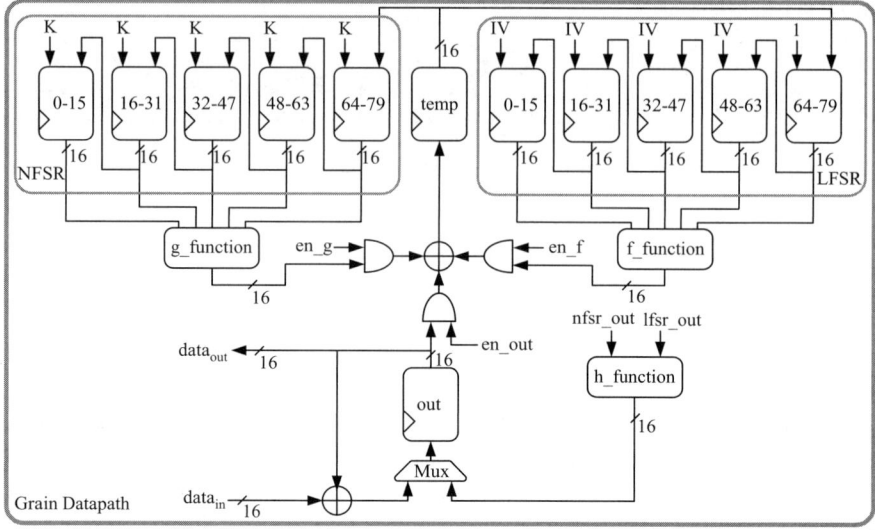

Fig. 13 Datapath of Grain

Table 6 Components of Grain datapath

Component	Min. area (radix-1)		Low-power (radix-16)	
Chip area	(GE)	(%)	(GE)	(%)
NFSR + LFSR registers (160 bits)	1,275	72.4	1,130	33.6
Temporary register	0	0	85	2.5
Output register	50	2.8	150	4.5
Combinational logic and misc.	315	17.9	1,835	54.6
Controller (FSM)	120	6.8	160	4.8
Total	1,760	100	3,360	100

4.4.2 Results of Grain Implementation

In comparison to the straight-forward, minimum-area implementation in radix-1 which has an 8-bit interface and requires 1,760 GEs, the area requirements for the low-power Grain implementation are 3,360 GEs. A detailed comparison of the components in these types of implementation can be seen in Table 6. The 160 bits for the NFSR and the LFSR require 1,130 GEs accounting for one-third the total chip area. The combinational part of the circuit has also a significant chip area compared to other algorithms. This is due to the resource-intensive feedback functions that are implemented in radix-16. The mean current consumption of the Grain module is $0.8\,\mu A$ at $100\,kHz$ and $1.5\,V$ supply voltage. After an initialization phase of 130 clock cycles encryption and decryption of a 128-bit data block requires 104 cycles.

Fig. 14 Pseudocode of
Trivium stream cipher

$$
\begin{aligned}
&\text{for } i = 0 \text{ to N-1 do} \\
&\quad t_0 = s_{65} + s_{92} \\
&\quad t_1 = s_{161} + s_{176} \\
&\quad t_2 = s_{242} + s_{287} \\
&\quad out_i = t_0 + t_1 + t_2 \\
&\quad t_0 = t_0 + s_{90}{\cdot}s_{91} + s_{170} \\
&\quad t_1 = t_1 + s_{174}{\cdot}s_{175} + s_{263} \\
&\quad t_2 = t_2 + s_{285}{\cdot}s_{286} + s_{68} \\
&\quad (s_0,s_1,...,s_{92}) = (t_2,s_0,...,s_{91}) \\
&\quad (s_{93},s_{94},...,s_{176}) = (t_0,s_{93},...,s_{175}) \\
&\quad (s_{177},s_{178},...,s_{287}) = (t_1,s_{177},...,s_{286}) \\
&\text{end for}
\end{aligned}
$$

4.5 Trivium

The developers of the stream cipher Trivium are Christophe De Cannière and Bart
Preneel [2]. This hardware-oriented synchronous stream cipher was designed to explore how simple a stream cipher could be designed without sacrificing its security.
It is possible to generate up to 2^{64} bits of key stream from an 80-bit key and an
initial value (IV) of 80 bits. The state consists of 288 bits which are denoted as
$s_0, s_2, ..., s_{287}$. The pseudocode in Fig. 14 shows how the algorithm uses 15 specific
bits of the state to generate three variables which are used to update the state and
which produce one bit of the output. During the initialization, which is not shown
in the figure, the key and the IV are loaded to the state and the same update function
is applied 1,152 times without using the output for the key stream. In the algorithm
description, N is used for the number of output bits of the stream cipher.

4.5.1 Architecture of Trivium Module for RFID Tags

The implementation of the Trivium module has the same 16-bit AMBA APB interface as the one for Grain. Implementing a radix-16 datapath is also motivated
by the low-power design technique. Figure 15 shows the details of the architecture.
The boxes denoted with *comb* are the combinational logic elements of the algorithm
that are used for updating the state according to the algorithm specification. The
288 flip-flops for the state are separated in registers of size 16 bits and smaller. Additionally, two temporary registers are necessary which store intermediate results.
The output register is again used for directly applying the XOR operation of the key
stream with the input value. Again, clock gating is used to clock only one register
per clock cycle. During initialization, the key, the IV, and the constants are loaded
into the registers. Then the combinational circuit is used to update the registers in
a kind of pipeline where the temporary registers are used to prevent overwriting of
values needed later. The generation of a 16-bit key stream after the initialization
phase requires 22 clock cycles.

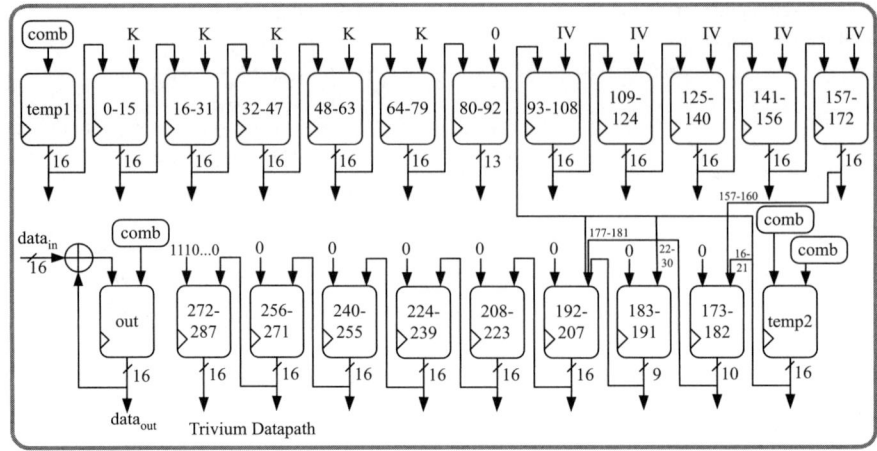

Fig. 15 Datapath of Trivium

Table 7 Components of Trivium datapath

Component	Min. area (radix-1)		Low-power (radix-16)	
Chip area	(GE)	(%)	(GE)	(%)
State registers (288 bits)	1,840	77.0	2,040	66.0
Temporary registers	0	0	200	6.5
Output register	50	2.1	150	4.9
Combinational and misc.	290	12.1	410	13.3
Controller (FSM)	210	8.8	290	9.4
Total	*2,390*	*100*	*3,090*	*100*

4.5.2 Results of Trivium Implementation

A comparison of the synthesis results of a minimum-area implementation and the low-power implementation for passive RFID tags can be seen in Table 7. The biggest component in Trivium is the 288-bit state register which requires 2,040 GEs. The total chip are for the Trivium module suitable for RFID tags is 3,090 GEs. The difference between the minimum-area variant and the low-power variant is not as big as for the Grain algorithm because the combinational circuit for radix-1 and radix-16 implementations do not differ as much. The low-power variant requires only slightly more area for the state registers, the additional temporary registers, the output register, and the combinational logic. The mean power consumption of the Trivium module is 0.68 µA. One drawback of this solution is the long initialization phase which requires 1,603 clock cycles. After that the encryption of a 128-bit block of data requires 176 cycles.

4.6 ECC

The implementation of asymmetric cryptography will be covered in chapters later in this book. At this point, only a short coverage of ECC will be given in order to classify the results achieved for symmetric cryptography. ECC is the only technology providing standardized asymmetric cryptography that could fulfill the strong requirements of passive RFID tags. ECC operates on word sizes of roughly 160 bits. Word sizes of $w = 113$ to $w = 192$ bits are proposed for use in RFID systems. Similar to the implementation of symmetric cryptography, the storage requirements of ECC mainly determine the circuit size of ECC. At least six registers of full word size w are needed to store intermediate results. In addition, some curve parameters of same size need to be stored in a ROM. In sum, an ECC circuit has to store roughly 1,500 bits. The combinational circuit of ECC is dominated by a multiplier. Most reported hardware implementations operate on the full word size using digit-serial multipliers. In order to keep the silicon area small, bit-serial multipliers are preferred. Multiplication dominates the timing of ECC computations. Bit-serial multipliers need roughly w cycles per multiplication. Up to ten multiplications are needed per point operation. $1.5w$-point operations make up an ECC point multiplication – the central operation of ECC. Thus, the total number of clock cycles is roughly $15w^2$. So ECC requires 200,000 – 600,000 clock cycles for computation. The computation time is much larger than those of symmetric algorithms. Also the power consumption is higher because in most clock cycles three w-bit registers have to be clocked. Table 8 gives results for ECC-192 and compares them to those results achieved by symmetric cryptography.

4.7 Comparison of Results

The results and a comparison of the different hardware modules of cryptographic algorithms can be seen in Table 8. A similar analysis is presented in [13]. The symmetric algorithms AES-128, TEA/XTEA, SHA-256, SHA-1, MD5, Grain, and Trivium are compared concerning their security level, power consumption, chip area, and required number of clock cycles. Algorithms were only analyzed that have a high level of security of at least 80 bit key size. In addition to the standardized algorithms AES, SHA-1, SHA-256, MD5, and ECC, where the security of the algorithm is well scrutinized, the new stream cipher algorithms Grain and Trivium were selected for comparison. Although there are some doubts about the security of TEA and XTEA these algorithms were implemented because it is often stated in literature that their implementation is very resource efficient. The mean current consumption in the third column is given in μA at a nominal clock frequency of 100 kHz and a supply voltage of 1.5 V. The current values were obtained by power simulations with Synopsys NanoSim. This is a near-spice level power simulator which can be used to simulate large digital designs in reasonable time. For power simulation the

Table 8 Results of symmetric crypto hardware modules for passive RFID tags

Algorithm	Security (bits)	I_{mean} (μA at 100 kHz)	Chip area (GE)	Clock cycles
AES-128	128	3.0	3,400	1,032
XTEA	128	3.86	2,636	705
TEA	128	3.79	2,633	289
SHA-256	128	5.86	10,868	1,128
SHA-1	80	3.93	8,120	1,274
MD5	80	3.16	8,001	712
Grain	80	0.80	3,360	(130)+ 104
Trivium	80	0.68	3,091	(1,603) +176
ECC-192	96	18.85	23,600	502,000

specific algorithm was executed several times using random input data. In addition to the current trace, NanoSim provides results for the mean current consumption as output. The simulation results have been verified by measurements in the case of AES-128 where a test chip on silicon is available. Thereby, a simple switched capacitor measurement approach was chosen to find out the mean current consumption. The chip area results are based on synthesis and are given in gate equivalents GEs. For the used 0.35-μm CMOS process technology one gate equivalent compares to a NAND2 cell of $55\,\mu m^2$. The number of clock cycles are stated for encryption of a 128-bit block of data for AES, XTEA, TEA, Grain, and Trivium. For the hash functions the number of clock cycles for hashing a message of size 512 bits are given. For the EC module the values for a point multiplication over ECC-192 is used.

The comparison shows that nearly all implemented algorithms achieve the required design goals for passive RFID tags. The comparison of the hash functions with the block cipher AES shows that they require more than twice as much chip area. This is due to the high number of registers required for the internal state. Also the power consumption in case of SHA-256, which has the same security level, is nearly twice as high while requiring a similar amount of clock cycles. A comparison of the block ciphers XTEA and TEA with the AES shows that it is not that advantageous to use these algorithms which do not have high confidence from the cryptanalytic point of view. The required hardware resources are only slightly smaller while having a little bit high-power consumption. The upcoming stream ciphers Grain and Trivium could be an interesting alternative to AES because of their outstanding low-power consumption while having nearly the same chip area. It depends on the application whether it is worth to use the, until now not standardized, algorithms where the security is not analyzed as intensive as for AES. ECC could be an interesting option in future when newer process technologies allow more hardware resources on RFID tags. The high number of required clock cycles should be considered here.

5 Conclusion

In this chapter, different hardware implementations of symmetric algorithms for RFID security have been analyzed. The motivation for security on RFID tags is to prevent attacks like cloning of tags by means of cryptography. The security assurances confidentiality, authentication, and data integrity are introduced and an overview about state-of-the-art crypto primitives has been presented. The most commonly used primitives like hash functions, block ciphers, and stream ciphers have been explained in detail. Many security protocols were published in the last years which use these primitives for authentication. Nearly all of us use challenge–response protocols to achieve one-way or mutual authentication which proves the origin of an RFID tag which is attached to a certain product.

The implementation of cryptographic primitives for passive RFID tags is challenging due to the stringent resource requirements. These requirements have been analyzed in detail and the differences between the HF and the UHF frequency range have been discussed. The low-power consumption of the tag is the most important issue because the operating range of an RFID tag is directly influenced by the power consumption. Different CMOS process technologies have been scrutinized concerning their power consumption and chip area requirements. Because chip size significantly influences the price of tags the required hardware resources for a crypto module have to be considered. This is of importance because of the high number of tags. Different to other hardware modules which require high throughput and high clock frequencies, in RFID technology the data rates are low and hence the algorithms could be made slow. This increases the energy consumed for an operation but the reduction of power consumption is more important for RFID tags where power is transmitted from the reader to the tag continously. It has been shown that the implementation security is an issue for passive RFID tags since the publication of successful power-analysis attacks against passive tags. In a separate section the design measures for low-resource requirements have been evaluated. This includes the discussion about the power consumption of a hardware module in general. Different possible measures have been listed and details about design issues for low-power consumption and low die size have been demonstrated. Clock gating is the most useful design measure for low-power consumption while sleep logic also reduces the glitching activity in the circuit.

The design guidelines for low-resource requirements have been used to implement different hardware implementations of symmetric crypto primitives. This survey presented details about the bock ciphers AES, XTEA, and TEA, the hash functions SHA-256, SHA-1, MD5, and the new upcoming stream ciphers Grain and Trivium which should be suitable for realization on passive RFID tags. The main design principles were to implement extensive clock gating on different word widths. For AES, a 8-bit datapath has been implemented, while XTEA, TEA, and the hash functions use 32-bit words. The stream ciphers Trivium and Grain have been realized using a 16-bit architecture. All modules have been implemented with a AMBA bus interface which allows to use the modules directly without any additional resources on an RFID tag. The presented results allow a fair comparison of the implemented modules because all algorithms have been implemented on the same CMOS standard-cell technology under the same conditions. The methodology was to use a standard-cell design flow starting from a high-level language implementation and getting a chip layout via HDL description, synthesis, and backend verification. Power consumption figures have been achieved using the transistor-level simulation tool NanoSim from Synopsys. The presented results show detailed current consumption figures in μA at a nominal clock frequency of 100 kHz and a supply voltage of 1.5 V. This allows to directly argument about the suitability of a certain algorithm for passive RFID tags. The chip area is given in gate equivalents which is commonly used to avoid the technology-depend chip size in μm^2. Although not of utmost importance, the number of clock cycles for the execution of the algorithms has also been presented.

The comparison of the implementation of the selected symmetric crypto algorithms has shown that the AES implementation for passive RFID tags is to favor due to the following reasons. In comparison to the hash functions SHA-256, SHA-1, and MD5 it has a lower chip area because of lower memory requirements. SHA-256 which has the same level of security but requires a higher power consumption. Although the block ciphers XTEA and TEA require slightly less chip area it is not worth the risk of using an algorithm which has potential security issues. Additionally, it has a higher power consumption. The new upcoming stream ciphers Grain and Trivium could be an interesting alterative to AES for RFID tags due to their lower power consumption of factor four. Until now they do not have high confidence from the cryptanalytic point of view. Hence, it depends on the application whether this potential risk can be taken or staying at the AES which has passed many security analyses. Public-key algorithms like ECC could also be interesting in near future when new process technologies allow more hardware resources for cheap RFID tags. Up to now, using the AES algorithm for securing passive RFID tags is the most promising solution due to the efficient implementation in terms of chip area and power consumption while having a high level of security.

References

1. S. Bono, M. Green, A. Stubblefield, A. Juels, A. Rubin, and M. Szydlo. Security Analysis of a Cryptographically-Enabled RFID Device. In *USENIX Security Symposium, Baltimore, Maryland, USA, July–August, 2005, Proceedings*, pp. 1–16, USENIX, 2005
2. C.D. Canniére and B. Preneel. *TRIVIUM Specifications*. eSTREAM, ECRYPT Stream Cipher Project (http://www.ecrypt.eu.org/stream), Report 2005/030, April 2005
3. C.D. Canniére and C. Rechberger. Finding SHA-1 Characteristics: General Results and Applications. In X. Lai and K. Chen, editors, *Advances in Cryptology - ASIACRYPT 2006, 12th International Conference on the Theory and Application of Cryptology and Information Security, Shanghai, China, December 3–7, 2006, Proceedings*, volume 4284 of *Lecture Notes in Computer Science*, pp. 1–20, Springer, Berlin, 2006
4. E.Y. Choi, S.-M. Lee, and D.H. Lee. Efficient RFID Authentication Protocol for Ubiquitous Computing Environment. In T. Enokido, L. Yan, B. Xiao, D. Kim, Y. Dai, and L.T. Yang, editors, *Embedded and Ubiquitous Computing – EUC 2005 Workshops, EUC 2005 Workshops: UISW, NCUS, SecUbiq, USN, and TAUES, Nagasaki, Japan, December 6–9, 2005,*

Proceedings, volume 3823 of *Lecture Notes in Computer Science*, pp. 945–954, Springer, Berlin, December 2005

5. L. Dadda, M. Macchetti, and J. Owen. The Design of a High Speed ASIC Unit for the Hash Function SHA-256 (384, 512). In *2004 Design, Automation and Test in Europe Conference and Exposition (DATE 2004), 16–20 February 2004, Paris, France*, volume 3, pp. 70–75, IEEE Computer Society press, Washington, DC, February 2004

6. J. Daemen and V. Rijmen. *The Design of Rijndael*. Information Security and Cryptography, Springer, Berlin, 2002. ISBN 3-540-42580-2

7. T. Dimitriou. A Lightweight RFID Protocol to Protect Against Traceability and Cloning attacks. In *First International Conference on Security and Privacy for Emerging Areas in Communications Networks (SecureComm 2005), Athens, Greece, 59 September 2005, Proceedings*, Athens, Greece, pp. 59–66, IEEE Computer Society Press, Washington, DC, September 2005

8. S. Dominkus. A Hardware Implementation of MD4-family Hash Algorithms. In *Ninth IEEE International Conference on Electronics, Circuits and Systems, Dubrovnik, Croatia, 15–18 September, 2002, Proceedings*, volume 3, pp. 1143–1146, IEEE, New York, NY, October 2002

9. ECRYPT. eSTREAM – The ECRYPT Stream Cipher Project Website. http://www.ecrypt. eu.org/stream/

10. M. Feldhofer. Comparison of Low-Power Implementations of Trivium and Grain. In *Workshop on The State of the Art of Stream Ciphers (SASC 2007), January 31–February 1, 2007, Bochum, Germany*, pp. 236– 246, ECRYPT, February 2007

11. M. Feldhofer and C. Rechberger. A Case Against Currently Used Hash Functions in RFID Protocols. In R. Meersman, Z. Tari, and P. Herrero, editors, *First International OTM Workshop on Information Security (IS'06), Montpellier, France, Oct 30–Nov 1, 2006. Proceedings, Part I*, volume 4277 of *Lecture Notes in Computer Science*, pp. 372–381, Springer, Berlin, October 2006

12. M. Feldhofer and J. Wolkerstorfer. Low-power Design Methodologies for an AES Implementation in RFID Systems. In *Workshop on Cryptographic Advances in Secure Hardware 2005 (CRASH05), September 6–7, Leuven, Belgium*, September 2005

13. M. Feldhofer and J. Wolkerstorfer. Strong Crypto for RFID Tagsa Comparison of Low-Power Hardware Implementations. In *IEEE International Symposium on Circuits and Systems (ISCAS 2007), New Orleans, USA, May 27–30, 2007, Proceedings*, pp. 1839–1842, IEEE, New York, NY, May 2007

14. M. Feldhofer, S. Dominikus, and J. Wolkerstorfer. Strong Authentication for RFID Systems. Using the AES Algorithm. In M. Joye and J.-J. Quisquater, editors, *Cryptographic Hardware and Embedded Systems – CHES 2004, Sixth International Workshop, Cambridge, MA, USA, August 11–13, 2004, Proceedings*, volume 3156 of *Lecture Notes in Computer Science*, pp. 357–370, Springer, Berlin, August 2004

15. M. Feldhofer, K. Lemke, E. Oswald, F.-X. Standaert, and J. Wolkerstorfer. *D.VAM.2 – State of the Art in Hardware Architectures*, August 2005

16. M. Feldhofer, J. Wolkerstorfer, and V. Rijmen. AES Implementation on a Grain of Sand. *IEE Proceedings on Information Security*, 152(1): 13–20, October 2005

17. T.S. Ganesh and T.S.B. Sudarshan. ASIC Implementation of a Unified Hardware Architecture for Non-Key Based Cryptographic Hash Primitives. In *International Symposium on Information Technology: Coding and Computing (ITCC 2005), 4–6 April 2005, Las Vegas, Nevada, USA, Proceedings*, volume 1, pp. 580–585, IEEE Computer Society Press, Washington, DC, April 2005

18. M. Hell, T. Johansson, and W. Meier. Grain – A Stream Cipher for Constrained Environments. eSTREAM, ECRYPT Stream Cipher Project (http://www.ecrypt.eu.org/stream), Report 2005/010, 2006. Revised version

19. D. Henrici and P. Müller. Hash-based Enhancement of Location Privacy for Radio-Frequency Identification Devices using Varying Identifiers. In *Second IEEE Conference on Pervasive Computing and Communications Workshops (PerCom 2004 Workshops), Orlando, FL, USA,*

14-17 March 2004, Proceedings, pp. 149–153, IEEE Computer Society Press, Washington, DC, March 2004

20. M. Hutter, S. Mangard, and M. Feldhofer. Power and EM Attacks on Passive 13.56 MHz RFID Devices. In P. Paillier and I. Verbauwhede, editors, *Cryptographic Hardware and Embedded Systems – CHES 2007, Ninth International Workshop, Vienna, Austria, September 10–13, 2007, Proceedings*, volume 4727 of *Lecture Notes in Computer Science*, pp. 320–333, Springer, Berlin, September 2007

21. IEEE. IEEE Standard 802.11i-2004: Wireless LAN Medium Access Control (MAC) and Physical Layer (PHY) specifications. Amendment 6: Medium Access Control (MAC) Security Enhancements. Available online at `http://standards.ieee.org/getieee802/`, July 2004

22. ISO/IEC. *Information technology – Security techniques – Hash-functions – Part 3: Dedicated Hash-Functions*. Available from `http://www.iso.org/` (with costs), 2004

23. P. Israsena. Securing Ubiquitous and Low-Cost RFID Using Tiny Encryption Algorithm. In *First International Symposium on Wireless Pervasive Computing (ISWPC 2006), Phuket, Thailand, 16–18 January, 2006, Proceedings*, IEEE, New York, NY, January 2006

24. Y. Lee and I. Verbauwhede. Secure and Low-Cost RFID Authentication Protocols. In *Second IEEE Workshop on Adaptive Wireless Networks (AWiN), November 28, 2005, St. Louis, MO*, 2005

25. S. Mangard, M. Aigner, and S. Dominikus. A Highly Regular and Scalable AES Hardware Architecture. *IEEE Transactions on Computers*, 52(4): 483–491, April 2003

26. S. Mangard, E. Oswald, and T. Popp. *Power Analysis Attacks – Revealing the Secrets of Smart Cards*, Springer, Berlin, 2007. ISBN 978-0-387-30857-9

27. National Institute of Standards and Technology (NIST). FIPS-46-3: Data Encryption Standard, October 1999. Available online at `http://www.itl.nist.gov/fipspubs/`

28. National Institute of Standards and Technology (NIST). FIPS-197: Advanced Encryption Standard, November 2001. Available online at `http://www.itl.nist.gov/fipspubs/`

29. National Institute of Standards and Technology (NIST). FIPS-180-2: Secure Hash Standard, August 2002. Available online at `http://www.itl.nist.gov/fipspubs/`

30. K. Rhee, J. Kwak, S. Kim, and D. Won. Challenge–Response Based RFID Authentication Protocol for Distributed Database Environment. In D. Hutter and M. Ullmann, editors, *Security in Pervasive Computing, Second International Conference, SPC 2005, Boppard, Germany, April 6–8, 2005, Proceedings*, volume 3450 of *Lecture Notes in Computer Science*, pp. 70–84, Springer, Berlin, April 2005

31. A. Satoh and T. Inoue. ASIC-Hardware-Focused Comparison for Hash Functions MD5, RIPEMD-160, and SHS. In *International Symposium on Information Technology: Coding and Computing (ITCC 2005), 4–6 April 2005, Las Vegas, Nevada, USA, Proceedings*, volume 1, pp. 532–537, IEEE Computer Society Press, Washington, DC, April 2005

32. S. Tillich, M. Feldhofer, and J. Großschädl. Area, Delay, and Power Characteristics of Standard-Cell Implementations of the AES S-Box. In S. Vassiliadis, S. Wong, and T. Hämäläinen, editors, *Sixth International Workshop on Embedded Computer Systems: Architectures, Modeling, and Simulation, SAMOS 2006, Samos, Greece, July 17–20, 2006, Proceedings*, volume 4017 of *Lecture Notes in Computer Science*, pp. 457–466, Springer, Berlin, July 2006

33. X. Wang, Y.L. Yin, and H. Yu. Finding Collisions in the Full SHA-1. In V. Shoup, editor, *Advances in Cryptology – CRYPTO 2005: 25th Annual International Cryptology Conference, Santa Barbara, California, USA, August 14–18, 2005, Proceedings*, volume 3621 of *Lecture Notes in Computer Science*, pp. 17–36, Springer, Berlin, 2005

34. S.A. Weis, S.E. Sarma, R.L. Rivest, and D.W. Engels. Security and Privacy Aspects of Low-Cost Radio Frequency Identification Systems. In D. Hutter, G. Müller, W. Stephan, and M. Ullmann, editors, *Security in Pervasive Computing, First Annual Conference on Security in Pervasive Computing, Boppard, Germany, March 12–14, 2003, Revised Papers*, volume 2802 of *Lecture Notes in Computer Science*, pp. 201–212, Springer, Berlin, March 2003

35. D.J. Wheeler and R.M. Needham. TEA, a Tiny Encryption Algorithm. In B. Preneel, editor, *Second International Workshop on Fast Software Encryption (FSE94), Leuven, Belgium, 14–16 December 1994, Proceedings*, volume 1008 of *Lecture Notes in Computer Science*, pp. 363–366, Springer, Berlin, 1995

36. J. Wolkerstorfer, E. Oswald, and M. Lamberger. An ASIC implementation of the AES SBoxes. In B. Preneel, editor, *Topics in Cryptology – CT-RSA 2002, The Cryptographers' Track at the RSA Conference 2002, San Jose, CA, USA, February 18–22, 2002, Proceedings*, volume 2271 of *Lecture Notes in Computer Science*, pp. 67–78, Springer, Berlin, 2002.

Hardware Implementation of a TEA-Based Lightweight Encryption for RFID Security

P. Israsena* and S. Wongnamkum

Abstract This chapter discusses hardware implementation strategies for employing the relatively lightweight Tiny Encryption Algorithm (TEA) in low-cost secure RFID systems. Low-cost RFID tags have stringent requirements, particularly in terms of cost related to silicon area, making conventional encryption unsuitable. Three different architectures implementing the TEA are evaluated and benchmarked with reference to designs that are area-optimized AES cores. It is found that using a customized digit-serial architecture, the TEA core has met the low-cost area requirement while consuming significantly less power than the AES equivalences. The core has an equivalent gate number of 3,872. Based on 0.35-μm CMOS technology, the complete layout has an area of $0.211\,\text{mm}^2$, suggesting a highly compact core solution.

1 Introduction

Recently, low-cost radio frequency identification (RFID) has been the topic of discussion within both academic and industry domains. RFID systems allow RFID-tagged products to be read simultaneously, require no line-of-sight, and provide a reading range superior to existing systems. RFID tags are expected to be used ubiquitously as a means to implement automated product identification systems. At present, there are a number of standards for RFID systems, such as ISO 11784-5 for animal identification, ISO15693 vicinity card, and ISO14443 for hi-end RFID tags used as contactless smartcards [1, 2]. In particular, the attention has been paid to electronic product code (EPC) Class-1 Generation-2 (C1G2) standard [3] proposed by EPCglobal for UHF tags. EPG Global is a joint venture between EAN international (Europe) and UCC (USA). The organizations are responsible for the

P. Israsena
National Electronics and Computer Technology Center (NECTEC), 112 Thailand Science Park, Klong Luang, Pathumthani, Thailand

P. Kitsos, Y. Zhang (eds.), *RFID Security: Techniques, Protocols and System-on-Chip Design,* © Springer Science+Business Media, LLC 2008

standardization of the present barcode systems, and C1G2 (or Gen-2) tags are proposed to replace or complement the existing barcode systems used at present [4].

In such an ubiquitous system, RFID tags have to be highly cost effective. That is, it is estimated that the total cost involved should not be more than 5 cents per tag [5]. As its size directly affects the manufacturing cost, so the design of tag's integrated circuit has to be highly silicon efficient. Even though RFID systems, similar to other wireless systems, are inherently susceptible to security and privacy related attacks [6], only a minimum level of security is identified in C1G2 in an attempt to make the concept as cost effective and commercially viable as possible. C1G2 standard supports functions for passive tags such as anticollision, 16-bit cyclic redundancy code (CRC) checksum, and 10-bit pseudorandom number generator (PRNG), along with ensuring consumer privacy with a simple Kill command [3]. After receiving the command, the tag becomes permanently unusable. The legitimacy of the reader that sends the Kill command to tag is ensured via a 32-bit PIN. Although this simple arrangement can prevent the tag from privacy issues like tracking, the advantages of using RFID such as dynamic data update and analysis can be lost.

Although this arrangement is justified on the short-term viability, it will become ineffective in the longer term. Inevitably, additional features will need to be included to provide better security measures. As such, there have been various researches on RFID security and privacy issues [7–12]. Some of the measures are compatible with C1G2 protocols, while others may require hardware and/or software modifications. Among the key issues remain, if one is to follow the more conventional but proven security procedures using private or pubic key infrastructure approaches successfully employed in more sophisticated systems, is that there are seemingly no appropriate hardware solutions for the encryption/authentication core. The available cores are generally too large and consume too much power. This is largely due to the complexity of the algorithms themselves, or the fact that they are usually designed for high-speed applications without having high restrictions on their silicon sizes. This chapter discusses the use of the Tiny Encryption Algorithm (TEA), developed by David Wheeler and Roger Needham at the Computer Laboratory of Cambridge University [13], as the main encryption/authentication engine for medium secure RFID systems. TEA is lightweight and is potentially a strong candidate. Different styles of TEA architectures are implemented, with detail analyses of results obtained from actual hardware implantations reported. They are compared to two alternative implementations of industrial standard Advanced Encryption Standard (AES) core based on area-optimized designs publicly reported [14, 15].

The chapter is organized as follows. In Sect. 2, characteristics and requirements for low-cost RFID tags are further discussed. Relevant issues regarding security measures in low-cost RFID systems are provided in Sect. 3. The emphasis is on security based on symmetric-key approach under which TEA algorithm and implementation strategies are discussed, respectively, in Sects. 4 and 5. Performance comparison with industrial standard, area-optimized AES cores is given in Sect. 6. Section 7 summarizes the chapter.

2 Low-Cost RFID Tag

A simple RFID system implements the basic function of item marking, i.e., it only handles bidirectional communication between a reader device and an RFID tag without any mechanism that defines authentication procedure [16].

A very basic passive tag is composed of an integrated circuit (IC) connected to an antenna coil used for receiving and transmitting wireless signals. Inside the IC there are three functional blocks; an RF front-end, a nonvolatile memory, and a microcontroller (or simple state-machine) used for data coding and implementation of protocol commands (Fig. 1). Optional functions, such as anticollision mechanisms, error detection (CRC), pseudorandom number generation (PRNG), or cryptographic coprocessor for accelerated performance, are usually only available in higher models. For the low-cost version of an RFID tag, recent research by MIT's Auto-ID Center [5] suggested a maximum cost per tag to be 5 cents, of which only 1–2 cents are available for IC manufacturing. To meet the target, the basic tag discussed seems an obvious choice. The arrangement is fine for a number of scenarios, but in situations where an RFID system has to avoid eavesdropping, traffic analysis, spoofing, and denial of service (DOS) [17], the system will inevitably need to include enhanced functionalities such as encryption/authentication to support security applications. Although the number of gates per layout area and the fabrication cost vary depending on the process technology used, it is generally considered that an average cost per mm^2 is roughly 4 cents [5]. With those higher security scenarios in mind, and assuming half of the layout is to be taken up by an encryption core, under the extreme case where only 1 cent is available for IC manufacturing per tag, the layout area available for that encryption core is only $0.125\,mm^2$. That rises to $0.25\,mm^2$ for a more typical case. For a typical 0.35-μm CMOS technology, the maximum layout areas allowed translate to equivalent numbers of gates from 2,000 to 4,000 gates [5].

A passive tag also draws its current from the reader, and for it to be functional the security block must not consume more than 30 μW of power for its security

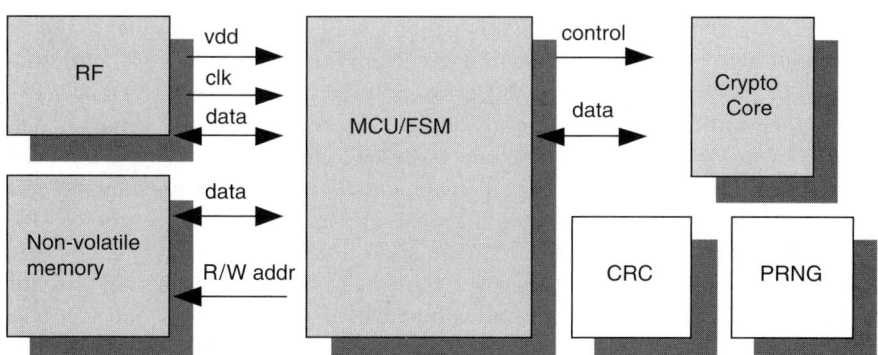

Fig. 1 Architecture of an RFID tag

Table 1 Hardware specifications of an encryption core for low-cost RFID tag

Area	Less than $0.25\,\text{mm}^2$ (or 4,000 gates)
Power	Less than $30\,\mu\text{W}$
Execution time	2.5 ms for encryption

operations [14], which is about half of the total power used by the tag. Multiple tags will also need to be able to access the same reader simultaneously in a way that the user of each tag will not notice or be annoyed by the time taken. This paper looks at relatively extreme situations where 100 tags are simultaneously accessing the same reader. Correspondingly, for all the tags to be processed within 1 s [14], the available time for encryption on each tag is approximately 2.5 ms. To summarize, the targeted specifications for a security block in a low-cost tag can be as in Table 1.

3 Security in Low-Cost RFID Systems

Security and privacy have increasingly become issues of concern for low-cost RFID systems. Similar to any communication systems, an RFID system is vulnerable to security breaches such as eavesdropping, traffic analysis, spoofing, and DOS [17]. Attacks can be made at different parts of the system, with the link between the tag and the reader considered to be the weakest. For example, a bogus reader can be used as a man-in-the-middle to gather information sent out wirelessly by the communication partners or even to track the tag to gather private information that can be useful for example in marketing or in logistic.

Conventionally, one of the best ways to provide security and privacy measures is through an authentication process. Authentication is an assurance of the identity of an entity at the other end of communication channel. It is performed to guarantee that information is only exchanged by authorized partners. When a reader authenticates a tag, it makes sure that the tag is not a forged one that may be used, for example, for spoofing. With tag authenticating a reader, it can make sure that secret information will not be sent to a wrong receiver. There are various authentication schemes or protocols [18]. Password protection, for example, is an example of a weak authentication. The more useful solutions, however, are to use strong authentications. A strong authentication scheme based on a challenge and response concept has already been discussed in Chap. 8 as a potential candidate for implementing security in RFID systems. In symmetric unilateral challenge response scheme [19], there are two partners, A and B. First B sends A a random number r_B. A encrypts r_B using key K and sends the encrypted value back to B. B can verify if A shares the same key K by encrypting r_B and compares the result with the one sent back by A. It is clear from this example that encryption is the main operation, and in this case a symmetric key, where both parties share the secret of a same key is used. Compared to asymmetric or public key alternatives, a symmetric key is generally less complicated, requires less number of operations, and can have the same security strength as its asymmetric equivalence using a key of smaller size. These facts make the symmetric key approach more suitable for limited resource RFID systems, although it should be noted that it is generally more difficult to manage the keys in a private key system [18]. For example, each reader will need to have an access to the database that keeps all the keys, making the authentication scheme more suitable for closed systems. Applications for such systems for example are engine immobilizer and airport luggage management.

Still, one possible bottleneck remains regarding the pursue of this approach for authentication in RFID systems, as it is generally believed that encryption core required to implement the authentication will not fit into the requirements in Table 1, especially in terms of the size of the design. Encryption cores tend to have sizes of more than 20 kgates, as they are mostly optimized for maximum speed. Given the limited resource available, a number of research works have therefore focused on various noncryptographic approaches [7–12]. These approaches have advantages and disadvantages, most of which have been widely discussed [12]. On the other hand, it remains inconclusive whether the cryptographic-based approach can also, in practice, be the answer to the problem of security in RFIDs. Recently work such as [14] has begun to look into implementation of the AES core in RFID tag. This is understandable as AES is the industry's present standard in symmetric key algorithm for data security [20]. Two works, in particular, have provided the results that are very positive. In [14], Feldhofer et al. discussed area-optimized design for the AES with reported gate count of 3,500 gates, while Pongjit's work [15] is also an area-optimized design with a particular target on higher-end smartcard applications. In this work, we will take a different route and we shall look at the possibility of using a different encryption algorithm to implement the encryption core and compare the results to the ones from [14, 15]. The candidate algorithm considered here is the lightweight tiny encryption algorithm (TEA) core discussed in the following sections.

4 TEA Algorithm

TEA [13] is thought to be one of the fastest and most efficient cryptographic algorithms. TEA is a Feistel cipher that uses only XOR, ADD, and SHIFT operations to provide Shannon's properties of diffusion and confusion necessary for a secure block cipher without the need for complicated P-boxes and S-boxes as required in algorithms such as Data Encryption Standard (DES) or AES. TEA operates on a 64-bit data block using a 128-bit key and can achieve complete diffusion after six rounds. TEA is believed to be as secure as the IDEA algorithm [21] that uses the same mixed algebraic groups technique, but is much simpler and faster. It is also in public domain, whereas IDEA is patented. For software implementation, the code is lightweight and portable and therefore particularly suits real-time applications. The original code of TEA is as shown in Fig. 2. Although TEA has a few weaknesses,

```
void code(long* v, long* k)
{
unsigned long y=v[0], z=v[1], sum=0, delta=0x9e3779b9, n=32;
      while (n-->0)
          {
              sum += delta;
              y += ((z<<4)+k[0]) ^ (z+sum) ^ ((z>>5)+k[1]);
              z += ((y<<4)+k[2]) ^ (y+sum) ^ ((y>>5)+k[3]);
          }
          v[0]=y; v[1]=z;
}
```

Fig. 2 TEA source code

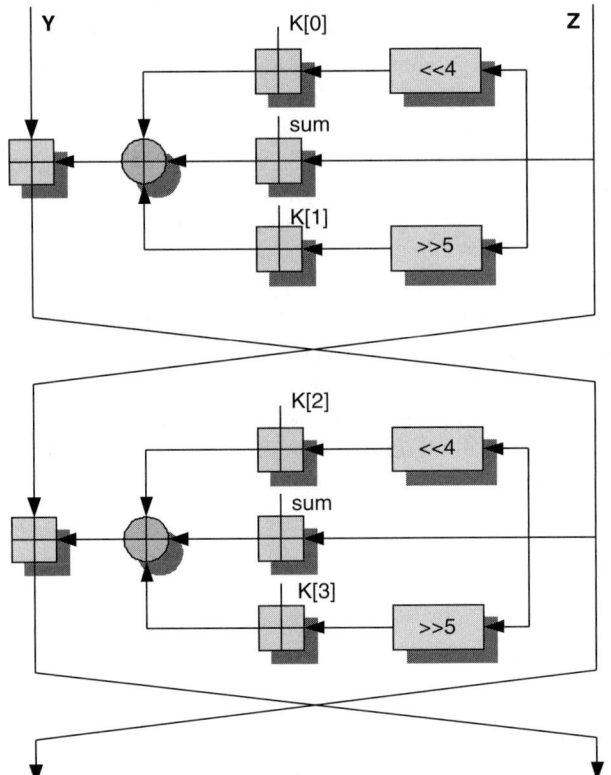

Fig. 3 TEA algorithm, cycle i

most notably from equivalent keys and related-key attacks [22,23], TEA should still provide good security for mobile systems such as RFIDs where hackers have neither time nor the right environment to break up the systems.

The block diagram of cycle i of TEA encryption is shown in Fig. 3. Each cycle consists of two rounds. At the beginning, the 128-bit key is divided into four equally sized subkeys, $k[0]$, $k[1]$, $k[2]$, and $k[3]$, respectively, from most significant bits (MSBs) to least significant bits (LSBs). The incoming 64-bit text is divided into two groups, *Y* is the 32-bit MSB and *Z* is the 32-bit LSB. During the first round, *Y* accumulates itself with the result of {(*4-bit left-shifted Z* plus *k[0]*) XOR (*Z* plus *sum*, which is the accumulated *delta*) XOR (*5-bit right-shifted Z* plus *k[1]*)}. This follows by the next round to calculate new *Z*, which is the result of *Z* accumulating its value with the result of (*4-bit left-shifted Y* plus *k[2]*) XOR (*Z* plus current *sum*) XOR (*5-bit right-shifted Z* plus *k[3]*)}. *Z* and *Y* then exchange their values to conclude one cycle. Similar procedure is observed for the following cycles, with TEA considered to be secure after 64 rounds (32 cycles) [13]. For hardware implementation, it is interesting to note that the structure is very simple and only hardware for addition, XOR, and registers is required. This is compared to other block ciphers such as DES or AES, for which bigger blocks such as the s-boxes are necessary.

5 TEA Architectures and Implementations

When implemented as part of a system, an application-specific co-processor core such as the encryption core considered will usually be controlled by a main processing unit. For the example of an RFID tag in Fig. 4, the main processing unit is in the form of an 8-bit microprocessor. The architecture has separate data and address buses that are used to provide the cryptographic processor with both the text and key information. Inside the cryptographic core there is an IO buffer used to preprocess the data before entering the cryptographic block. Three different styles of architectures for the core that are parallel, sequential, and digit-serial designs are considered here as discussed in the following sections.

5.1 The Parallel Architecture

Investigating the hardware style suitable for TEA implementation, the TEA encryption algorithm is implemented using three different architectures. The first architecture, as shown in Fig. 5, is a relatively straight-forward mapping from the functional

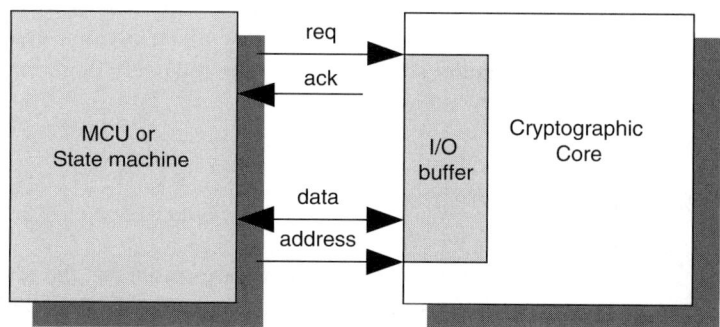

Fig. 4 Communication between MCU and cryptographic core

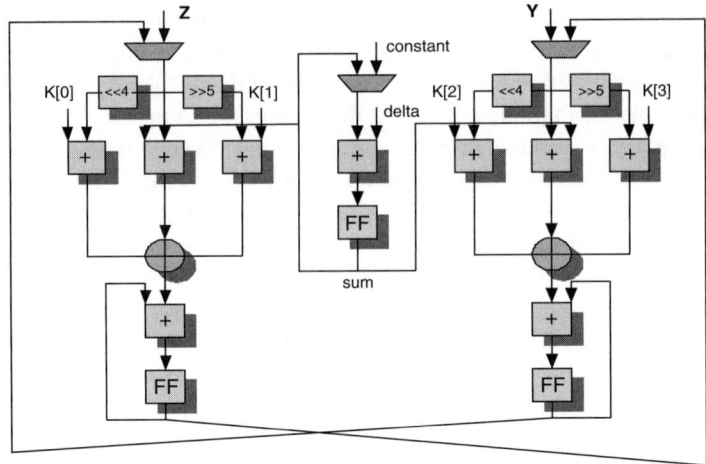

Fig. 5 Parallel architecture

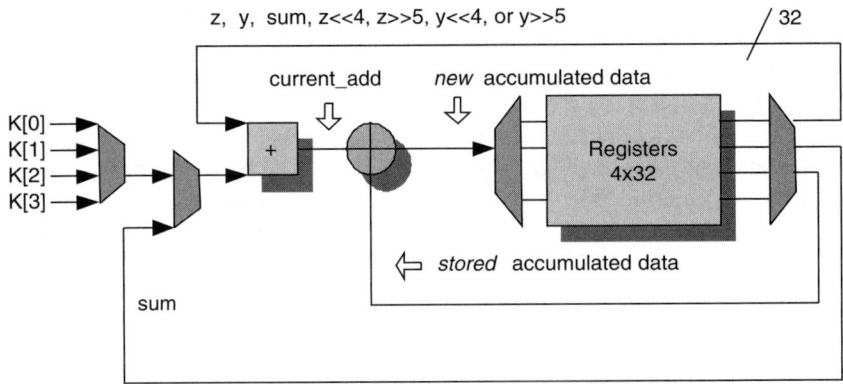

Fig. 6 Sequential architecture

view of Fig. 3. Multiple 32-bit Ripple Carry Adders (RCA) are used concurrently to perform operations needed for one encryption cycle. Synchronization with both rising and falling edges is observed, respectively, for the first and second rounds, resulting in an architecture that processes one cryptographic cycle at every clock cycle. Intuitively, this fully parallel structure should be relatively large in terms of hardware area but should also be inherently fast.

5.2 The Sequential Architecture

In contrast, as a conventional way to use less hardware resources, the second architecture performs TEA operations sequentially using a single 32-bit adder. The block diagram is as shown in Fig. 6. The architecture is in many ways similar to a software-based approach where designer writes assembly codes to program an MCU. Reduction in hardware is partially offset by the additional control and storage units needed. Here the architecture includes a 4×32-bit register array used to hold temporary data. Speed is sacrificed as the number of clocks per cycle increases. Using both clock edges, as in the parallel design case, the design needs nine clocks for each round. Communications with the connecting MCU is also arranged in a similar fashion to the parallel case.

5.3 The Digit-Serial Architecture

In the third design, to take advantage of customized architecture offered via the application-specific (ASIC) hardware approach, the architecture employs 8-bit digit-serial adders as main engines for its sub-blocks. The design consists of five main sub-blocks. Sub-block KLS, KRS, DDL are, respectively, used to calculate additions between key and left-shifted data, key and right-shifted data, and data and accumulated *delta*. Each block has two-stage multiplexers to select the appropriate 8-bit inputs that are to be added up using the serial/parallel adder. The output of the three blocks are XORed together, with the result fed to the OAD sub-block for final output calculation. Intermediate and output values calculated are stored at corresponding addresses in the register array. All timing and control are done by the control logic block. In all, the design needs eight clock cycles to calculate one cycle of TEA encryption (Figs. 7 and 8).

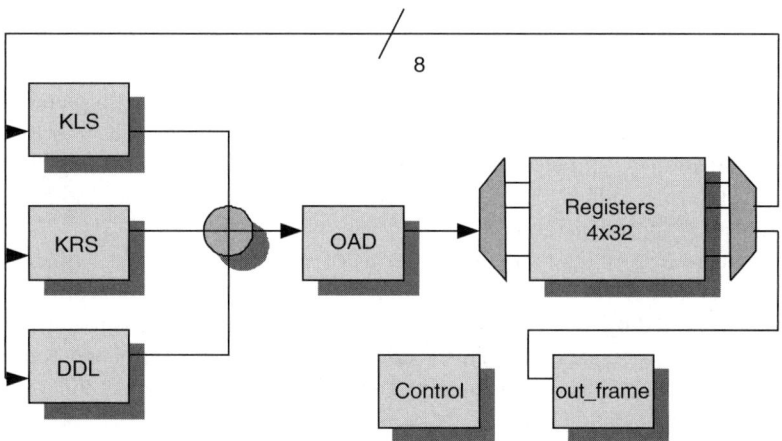

Fig. 7 Block diagram of the digit-serial TEA

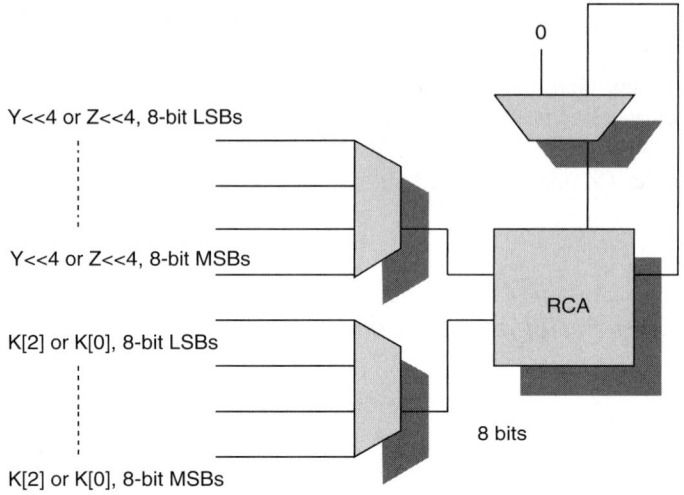

Performs (Y<<4 + K[2]), or (Z<<4+K[0])

Fig. 8 Sub-block architecture (showing KLS)

5.4 TEA Implementation

Although the sequential and digit-serial designs are aimed for low-area solutions, because of the overhead in terms of control, datapath selection (i.e., multiplexing/demultiplexing), and data storage, their effectiveness remains to be confirmed. Also the performance in terms of speed and power consumption will need to be compared and evaluated against the requirements set in Table 1. The TEA encryption cores based on the three architectures discussed are implemented using Verilog HDL and are first functionally verified using Mentor Graphic's ModelSim for Xilinx FPGA implementation. The designs are synthesized for ASIC implementation using Mentor Graphic's LeonardoSpectrum targeting 0.35-μm CMOS technology. Power dissipation is estimated based on observing switching activities together with information provided in [24]. The estimated results for all area, time, and power parameters are as shown in Table 2.

It can be seen from the table that, in terms of area utilization, the sequential and digit-serial architectures are equally small. It may also be noted that the effect of area reduction provided by the sequential and digit-serial architectures is not as much as what might have been expected. This is due to the additional control and data selection logics necessary. The results obtained are consistent with the fact that overhead is more significant in smaller circuits, which is the case of the TEA cores considered here. Interconnect routings also affect the overall size, as will be discussed more in Sect. 6. Although all three architectures meet the typical area specification given in Table 1, for extremely cost-sensitive applications only the sequential and digit-serial architectures are small enough. This can be very important in providing cost advantage when commercialized.

Table 2 TEA implementation results

	Parallel	Sequential	Digit-serial
Prerouted area (mm^2)	0.207	0.124	0.127
Maximum clock speed (MHz)	53	70.3	80.9
Number of clocks per cycle (double edges)	1	9	8
Number of clocks per 32 cycles (double edges)	32	288	256
Maximum throughput rate (Mbps)	106	15.6	20.225
Power (µW at 51.2 kbps)	7.37	39.0	38.4
Current (µA at 51.2 kbps)	2.23	11.8	11.6

In terms of the maximum speed, the digit-serial design, with its short critical path, is able to operate at the highest clock speed. The fact that it requires multiple clocks per cycle, however, results in the maximum throughput for an encryption cycle being less than that of the parallel design. A similar argument also goes for the sequential design, with more severe effect given its longer critical path and more clock cycles required. The throughput can be calculated from the equation

$$(\text{Data size}) \times (\text{max. clock speed}) \div (\text{no. of clock per cycle}) \div (\text{no. of cycle}). \quad (1)$$

The throughput for each design is shown in Table 2. It can be seen that the parallel design is clearly superior to the other two. To successfully use the TEA core in secure RFID systems, however, only 2.5 ms per tag for encryption is required as already discussed. Given that 32 cycles are performed for one encryption (64 bits), and assuming a 128-bit data (64 cycles) is to be encrypted, the corresponding encryption rate required is only 51.2 kbps or 25.6 kcycle per second, meaning that all three designs can easily meet the timing requirement. It is interesting to see also that the parallel design, which is the largest, dissipates the least power. Further analysis on power dissipation reveals that, although the sequential and digit-serial designs require multiple clocks per cycle, under the same throughput the powers contributed by their registers (FFs) are essentially similar to that of the parallel design, which is as it should be given that they perform similar operations. The increase in power in the latter two designs is, however, contributed from the significant increase in control logics that switch much closer to the clock speed. In the digit-serial case, although some of the glitches are removed from the adders as a result of their shortened critical path, the overhead in terms of data control logic is more severe and adds up to the total power consumption. The power values shown in Table 2 (with 3.3 V supply) are for 51.2 kbps throughput, which is the requirement for 2.5-ms process time. It can be seen that only the parallel design easily meets the specification, although with only further modifications in design or technology, such as using a specific low-power library, the other two designs should be able to meet the power consumption specification too. It is noted also that the process time targeted is for extremely demanding applications. In other applications, power can be reduced further to meet the specification by relaxing the process time so that a slower clock can be used.

Given the results, the decision about which architecture is to be used depends largely on the performance criteria targeted. As design bounds, the parallel and digit-serial architectures are selected for detailed comparison with equivalent AES cores in Sect 6.

6 Comparisons Between TEA and AES

As a benchmark to evaluate TEA's effectiveness, complete layout designs are compared with their AES equivalences. Since 2001, The National Institute for Standards and Technology (NIST) had selected the Rijndael algorithm as the official Federal Information Processing standard (FIPS) for AES [20], replacing the DES [25] which expired in 1998. The AES is a block cipher that operates on a 128-bit block of data (in original Rijndael's proposal, the block size can be 128, 192, or 256 bits). The key length used can be 128, 192, or 256 bits with the corresponding 10, 12, 14 rounds of operations required, respectively. Each data block is arranged into an array-of-byte format called the state array, on which all the AES operations are performed. As AES is used in various systems, from Gbps high-speed wireless communication systems to personal systems such as the smartcard [14,15,26–29], there are different types of architectures possible. Here, with a similar key size, only the 128-bit key version (4 × 4 state array, 10 rounds) is discussed. The encryption and decryption flows are as shown in Fig. 9.

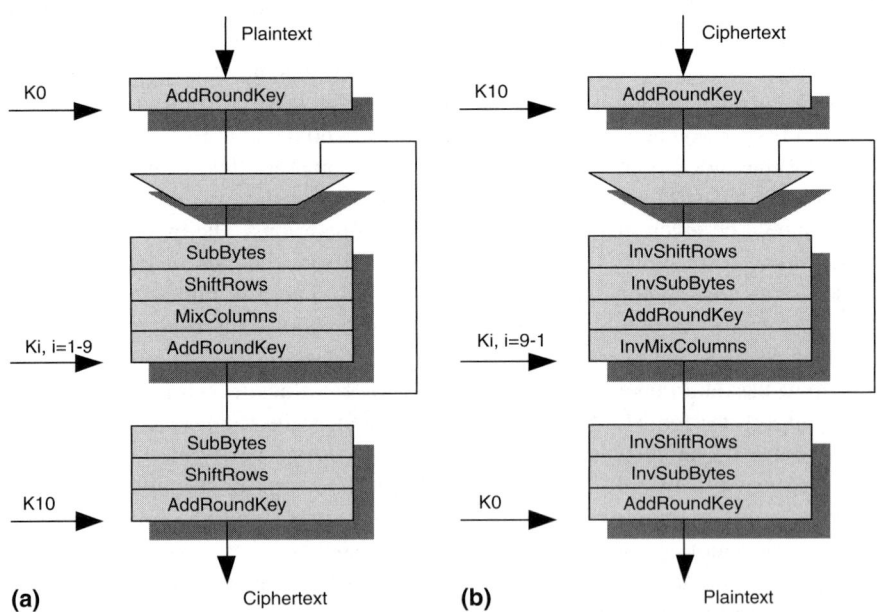

Fig. 9 AES operations; (**a**) Encryption (**b**) Decryption

SubBytes(). SubBytes transformation (nonlinear) is performed on each byte of the state array using a substitute table (s-box). The s-box is constructed by first computing the multiplicative inverse for each element of $GF(2^8)$ with irreducible polynomial $p(x) = x^8 + x^4 + x^3 + x + 1$ (element $\{00\}$ mapped to itself), followed by an affine transformation. In terms of implementation, a well-designed s-box is critical for the overall performance of the AES. Conventional method uses look up table. For high performance solutions, complicated modifications in finite or composite fields GF() have been proposed [19, 27, 28].

ShiftRows(). The rows in the array are cyclically left-shifted. The top row (row0) is not shifted. Rows 1, 2, and 3 are cyclically shifted by one, two, and three bytes, respectively.

MixColumns(). The MixColumns transformation is performed on all four the state array columns, column-by-column. Each column is regarded as a four-term polynomial of $GF(2^8)$ multiplied by $c(x)$ modulo $(x^4 + 1)$ where

$$c(x) = x^3 + \{01\}x^2 + \{01\}x + \{02\}.$$

Key expansion. For each round of operation, a round key is derived from a key expansion algorithm [20]. The operations include rotating the bytes in the right-most column, SubBytes transformed, XOR with round constant RCON, and finally XOR the result with the previous round key to generate the next key.

Decryption. Once the cipher text is received, the decryption algorithm as in Fig. 9b is used. InvShiftRows is similar to ShiftRows only that the state bytes are shifted in opposite directions. The InvSubBytes is inverse-affine transformation followed by multiplicative inverse in $GF(2^8)$. InvMixColumns simply involves modular multiplication with different value of $c(x)$. One major difference, however, is that the round key is in the reverse order, i.e., the first decryption round uses the last encryption round key and so on. Careful design consideration, about when to generate and where to keep the round keys, is usually required.

TEA is to be compared to two area-optimized AES designs. The first design, termed AES-1 here, was first reported by Feldhofer et al. in [14]. The design is arguably at present the main reference in AES implementation targeting the RFID tag. It was compared favorably with other designs reported such as that of Mangard [29] or Verbauwhede [19]. Among many detailed optimizing schemes, AES-1 is designed purposely to implement AES encryption only, as opposed to both encryption and decryption. It is argued in [14] that it is possible to implement challenge–response authentication based on encryption only, and that is the assumption made for TEA-based systems as well. AES-2, on the other hand, is recreated based on Pongjit's work in [15] (Fig. 10). The design is an area-optimized encryption/decryption core designed for smartcard applications. AES-2 design optimization techniques include composite field $GF((2^4)^2)$-based s-box, datapath sharing for encryption/decryption, on-the-fly key expansion, and logic optimization [15]. The two designs are compared to parallel and digit-serial TEA cores discussed earlier. Similar to AES-1 reported in [14], AES-2 and the two TEA cores are implemented

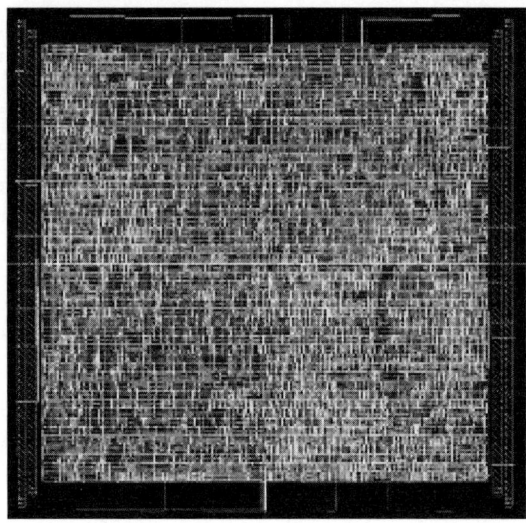

Fig. 10 AES-2 layout $(0.337\,\text{mm}^2)$

Table 3 Comparisons between AES and TEA

	AES-1 [14]	AES-2 [15]	TEA parallel	TEA digit
Key (bit)	128	128	128	128
Data size (bit)	128	128	64	64
Round per data	10	10	32	32
Clock per 128-bit data encryption	1016	226	64	512
Function	Encryption	Enc/Dec	Encryption	Encryption
Equivalent gate count	3,595	6,177	6,918	3,872
Layout area (mm^2)	0.25[a]	0.337	0.378	0.211
Maximum throughput (routed) Mbps	9.9[a]	11.33	48.0	6.25
Current µA at 12.8 kbps	8.15	11.1	0.55	2.90
CMOS technology	0.35 µm	0.35 µm	0.35 µm	0.35 µm

[a] Additional data from [30]

using 0.35-µm CMOS technology for direct comparison. The designs are Verilog coded, with Cadence Design Suites used for synthesis and place and route of the layouts. The results are shown in Table 3. Since AES operates on a 128-bit data, while TEA is on 64-bit, the performances reported are based on the assumption that TEA encrypts two set of 64-bit data as opposed to AES's one set of 128-bit data.

It can be seen that digit-serial-based TEA has a comparable gate count to the best AES design report (AES-1) while consumes much less current (power). The current values shown in the table are for the encryption cores operating at the throughput rate of 12.8 kbps, which corresponds to clock frequency of 100 kHz for AES-1, as used in [14]. At this rate, all the designs meet current consumption requirement, except AES-2 which is only slightly higher. It is noted here that the current values for the TEAs are exclusive of IO buffer circuits of which currents vary depending on the communication protocols. Also, even though meeting the current requirement is significant, lower current consumption can still be critical in terms of tag's reliability and read range. In the case where lowest power consumption is the design target, then the parallel-based TEA can be the solution at the expense of silicon area cost. When fully routed, with complete IOs, the parallel TEA has the layout area of $0.378\,\mathrm{mm}^2$, exceeding the design criteria set in Table 1. Routing also has an effect on the speed, as can be seen from the throughputs in Table 3. Among the four designs, only AES-2 fails to meet the requirement in terms of current consumption, consuming the amount of current slightly above the $10\,\mu\mathrm{A}$ target. Although still considerably small compared to other AES designs reported, the AES-2 scarifies silicon cost for flexibility in terms of providing both encryption and decryption functions.

In all, TEA-based cryptographic core has been shown to be able to meet all the stringent requirements in terms of silicon area, current drawn, and execution time. A cryptographic core based on TEA algorithm is similar to the best optimized AES core in terms of silicon cost, but generally consumes less current. Selection of architecture for implementing the TEA algorithm depends on the target criteria, but in general it can be said that the digit-serial design is the architecture of choice for its all round performance. The parallel design should be considered when extremely low-power consumption is required (Fig. 11).

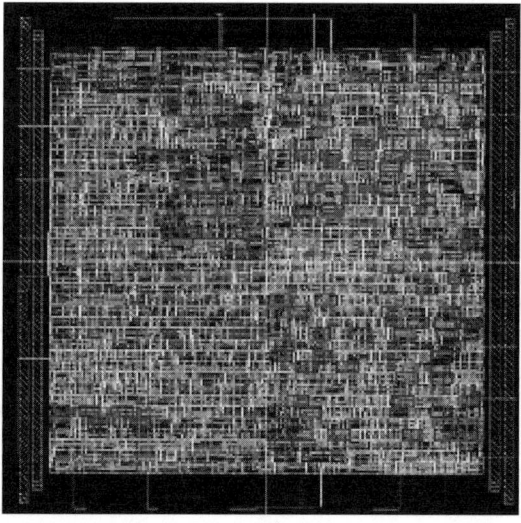

Fig. 11 Layout of digit-serial TEA $(0.211\,\mathrm{mm}^2)$

7 Conclusion

This chapter discusses a potential solution for low-cost secure RFID based on TEA encryption. One of the challenges when employing authentication scheme for secure RFID systems is to find a suitable encryption algorithm that meets the all stringent demands in terms of current consumption, silicon area used, and the processing time. The TEA algorithm has been shown to be suitable for hardware implementation meeting all design targets. Three different hardware architectures of the TEA are considered, with the digit-serial design considered to be best all-round. For analysis and evaluation purposes, the TEAs are benchmarked with equivalent area-optimized AES cores, with encouraging results.

Acknowledgments The authors would like to express their gratitude for the advice and support they receive from their colleagues at Thailand IC Design and Innovation lab (TIDI) of the Thai Microelectronics Center (TMEC), NECTEC.

References

1. Finkenzeller K. (2003) RFID-Handbook, Fundamentals and Applications in Contactless Smart Cards and Identification, 2nd edn. Wiley, New York, NY
2. Rankl W. (2000) Smart Card Handbook, 2nd edn. Wiley, New York, NY
3. EPC Global (2005) EPC Radio-Frequency Identity Protocols Class-1 Generation-2 UHF RFID Protocol for Communications at 860 MHz–960 MHz. http://www.epcglobalinc.org. Accessed 21 February 2008
4. EPC Global (2004). http://www.epcglobalinc.org
5. Sarma S.E. (2001) Towards the 5 ¢ Tag. Technical Report MIT-AUTOID-WH-006. MIT, Auto-ID Center
6. Weis S.A. (2004) RFID Privacy Workshop. IEEE Security and Privacy Magazine, 2(2): 48–50
7. Juels A., Rivest R.L., Szydlo M. (2003) The blocker hag: Selective blocking of RFID tags for consumer privacy. In: Tenth ACM Conference on Computer and Communication Security, pp. 103–111
8. Duc D.N., Park J., Lee H., Kim K. (2006) Enhancing security of EPCglobal GEN-2 RFID tag against traceability and cloning. In: Symposium on Cryptography and Information Security, Japan
9. Juels A. (2005) Strengthening EPC tag against cloning. In: Jakobsson M., Poovendran R. (eds.) ACM Workshop on Wireless Security (WiSe), pp. 67–76
10. Karthikeyan S., Nesterenko M. (2005) RFID Security without extensive cryptography. In: Third ACM Workshop on Security of Ad Hoc and Sensor Network, pp. 63–67
11. Weis S.A., Sarma S.E., Rivest R.L., Engels D.W. (2003) Security and privacy aspects of low-cost radio frequency identification systems. In: First International Conference on Security in Pervasive Computing. Lecture Notes in Computer Science 2802, pp. 201–212
12. Juels A. (2006) RFID security and privacy: A research survey. IEEE Journals on Selected Areas in Communications 24(2): 381–394
13. Wheeler D.J., Needham R.M. (1994) TEA, a tiny encryption algorithm. In: Second International Workshop on Fast Software Encryption. Lecture Notes in Computer Science 1008, pp. 363–366
14. Feldhofer M., Dominikus S., Wolkerstorfer J. (2004) Strong authentication for RFID systems using the AES algorithm. In: Cryptographic Hardware and Embedded Systems. Lecture Notes in Computer Science 3156, pp. 357–370

15. Pongjit J. (2003) Power–Area efficient advanced encryption standard IP core targeting smart card applications. Master Thesis, Asian Institute of Technology
16. International Organization for Standardization (2003) ISO/IEC 18000-4 Information Technology AIDC Techniques – RFID for Item Management
17. Schneier B. (1996) Applied Cryptography, 2nd edn. Wiley, New York, NY
18. Menezes A.J., van Oorschot P.C., Vanstone S.A. (1997) Handbook of Applied Cryptography. CRC, Boca Raton, FL
19. Verbauwhede I., Schaumont P., Kuo H. (2003) Design and performance testing of a 2.29 Gb/s Rijndael processor. IEEE Journal of Solid-State Circuits, March: 569–572
20. National Institute of Standards and Technology (NIST) (2001) FIPS-197: Advanced Encryption Standard
21. Lai X., Massey J.L. (1990) A proposal for a new block encryption standard. In: Advances in Cryptography-EUROCRYPT 1990. Lecture Notes in Computer Science 473, pp. 389–404
22. Kelsey J., Schneier B., Wagner D. (1996) Key-schedule cryptanalysis of IDEA, G-DES, GOST, SAFER, and Triple-DES. In: Advances in Cryptology-CRYPTO '96. Lecture Notes in Computer Science 1109, pp. 237–251
23. Kelsey J., Schneier B., Wagner D., (1997) Related-key cryptanalysis of 3-WAY, Biham-DES, CAST, DES-X NewDES, RC2, and TEA. In: First International Conference on Information and Communication Security. Lecture Notes in Computer Science 1334, pp. 233–246
24. Austria Micro Systems (2003) 0.35 μm CMOS Digital Standard Cell Databook
25. Ehrsam W.F. et al. (1976) Product block cipher system for data security, U.S. Patent 3,962,539
26. Kau H., Verbauwhede I. (2001) Architectural optimization for a 1.82 Gbits/sec VLSI implementation of the AES Rijndael algorithm. In: CHES 2001. Lecture Notes in Computer Science 2166, pp. 51–64
27. Satoh M.S., Takano K., Munetoh S. (2001) A compact Rijndael hardware architecture with s-box optimization. In: ASIACRYPT 200. Lecture Notes in Computer Science 2248, pp. 239–254
28. Lin T.F. et al. (2002) A high throughput low-cost AES cipher chip. In: Third IEEE Asia-Pacific Conference on ASIC, pp. 85–88
29. Mangard S., Aigner M., Dominikus S. (2003) A highly regular and scalable AES hardware architecture. IEEE Transactions on Computers, 52(4): 483–491
30. Feldhofer M., Wolkerstorfer J., Rijmen V. (2005) AES implementation on a grain of sand. IEE Proceedings on Information Security 152(1): 13–20

Index

Printed in the United States of America